Lecture Notes in Computer Science 3262

Commenced Publication in 1973
Founding and Former Series Editors:
Gerhard Goos, Juris Hartmanis, and Jan van Leeuwen

Mário Marques Freire   Prosper Chemouil
Pascal Lorenz   Annie Gravey (Eds.)

Universal
Multiservice Networks

Third European Conference, ECUMN 2004
Porto, Portugal, October 25-27, 2004
Proceedings

 Springer

Volume Editors

Mário Marques Freire
University of Beira Interior
Department of Informatics
Rua Marquês d'Ávila e Bolama, 6201-001 Covilhã, Portugal
E-mail: mario@di.ubi.pt

Prosper Chemouil
France Telecom
Research and Development, CORE/SPP
38-40 rue du Général Leclerc, 92794 Issy-les-Moulineaux Cedex 9, France
E-mail: prosper.chemouil@francetelecom.com

Pascal Lorenz
University of Haute Alsace, IUT
34 rue du Grillenbreit, 68008 Colmar, France
E-mail: pascal.lorenz@uha.fr

Annie Gravey
ENST-Bretagne
Département Informatique
CS 83818, 29238 Brest Cedex 3, France
E-mail: Annie.Gravey@enst-bretagne.fr

Library of Congress Control Number: 2004113912

CR Subject Classification (1998): C.2, H.4, H.3, H.5.1

ISSN 0302-9743
ISBN 3-540-23551-5 Springer Berlin Heidelberg New York

Springer is a part of Springer Science+Business Media

springeronline.com

© Springer-Verlag Berlin Heidelberg 2004
Printed in Germany

Typesetting: Camera-ready by author, data conversion by PTP-Berlin, Protago-TeX-Production GmbH
Printed on acid-free paper   SPIN: 11336310   06/3142   5 4 3 2 1 0

Preface

On behalf of the Organizing and Program Committees of the 3rd European Conference on Universal Multiservice Networks (ECUMN 2004), it is our great pleasure to introduce the proceedings of ECUMN 2004, which was held during October 25–27, 2004, in Porto, Portugal.

In response to the Call for Papers, a total of 131 papers were submitted from 29 countries. Each paper was reviewed by several members of the Technical Program Committee or by external peer reviewers. After careful assessment of the reviews, 53 papers were accepted for presentation in 13 technical sessions: half of them originated from countries outside Europe (mainly Asia). This illustrates the strong interest of this conference beyond its original geographical area.

The conference program covered a variety of leading-edge research topics which are of current interest, such as wireless networks, mobile ad hoc networks, sensor networks, mobility management, optical networks, quality of service and traffic, transport protocols, real-time and multimedia, Internet technologies and applications, overlay and virtual private networks, security and privacy, and network operations and management. Together with three plenary sessions from France Télécom, Siemens, and Cisco Systems, these technical presentations addressed the latest research results from the international industry and academia and reported on findings on present and future multiservice networks.

We thank all the authors who submitted valuable papers to the conference. We are grateful to the members of the Technical Program Committee and to the numerous reviewers. Without their support, the organization of such a high-quality conference program would not have been possible. We are also indebted to many individuals and organizations that made this event happen, namely IEEE, EUREL, SEE, Order of Engineers, Institute of Telecommunications, France Télécom and Springer. Last but not least, we are grateful to the Organizing Committee for its help in all aspects of the organization of this conference.

We hope that you will find the proceedings of the 3rd European Conference on Universal Multiservice Networks in Porto, Portugal a useful and timely document that presents new ideas, results and recent findings.

October 2004

Mário Freire, Prosper Chemouil, Pascal Lorenz and Annie Gravey

Conference Committees

General Co-chairs

Prosper Chemouil (France) – France Télécom R&D
Annie Gravey (France) – Groupement des Écoles des Télécommunications/ÉNST
    Bretagne
Pascal Lorenz (France) – University of Haute Alsace
Mário Freire (Portugal) – University of Beira Interior/IT

Steering Committee

Prosper Chemouil (France) – France Télécom R&D
Annie Gravey (France) – Groupement des Écoles des Télécommunications/ÉNST
    Bretagne
Pascal Lorenz (France) – University of Haute Alsace
Jean-Gabriel Rémy (France) – Cegetel
Sylvie Ritzenthaler (France) – Alcatel
Pierre Rolin (France) – France Télécom R&D

Technical Program Committee

P. Bertin (France) – France Télécom R&D
F. Boavida (Portugal) – University of Coimbra
S. Bregni (Italy) – Politecnico Milano
P. Brown (France) – France Télécom R&D
E. Carrapatoso (Portugal) – University of Porto
P. Castelli (Italy) – Telecom Italia Labs
T. Chahed (France) – INT
J. Craveirinha (Portugal) – University of Coimbra
M. Diaz (France) – LAAS-CNRS
P. Dini (USA) – Cisco
N. Fonseca (Brazil) – Campinas University
Y. Gourhant (France) – France Télécom
A. Jamalipour (Australia) – University of Sydney
H.-K. Kahng (Korea) – Korea University
N. Kamiyama (Japan) – NTT
S. Karnouskos (Germany) – Fraunhofer FOKUS
L. Lancieri (France) – France Télécom R&D
K. Lim (Korea) – Kyungpook National University
M. Maknavicius Laurent (France) – INT
Z. Mammeri (France) – University of Toulouse

E. Monteiro (Portugal) – University of Coimbra
S. Oueslati (France) – France Télécom R&D
G. Petit (Belgium) – Alcatel
M. Pioro (Poland) – Warsaw University of Technology
S. Ritzenthaler (France) – Alcatel
A. Santos (Portugal) – University of Minho
R. Valadas (Portugal) – University of Aveiro
M. Villen-Altamirano (Spain) – Telefonica I+D
J. Yan (Canada) – Nortel

Organizing Committee

C. Salema (Portugal) – IT/Order of Engineers
L. Sá (Portugal) – University of Coimbra/IT
M. Freire (Portugal) – University of Beira Interior/IT
H. Silva (Portugal) – University of Coimbra/IT
J. Rodrigues (Portugal) – University of Beira Interior/IT
F. Perdigão (Portugal) – University of Coimbra/IT
R. Rocha (Portugal) – IST/IT

Table of Contents

Wireless Networks

Quality of Service

Optical Networks

Mobility Management

Transport Protocols

Mobile Ad Hoc Networks

Real Time and Multimedia

Traffic

Network Operations and Management

Wireless and Sensor Networks

The 4GPLUS Project, Overview and Main Results

Dirk-Jaap Plas[1], Mortaza Bargh[2], Jan Laarhuis[3], Jeroen van Vugt[3], Jacco Brok[1], and
Herma van Kranenburg[2]

[1] Lucent Technologies, Bell Labs Advanced Technologies,
Capitool 5, 7521 PL Enschede, The Netherlands
{dplas, brok}@lucent.com
[2] Telematica Instituut, P.O. Box 589, 7500 AN Enschede, The Netherlands
{Mortaza.Bargh, Herma.vanKranenburg}@telin.nl
[3] TNO Telecom, P.O. Box 5050, 2600 GB Delft, The Netherlands
{J.H.Laarhuis, J.M.vanVugt}@telecom.tno.nl

Abstract. A 4G-environment is based on the integration of 3G mobile and
other wired and wireless technologies into an all-IP network environment. Central to such an environment is a 4G-service platform, which offers services such
as AAA, mobility management and session control, and also mediates between
the users and providers of these services. Each service platform and its associated users, service providers and access networks constitute a service platform
domain. Extension of offered functionality and expansion of coverage is obtained by federation between multiple service platform domains. In the
4GPLUS project a conceptual framework is developed for 4G-environments,
which specifies the functionality and structure of the 4G-service platform and
the concept of federation. Several key aspects of this framework have been refined, implemented and prototyped within this project. This paper provides an
overview of the 4G-concepts developed within the project and the main results
achieved.

1   Introduction

Current communication infrastructures comprise fixed networks (e.g. Ethernet and
ADSL), mobile Wide Area Networks (e.g. GPRS and UMTS), Wireless Local Area
Networks (WLAN) and their associated terminals. Services supporting all kinds of
media, such as voice, video and data, are offered over these infrastructures by an increasing number of providers. Examples are voice over IP provided by telephony providers, conferencing services and multimedia streaming. Though these communication infrastructures have capable functionality, they fail to meet obvious upcoming
user requirements: users are *technology agnostic*, they want to have *universal access*
to their services, and their *services should be adapted* according to the context and/or
their personal preferences [16]. In order to meet these user requirements, the following system requirements must at least be imposed on a supporting infrastructure. The
first is *seamless mobility,* which includes hand-overs to other terminals, to other access networks as well as to other administrative domains. The second is *generic access* including authentication and authorization for both network access and service

M. Freire et al. (Eds.): ECUMN 2004, LNCS 3262, pp. 1–11, 2004.

access. The third is *session control* including session negotiation as well as session adaptation.

The design of a supporting infrastructure meeting these requirements is the topic of the 4GPLUS project [1,4]. Such an infrastructure is generally denoted a 4G-environment. The design includes a framework that integrates network and service architecture. Central to the service architecture in 4GPLUS is the service platform, whose (distributed) generic software components provide feasible technological solutions for meeting the user requirements mentioned above. To implement our architectural framework, we combine and extend existing technologies. Also, we introduce new business opportunities without taking away assets from current operators and a new enterprise model that supports our solution.

In this paper we will explain our vision and solutions on facilitating software infrastructures for the 4G-environment and highlight results that we achieved within the 4GPLUS project. The remainder of this paper is structured as follows. A concise overview is given of the 4G-environment and the impact that the introduction of this 4G environment has on the current enterprise model. An end-user scenario illustrates the issues that have been addressed by the 4GPLUS project. These issues, i.e. mobility management, session control, federation and service provisioning, are described with references to papers, presentations and reports [20-25] published earlier. Our conclusions are given in the final section.

2   4G-Environment

A 4G-environment in the context of the 4GPLUS project consists of: (a) access networks of different technology types including wired (e.g. xDSL) and wireless networks (e.g. UMTS and Wi-Fi™); (b) next generation mobile terminals with a wide range of communication, computing and storage capabilities; (c) a rich set of (3rd party) services that offer value added services to mobile end users across these heterogeneous networks and terminals, and; (d) service platforms for development and provisioning of services.

Functionally, the 4G-environment adheres to a 3-layered model consisting of an application layer, a service control layer and a transport layer [24]. The transport layer consists of heterogeneous access networks and core networks. The application layer contains all application logic needed to provide services from 3rd Party Service Providers (3PSPs) to end-users. The service control layer is logically located between the application and transport layers and shields the network heterogeneity for the different parties. The service control layer is made up of service platforms interoperating through federation [2,5,9,24].

Fig. 1 shows a schematic structure of the 4G-environment, including the three layers mentioned above. There is global IP-connectivity between all networks and all end-to-end communication is IP-based. The IP packets exchanged between access networks and the core network are assumed to have globally routable IP addresses. End users, 3PSPs and service platform operators are the end points connected to these access networks. The services provided in the application layer range from user-to-user services, such as telephony, to user-provider services, such as content retrieval services.

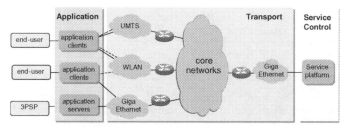

Fig. 1. The physical structure of a 4G-environment

The service platform(s) in the control layer offer(s) service control functions that enable end users to easily gain and maintain access to (new) services, while roaming between different access networks and terminals. For 3PSPs, a service platform acts as a one-stop-shop for providing their services to end-users and hides the changes of access networks and terminals due to roaming of end-users. For access networks, a service platform provides control functions for transport outsourcing services. Ref. [5] describes how service platforms and the envisioned federation among them realize the service control functionality.

3   4G-Enterprise Model

The current enterprise model for Internet access business is inadequate for 4G-environments, and therefore needs a revision. We first analyze in what respects the 4G-environment differs from the environment of current Internet access business. Based on these observations, an enterprise model for 4G-environments (4G-EM) is proposed. Finally we elaborate on the central role in this new enterprise model, i.e. the Service Platform Provider (SPP).

The current enterprise model for Internet access is shown in Fig. 2.a. The role of the Internet Service Provider (ISP) needs some explanation. The currently known ISP is not an atomic role. Rather it is a functional composition of three roles [18]. By offering services such as e-mail, web hosting, virus scanning and the like, the ISP performs the role of an Application Service Provider (ASP). Since the ISP has, in most cases, contact with the customer and sends the bills, it also performs the role of the packager. Finally, the ISP takes care of naming and addressing and thus of IP-address allocation. Therefore, it also plays the role of Internet Communication Provider (IP-CP). It is assumed that each ISP has its own range of public IP-addresses.

Opposed to the current Internet access business, the 4G-environment is in many cases an open service environment. By exporting interfaces to 3PSPs in a standardized way, the SPP introduces a new kind of functionality. Notice that the 3PSP role is similar to the ASP role. The difference is just a matter of parlance. Also, both personal and terminal mobility are supported by a 4G-environment. This means that the association between a customer and an access network, and thus the IP-edge, is a dynamic one rather than a permanent one.

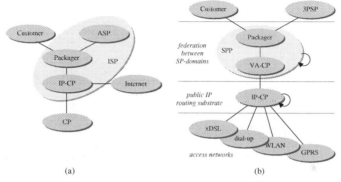

Fig. 2. Enterprise models for (a) the current Internet access business, and for (b) the 4G-environment

The 4G-EM shows some salient aspects of 4G-environments as defined within the 4GPLUS project, namely the incorporation of all kinds of access networks, the use of public IP as the common transport technology, the federation between SP-domains, and the various 3PSPs that take the role of ASPs. The SPP role entails three major aspects, as shown by the decomposition of this role in Fig. 2.b:

1. *value-added IP-connectivity provider (VA-CP)* – In addition to simply issue IP-addresses and process AAA, the SPP can provide mobility, session control, profile management, personalization and QoS-management.
2. *packager* – The SPP performs the packager role and thus offers the services of the 3PSPs and the access network providers within its domain to the customers. The relation of an SPP with multiple access networks and multiple 3PSPs makes it a one-stop-shop for customers.
3. *one-stop-shop for 3PSPs* – Due to federation with other service platform domains and its role as a packager, the SPP is a one-stop shop for 3PSPs. 3PSPs need to have a relation with only one SPP in order to get a much larger 'audience'.

4   End User Scenario

Following the descriptions of the 4G-environment and the 4G-enterprise model we now introduce an end-user scenario that provides links to the concepts that have been developed within the 4GPLUS project.

John is an end-user (customer), who has a subscription with a public SPP, and who (also) has IP-connectivity via his corporate domain in his employee-role. The public SPP domain and corporate domain, including the involved parties, are depicted in the scenario environment of Fig. 3. The SPPs provide John with access to a number of different access networks and services, offered by different CPs and 3PSPs. In the fol-

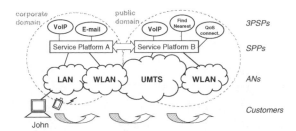

Fig. 3. Overview of the scenario environment

lowing scenario description this is elaborated upon. Further, with superscripts we refer to the sections that follow, where we describe the specific 4GPLUS concepts and results, being:

A. federation: enabling access to services across administrative domains;
B. mobility management: handling the mobility of the user (equipment);
C. session control: controlling sessions in a heterogeneous network environment; and
D. services: providing support for value added services.

John is planning to go home when he receives a video call from his wife[C.1, B.1]. He answers the call using his fixed videophone. Since he was just about to leave the office, he transfers the call[B.2] to his PDA that is connected to the corporate WLAN network[A.3,B.3]. According to company policy only audio calls are allowed[C.2] and the video is dropped[C.1]. Meanwhile, John can continue[B.2] the conversation with his wife; she reminds him on her mother's birthday coming up and asks John to buy a gift on his way home. While John leaves the office premises his call is seamlessly handed over to the UMTS network[A.1,A.2,A.3,B.2,B.3].

He walks over to the railway station where a Wi-Fi™ hotspot is available[A.1,A.3]. When he accesses the network[B.3] he is notified of the cost. A local service [D] provided by the railway operator informs him[A.3,D] on the delay of certain trains. This provides John with sufficient time for buying a present. He finds a nearby gift shop using the FindNearest service[A.3,D]. John's preferences indicate he likes to be notified about interesting offers while being in a shopping center and he receives a number of advertisements before arriving at the gift shop[A.3,D]. Once there, he cannot make up his mind on the best present. On the spot he sends three picture-messages[C] to his wife and she – knowing her mother best, selects the appropriate gift.

The railway service reminds him to get back to the railway station to catch his train. In the 45-minutes train trip John spends his time on reading the latest news and his private e-mail. He also views the video clips that were recorded by his personal video recorder during the day[C]. Suddenly he remembers he forgot to send an important e-mail message to his colleague Jan. Although he is not in his corporate domain, he has access[A.1,A.3] to all corporate services and is able to send the e-mail[A.2]. When he gets home, his PDA automatically obtains access to his private WLAN[A.3,B.2,B.3], which now is being used to receive all private telephone calls[B.1].

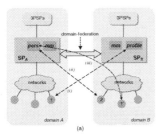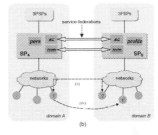

Fig. 4. Federation in a 4G-environment: (a) domain federation, and (b) service federation

5   The 4GPLUS Results

A   Federation

In [5,24] the service platform domain is defined as the service platform plus all entities it maintains administrative relations with. Federation between service platform
domains in a 4G-environment is a key concept. Federation is a primary necessity for
inter-domain mobility or roaming. Notice, however, that mobility applies to the intra-
domain case as well.

1. Domain Federation. Several service platform domains can create a domain
federation, as shown in Fig. 4.a between domains A and B. Through federation, the
service capabilities of SP_B be-come available for the end-users of SP_A, and vice versa.
This is illustrated as end-user 1 being able to access the profile capability of SP_B (i)
and end-user 2 being able to access the personalization capability of SP_A (ii). In
addition, when end-user 1 is connected to the networks within SP_B, indicated by end-
user 1', the domain federation enables end-user 1 to still access the personalization
capability (iii). Here it is assumed that terminal mobility is not solved at the IP-
network level, and that through federation, end-user 1' is granted IP-access as a
visitor by SP_B.

2. Service Federation. SP-providers can also be engaged in service federation. This
is illustrated in Fig. 4.b. If SP_A and SP_B offer similar functions, in this example
mobility management and session control, these functions can be extended over these
domains. In a first example (iv), end-user 1 has, in addition to the capabilities offered
by domain federation also (SIP-based) terminal mobility across the federated
domains. This means that the location of this end-user is tracked, and therefore that
end-user 1 not only is able-to-reach, but also is reachable by others, and maintains
active sessions during hand-over. The second example of service-federation in Fig.
4.b is session control (v). Both SP_A and SP_B have session control capabilities. Due to
federation of these capabilities, end-user 1 in domain A can set up a session with end-
user 2 in domain B.

3. User Identification and Authentication. In order to support seamless roaming in a heterogeneous network environment, the mobile terminal should be able to authenticate to all available networks without requiring user interaction. Assuming a subscription with a SPP, authentication requests must be terminated by the SPP and not the access networks. For Wi-Fi™ and even LAN networks, the IEEE 802.1x framework in combination with RADIUS supports such an authentication scheme. One authentication variant supported by IEEE 802.1x is EAP-TLS, based on client certificates containing a user identifier (user@domain) and credentials (private and public keys). Another variant is EAP-SIM, where the authentication triplets – commonly used in GSM networks – from a SIM-reader and a SIM card allow Wi-Fi™ and LAN authentication. Both have been implemented and validated in the 4GPLUS solutions [13].

In addition to network authentication, users may need authentication to (web-based) services as well. Rather than registering for every service, the user's subscription with the SPP enables single sign-on to 3rd party services. A prerequisite for this to work is a trust relationship between SPP and 3PSP. The 4GPLUS service platform has adopted Liberty Alliance concepts, and in particular SAML [12].

B   Mobility Management

The Mobility Management functionality of the service platform provides three functions: (1) Location Management to locate the network attachment points of end-users for session initiation, (2) Handover Management to modify the IP routes of the ongoing sessions of end-users and (3) Network Access to establish IP connectivity to different access networks. Ref. [10] and Ref. [11] describe a high-level architecture of the Mobility Management component that uses Mobile IP and SIP to enable IP-level user and terminal mobility in 4G-environments. In Ref. [3] we motivate the idea of decoupling Location Management and Hand-over Management in 4G environments, where a mobile host is equipped with wireless WAN and wireless LAN interfaces (such as UMTS and Wi-Fi™). The following subsections briefly elaborate upon our implementation of the Mobility Management functions.

1. Location Management. The Location Management function keeps track and discovers *the current IP-addresses* associated with mobile end-users' terminals. The current IP addresses associated with a mobile end-user are dynamically maintained in a corresponding Home Agent (when using Mobile IP) or a registrar at a corresponding Service Platform (when using SIP) [10,11]. Every other end-user contacts the Home Agent or the SIP registrar, when initiating a new session towards the mobile end-user. [17] describes in detail our SIP based roaming solution that realizes also session control aspects.

2. Hand-over Management. The Hand-over Management function maintains an end-user's ongoing sessions as the corresponding IP-addresses change due to terminal mobility or terminal change. Within the 4GPLUS project, we have combined SIP and Mobile IP for Handover Management [10,11]. As a single point of control on the mobile terminal, a mobility manager manages the handover process (as well as network selections) and hosts a Mobile IP client. Mobility-aware applications can register with the mobility manager to monitor and request network handovers, with the purpose to realize session mobility with SIP [17]. In Ref. [15], we focus on the mobility management functionality residing on the mobile host and describe the implementation of a mechanism to systematically exploit the availability of multiple mobility protocols and network interfaces. The mechanism provides the applications running on the mobile host with information about the state of the lower-layer mobility management protocols as well as the state and characteristics of the available network resources. Applications may consecutively adapt their behavior depending on this mobility process information or use SIP to deal with some mobility issues (e.g., route optimization).

3. Network Access. IP connectivity can be arranged to one or more access networks simultaneously in 4G-environments. Being a prerequisite to Location Management and Hand-over Management, the Network Access function enables a mobile end-user's terminal to send and receive IP traffic by detecting network interfaces and available access networks, selecting access networks and configuring terminal interfaces. Access network selection, implemented as part of the mobility manager on the mobile terminal, includes a decision process that uses input parameters such as availability of interfaces and networks, signal quality and bandwidth [6,7,25]. In 4GPLUS, we have extended the decision process by including user preferences, operator policies, network cost and local application requests as well [25].

C   Session Control

In the 4G-environment, the availability and almost constant access to networks with increasingly higher bandwidth will stimulate the use of multimedia services. For these types of services the ability of seamless roaming is even more critical. In addition to the proper configuration at the network level, special care should be taken to control the ongoing multimedia sessions.

1. Session Adaptation. Besides the ability of reconfiguring the appropriate network settings due to changes in the network environment, the SIP protocol also enables the adaptation of multimedia sessions that is required in such circumstances. This means that also the service itself can be adapted based on the available network resources. In 4GPLUS, we have created a multimedia application supporting video conferencing sessions over Ethernet or Wi-Fi™ networks, while dropping video – maintaining audio – when switching to GPRS. Other characteristics that could be changed are for instance the frame rate of the video, the codecs to be used, etc. This mechanism allows services to achieve the best user experience and enables efficient use of network resources, in an environment where users are using different services across a wide range of different access networks. We have tested a number of scenarios and also combined the use of Mobile IP for mobility management and SIP for session control, for which the results can be found in this paper [17].

2. Controllable Gateway. A Controllable Gateway (CGW) in the access network functions as a dynamic firewall that controls admission to end-user service sessions

[8]. The CGW resides on the IP path at the IP-edge between the access network and the rest of the world. While the access network provider may own the CGW, the SPP controls the CGW, possibly via another SPP through federation. All IP traffic from and to a user connected to, for example, a Wi-Fi™ hotspot must go through the CGW belonging to this access network. The basic functionality of the CGW can possibly be extended with e.g. QoS control, network security, metering (for billing purposes) and legal intercept. The CGW is a suitable point to implement these extra functionalities, especially legal intercept, since all traffic flows are available here. The SPP remains the controlling entity for these extra functionalities.

D   Value Added Services

The technologies and solutions described so far enable roaming in a heterogeneous network environment. In 4GPLUS, we have enhanced the end-user experience with additional service platform services accessible from the mobile terminal. Besides required services such as support and contact information, added value can be achieved with Service Platforms delivering lists of Wi-Fi™ hotspots and their locations, location-based services and personal and local portals. To support location-based services, the service platform offers a single standardized Parlay X interface – embedded in the SAML security context – for 3PSPs to retrieve user location information independent of access networks. The local portal concept allows end-users to access a web-based service portal for the access network currently in use. The resulting channel to advertise various services provides an incentive for SMEs to offer public Wi-Fi™ hotspot connectivity.

6   Conclusion

Our 4G Service Platform is a (distributed) software infrastructure, supporting applications and services with functionality for, e.g., network and service authentication, profiling, session control, seamless mobility and charging.

Through co-operation of the distributed components, the interoperability issues between the various (network) technologies and administrative domains are solved. Seamless mobility, i.e., the capability of end users to roam through arbitrary environments while automatically maintaining connectivity to their application, is realized by federated mobility management functions. Various kinds of access control, mobility management, inter-domain AAA are put in place to make this work. Also session adaptation to the changing environment is realized, as are appealing end-user services. Important key protocols used in the 4GPLUS solutions include Mobile IP, SIP, SAML and IEEE 802.1x with different EAP variants.

The result is that, irrespective of the network or administrative environment the user is in, the services and applications are made available, tailored to the specific environment and user preferences, and maintained in a seamless and transparent manner while roaming. In other words, our 4GPLUS solution meets the emerging user requirements of technology agnosticism, universal and secure service access and service adaptation.

Acknowledgements. The work described in this paper is part of the Freeband 4GPLUS project (http://4gplus.freeband.nl) and the Freeband AWARENESS project (http://awareness.freeband.nl). Freeband is sponsored by the Dutch government under contract BSIK 03025. The authors like to thank Frank den Hartog from TNO Telecom and Jeroen van Bemmel from Lucent Technologies for reviewing this manuscript.

References

[1]   4GPLUS (4th Generation Platform Launching Ubiquitous Services) project. http://4gplus.freeband.nl.

[2]   Herma van Kranenbrug, "4G Service Platform", Proceedings of EU workshop on "Interconnecting Heterogeneous Wireless Testbeds", 26 March 2003, Brussels, Belgium, also published on EU web site: http://www.cordis.lu/ist/rn/wireless-testbed-ws.htm

[3]   Mortaza S. Bargh, Hans Zandbelt, Arjan Peddemors, "Managing Mobility in Beyond 3G-Environments", to appear in the proceedings of the 7th IEEE international conference on High Speed Networks and Multimedia Communications (HSNMC 2004), June 30-July 2, 2004, Toulouse, France.

[4]   Jeroen van Bemmel, Harold Teunissen, Dirk-Jaap Plas, Bastien Peelen, Arjan Peddemors, "A Reference Architecture for 4G Services", Wireless World Research Forum (WWRF) # 7, 3 and 4 December 2002, Eindhoven, The Netherlands.

[5]   Mortaza S. Bargh, Jan H. Laarhuis, Dirk-Jaap Plas, "A Structured Framework for Federation between 4G-Service Platforms", PIMRC 2003, Beijing, September 2003.

[6]   Ronald van Eijk, Jacco Brok, Jeroen van Bemmel, Bryan Busropan, "Access Network Selection in a 4G Environment and the Roles of Terminal and Service Platform", WWRF #10, 27-28 October 2003, New York.

[7]   Bryan Busropan, Jan van Loon, Frans Vervuurt, Ronald van Eijk, "Access Network Selection in Heterogeneous Networks and the Role of the Operator", WWRF #9, 1 and 2 July 2003, Zurich, Switzerland.

[8]   B.J. Busropan, M.S. Bargh, J.M. van Vugt, "Controllable Gateways for Network-based Session Control", WWRF #10, 27-28 October 2003, New York.

[9]   Herma van Kranenburg, Ronald van Eijk, Mortaza S. Bargh, Arjan Peddemors, Hans Zandbelt, Jacco Brok, "Federated Service Platform Solutions for Heterogeneous Wireless Networks", Proceedings DSPCS'2003 (7th Int. Symp. on Digital Signal Processing and Communication Systems), Australia, 8 11 December 2003.

[10]   Mortaza S. Bargh, Hans Zandbelt, Arjan Peddemors, "Managing Mobility in 4G Environments with Federating Service Platforms (an overview)", EVOLUTE workshop, Surrey University (Guildford, UK), November 10, 2003.

[11]   Mortaza S. Bargh, Dennis Bijwaard, Hans Zandbelt, Erik Meeuwissen, Arjan Peddemors, "Mobility Management in beyond 3G-Environments", WWRF #9, 1 and 2 July 2003, Zurich, Switzerland.

[12]   Jeroen van Bemmel, Harold Teunissen, Gerard Hoekstra, "Security Aspects of 4G Services", WWRF #9, 1 and 2 July 2003, Zurich, Switzerland.

[13]   Hong Chen, Miroslav Živkovi , Dirk-Jaap Plas, "Transparent End-User Authentication Across Heterogeneous Wireless Networks", Proceedings of IEEE VTC 2003 Fall conference, October 2003.

[15]   A.J.H. Peddemors, H. Zandbelt, M.S. Bargh, "A Mechanism for Host Mobility Management supporting Application Awareness", Accepted to MobiSys2004, Boston, Massachusetts, USA, June 6-9, 2004.

[16] Herma van Kranenburg (editor), Johan Koolwaaij, Jacco Brok, Ben Vermeulen, Inald
      Lagendijk, Jan van der Meer, Peter Albeda, Dirk-Jaap Plas, Bryan Busropan, Henk Eer-
      tink, "Ambient Service Infrastructures - Supporting tailored mobile services anytime,
      anywhere", Freeband Essentials 1, Telematica Instituut, Enschede, January 2004, avail-
      able at http://www.freeband.nl/essentials.
[17] Willem A. Romijn, Dirk-Jaap Plas, Dennis Bijwaard, Erik Meeuwissen, Gijs van Ooijen,
      "Mobility Management for SIP sessions in a Heterogeneous Network Environment", Bell
      Labs Technical Journal (BLTJ), BLTJ Vol. 9 No. 3, 2004.
[18] Jan Laarhuis, "Towards an Enterprise Model for 4G-Environments", submitted to IEEE
      Transactions on Networking.
[20] 4GPLUS, deliverable D3.1-4.1, "SOTA Interworking of heterogeneous network tech-
      nologies and Mobile Multimedia Session Management", November 2002, available at
      http://4gplus.freeband.nl.
[22] 4GPLUS, deliverable D3.2-4.2, "Architectural Specification of Interworking of Hetero-
      geneous Network Technologies and MultiMedia Session Control", April 2003, available
      at http://4gplus.freeband.nl.
[23] 4GPLUS, deliverable D2.3-3.5-4.5, "Operational aspects of the Federated Service Plat-
      form", December 2002, available at http://4gplus.freeband.nl.
[24] 4GPLUS, deliverable D2.2v3, "Functional specifications of the federated platform archi-
      tecture", June 2003, available at http://4gplus.freeband.nl.
[25] 4GPLUS, deliverable D5.3v2, "Architectural Specification of the demonstrator", Febru-
      ary 2004, available at http://4gplus.freeband.nl.

Providing Real Time Applications with a Seamless Handover Using a Combination of Hierarchical Mobile IPv6 and Fast Handover

David García Ternero*, Juan A. Ternero, José A. Gómez Argudo, and
Germán Madinabeitia**

Área de Ingeniería Telemática
Departamento de Ingeniería de Sistemas y Automática
Escuela Superior de Ingenieros
Sevilla, Spain
jternero@trajano.us.es

Abstract. Users of real time applications expect to roam between fixed
and different wireless networks without notice, therefore handovers are
expected to be seamless. The use of IP, and specifically Mobile Ipv6, as
a transport technology solves some of the internetworking problems, but
not the seamless handover. Although some extensions, as Hierarchical
Mobile Ipv6 and Fast Handover, were introduced, these extensions, and
combinations of them, do not offer an optimal handover performance.
In this article, a new combination of these two extensions is proposed.
This solution tries to solve the problems of seamless handoff for real time
applications.

1 Introduction

Current trends in communication networks point to an aggregation of all kinds
of traffic (data, voice, etc) in the same transport technology, and this is valid for
both fixed and wireless networks. The use of IP as a transport technology solves
some of the internetworking problems between different technologies, but at the
same time, the IP Quality of Service mechanisms [2] are not fully adapted to the
mobility between different technologies.

In the last few years, there has been a fast development of real-time applications, and customers are demanding an efficient solution to the problems
emerging from this kind of applications. Ideally, a real time application should be
able to roam between a fixed network, a wireless LAN [3] or a UMTS provider.
Multiple and different handovers have to be established in this type of environment: between technologies (vertical handover) and between cells of the same
technology (horizontal handover). These handovers are expected to be seamless,

---

* Part of his work was developed at [1], under Socrates-Erasmus Program
** The work leading to this article has been partly supported by CICYT and the EU
under contract number TIC2003-04784-C02-02

M. Freire et al. (Eds.): ECUMN 2004, LNCS 3262, pp. 12–19, 2004.

that is, undetected by the users. To make it transparent to the user, these handovers have to be done in a short period of time, without disturbance in the established communication.

As both the usage of real time applications and the development of wireless devices grow more and more, a protocol being able to support a seamless handover is required. A protocol that was able to make a handover between different IP networks, Mobile IPv4, appeared a few years ago. However, this protocol lacked a method to establish a seamless handover.

A new protocol, trying to overcome the deficiencies of Mobile IPv4, and fully adapted to the new generation of mobile devices and protocols, appeared. This protocol is Mobile IPv6 [4], based on the IPv6 protocol. Nevertheless, the seamless handover problems had not been solved yet, and some extensions trying to come up with the definite solution were introduced. The most important extensions to Mobile IPv6 protocol are Hierarchical Mobile IPv6 and Fast Handover. The first one tries to minimize the communication with the Home Agent, which may be situated far from the Mobile Node. In this case, this extension opts for reducing the number of (home) network registrations, using a hierarchical scheme. The other important extension is Fast Handover, that tries to minimize the address resolution time.

It has been proven that these extensions do not offer an optimal handover performance, so a few combinations of these protocols have been presented so far. In this article, a new combination of these two extensions, fully adapted to the new generation of mobile devices and applications is proposed. This solution tries to solve the problems of seamless handoff for real time applications.

2   Mobile IPv6

Mobile IPv6[5] protocol was principally designed to solve the mobility problem caused by the IP address architecture. A node is uniquely identified by its IP address, which is composed of two different parts. The first part is the network identifier, a part used to designate the network in which the node is located. The second part is the host identifier, which identifies a single host within a certain network. Due to this architecture, when a Mobile Node moves to another network, it must change its IP address, to get another one with the network prefix corresponding to the new network. This process cannot be done without an interruption in a possible ongoing communication, with the corresponding interruption in the application level.

Mobile IPv6 tries to overcome this problem introducing a level of indirection at the network layer. A special node, called Home Agent, located in the home network, intercepts the packets addressed to the Mobile Node while it is away from the home network, sending the data to the current location of the Mobile Node. To do this, the Mobile Node sends a Binding Update message to the Home Agent, informing about the new IP address, called care-of address, topologically correct, that the Mobile Node has obtained from the new network, normally using the method described in [6]. The Mobile Node can also send the Binding Update message to the Correspondent Node in ongoing communication, to achieve a more direct communication, without the intervention of the Home Agent. All this process is transparent to the user, and to the higher layers, allowing the ongoing communication to continue without any disruption.

However, although this protocol solves the mobility problem, the handoff is not seamless at all. Several things must be taken into account. First of all, the Mobile Node, already in the new network, must register the new care-of address in the Home Agent. This process can take a long time because the home network can be far from the Mobile Node. In this period of time, losses of packets can occur. Secondly, the process of discovering and forming a new care-of address in the new network can also be very time consuming, and packet losses can occur during the process. New extensions trying to solve these two different problems have been proposed in the literature[7,8].

2.1   Hierarchical Mobile IPv6

This extension[7] tries to solve the first problem mentioned above by reducing the number of home network registrations when the Mobile Node moves between a number of networks, thanks to the use of a hierarchical scheme. A new node is presented in this extension, a node called MAP (Mobility Anchor Point). This node defines a regional domain, and all the Mobile Nodes within this domain have a new care-of address, named a regional care-of address, only valid in this regional domain.

So, when a Mobile Node hands off to a new network that belongs to the same regional domain, it must obtain a new local care-of address, using the method described in [6], and register it in the MAP, binding it with the regional care-of address. The MAP intercepts all the packets directed towards the regional care-of address of the Mobile Node and sends them to the current local care-of address.

In the case of a handoff between networks belonging to different MAP domains, the Mobile Node must obtain a new regional care-of address and register it in the Home Agent, which intercepts the packets sent to the home address and redirect them to the regional care-of address.

Therefore, using the hierarchical structure, only in the case of a handover between two MAP domains, a Binding Update to the Home Agent is required.

2.2   Fast Handover

The other extension, presented in [8], tries to minimize the address resolution time forming a new care-of address while still connected to the old access router. Several new messages are utilized in this extension. Upon receiving a layer 2 trigger, informing that the Mobile Node is going to perform a handover, the Mobile Node sends the Router Solicitation Proxy message to the old access router, informing about the imminent handover. The old access router responds with the Proxy Router Advertisement message, indicating either the new point of attachment is unknown, or known but connected through the same access router, or known and connected to a new access router, including, in the latter case, the necessary prefix to form a new care-of address to be used in the new network, using the method described in [6].

In addition to this message, the old access router sends the Handover Initiation message to the new access router, with the newly formed care-of address. In response, the new access router sends the Handover Acknowledgement message, either accepting or rejecting the new care-of address. The Mobile Node, before proceeding with the handoff, sends a Fast Binding Update message to the old access router, binding the old care-of address with the newly-formed care-of address. At this moment, the old access router starts the redirection of the packets directed towards the Mobile Node to the new network, anticipating the packet flow to the actual movement of the Mobile Node.

In response to the Fast Binding Update message, the old access router sends the Fast Binding Acknowledgement message to the Mobile Node, via the new or the old network, and either accepting or rejecting the previous binding. Once the Mobile Node is in the new network, it sends the Neighbor Advertisement message to initiate the packet flow from the new access router to the Mobile Node. After this, the normal Mobile IPv6 procedures are followed, sending the Binding Update to the Home Agent.

3   Previous Work

These two protocols do not offer an optimal performance in the handover process, and a period of packet loss occurs. For this reason, in recent times, a number of attempts to minimize the period of packet loss, using a combination of the two protocols, have appeared.

The scheme proposed in [9] presents a simple superimposition of the two protocols, one after the other. First, it proceeds with the normal Fast Handover mechanism, resulting in the redirection of the packets from the old access router to the Mobile Node, located in the new access router domain. Once in the new network, the Mobile Node sends a Binding Update to the MAP, changing the data flow from the old access router to the new access router.

However, although this method introduces a great advantage over the use of one or no extension, the handover process cannot be considered seamless. A simulation using the NS-2 simulator [10] was carried out, and showed a handover period time of more than 300 milliseconds. The simulation was performed using TCP over Mobile IPv6, and several problems appeared. Some packets retransmitted from the old access router to the new access router, arrived later than some packets directly sent from the MAP to the new access router, the former being chronologically older than the latter. This causes TCP to interpret the packets that arrived later, and chronologically older, as missing packets, starting the congestion avoidance mechanisms and retransmitting such packets.

A new scheme that solves this problem was presented in [11]. It starts doing the same as in the previous scheme, with the fast handover redirection mechanism, from the old access router to the new access router. However, in this case, immediately after the old access router receives the Fast Binding Update, it sends a packet to the MAP indicating that a handoff will occur shortly. MAP proceeds to redirect the packet flow to the new access router, continuing to send packets to the old access router. Doing this, if finally the Mobile Node does not hand off, it will be able to receive the information properly.

In this scheme, two redirection points are introduced, the old access router and the MAP, but TCP retransmission mechanisms are not triggered because the two redirection flows are marked as coming from different nodes. Therefore, the Mobile Node will first receive packets from the old access router, which probably arrived later to the new access router than some packets from the MAP, but which were stored in a different buffer, thanks to the flow differentiation. Consequently, the buffer containing the packets from the old access router is emptied first, and after the completion of the process, it proceeds to do the same with the other buffer, containing the packets coming from the MAP. Simulation carried out using the NS-2 simulator , with TCP Tahoe over Mobile IPv6 showed that with this method the handover is almost seamless.

It is important to note that all these previous schemes and simulations have been performed bearing TCP in mind, a protocol that produces a great deal of complications in the handover process, due to the retransmission mechanism. However, this paper is focused on real time applications, such as voice, where a seamless handover can offer a definitive advantage. In this case, the use of TCP is not suitable. In this kind of applications, the most important thing is for packets to arrive on time, within the delay limits established by a specific application, even with the possible loss of some packets. For this reason, the use of TCP and its retransmission mechanism is counterproductive, and the utilization of UDP is recommended.

4 Proposed Solution

In figure 1, the proposed signaling scheme is shown. The handover process starts with the arrival of a layer 2 trigger, informing that a new network is available at that moment. The Mobile Node responds sending the Router Solicitation Proxy (RtSolPr) message to the old access router, informing this node that the Mobile Node is going to perform a handoff, and indicating the network which it is going to move to. At this moment, the old access router has the knowledge of the subnet prefix of the network the Mobile Node is going to move to, along with the host prefix of the Mobile Node.

In response to the RtSolPr, the old access router sends the Proxy Router Advertisement (PrRtAdv) message, indicating if the network the Mobile Node is going to move to is unknown, known but attached to the same access router, or known and attached to another access router, in which case, this message provides the subnet prefix necessary to form a new care-of address. After this, the old access router forms a new care-of address on behalf of the Mobile Node using [6], and sends a message called Initiation (Init) message to the MAP serving the network to which the Mobile Node is going to move. This message contains

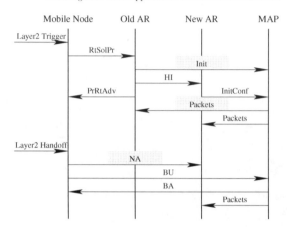

Fig. 1. Signaling scheme of the proposed handoff protocol

the newly-formed care-of address and the care-of address the Mobile Node is using at this moment.

Once the MAP has received this message, it knows that the Mobile Node with the care-of address included in the Init message is going to move to another network, in which the Mobile Node will try to use the second care-of address, the one formed by the old access router. Therefore, the MAP is informed that, very soon, it will have to change the packet flow of the Mobile Node to another network, going through the new access router.

After this, the old access router sends the Handover Initiation (HI) message to the new access router, indicating that the Mobile Node is going to perform a handover to a network that depends on the new access router. Two care-of addresses are included in this message, the care-of address newly-formed by the old access router, with the topologically correct subnet prefix of the new network, and the care-of address the Mobile Node is using at this time.

The new access router checks if the new care-of address is valid and although it is possible to perform Duplicate Address Detection [6] with the new address, it may take a long time, delaying the handover process. The chance of the new address being already in use in the new network is low, as the new care-of address has been formed using a unique identifier, based on the MAC address of the Mobile Node. For this reason, we can skip this detection method.

With the result of the checking, the new access router sends a message, called Initiation Confirmation (Init Conf), informing the MAP of which care-of address the Mobile Node will use in the new network. If the new care-of address is valid, the MAP will start a new packet flow to the new care-of address, going through the new access router. If the new care-of address is not valid, the MAP will set up a temporary tunnel to the new access router, using the old care-of address. The moment the MAP receives the Init Conf message, it starts to redirect the packet flow of the Mobile Node to the new care-of address, but still maintains the packet flow to the old access router. This is done to allow the continuation of the communication in case the Mobile Node does not perform a handoff in the end.

Once the layer 2 handover is performed, the Mobile Node sends the Neighbor Advertisement (NA) message to the new access router, indicating that the Mobile Node is currently attached to this new network. The new access router starts sending the packets that it is already receiving from the MAP. After this, the Mobile Node sends a Binding Update (BU) message to the MAP, to confirm that the Mobile Node is using the new care-of address correctly. When the MAP receives this message, it stops the communication with the old access router. If after a certain period after the MAP has received the Init Conf message, it has not received the Binding Update message, it stops sending the packet flow to the new access router, meaning that, in the end, the Mobile Node did not perform the handoff. The final message is the Binding Acknowledgement (BA), with the same use as in [7].

5   Conclusions

This paper assumes that real time applications use UDP, so no further studies about the impact of the retransmission mechanism of TCP on the proposed scheme have been carried out.

The proposed scheme offers some advantages. The most important one is that only one redirection point is used, the MAP, not two as in other schemes, the MAP and the old access router. Besides, in our scheme, the MAP could not be considered a redirection point, because it is part of the final path to the Mobile Node. Doing this, we achieve a much more simplified scheme, with no need for either double buffering or flow identification, and preserving all the efficiency required for a seamless handoff. The decision to move the redirection point from the old access router, as in [7,9,11], to the MAP was based on the idea that the redirection through the old access router adds an extra delay to the end-to-end delay experienced by the packets. This fact can be unacceptable in many cases.

This scheme takes advantage of two key points. The first one is that when the old access router knows the identity of the Mobile Node and the new network is going to move, it has all the information necessary to form the new care-of address. At this point, the old access router informs the new access router and the MAP about the possible change of location of the Mobile Node. Secondly, the last decision about the possible use of the new care-of address is taken by the new access router. Inmediatly after this address has been approved by the new access router, it sends the signal to the MAP to initiate the packet redirection. It is not necessassary to receive the Fast Building Update message from the Mobile Node because it has no further information about the actual change of network, only the previously received Layer 2 Trigger. For this reason, the starting of the packet redirection can be done even before the Mobile Node receives the Proxy Router Advertisement message.

Consequently, this protocol is able to perform a seamless handover and provide full support to real time applications using a very simple signaling scheme and reducing the complexity and resources of other proposals.

In order to demonstrate the applicability of this new scheme we are developing a new network simulator. This simulator uses object oriented programming techniques, and the main aim is to obtain more reliability than previous work with other simulators.

References

1. The Assar Project: Looking at real time properties using Hierarchical Mobile IPv6 and Fast Handover. Communication System Design, KTH, Stockholm, Sweden. http://2g1319.ssvl.kth.se/~csd2002-mip/
2. Marques, V., Aguiar, R., Fontes, F., Jaehnert, J., Einsiedler, H.: Enabling IP QOS in Mobile Environments. IST Mobile Communications Summit 2001, Barcelona, Spain, September 9-12, 2001, pp. 300-305.
3. Köpsel, A., Wolisz, A.: Voice transmission in an IEEE 802.11 WLAN based access network. Proceedings of the 4th ACM international workshop on Wireless mobile multimedia. Rome, Italy 2001 Pages: 23 - 32
4. Willén, J.: Introducing IPv6 and Mobile IPv6 into an IPv4 wireless Campus network. M.Sc. Thesis. KTH, Stockholm, Sweden. 27th February 2002
5. Johnson, D., Perkins, C., Arkko, J.: Mobility Support in IPv6. Internet Draft, IETF, June 2003. Work in progress.
6. Thomsom, S., Narten, T.: IPv6 stateless address autoconfiguration. RFC 2462, IETF, December 1998.
7. Soliman, H., Castellucia, C., Malki, K., Bellier, L.: Hierarchical MIPv6 mobility management. Internet Draft, IETF, June 2003. Work in progress.
8. Koodli, R.: Fast Handovers for Mobile IPv6. Internet Draft, IETF, October 2003. Work in progress.
9. Hsieh, R., Sereviratne, A., Soliman, H., El-Malki, K.: Performance Analysis on Hierarchical Mobile IPv6 with Fast-Handoff over TCP. Proceedings of Globecom, Taipei, Taiwan, 2002.
10. The ns-2 Website: The Network Simulator ns-2 http://www.isi.edu/nsnam
11. Hsieh, R., Zhou, Z.G., Seneviratne, A.: S-MIP: A Seamless Handoff Architecture for Mobile IP. Proceedings of INFOCOM, San Francisco, USA, 2003.

Fast Handoff Algorithm Using Access Points with Dual RF Modules

Chun-Su Park[1], Hye-Soo Kim[1], Sang-Hee Park[1],
Kyunghun Jang[2], and Sung-Jea Ko[1]

[1] Department of Electronics Engineering, Korea University, Anam-Dong,
Sungbuk-Ku, Seoul 136-701, Korea.
Tel: +82-2-3290-3672
{cspark, hyesoo, jerry, sjko}@dali.korea.ac.kr
[2] i-Networking Lab. Samsung Advanced Institute of Technology, Suwon, Korea.
khjang@samsung.com

Abstract. With the spread of portable computers, mobile users rapidly
increase and have growing interests in wireless LAN (WLAN). However,
when a mobile node (MN) moves, handoff can frequently occur. The
frequent handoff makes fast and seamless mobility difficult to achieve.
Generally the handoff delay is so long that a provider can not support
realtime services sufficiently. In this paper, to address the link layer (L2)
handoff delay problem, we propose a fast handoff algorithm using access
points (APs) with dual radio frequency (RF) modules. The proposed
handoff algorithm is based on the modified neighbor graph (NG). Experimental results show that the proposed method reduces the L2 handoff
delay drastically. Furthermore, the handoff delay in the network layer
(L3) can be also reduced simultaneously, since the proposed algorithm
can support L2 triggers.

1 Introduction

There has been considerable interest recently in WLAN. The main issue in
WLAN is handoff management between APs [1][2]. As a moving MN may ir-
regularly need to change the associated AP, the APs must be identified and
the target AP must be selected. When this process is finished, the connecting
process begins [3]-[5]. The whole handoff procedure can be divided into three
distinct logical phases [6]: scanning, authentication, and reassociation. During
the first phase, an MN scans for APs by either sending *ProbeRequest* messages
(Active Scanning) or by listening for *Beacon* messages (Passive Scanning). After scanning all channels, an AP is selected by the MN using the received signal
strength indication (RSSI), link quality (LQ), and etc. Then, the MN exchanges
IEEE 802.11 authentication messages with the selected AP. Finally, if the AP
authenticates the MN, an association moves from an old AP to a new AP as
following steps:

(1) An MN issues a *ReassociationRequest* message to a new AP. The new AP
must communicate with the old AP to determine that a previous association
existed;

M. Freire et al. (Eds.): ECUMN 2004, LNCS 3262, pp. 20–28, 2004.
© Springer-Verlag Berlin Heidelberg 2004

(2) The new AP processes the *ReassociationRequest*:

(3) The new AP contacts the old AP to finish the reassociation procedure with inter access point protocol (IAPP) [7][8]:

(4) The old AP sends any buffered frames for the MN to the new AP:

(5) The new AP begins processing frames for the MN.

The delay incurred during these three phases is referred as the L2 handoff delay, that consists of probe delay, authentication delay, and reassociation delay. Mishra [9] showed that scanning delay is dominant among the three delays. Thus, to solve the problem of the L2 handoff delay, the scanning delay has to be reduced or abbreviated.

In this paper, to minimize the disconnected time while an MN changes the associated AP, we propose a fast handoff algorithm using APs with dual RF modules. The proposed handoff algorithm is based on the modified NG. By adding to AP an RF module that can only receive signals (SNIFFER), the L2 handoff delay is reduced drastically. The proposed algorithm can eliminate the scanning phase of the MN.

This paper is organized as follows. We briefly review the NG and introduce the modified NG in Section 2. In Section 3, the AP with dual RF modules is presented. Then, Section 4 describes a fast handoff algorithm using AP with dual RF modules. Finally, Section 5 shows the results experimented on our test platform and presents brief conclusion comments.

2   Modified NG

Before introducing our proposed method, we briefly review the NG. In this section, we describe the notion and motivation for neighbor graphs, and the abstractions they provide. Given a wireless network, an NG containing the reassociation relationship is constructed [10].

Reassociation Relationship: Two APs, ap_i and ap_j, are said to have a reassociation relationship if it is possible for an MN to perform an 802.11 reassociation through some path of motion between the physical locations of ap_i and ap_j. The reassociation relationship depends on the placement of APs, signal strength, and

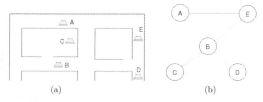

(a)   (b)

Fig. 1. Concept of neighbor graph. (a) Placement of APs. (b) Corresponding NG.

other topological factors and in most cases corresponds to the physical distance (vicinity) between the APs.

Data Structure (NG): Define an undirected graph $G = (V, E)$ where $V = \{ap_1, ap_2, \cdots, ap_n\}$ is the set of all APs constituting the wireless network. And the set E includes all existing edges e_{ij}'s where $e_{ij} = (ap_i; ap_j)$ represents the reassociation relationship. There is an edge e_{ij} between ap_i and ap_j if they have a reassociation relationship. Define $N(ap_i) = \{ap_{i_k} : ap_{i_k} \in V, e_{ik} \in E\}$, i.e., the set of all neighbors of ap_i in G.

While, the NG can be implemented either in a centralized or a distributed manner. In this paper, the NG is implemented in a centralized fashion, with correspondent node (CN) storing all the NG data structure (see Fig. 6). The NG can be automatically generated by the following algorithm with the management message of IEEE 802.11.

(1) If an MN associated with AP_j sends Reassociate Request to AP_i, then add an element to both $N(ap_i)$ and $N(ap_j)$ (i.e. an entry in AP_i, for j and vice versa);
(2) If e_{ij} is not included in E, then creates new edge. The creation of a new edge requires longer time and can be regard as 'high latency handoff'. This occurs only once per edge.

The NG proposed in [10] uses the topological information on APs. Our proposed algorithm, however, requires channels of APs as well as topological information. Thus, we modify the data structure of NG as follows:

$$
\begin{aligned}
&G' = (V', E), \\
&V' = \{v_i : v_i = (ap_i, channel), v_i \in V\}, \\
&e_{ij} = (ap_i, ap_j), \\
&N(ap_i) = \{ap_{i_k} : ap_{i_k} \in V', e_{ik} \in E\},
\end{aligned}
\tag{1}
$$

where G' is the modified NG, and V' is the set which consists of APs and their channels. In Section 5, we develop a fast handoff algorithm based on the modified NG.

3   Improved Network Architecture

The MN scans all channels from the first channel to the last channel, because it is not aware of the deployment of APs near the MN. The long L2 handoff delay caused by scanning phase makes internet services such as VoIP and gaming not realizable. The fast and reliable handoff algorithm is a critical factor for providing seamless communication services in WLANs.

In general, the AP contains an RF module which can receive and transmit signals by turns in the allotted channel (see Fig. 2 (a)). By adding an RF module which can only receive signals to the AP (SNIFFER), i.e., the AP has two RF modules, the AP can eavesdrop channels of its neighbor APs. Figure 2 (b) shows the proposed network architecture. If the MN enters the cell range of a new AP,

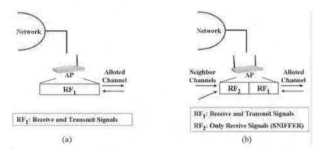

(a)   (b)

Fig. 2. (a) General AP with single RF module. (b) Proposed AP with dual RF modules.

the SNIFFER of the new AP can eavesdrop the MAC frame. Thus, by examining the MAC frame of incoming MN, the new AP can get the address of the AP associated with the MN.

As shown in Fig. 3 (a), the Basic Service Set (BSS) can use only three channels at the same site because of interference between adjacent APs. For example, in the USA, channels 1, 6, and 11 are used. If the SNIFFER knows the channels of neighbor APs, it does not need to receive frames in all channels. In Fig. 4, for example, the SNIFFER of the AP3 receives frames in channels 1 and 6 selected by the modified NG. Assume that the destination address in the received frame is the address of AP2. Then, AP3 informs its neighbor AP (AP2) that the MN associated with neighbor AP (AP2) enters the cell range of AP3.

The proposed network does not require any change of the MN. Existing wireless network interface card (WNIC) does not require extra devices but can be serviced simply by upgrading software. This is very important factor to the vender. In the next section, we will show in detail the difference between standard network and the proposed network architecture.

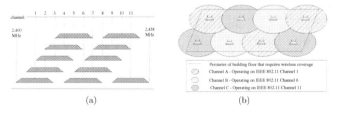

(a)   (b)

Fig. 3. Selecting channel frequencies for APs. (a) Channel overlap for 802.11b APs. (b) Example of channel allocation.

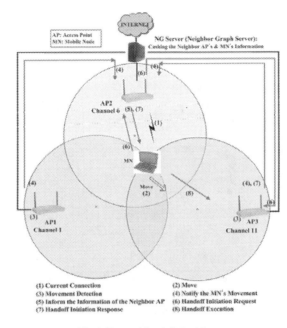

Fig. 4. Proposed handoff algorithm.

4   Proposed Handoff Algorithm

Before handoff, the MN or network should first decide the handoff depending on certain policy. Handoff policies can be classified into three categories [11]: network controlled handoff, network controlled and MN assisted handoff; MN-controlled handoff. In IEEE 802.11 WLAN, a handoff is entirely controlled by the MN.

In this paper, we propose a new type of fast handoff algorithm that is MN-controlled and network assisted. Network encourages the MN to handoff by providing it with information on the new AP. And the MN decides whether the network situation matches the handoff criterion. If it matches the handoff criterion, the MN performs handoff according to received new AP information. Figure 4 shows the proposed handoff procedure:

(1), (2) The MN associated with AP2 moves toward AP3:
(3) The SNIFFERs of AP1 and AP3 receive the MN's MAC frames:

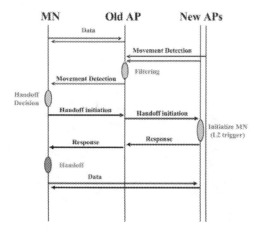

Fig. 5. Flowchart.

(4) Using the destination address in the MAC frame of the MN, SNIFFERs of AP1 and AP3 can know that the MN has been associated with AP2. Let us denote an RSSI between APi and the MN by $RSSI_{APi}$. If the $RSSI_{AP3}$ is over the threshold, SNIFFERs of the AP3 send AP2 a message including handoff information such as the measured RSSI, the MAC address of the MN, and so on:

(5) As the WLAN is very scarce resource, AP2 must not relay all messages from other APs to the MN. After removing redundant messages, AP2 relays messages to the MN:

(6) The MN decides handoff if the $RSSI_{AP3}$ in the received message exceeds the $RSSI_{AP2}$ and the $RSSI_{AP2}$ is below a threshold T (handoff if $RSSI_{AP3}$ > $RSSI_{AP2}$ and $RSSI_{AP2}$ < T). After deciding handoff, the MN sends a handoff initiation message to AP3 through AP2. If the MN does not need handoff, it does not issue a handoff initiation message:

(7) AP3 send a response message to the MN. In this phase, AP3 can supply L2 triggers to L3:

(8) The MN performs handoff from AP2 to AP3.

The proposed handoff algorithm omits the scanning phase by using the SNIFFER. And the MN can be authenticated and associated before handoff [12]. If network allows this preprocess, the proposed handoff algorithm producing zero delay can be implemented in L2. Furthermore, as the proposed algorithm supplies L2 triggers, the L3 handoff delay can be diminished drastically.

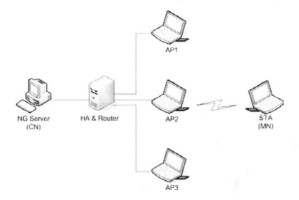

Fig. 6. Experimental platform.

5   Experimental Results and Conclusion

Figure 6 shows our experimental platform consisting of an MN, APs, router, and Correspondent Node (CN). All machines we used are SAMSUNG SENSV25 laptops with Pentium IV 2.4 GHz and 1,024 MB RAM. All machines are running RedHat Linux 9.0 as the operating system. To exchange the NG information, socket interface is used, and Mobile IPv6 is applied to maintain L3 connectivity while experimenting. The device driver of a common WNIC was modified so that the MN operates as an AP that can support handoff initiation message. For the proposed mechanism, we developed three programs: NG Server, NG Client, and SNIFFER. The NG Server manages the data structure of NG on the experimental platform and processes the request of the NG Client that updates the NG information of the MN after the MN moves to the another AP. Using destination address in the MAC frame of the MN, the SNIFFER can be aware of the AP associated with the MN. If the RSSI of the received frame is over the threshold, the SNIFFER sends the old AP a message including handoff information such as the measured RSSI, the MAC address of the MN, and so on.

In general, the wireless MAC protocol can be implemented using device driver and firmware. The wireless MAC protocol can not be easily extended or modified because wireless MAC protocol is implemented in dedicated firmware. But, in our experiment, we measured RSSI, LQ, and L2 handoff delay in both firmware and device driver.

In the proposed handoff scheme, the handoff delay is defined by the duration between the first ReassociationRequest message and the last ReassocationResponse message. But, the device driver of the WNIC starts handoff with Join process. So, we measure the L2 handoff delay between Join and ReassocationRe-

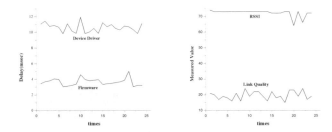

Fig. 7. L2 Handoff delay and the other parameters.

sponse message. To diminish the handoff delay, we also modified Join process in firmware. The handoff delay experienced by a single mobile device is presented in Fig. 7 with RSSI and LQ. In Fig. 7, the RSSI value is reported on chip set and is linear with signal level in dB. We can simply get signal level in dBm on a radio by

$$RSSI(dBm) = RSSI(measured\ value) - 100. \qquad (2)$$

The LQ is available on handoff algorithm in many ways. This measure is the SNR in the carrier tracking loop and can be used to determine when the demodulator is working near to the noise floor and likely to make errors.

Since the handoff delay in the device driver includes message processing time in addition to receiving and transmitting time, the handoff delay in the device driver is longer than the handoff delay in the firmware. Our experimental results show that the L2 handoff delay is about 11msec in device driver. But, if we optimize our experimental platform, the L2 handoff delay can be reduced to measured delay in firmware. Thus, by comprising the other L3 solution, realtime seamless multimedia service can be realized sufficiently.

To address the L2 handoff delay problem, we proposed a fast handoff algorithm using AP with dual RF modules that can eliminate the scanning phase of the MN. Since proposed algorithm enables the MN to be authenticated and associated before handoff, it produces zero delay in L2. But careful attention has to be paid to the design of L3 layer protocol to minimize total handoff delay. In our next paper, we will provide details of the L3 protocol to support the fast handoff.

Acknowledgement. The work was supported by the Samsung Advanced Institute of Technology (SAIT).

References

1. Koodli, R.: Fast Handovers for Mobile IPv6. IETF Draft (2003)
2. Ohta, M.: Smooth Handover over IEEE 802.11 Wireless LAN. Internet Draft: draft-ohta-smooth-handover-wlan-00.txt (2002)
3. Koodli, R., Perkins, C.: Fast Handovers and Context Relocation in Mobile Networks. ACM SIGCOMM Computer Communication Review **31** (2001)
4. Cornall, T., Pentland, B., Pang, K.: Improved Handover Performance in wireless Mobile IPv6. ICCS. **2** (2003) 857–861
5. Montavont, N., Noel, T.: Handover Management for Mobile Nodes in IPv6 Networks. IEEE Communications Magazine (2002) 38–43
6. IEEE: Part 11: Wireless LAN Medium Access Control (MAC) and Physical Layer (PHY) Specifications. IEEE Standard 802.11 (1999)
7. IEEE: Draft 4 Recommended Practice for Multi-Vendor Access Point Interoperability via an Inter-Access Point Protocol Across Distribution Systems Supporting IEEE 802.1f/D4 Operation. IEEE Standard 802.11 (2002)
8. IEEE: Draft 5 Recommended Practice for Multi-Vendor Access Point Interoperability via an Inter-Access Point Protocol Across Distribution Systems Supporting IEEE 802.1f/D5 Operation. IEEE Standard 802.11 (2003)
9. Mishra, A., Shin, M. H., Albaugh, W.: An Empirical Analysis of the IEEE 802.11 MAC Layer Handoff Process. ACM SIGCOMM Computer Communication Review **3** (2003) 93–102
10. Mishra, A., Shin, M. H., Albaugh, W.: Context Caching usingG Neighbor Graphs for Fast Handoff in a Wireless. Computer Science Technical Report CS-TR-4477 (2003)
11. Akyildiz, I.F., et al.: Mobility management in next-generation wireless systems. Proceedings of the IEEE **87** (1999) 1347–1384
12. Park, S. H., Choi, Y. H.: Fast Inter-AP Handoff Using Predictive Authentication Scheme in a Public Wireless LAN. IEEE Networks ICN. (2002)

UMTS-WLAN Service Integration at
Core Network Level

Paulo Pinto[1], Luis Bernardo[1], and Pedro Sobral[2]

[1] Faculdade de Ciências e Tecnologia, Universidade Nova de Lisboa,
P-2829-516 Caparica, Portugal, {pfp,lflb}@uninova.pt
[2] Faculdade de Ciência e Tecnologia, Universidade Fernando Pessoa,
Porto, Portugal, pmsobral@ufp.pt

Abstract. The integration of wireless LANs (WLANs) and 3G systems performed at core network level requires very little modifications to the current
3GPP architecture and provides a large set of benefits: seamless service integration, exploration of user mobility by applications, seamless use of different radio access network (RANs), easy availability of current services such as Short
Message Service (SMS) in other RANs, etc. This paper describes an architecture that has the GPRS as the primary network. Each of the other networks has
a 3G core-level component to manage it and to perform the integration. Vertical handovers between RANs are not needed and secondary networks are used
on an availability basis. Users can have at least one session per RAN that is
maintained even when they are moving in dark areas of that RAN (and the
communication is still possible via the primary network). Our proposal does
not require the system to be *all-IP*, but simply *IP-enabled*.

1 Introduction

The traditional approaches to the integration of wireless LANs (WLANs) and 3GPP
systems have avoided any changes to the core 3GPP network. WLANs can either
appear as 3GPP cells (known as the *tightly coupled* approach [1], [2]) or interact with
the 3GPP at IP level (the *loosely coupled* approach [1], [3]). If one considers that it
might be possible to change the core network (basically minor software upgrades) the
integration of wireless systems can become powerful and attractive at various levels.

We assume that the GPRS (General Packet Radio Service) network is ubiquitous and
forms the primary radio access network (RAN)[1]. Users can have sessions over the
different RANs and these sessions persist over out of range periods. Consequently,
users can always be contacted (either using that RAN, or the GPRS). All non-primary
RANs are used as a complement to the GPRS making vertical handovers less critical.
In fact, there is not really any vertical handover as users maintain the GPRS connectivity all the time.

---

[1] In the rest of the paper, we consider GPRS as a packet service in both 2.5G and 3G 3GPP
systems

M. Freire et al. (Eds.): ECUMN 2004, LNCS 3262, pp. 29–39, 2004.

A second aspect refers to the management of service contexts (i.e. the possibility for a service to use one RAN or another). If the core has some awareness of different RANs it allows the exploration of the user mobility in such aspects as connection availability in other RANs and decisions for forwarding traffic through certain RANs. Currently, user mobility in 3GPP systems is mainly concerned with maintaining the bearer services (control and data channels) to enable communication. The core just pushes packets through without any high level concern such as different cell capacity. The current work in 3GPP for interworking with WLANs leaves most of these issues to further work. This paper is also a contribution with a different angle to the problem. Lastly, we stress a view of more autonomy to WLAN owners. They trust on the 3G system to authenticate users but remain with all the power to authorize them. The 3GPP has, naturally, a more 3G centric view of the problem.

Our envisaged application scenarios are an extension of the infostation model [4] with cellular network integration. Imagine a user landing on an airport. Once there, he starts a session using the airport's WLAN. He wants to download a report. Next he takes a taxi to the hotel. On his way, any WLAN will be used to send parts of the report (semaphores, etc.). Inside the taxi, in areas not covered by WLAN, the user can still be contacted using the GPRS RAN. Eventually all the report will be transferred by WLAN. When he arrives at the hotel (which has also a WLAN) the same session is still on. We assume that UEs (User Equipments) are equipped with two, or more, wireless interfaces working simultaneously.

2 Hotspot Integration

In the future, 3GPP cells will be smaller and will have more bandwidth. However, extremely high rates will not be necessary everywhere, but just in small hotspots [5]. How will these hotspots be integrated?

2.1 Homogeneous Approach

One possibility is that these new cells will make WLAN integration useless because they will have the same characteristics – they are 3G cells. There are some drawbacks though. The network would have to predict the user movement (using cell information) to schedule data when the user enters in this kind of cells. It is a hard task to be performed at network level because it needs knowledge of the application. Such seamless environment is also difficult for users because they can step out of the cell and feel a drop in the bandwidth. The *tightly coupled* approach [1], [2] of integrating WLANs is somehow similar to this homogeneous approach: WLAN cells behave like ordinary cells offering an interface compatible with the 3GPP protocols. This approach has further disadvantages: (a) the WLAN must be owned by the 3GPP operator (due to strong exposure of core network interfaces); (b) it is hard to incorporate a WLAN cell because cell displacement demands carefully engineered network planning tools and because a great deal of control procedures are based on configuration parameters (CellID, UTRAN Registration Area (URA), Routing Area (RA), etc.); (c) paging procedures, registration updates and handovers (including vertical handovers) have to be defined and some technologies (e.g. IEEE 802.11) are not so optimized to make them fast enough; and (d) high-rate traffic has to pass through the current 3GPP core network.

2.2 Heterogeneous Approach

Another way to integrate WLANs is the *loosely coupled* approach [1][3]. It assumes there is a WLAN gateway on the WLAN network with functionalities of Foreign Agent, firewall, AAA (Authentication, Authorization and Accounting) relaying, and charging. The connection to the 3GPP uses the GGSN (Gateway GPRS Support Node) having a functionality of Home Agent. It only makes sense to use this option with multi-mode UEs because a vertical handover to WLANs would disconnect the UE from all the functionality of the cellular networks (paging, etc.). One advantage is that high-speed traffic is never injected into the 3GPP core network. A major disadvantage is the degree of integration. WLAN networks are handled independently and will be used on an availability basis by the users, whom have to stay within the same coverage. WLAN access to any service provided by the 3GPP (e.g. SMS) has to consider the cellular system's internet interface becoming more complex. Any exploitation of the UE's mobility (both in the cellular system and inside the WLAN aggregation of cells) is hidden by the mechanism of Mobile IP, for instance. From an application point of view, the UE is stationary, placed inside a big cloud called GPRS (or WLAN). I.e. it has a stable IP address and any mobility inside the 3GPP network is not seen from the exterior.

2.3 Core-Level Approach

Yet another possibility is that high bandwidth cells are seen as *special cells*, not integrated in the cellular system and having a special (direct) connection to a packet data network. The user knows he is using a different interface and stepping out of coverage is easy to detect. This possibility is easy to implement if the integration is performed somewhere in between the tightly and the loosely coupled approaches – at core network level. The packet data network of these *special cells* is added to the current 3GPP core network and can communicate with the current elements (SGSN, HSS, etc.). Any communication from the core to the UE (regardless of the RAN that is used) does not have to leave the core. This makes some fundamental differences towards the loosely coupled approach as we will see. A first one is related with authentication and authorization: in 3GPP, users become valid after an AAA procedure with the core and can use any available service. Having an AAA procedure in WLANs identical to the one used in UMTS allows the delivery of any packet from the core (either belonging to GPRS or to other WLANs) using any RAN. At a certain layer in the core there is no notion of services, but only packets. A second difference is related with mobility management: a cellular network has its own model to handle mobility (i.e. below the IP level with its own authentication procedures and control nodes). The current state-of-the-art in the Internet is Mobile IP where care-of-addresses and tunnels are used to hide mobility. In the heterogeneous approach the GGSN is overloaded with these tasks. The introduction of a control node inside the core for WLANs avoids such complexity and unifies the mobility management model with the one used in 3GPP. This control node can also offer a standard and protected programming interface for developing new services that are aware of both mobility and available RANs. In summary, this core-level approach allows the use of WLAN as a complement to the GPRS network.

3GPP defined requirements for six scenarios [6] of increasing levels of integration between 3GPP systems and WLANs. Scenario 3 addresses access to 3GPP packet services including access control and charging; [7] specifies how it should be done. An IP access to 3G services was adopted (loosely coupled approach)

3 Architecture

3.1 Primary and Secondary RANs

The GPRS forms the primary network. It is the only one to have control services (paging, compulsory registration updates, etc.) and the user is always attached to it. All other networks (secondary networks) are simpler and are basically data networks (if they have a paging facility to save battery life, it is only an internal optimization not seen at core network level). Most of the works in internetworking [8][9][10] assume that all control features exist in all networks and are seen at core level. IDMP [11] is one of the exceptions stating that they should be customized. The most similar approach to ours was taken by MIRAI [12]. They also have a primary network but it is mostly concerned about control features – the Basic Access Network (BAN). Its most important task is to help users in choosing a RAN. The choice is based on a list provided by the BAN considering user location and preferences. Although the authors consider a long list of issues to help the UE choose the RAN, some too low level or "external" reasons (e.g. battery life) can lead to unexpected choices from the applications' point of view. The control features of the BAN are very similar to the ones in UMTS. It could have been implemented by the 3G system (as also stated in [12]) but MIRAI authors decided to implement a new radio interface.

Other works consider all RANs at the same level. [13] defines a flow router at the core that uses all RANs. This will lead to the existence of control functions in all of them. If only one is chosen to have these features the system will fall back to ours. Moreover, with a monolithic core it would be more difficult to add a new RAN.

3.2 General Description

Secondary RANs are not ubiquitous. Sets of cells form islands and the group of islands belonging to a certain technology (e.g. 802.11, Hyperlan) is seen as a *Hotspot*

Network (HN) (Figure 1). Each island is controlled by a local Island Manager (IM), and a component at the core, called HNAC (Hotspot Network Area Controller), is responsible for one, or more, islands of the same HN. One task of HNAC is also to maintain a user session regardless of the connection status of the user at a certain moment.

Figure 2 shows the main components responsible for the data traffic. The novel parts at core level are the connection between the SGSN (Serving GPRS Support Node) and the HNAC; the ability of the HNAC to access information stored at the HSS (Home Subscriber Server); and a component called GHSN (Gateway Hotspot network Support Node) which is responsible for context management and Internet access (just the way GGSN is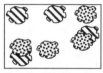

Fig. 1. WLANs form islands

for GPRS). The thicker lines (at the right of the IM) belong to the core but they are not present in the current 3G core. All the high speed traffic goes through them not overloading the current 3G core.

The 3GPP specification for scenario 3 [7] has a component called PDG (Packet Data Gateway) that gives IP access to 3G services (including external Internet access). All data passes through the PDG and there is no direct contact with the 3G core as all interactions are at IP level.

An UE has a unique identification at core level in the form of its IMSI (Int. Mobile Subscriber Identity). The idea is that when a packet is destined to an IMSI UE it can travel through the UTRAN to the UE identified by the IMSI, or through the WLAN RAN to the same UE now identified by an HN dependent identifier. In figure 2 an IP address, IPa, was chosen as this dependent (local) identifier. In the sequel it will be seen that it is not relevant that this address has to be an IP address. It is only necessary that it remains stable at IM interface.

As stated above, the UE is always attached to the GPRS network. It can create a session (PDP context) and define a stable IP address at GGSN (IP_1 in Fig. 2). On the

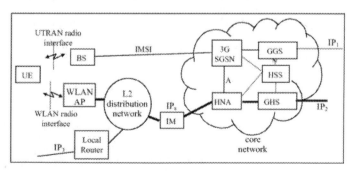

Fig. 2. Data traffic in the hybrid network

other hand, HN sessions can be established in two ways: directly over the WLAN RAN, or indirectly over the UTRAN.

The first way happens when an UE senses a WLAN and discovers it has a 3GPP agreement. It performs a two phase attachment procedure with secure authentications – first to the local WLAN and second to the 3GPP system. The first phase is not covered by 3GPP standards. Our solution is a challenge/response procedure between the UE and the IM (with the assistance of the 3G system). The WLAN trusts the 3G network in terms of authentication but remains with the power to authorize the user to use local services. The main operations in this first phase are the following: the IM asks the HNAC for a challenge based on authorization vectors of the 3GPP. The IM relays the challenge to the UE, and the response back to the HNAC. The HNAC informs the IM of the UE identity (e.g. its MSISDN, as this is not a sensitive piece of information). The IM will then use the MSISDN to verify in its authorization database if the UE can use local services. If it does, the address IP3 is defined by the Local Router (the functionality of the Local Router can reside in the IM) and session keys are provided by the IM. The second phase provides access to the 3G network. It can be done regardless of the outcome of the first phase. This second phase follows the lines of [7]. Once registered in the HN the UE can create a context with the GHSN defining a stable and routable IP address, IP_2.

The second way (HN session establishment via UTRAN) can happen during dark areas and the UTRAN is used to connect the UE to one HNAC (see below).

3.3 Overview of the Core Network Interactions

The HSS has information about the UEs (identity and routing information, etc.). It must be enlarged with information about HN related status (registered, reachable, current identification, serving HNAC, etc.). HNAC will go to HSS to get updated information in the same way as the SGSN goes nowadays. The HSS also provides authentication vectors, subscriber profiles, and charging information to HNACs.

We will describe two working scenarios: the smooth one that requires very little changes to the current core network; and the abrupt one demanding more modifications. In the smooth one, the PS domain works as today and the HNAC can access the UE directly or via SGSN (line A in fig. 2). An easy way to implement this indirect via is to allow the HNAC to establish a PDP context with the SGSN (similar to the PDP context of the GGSN). A GTP-U (GPRS Tunneling Protocol – for User Plan) is established between the HNAC and the serving RNC (Radio Network Controller) and used at HNAC discretion. The second scenario is more interesting and both SGSN and HNAC can convey their traffic through the other one if they see some advantage. One first change is the partition of the current GTP-U tunnel between the GGSN and the SRNC into two half-tunnels: GGSN-SGSN and SGSN-SRNC. It is very much a return to the original GPRS specification. The latter second half-tunnel ise replaced by a SGSN-IM connection via HNAC. So, the major modification is the ability of the SGSN to use the HNAC to reach the UE. Before describing in functional terms the new interfaces let's see how the communication between the core and the UE takes place.

Figure 3 shows the protocol stack in the UE. There is a Connection Manager (CM) that manages the status of both connections and offers a unique upward interface to both RANs. The Delivery Service (DS) operating over WLANs is a confirmed service and switches to the UTRAN if it senses a failure. The DS operating over UTRAN is not confirmed because the UTRAN is ubiquitous. If more than one RAN is active the default one for each message is used, unless otherwise indicated by the application. The CM can signal the applications (and be queried by them) about the current status of a specific connection. With this information, applications can avoid using the link if the proper interface is not active (transferring only urgent or control information, for instance). The CM is able to contact each of the core network components (SGSN or HNAC) either directly or via the other RAN (for link maintenance messages, etc.). The session control layer is responsible for session survival when the UE is in dark areas.

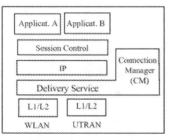

Fig. 3. Protocol stack at a UE

In figure 3 the DS of the core component (SGSN or HNAC) communicates with the DS of the UE using L1/L2 protocols. VLANs could be used. However, IP-tunnels can as well be used. In this case the DS establishes a tunnel to the peer DS, and the protocol stack would look a little bit different from the one pictured in the figure.

This setting (core-level approach) is functionally similar to the tightly coupled approach [1], without the need to expose the core network interfaces (Iu). Instead, the HNAC has to be introduced in the core.

3.4 Interfaces in the Core Network

For the second scenario (abrupt) the following interactions are needed:

(a) HSS must store information about the HN activity of UEs. It must provide this information to HNACs as well as authorization vectors, registration procedures, updates, etc. It is assumed that HNACs have the functionality of an AAA proxy;

(b) SGSN must perform the relaying of packets between the HNAC and the UE (this is better done by the DS than by creating an IP tunnel);

(c) HNAC must perform the relaying of packets between the SGSN and the UE (again, the DSs will cooperate to achieve this). (both (b) and (c) allow core components to communicate with the UE via the other RAN);

(d) logic must exist in the SGSN to enable it to use the WLAN in advantageous circumstances;

(e)   SGSN must have an event service to notify any interested core component (namely HNAC) about relevant events – "UE availability", "cell update", "routing area update", "positive cell identification" and "undefined cell identification";

(f)   SGSN must provide access to its mobility management information (cell identification if in GPRS state *ready*, or routing area identification, otherwise) - it can be useful for the HNAC if it has a relation between CellIDs and WLAN placements (it can force the WLAN interface to switch off if no islands are known in a certain routing area, for instance). It is also important because HNAC change of responsibility can happen when the UE performs a routing area update.

4 Discussion and Performance Evaluation

In our system there is no need for vertical handovers because the GPRS session is always on and the other RANs are used as a complement. Communication to the UE can use indistinguishably any available RAN. As the choice of RANs is performed by the core components, no information is ever lost. In systems with traditional vertical handover, the dominant factor is the time the UE takes to discover that it has moved in/out of coverage (i.e. the cell has to become active or inactive) [14]. Figure 4 shows the procedures when an UE loses and finds coverage (thin and blank lines represent no bandwidth available). The lost of coverage is the most critical of the two [14].

In the tightly coupled approach, when the UE loses coverage a new RAN has to be sensed and the secure associations must be re-established. As the new RAN behaves like the old one everything is in order after that.

In the loosely coupled approach the same thing happens but an extra phase of mobility management must exist (tunnels, care-of-address, etc.). In our approach the new RAN is always ready to be used and the core just starts to use it (if not any other RAN, the UTRAN is always ready).

Considering multi-mode UEs, the sensing and authorization phases could be made in advance and the tightly approach will be similar to ours. However, it is questionable if a new AAA establishment with another cell could be done while the current one is still valid (in the current 3GPP specification

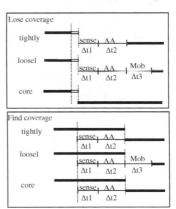

Fig. 4. Handovers between RANs

this is not valid). In the loosely coupled approach these two phases could be done in advance but the mobility management phase cannot.

Another aspect is the requirement of either an *all-IP* architecture or an *IP-enabled* one. The most critical part in our system is the communication between DSs. It can be performed using VLANs and no IP requirements exist, or it can be performed using IP-tunnels. If the tunnel is only established between the core component (SGSN or HNAC) and the IM, the destination of the packets is the UE identified by a stable identifier. The identifier can be IP, or not (in figure 2 is IPa). If the tunnel goes directly from the core to the UE, then the UE has to have an IP local identifier (IPa).

Note that IPa is a local identifier not seen by the applications. For the applications, the UE can either be working with the IP_1 or the IP_2 addresses. These are the working addresses at IP layer of figure 3. The Delivery Service is responsible for forwarding the packets between the IP layer of the core component (HNAC or SGSN) and the IP layer of the UE using the IMSI or the other identifier (IPa in this paper). Therefore, there are no assumptions about the necessity of using Mobile IP, for instance. The mobility management inside an island is performed by the IM without any knowledge of the core components. It can be based on IP, use VLANs, etc. So, if IPa is used in the routing process, or not, is not a requirement either. The stability of IPa is needed to avoid resolution towards the IMSI every time the core wants to use it.

HNACs can play an important role in user mobility management. They can have a standard (and protected) programming interface to be used by third-party organizations to build services that take advantage of the information gathered at the core (HSS) not explored fully today (e.g. if the user has a HN session, if it is under coverage, etc.). Most of the mobility management nowadays in 3GPP systems is only concerned on maintaining RABs (Radio Access Bearers). HNACs can store data (locally, or in components inside the core) in cooperation with the application, to defer its transfer until the UE gets into an island again (having the data nearby can be decisively if connection times are very short).

5   Examples of Service Integration

A good example that shows how this architecture can simplify the integration of wireless systems a great deal is taken from [7]. Figure 5, from [7], shows how 3GPP plans to support SMS over WLANs (providing IP bearer capability).

A service specific gateway, called IP-SM-GW, must exist and offer an interface similar to an MSC or an SGSN (interfaces E or Gd) to the GMSC/SMS-IWMSC. The address of this gateway is returned by the HSS in the "send routing information for short message" primitive. This gateway has a private database to associate MSISDN to IP addresses. UEs in WLAN have to specifically register and specifically authenticate for SMS services and have secure associations to the gateway. The gateway communicates with the UE via Internet (PDG, etc.).

In our system (second scenario), the SMS service could be provided without any need for service-specific extra components, service-specific private databases, or

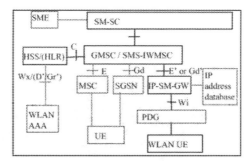

Fig. 5. Support of SMS over WLAN (3GPP version)

service-specific registration by the UEs. The SGSN just gets the message from the GMSC/SMS-IWMSC as before and can use the HNAC to convey the message to the UE, using the secure associations that are already in place between the HNAC and the UE.

6 Conclusions

The internetworking of wireless infrastructures performed at core level with a pivot network seems a simple and executable model with many advantages:

  (a) as most of the control features already exist in the 3GPP network, they can be absent in other networks;
  (b) certain details (such as micro-mobility) are not managed at core level;
  (c) it defines an environment where new features and services can be added to the core;
  (d) it does not impose an *all-IP* architecture;
  (e) there is no critical dependence on vertical handovers; and
  (f) it does not create extra load to the current 3GPP core network.

The addition of new modules at core level with standard (and protected) programming interfaces can open up new possibilities to explore terminal mobility (a topic that is absent today).

Topics relevant for further work include the algorithms to be used on top of the HNACs to explore the mobility of UEs and their connection periods, and the viability of service continuity using this type of handovers.

References

1. Salkintzis, A., et al., "WLAN-GPRS Integration for Next-Generation Mobile Data Networks", IEEE Wireless Comm., vol. 9, pp. 112-124, Oct 2002
2. Bing, H., et al., "Performance Analysis of Vertical Handover in a UMTS-WLAN Integrated Network", IEEE PIMRC 2003.
3. Buddhikot, M., et al., "Design and Implementation of a WLAN/CDMA2000 Interworking Architecture", IEEE Comm. Mag., pp 90-100, November 2003.
4. Frenkiel, R., et al, "The Infostations Challenge: Balancing Cost and Ubiquity in Delivering Wireless Data", IEEE Personal Comm., v 7, pp. 66-71, Ap 2000
5. Frodigh, M, et al, "Future-Generation Wireless Networks", IEEE Personal Communications, vol. 8, pp. 10-17, October 2001
6. 3GPP, Group Services and System Aspects; Feasibility study on 3GPP system to WLAN internetworking; (Release 6), TS 22.934.v.6.1.0, Dec. 2002
7. 3GPP, Group Services and System Aspects; 3GPP system to WLAN internetworking; System Description (Release 6), TS 23.234.v.6.1.0, Jun. 2004
8. Campbell, A., et al, "Design, Implementation, and Evaluation of Cellular IP", IEEE Personal Communications, vol. 7, pp. 42-49, August 2000
9. Ramjee, R., et al, "IP-Based Network Infrastructure for Next-Generation Wireless Data Networks", IEEE Personal Communications, vol 7, pp. 34-41, Aug 2000
10. Das, S., et al, "TeleMIP: Telecommunications-Enhanced Mobile IP Architecture for Fast Intradomain Mobility", IEEE Personal Communications, vol. 7, pp. 50-58, August 2000
11. Misra, A., et al, "Autoconfiguration, Registration, and Mobility Management for Pervasive Computing", IEEE Personal Comm., vol. 8, pp. 24-31, August 2001
12. Wu, G., Havinga, P., Mizuno, M., "MIRAI Architecture for Heterogeneous Network", IEEE Commun. Mag., pp. 126-134, Feb. 2002
13. Tönjes, R., "Flow-control for multi-access systems", PIMRC'02, 13th IEEE Int. Symp. On Personal and Mobile Radio Communication, pp. 535-539, Sep. 2002, Lisboa, Portugal.
14. Steem, M., Katz, R., "Vertical handoffs in wireless overlay networks", Mobile Networks and Applications, vol. 3, pp 335-350, 1998

End-to-End Quality of Service Monitoring Using ICMP and SNMP

Yong-Hoon Choi[1], Beomjoon Kim[2], and Jaesung Park[1]

[1] RAN S/W Group, System S/W Dept.,
System Research Lab , LG Electronics Inc.,
LG R&D Complex, 533 Hogye-dong, Dongan-gu, Anyang-City, Kyongki-do,
431-749, Korea
dearyonghoon@lge.com
[2] Standardiztion & System Research Group (SSRG),
Mobile Communication Technology Research Lab., LG Electronics Inc.,
beom@lge.com
[3] Architecture Group, System Lab., LG Electronics Inc.,
4better@lge.com

Abstract. In this paper, we propose a new management information base (MIB) called Service Monitoring (SM) MIB. The MIB enables a network manager to gather end-to-end management information by utilizing special test packet. The special packet is an Internet control message protocol (ICMP) application that can be sent to a remote network element to monitor its Internet services. The SM MIB makes end-to-end management possible without using any special measurement architectures or equipments. Real examples show that the proposed SM MIB is useful for end-to-end quality of service (QoS) monitoring. The accuracy of obtained data, priority and security issues are discussed.

1 Introduction

The ability to assess the end-to-end QoS of real-time application is important for billing and provisioning purpose. Current network management system uses simple network management protocol (SNMP) for handling managed objects of IP networks. However, SNMP-based management platforms basically cannot handle the end-to-end management of Internet. This typical approach suffers from lack of scalability and scope restriction. If an end-to-end user flow traverses multiple Internet service provider (ISP) domains, then a network management system (NMS) could obtain information from multiple agents through local or manager-to-manager interfaces to retrieve the customer's end-to-end view of their service. This typical approach may cause management traffic and data retrieval time to increase because of querying the various hops and manager-to-manager interactions.

The problems associated with centralized management architecture, lack of extensibility, and poor support for end-to-end management has been identified and addressed by many researchers. Over the last few years, mobile agent/code

M. Freire et al. (Eds.): ECUMN 2004, LNCS 3262, pp. 40–49, 2004.

approach [1] has achieved a widespread interest to decentralize network management tasks. Autonomous delegated agent, which moves network element dynamically, makes end-to-end management feasible. It gives much flexibility and scalability to a network management system by relieving it from periodic polling. Performance management using management by delegation (MbD) has been studied by Cherkaoui et al. [2] and Bohoris et al. [3].

The concept of MbD influences IETF distributed management (DISMAN) working group to integrate this concept to its management framework. IETF efforts to standardize distributed management have focused on SNMP compliant MIBs and Agent. There are two types of developments with respect to distributed management. One is agent extensibility (AgentX) protocol (RFC 2741, RFC 2742) and the other is distributed Internet management with functional MIBs such as RFC 2925, RFC 2981, RFC 2982, RFC 3014, RFC 3165, and RFC 3231. The goal of AgentX technology is to distribute agent functions to sub-agents. In AgentX, however, relatively complex operations on the master agent side are required in order to realize SNMP lexicographic ordering and access control efficiently. The goal of the functional MIBs is to delegate functionality to network elements, still assuming that the control and coordination of a management task ultimately resides within a management program communicating with one or more agent. In spite of widespread interest of researchers, a mobile agent-based distributed management is still quite complex both to design and maintain, in contrast with the relative simplicity of a centralized management system.

RMON (Remote Network Monitoring) [4] is a major step towards decentralization of monitoring statistical analysis functions. It provides an effective and efficient way to monitor subnetwork-wide behavior while reducing the burden on other agents and on management stations. It can monitor the total traffic within a LAN segment such as number of collisions and number of packets delivered per second. RMON2 [5] has the capability of seeing above the MAC layer by reading the header of the enclosed network-layer protocol. This enables the agent to determine the ultimate source and destination and allows the network manager to monitor traffic in great detail. Although RMON MIB provides network-wide information, the data collection is quite a CPU and memory intensive task because RMON operates in promiscuous mode, viewing every packet on a subnetwork. End-to-end characteristics such as delay and jitter cannot be measured by passive packet probe.

Active measurement approach is one of the most promising schemes in terms of end-to-end QoS monitoring. The examples of active measurement architecture in practice today include [6], [7], [8]. IETF IP Performance Metrics (IPPM) working group has proposed measurement architecture and protocols: RFC 2330, RFC 2679, and RFC 2680. Recently, IPPM has proposed a management interface named IPPM-reporting-MIB [9], which is work in progress, for comprehensive measurement reporting based on their measurement architecture. The active measurement approaches that we mentioned above need special devices or protocol implementations at both sides of measurement point. The works using ICMP extensions such as [10] are useful for performance monitoring. They can give monitoring information to both operators and end-users without any additional protocol implementations unless intermediate routers discard ICMP packets. Our approach is focused on in-service end-to-end flow monitoring without using any special measurement framework. Our work is on the same category as [9] since the goal is to propose a management interface that gives network operators and end users remote management information.

Our approach is to utilize dynamic packet rather than dynamic agent (e.g., MbD and AgentX) or passive packet probe (e.g., RMON). For the purpose of management, we designed and implemented test packet and Service Monitoring MIB (SM MIB). Test packet is a measurement packet and it is an integral part of SM MIB. Instead of using our proposed ICMP extensions, different works such as RFC 2679 and RFC 2680 are also useful for a test packet. SM MIB provides a network manager with dynamic end-to-end management information, in particular, QoS information by utilizing test packets, which move through network elements. The MIB gives the network manager the ability to obtain end-to-end information, reduces the need to distribute MIBs throughout the network, and cuts the amount of management-related traffic.

The remainder of this paper is organized as follows. We describe the design and implementation details of the proposed test packet and SM MIB. The real examples that we conducted on various network with SM MIB are shown together with accuracy and security discussions. Finally, conclusions follow.

2   Test Packet Capabilities

Most of the Internet real-time multimedia services use user datagram protocol (UDP) as a transport protocol. At receiver side, absolute packet loss measure of UDP application is basically impossible because sequence number field is not defined in the UDP header. Since both ICMP and UDP do not support flow control mechanism and retransmission algorithm (i.e., they are directly encapsulated into IP datagram.) ICMP and UDP will experience the same loss probability if intermediate routers process them as the same priority. In order to extend ICMP message to sophisticated QoS monitoring tool, we developed test packet. It gives capabilities for an IP router or a host to measure quality of service metrics of user flows by dispatching test packets periodically. Test packets are generated periodically and circulate on a user flow. They are interspersed with user traffic at regular intervals to collect information such as network throughput, packet loss, delay, and delay variation from the routers along user flow. They have same packet length, type of service (TOS) value (DS codepoint field for IPv4 and Traffic Class field for IPv6), and followed the same route - They experience the same QoS as user packets. A test packet is generated with following information.

– MPSN: Monitoring packet sequence number.
– TOS value: Type of Service value of test packet.
– Time Stamp: The time at which the test packet was inserted.

As the packets are received at the destination endpoint, the test packets are replied. They can measure the following QoS metrics.

- RTT (Round Trip Time), the end-to-end delay of a flow: test packet is able to calculate the round-trip time (RTT) by storing the time at which it sends the echo request in the data portion of the message.
- Loss, the end-to-end packet loss of a flow: The end-to-end packet loss ratio can be estimated as total number of replied test packets over total number of requested test packets.
- Jitter, the variation of latency between packets: The end-to-end packet delay variation can be estimated as test packet delay variation.

We implemented test packet that works with both IPv4 and IPv6 based on the publicly available ping program source.

3   MIB Design and Implementation

Service Monitoring MIB (SM MIB) is an SNMP-compliant MIB for the delegation of dynamic packets to a destination endpoint and for gathering end-to-end QoS management information. Fig. 1 shows the logical structure of the SM MIB. It consists of three tables: one control table (named smControlTable) which specifies the destination and the details of sampling function, and two data tables (named smRequestTable and smReplyTable), which record the data.

For each of the K rows of smControlTable, there is a set of rows of smRequestTable and smReplyTable. For information of activated test flow specified by the row in the smControlTable, the data tables contain one row for each QoS information delivered by one test packet. Thus, as long as the control table information is not changed, one row is added to the smReplyTable each time a test packet arrives to the agent. The data tables are indexed by smRequestIndex (or smReplyIndex) and smRequestSeqNum (or smReplySeqNum). smControlMaxRows, one of the control table objects limits the size of the data table. The number of rows in the data table can be expressed as $\sum_{i=0}^{N}$ smControlMaxRows(i) assuming that all test replies are received by agent. Where smControlMaxRows(i) is value of smControlMaxRows for row i of the smControlTable, and N is the number of rows in the smControlTable.

After a new row is defined in the smControlTable, a new test flow starts. In other words, the network manager starts a test flow by writing an smControlEntry. Whenever a new test packet is captured by local network element, a new row is added to the data table. Once the number of rows for a data table becomes equal to smControlMaxRows, the set of rows for that test flow functions as a circular buffer. As each new row is added to the set, the oldest row associated with this test flow is deleted. The object is to prevent resource abuse; however, if a network manager sets smControlMaxRows too small, it may cause problems such that management information might be deleted before the manager obtains the information using Get/GetBulk request message. Each instances of the data table associated with smControlRowStatus will be deleted (associated test flow is deactivated) by the agent if this smControlRowStatus is destroy (6).

Fig. 1. Logical structure of the SM MIB. The entries represent the three tables of the MIB. Table index components used to identify entries are indicated by an arrow.

A prototype of the SNMP agent including SM MIB was built using ucd-snmp-4.2.6 package. If the SM MIB is deployed in DiffServ platform, the TOS field is interpreted in conformance with the codepoint allocation defined in RFC 2474.

4   Experimental Results

We have performed a number of measurement experiments on Internet. We varied the multimedia sources (e.g., Live TV sites, VoIP services, video conferencing, and VoD/AoD). We evaluate service quality in terms of RTT, jitter, and packet loss rate.

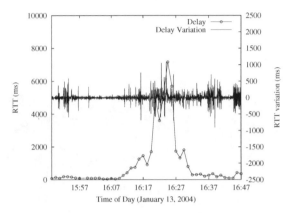

Fig. 2. RTT and RTT variation between site at Yonsei University in Seoul, Korea and IP Telephony site (www.dialpad.co.kr) in Korea. In-Service Monitoring during 15:47 (PM) January 13, 2004 - 16:47 (PM) January 13, 2004.

4.1   In-Service Monitoring

We sent test packets with 36-byte payload (which is the same to the UDP payload size of user packets) at 2 s regular intervals to Internet telephony site, www.dialpad.co.kr during call holding time, 60 minutes. Fig. 2 and 3 show the value of data table instances of SM MIB as graphs. For RTT, jitter and packet loss, the measured data are sampled with rate of 30/min during call holding time 60 min. We obtained 1800 samples for this flow. During the experimentation, from 16:10pm to 16:31pm, we offered high background load to monitor delay and packet loss.

RTT and Jitter. The circled line of Fig. 2 shows the RTT versus each monitoring time. The RTT values fluctuate from 12 ms to 7789 ms and the average RTT and standard deviation are 837.951 ms and 1501.671 ms, respectively. As shown in Fig. 2, RTT values increase when we offered high background load. Overloaded 20 minutes of poor performance lead to dissatisfaction. The solid line of Fig. 2 shows the jitter versus monitoring time. The jitter fluctuates from -1007 ms to 1073 ms and the average jitter is 0.907 ms. We observed that above 400 ms or below -400 ms jitter VoIP speech becomes irritating, and sometimes, speakers become unable to communicate.

Data Accuracy. For end-to-end delay variation, the measured data are sampled with the rate of λ_i (test packet generation rate) during call holding time $1/\beta_i$. We can get λ_i/β_i samples for user flow i. Suppose that $X_1, \cdots, X_n$ is a sample

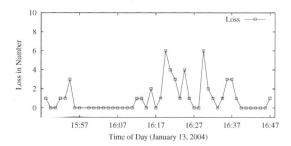

Fig. 3. Number of test packet losses between site at Yonsei University in Seoul, Korea and IP Telephony site (www.dialpad.co.kr) in Korea. In-Service Monitoring during 15:47 (PM) January 13, 2004–16:47 (PM) January 13, 2004.

from a normal population having unknown real traffic mean end-to-end delay variation μ and variance σ^2. It is clear that $\sum_{i=1}^{n} X_i/n$ is the maximum likelihood estimator for μ. Since the sample mean $\overline{X}$ does not exactly equal to μ, we specify an interval for which we have a certain degree of confidence that μ lies within. To obtain such an interval we make use of probability distribution of the point estimator. For example, 95% confidence interval for μ are:

$$\left(\overline{X} - 1.96\sigma/\sqrt{\lambda_i/\beta_i}, \ \ \overline{X} + 1.96\sigma/\sqrt{\lambda_i/\beta_i} \right) \tag{1}$$

assuming that the delay variation samples have normal distribution and the number of measured samples $\lambda_i/\beta_i \geq 30$. The 95% confidence interval of the average jitter is (0.906993 ms, 0.907007 ms).

Packet Loss. We have monitored 48 test packet losses from 1800 samples during 60 min. By the results, we can get 2.67% packet loss ratio for this application. Small number of packet drops was monitored under low background load. In this site, we observed that the packet loss behavior is affected by delay and delay variation. We observed that the occurrence of long delays of 3 seconds or more at a frequency of 5% or more packet loss is irritating for VoIP speech.

4.2 Long Time Scale Service Estimation

Every hour, we sent 11 test packets with 1000-byte payload to Live TV site, www.kbs.co.kr at 2 s intervals. This is repeated for a week. This site is located at a distance of 9 hops from our workstation. In order to experience the same QoS as the user packets, test packet size is set to 1,000 bytes, which is the same to the UDP payload size of user packets. The RTT, jitter, and the number of test packet losses are recorded at SM MIB.

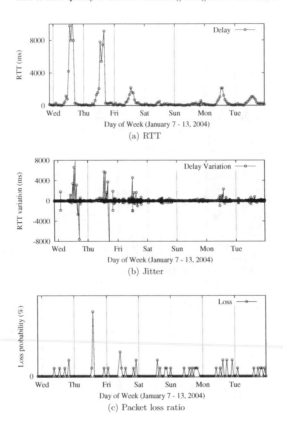

(a) RTT

(b) Jitter

(c) Packet loss ratio

Fig. 4. RTT, jitter, and packet loss ratio between site at Yonsei University in Seoul, Korea and Live TV site (www.kbs.co.kr) in Korea. Weekly plots during 22:16 (PM) January 7, 2004–21:24 (PM) January 13, 2004.

Fig. 4 shows the round trip time (RTT) and jitter versus each monitoring time. As soon as the control table entry was created, the first `smRequestTable` row was created right away, and the first `smReplyTable` row was created at t = 135 ms. The average time gap between request table row creation and relevant reply table row creation was 717.594 ms for this flow. For RTT, jitter and packet loss, the measured data was sampled with rate of 10/h during a week. We obtained 1710 samples for this flow.

RTT and Jitter. Fig. 4-(a) shows the RTT versus each monitoring time. The RTT values fluctuate from 11 ms to 10366 ms and the average RTT and standard deviation are 717.594 ms and 1887.739 ms, respectively. As shown in Fig. 4-(a), RTT values decrease during the weekend. Fig. 4-(b) shows the jitter versus each monitoring time. The jitter fluctuates from -10776 ms to 8255 ms and the average jitter and standard deviation are -0.00058 ms 618.944 ms, respectively. If we assume that the delay variation follows normal distribution, 95% confidence interval of the average jitter is (-0.00055 ms, 0.00061 ms).

Packet Loss. Fig. 4-(c) shows the packet loss versus each monitoring time. We have monitored 54 test packet losses from 1710 samples for a week. By the results, we can get 3.158% packet loss ratio.

Limitations. The collected data would be useful for postmortem analysis and future use of network provisioning. We collected samples at periodic intervals. Currently our SM MIB does not support other sampling methods such as Poisson and Geometric sampling. Periodic sampling is attractive because of its simplicity, but it may suffer from potential problem: If the metric being measured itself exhibits periodic behavior, then there is a possibility that the sampling will observe only part of the periodic behavior if the period happens to agree.

Security Issues. Since we use active measurement approach, injecting too much test traffic into the network may cause to distort the results of the measurement, and in extreme cases it can be considered as denial-of-service attacks by routers. To prevent this attack, some sites limit the amount of ICMP-echo traffic: the effect is that ICMP messages can get blocked so packet loss can look bad even though the network is OK [10]. The test packet generation rate in this paper was small enough not to experience drop by rate limiting. Routers using firewall may discard unauthorized packets to prevent attack. ICMP-based test packet as well as other types of measurement traffic will be discarded in this case.

Priority Issues. If routers can recognize test traffic and treat it separately, the results will not reflect actual user traffic. The test packet that we design has the same packet length and TOS value (DS codepoint field for IPv4 in DiffServ) as user traffic. Because both of them are directly encapsulated into IP datagram, ICMP-based test packet and UDP-based user traffic will experience the same loss probability if intermediate routers process them as the same priority. The security and priority issues are open problems in active measurement studies.

5   Concluding Remarks

Our goal is to suggest a practical solution for end-to-end management, still maintaining typical manager-agent paradigm. In this paper, we proposed a new MIB approach, called SM MIB. It offers a convenient solution for end-to-end QoS management. It provides a network manager with end-to-end management information by utilizing special test packets, which move through network elements dynamically. The MIB gives the network manager the ability to obtain end-to-end information, reduces the need to distribute MIB's throughout the network, and cuts the amount of management-related traffic. The implementation and maintenance of the MIB are simple because the method does not requires any special management functions at every intermediate router in the network, but just requires us to implement SM MIB at edge routers/end hosts only. The real examples showed that the SM MIB can be used as real-time in-service monitoring tool as well as long time scale load analysis tool. The accuracy of obtained data depends on the number of samples. We specified an interval for which we have 95% degree of confidence that measured data lies within.

Finally, we address the weakness of our work. SM MIB does not guarantee 100% accuracy due to the unavoidable errors of periodic sampling method and the limitations of test packet capabilities. The inaccuracy of the proposed test packet capabilities can be overcome by improving functionality of the packet, such as in RFC 2679 and RFC 2680. We expect that the SM MIB will be useful for end-to-end service evaluation as well as QoS monitoring.

References

1. G. Goldszmidt: Distributed Management by Delegation. Ph. D. Dissertation, Columbia University (1996).
2. O. Cherkaoui et al.: QOS Metrics tool using management by delegation. IEEE Network Operation and Management Symposium (NOMS'98) New Orleans LA, USA, (1998) 836-839.
3. C. Bohoris et al.: Using Mobile Agents for Network Performance Management. IEEE Network Operation and Management Symposium (NOMS'2000), Honolulu Hawaii, USA, (2000) 637-652.
4. S. Waldbusser: Remote Network Monitoring Management Information Base. RFC 2819, (2000).
5. S. Waldbusser: Remote Network Monitoring Management Information Base Version 2 using SMIv2. RFC 2021, (1997).
6. The Skitter Project, Online: http://www.caida.org/tools/measurement/skitter/
7. The Surveyor Project, Online: http://www.advanced.org/surveyor and http://betelgeuse.advanced.org/csg-ippm/
8. V. Paxson et al: An Architecture for Large-Scale Internet Measurement. IEEE Comm. Mgz., 36 (8) (1998), 48-54.
9. E. Stephan and J. Jewitt: IPPM reporting MIB. Internet Draft: draft-ietf-ippm-reporting-mib-05.txt, (2004).
10. W. Matthews and L. Cottrell: The PingER Project: Active Internet Performance Monitoring for the HENP Community. IEEE Comm. Mgz., (2000) 130-136.

Expedited Forwarding Delay Budget Through a Novel Call Admission Control

Hamada Alshaer and Eric Horlait

Lip6, UPMC, 8 rue Capitaine Scott,
75015 Paris, France
{hamada.alshaer,eric.horlait}@lip6.fr
http.//www-rp.lip6.fr

Abstract. Call admission control is a principal component for QoS delivery in IP networks. It determines the extent to which network resources are utilized. It determines also whether QoS are actually delivered or not. Continuing on the steps of the existing approaches, we introduce a distributed and scalable admission control scheme to provide end-to-end statistical QoS guarantees in Differentiated Services(DiffServ)networks. This scheme is based on the passive monitoring of aggregate traffic at the core routers, and active probing of network state between the different edge routers. Thus, we have simulated a DiffServ network where different users are provided different service classes. In this network, the back ground (BG or BE) flows are characterized by Poisson process, however the Expedited Forwarding (EF) flows are generated by Exponential and Pareto ON-OFF sources. Afterward, we have evaluated the admission mechanism employed at the different network ingresses by measuring the network utilization at the network bottlenecks, and the effect of this mechanism on the accepted number of EF flows and their delay budget.

1 Introduction

The fast growing of the Internet network has created the need for IP-based applications requiring guaranteed QoS characteristics. The integrated service(IntServ)and DiffServ have been proposed to address QoS. While IntServ operates on a per-flow basis and hence provides a strong service model that enables strong per-flow QoS guarantees, it suffers from scalability problem. However, DiffServ keeps per flow state only at the ingress or egress of a domain and aggregates flows into a limited set of traffic classes within the network, resolving the scalability problem at the expense of looser QoS guarantees.

Beside standardized functionality of the IP layer, considerable amount of work has been devoted to architectures and functions to deliver end-to-end (e2e) QoS. These functions can be classified into service management functions, and traffic engineering functions. Traffic engineering functions are principally concerned with the management of network resources for accommodating optimally offered traffic. Service management functions deal with the handling of the different service users connections, trying to maximize the number of

M. Freire et al. (Eds.): ECUMN 2004, LNCS 3262, pp. 50–59, 2004.
© Springer-Verlag Berlin Heidelberg 2004

connections while respecting the signed contracts between the service providers (SP's), and users. In order to guarantee the agreed QoS signed in the contracts, service management needs a call admission control(CAC) to avoid the network overloading [1].

Recently, different CAC's have been developed to provide deterministic and statistical QoS guarantees in DiffServ network [2,3,4]. In [2] Zhang et al. introduce a bandwidth broker(BB)to manage the network status. It implies that admission control decisions are made at a central node for each administrative domain. Although the cost of handling service requests is significantly reduced, it is unlikely approach scales for large networks. Furthermore, it estimates the available network resources assuming the worst case, which can't efficiently utilize the network resources. In order to cope with the scalability, most relevant studies adopt distributed admission control schemes, which are further categorized into model-based and measurement-based approaches. Both of these approaches assess QoS degradation probability upon service request arrivals: model based approaches[5] maintain state information for active and employ mathematical models, whereas measurement based approaches rely on either passive[7]or active[8]aggregate measurements.

In an attempt to provide statistical QoS guarantees, several end point admission control(EPAC) algorithms such as [2,3] have been developed in which end hosts probe a network and record the performance metrics of the probing packets. Then, if the metrics is below a threshold, the hosts admit the connection and start the data transmission phase: otherwise, the connection is rejected. An advantage of these schemes is that, no network control is required and all functions are performed by the end hosts. However, they are also suffering from long probing time, and they provide imprecise network measurements.

In this paper, we analyze a distributed admission control, and evaluate it through the NS in the DiffServ network shown in Fig.1. The connections are accepted or rejected at the different network ingresses according to the feedbacks of the actual network status. However, the number and measurement times of the probing packets which are used through the simulations to infer the network status are less and shorter respectively than those used in the EPAC algorithms. This distinguishes in turn our admission control than those introduced before.

The remainder of this paper is organized as follows: In section 2, and 2.1 we describe the simulation environment, different traffic classes, and its characteristics. In section 3, and 3.1 we analyze our proposed admission conrol. In section 5, we present the results of several experiments intended to demonstrate the ability of the proposed call admission control to guarantee the EF delay bound while maximizing the network utilization. Finally, section 6 will draw a conclusion to this paper, and we will provide some perspective points.

2   Network Model

In order to evaluate the proposed call admission control we have designed and simulated the DiffServ network shown in Fig. 1 using NS. Since, we aim at eval-

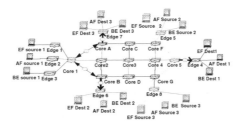

Fig. 1. Differentiated Service Topology.

uating distributed CAC, we see that through simulation the network topology shown in Fig. 1 we can realize our objectives, because it includes many edges through which different users can inject traffic in the network, and at which they can be controlled. Furthermore, we can evaluate the CAC on multiple bottlenecks. Also, we have two traffic sources$_{2,3}$ inject traffic in opposite directions, however we will see in the simulation section how the CAC does exploit the available bandwidth when they send their traffic in the opposite and the same directions. All the routers output links have the same capacity "C_0", except the network bottlenecks which are circled on Fig. 1 have a capacity equivalent to $\left\lfloor \frac{3}{4} * C_0 \right\rfloor$ [6].

2.1 Network Traffic

Network traffic is divided into three classes $\{C_{EF}, C_{AF}, C_{BE}\}$, where C_{EF}, C_{AF}, and C_{BE} are the traffic generated from EF, AF, and BE flows respectively. Nine users divided into three groups according to their Edge routers are allowed to inject these different traffic classes in the DiffServ network shown in Fig. 1. They are serviced at the different output links of the routers by the static priority scheduling service discipline (SPS) [14]. Therefore, each class has its own priority queue, and reserved bandwidth. However, since we are concerned with the QoS offered by the EF service class, hence the e2e delay bound and number of the different EF flows that are injected by the different EF sources shown in Fig. 1 are monitored, and calculated from their sources$_{1,2,3}$ until their destinations$_{1,2,3}$ respectively, in order to evaluate the efficiency of the CAC installed at the edge routers. At these ingresses EF flows are controlled by a dual leaky bucket controller (DLBC) with parameters(ρ, π, σ) where ρ is the average traffic rate, π is the peak traffic rate, and σ is the burst size. Therefore, the amount of EF traffic that arrives to a core router over a time interval(t_1, t_2) is denoted by $A(t_1, t_2)$ and determined through the following formula :

$$\Lambda(t_1, t_2) \leq min\left\{\rho(t_2 - t_1) + \sigma, \pi(t_2 - t_1)\right\} \tag{1}$$

[9,10]. Nevertheless, the CAC is configured such that the e2e delay bound of EF flows is statistically guaranteed through the following relation :

$$Prob(d_j \succ D_{EF}) \prec \epsilon \tag{2}$$

Where D_{EF} is the determined e2e delay bound to the EF traffic class in the CAC, d_j is the EF flow's e2e delay bound suffering from in the network, and ϵ is the probability of EF flows'e2e delay bound violation .

3   Statistical Multiplexing Based on Bandwidth

Statistical multiplexing improves the network resources sharing and raises its links utilization. This mechanism is used to allocate the bandwidth for each connection in IntServ network and for the different traffic classes in DiffServ network. This allocation is based on the traffic characteristics: such as, peak rate "ρ", and burst"σ". For example, EF class in DiffServ network shown in Fig. 1 is assigned at each node's output link a portion of bandwidth C_{EF} equivalent to its peak rate ρ_{EF}, provided that, $\left\{ \rho_{EF} = \sum_{i=1}^{N_{EF}} \rho_i \right\} \leq C_{EF}$. Furthermore, EF class is assigned a queue of length L at each network node, where, L is related to the required delay bound of EF class at each node through this simple formula : $L = C_{EF} * d_{req}$. This formula helps us to calculate the different queues' length that support the SPS service discipline.

However, when C_{EF} is consumed by the different EF flows, then the queue that is assigned for EF in SPS service discipline is nearly filled up, which results in degradation of the EF service class. But, when the number of EF flows is reduced, then EF queue is nearly empty, and EF traffic suffers from short delay and low packets loss. IP networks suffer from the bandwidth clustering, whereby the bandwidth is available at some network nodes, while heavy congestions occur at others due to lack of bandwidth. Therefore, measuring the available bandwidth at the output interfaces of each network node in DiffServ network is important for the CAC's that work on its edge routers to decide whether to accept more EF flows or not, such that the QoS offered through EF service class is guaranteed.

3.1   Local Measurement and Available Bandwidth Estimation

Through measuring the envelope of the multiplexed EF aggregate flows, the short time scale burstiness of the EF traffic can be captured, and then it can be employed for bandwidth reservation, thereby CAC builds its decision to accept new EF flows or not. Furthermore, the available bandwidth which is remained from the reserved bandwidth for EF aggregate C_{EF} of the output links to a router is estimated in terms of the EF aggregate arriving average and its variance over the different time intervals [2]. Let $A(t_1, t_2)$ denotes arriving traffic from EF aggregate in the time interval (t_1, t_2). This time interval is divided into M windows, where each is divided into k time slices of length τ. Then, for the past

m^{th} window from the current time t, the maximum rate $\overline{R_k}$, and variance σ_k^2 of the EF aggregate arriving rate over the time interval $K\tau$ can be computed as follows :

$$R_k^m = \frac{1}{k\tau} \max_{0 \leq i \leq T-K} A[(t - mT\tau) + i\tau,$$
$$(t - mT\tau) + (i + k)\tau] \tag{3}$$

$$R_k = \frac{1}{M} \sum_{m=1}^{M} R_k^m, \sigma_k^2 = \frac{1}{M-1} \sum_{m=1}^{M} (R_k^m - \overline{R_k})^2 \tag{4}$$

Where, k = 1,2,...., T , and m = 1,2,......,M.

However, After EF traffic flows are multiplexed with other traffic classes at the different routers in the DiffServ shown in Fig. 1, they are subjected to lose their characteristics, and get distorted. Furthermore, if there are more users offered EF service class, then EF flows will grow exponentially in the network core. Thus, we use the central limit theorem [13,14] as a statistical approach to characterize the backlog distribution that results from the EF traffic flows at the different network core routers; because it doesn't require any knowledge about the original EF flows distributions.

Hence, the statistical distribution of EF traffic backlog B($k\tau$) along the servers' busy period $k\tau$ approaches the normal distribution $N(k\tau\mu, k^2\tau^2\sigma^2)$. The probability of delay bound violation of EF flows can be expressed statistically as follows:

$$Prob(D \succ d) \approx \max_{0 \leq k\tau \prec \beta_{EF}} Prob((B(k\tau) - ck\tau) \geq c.d) \tag{5}$$

Thereby, an optimization problem arises which requires the computation of the minimum bandwidth eb. This is assigned to EF flow at any network core router, so that, its delay bound violation doesn't exceed the predetermined probability of EF service class delay bound violation ϵ, which is determined in the employed CAC at the different edge routers.

$$Prob((B(k\tau) + ebk\tau - ck\tau) \geq c.d) \leq \epsilon \tag{6}$$

then, we solve this for "eb" as follows :

$$Prob((B(k\tau) + ebk\tau - ck\tau) \geq c.d) = 1 - Prob$$
$$((B(k\tau) \leq (c(d + k\tau) - ebk\tau)) \tag{7}$$

$$1 - Prob((B(k\tau) \leq (c(d + k\tau) - ebk\tau)) = 1 -$$
$$\phi(\frac{c(d + k\tau) - (ebk\tau) - (k\tau\overline{R_k})}{k\tau\sigma_k}) = 1 - \phi(z) \tag{8}$$

$$eb = \frac{1}{k\tau}(c(d + k\tau) \quad (k\tau\overline{R_k}) - zk\tau\sigma_k) \tag{9}$$

Where, ϕ is the accumulative Gaussian distribution function that characterizes $B(k\tau)$ over the busy period $k\tau$. And, z is approximated in [13] by $\sqrt{|\log(2\pi\epsilon)|}$.

So fare, we have solved statistically the problem described in 6. However, this problem in [12,11] was solved stochastically for a buffer bound, and the critical time at which buffer overflow happens. In our case, the critical time is the delay bound d_{req} required for the EF flow at a node. However, we can allow limited delay bound violation ϵ due to the deviation from d_{req} by d^*. Nevertheless, each flow in EF aggregate is assigned effective bandwidth eb. But, here we come back to the same problem in [12], since we need to solve for the critical time d^* at which the reserved capacity for EF class is consumed by the current backlogged EF flows. Where, d^* is measured by Kullback-Leibler distance $\Delta(p_2(d^*, eb)||p_1(d^*))$. This measures in turn how unlikely is for a Bernoulli random variable with mean $p_1(d^*)$ to become another Bernoulli random variable with mean $p_2(d^*, eb)$.

Algorithm 1 : CAC based on the effective bandwidth(EB)measurements

---

1- An ingress router sends three probing packets to an egress router.

2- After the egress router receives the probing packets, it sends them back to the ingress router.

3- The ingress router calculates d_{e2e}, ϵ_{e2e}, and eb_{e2e} as follows :

$$d_{e2e} = \sum_{m=1}^{N} d_m + \sum_{m=1}^{N} \omega_m, \; \epsilon_{e2e} = 1 - \prod_{m=1}^{N}(1 - \epsilon_m), \; eb_{e2e} = min\left\{eb_{m=1}^{N}\right\}$$

Where, N, d_m, ϵ_m, and eb_m are the number of routers, the delay, delay violation probability, and available effective band width respectively at each router along the connection's path. And ω_m is the delay propagation.

   if (connection peak rate(P_{cr}) $\leq eb_{e2e}$ && $d_{e2e} \leq$ connection delay bound (d_{cb} or D_{EF}) && $\epsilon_{e2e} \leq \epsilon_{adm}$) **then**

      Connection is admitted

   else if ($P_{cr} \geq eb_{e2e} \,||\, d_{e2e} \succ (d_{cb})$) **then**

      new flow is reshaped at eb_{e2e} rate at the ingress router. And, the d_{e2e} is recalculated through the following expression:

$$d_{e2e} = d_{resh} + \sum_{m=1}^{N} d_m + \sum_{m=1}^{N} \omega_m$$

      Where,

$$d_{resh} = \sup_{0 \prec t \leq \beta_{EF}} \left\{inf\left\{T : T \succ 0, A(\tau) \leq S(\tau + T)\right\}\right\}$$

$$\beta_{EF}(worst\ busy\ period) = \frac{\sigma_{new_{EF}} + \sum_{m=1}^{N_{accepted_{EF}}} \sigma_m}{C_{EF}}$$

      $S(t)(Ingress\ Service\ Curve) = eb_{e2e} * t. \;\; \forall\, t \succ 0$

      $A(\tau)$ is computed from the expression 1

      if ($d_{e2e} \succ (d_{cb})$) **then**

         connection is rejected

      end if

   else

      Connection is admitted

   end if

---

4   Admission Control Functions and Qos Adaptation

The admission control function depends on active probing. For example, when a connection request arrives from EF source1 to the edge 1 on Fig.1. This determines to which service class the connection request belongs and its corresponding egress router, which is Edge 4 in this case. Then, it transmits 3 probing packets to the edge router 4 and starts a probing timer whose length given by the double of EF class delay bound. If edge 1 does not receive a response from edge router 4 along the probing timer, then edge 1 rejects the EF source 1 connection.

The admission control function depends also on passive monitoring. For example, during the previous example, the core routers$_{1,2,3,4,5}$ mark the probing packets with their status, such as the available bandwidth at its output interface links and packets loss value.

Upon probing packets arrivals to the edge 4, the probing packets are echoed back to the edge 1. This uses the collected status information in the probing packets to decide whether to accept the new connection or to do QoS adaptation by reshaping the connection's peak rate according to the minimum available bandwidth along its path. As a result, the minimum available bandwidth eb_{e2e} along the different paths between the ingress-egress routers is included in the rule base of the CAC illustrated in algorithm 1. Then, the rule base is used by the CAC inference mechanism to accept, adapt or reject a new EF flow at the different edge routers in Fig 1.

5   Simulation and Results

We have conducted three experiments to study the performance of the proposed CAC. The primary objectives of these experiments are to study the efficiency of the CAC's at the different edges. This is shown through the ability of each CAC to exploit the available bandwidth for accepting more EF flows in the face of BE or other competitors EF flows, which are generated by other users. The second objective is to study the blocking probability η_{EF} of the CAC through the following relation

$$\eta_{EF} = \frac{N_{rejected}}{N_{EF}} \tag{10}$$

$$\mu_{EF} = \frac{\sum_{i=1}^{N_{accepted}} r_i}{C_{EF}} \tag{11}$$

Where, r_i is the average rate of each admitted EF flow on the different bottlenecks.

In the first experiment, we measured through the relation described in 11 the average network utilization μ_{EF} of EF flows at the network bottlenecks in Fig 1. The capacity of these bottleneck links is 6Mbps, and the capacity of other links

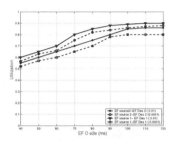
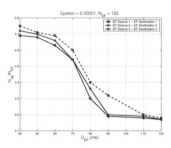

Fig. 2. Call Admission Region For EF Flows Generated By Sources 1-2.

Fig. 3. Call Admission Control Blocking Probability At Edges(1,8,5).

is 8Mbps. The capacity is shared between the supported traffic classes as 35% for the EF class, which is equivalent to 2,1Mbps of the bottleneck links' capacity, and 2,8Mbps of other links. And, 50 % of the links' capacities is reserved for AF class. And, the remained capacity can be used by BE flows. The propagation delay over all the network links is set to 4 ms.

The EF source$_1$ is characterized by ON-OFF voice traffic generator. The On and OFF states are characterized by an exponential random variable with mean 400ms, and 600ms respectively. In the on state, EF source$_1$ generates voice traffic at a peak rate of 64kbps. Meanwhile, EF source$_2$ and EF source$_3$ are characterized by Pareto ON-OFF video traffic generator. The average of ON state of EF source$_2$, and EF source$_3$ is set to 360ms, and the average of OFF state of these EF sources is set to 400 ms. In the on state EF sources$_{2,3}$ generate video traffic at a peak rate of 200kbps. All ON-OFF packets have size equivalent to 567 bytes. Best effort traffic is generated by 10 FTP connections characterized by Poisson process, and they are active a long the simulation run time which is set to 90 s. The size of packets generated by the FTP connections is 1000 bytes. Traffic shapers (DLBC) are installed at the network ingress routers {Edges$_{1,5,8}$} shown in Fig. 1.They are characterized with these parameters($\sigma = 4$ packets, $\rho = 115kbps, \pi = 200kbps$) at Edge routers$_{5,8}$, and with ($\sigma = 2$ packets, $\rho = 40kbps, \pi = 64kbps$) at edge router$_1$. The arrival to service ratio r is kept at 0.9.

Fig. 2 shows that at the lower delays the average utilization of exponential source is higher than that of Pareto ON-OFF for its long range dependent property. However, for the larger delays, the average utilizations of Exponential, and Pareto On-OFF sources are quiet close due to the buffering. Furthermore, it indicates that the CAC utilizes statistical multiplexing gain to achieve higher network utilization for bursty traffic flows when the probability value of delay bound violation of EF class is increased from $\epsilon = 0,0001$ to ϵ 0.01.

In the second experiment, we fixed ϵ at 0.00001, and number of EF flows of each EF source at 150, and we changed the D$_{EF}$ in the CAC from 40ms to 120ms. Consequently, Fig. 3 shows that the number of rejected EF flows

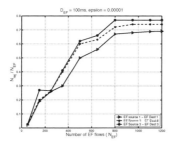

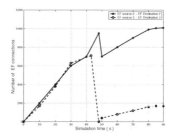

Fig. 4. Blocking probability vs number of EF flows gnerated by sources$_{1,2,3}$.

Fig. 5. Superposition of accepted EF flows along the simulation run time.

decreases as long as the D_{EF} increases. However, this decrease stabilizes in the interval($85ms, 120ms$) around the average delays of EF flows of the different EF sources. Furthermore, the number of rejected EF flows by CAC at Edge$_8$ is greater than that at edge$_5$, since EF source$_2$ starts to generate traffic before EF source$_3$. On the other hand, the number of EF flows of source$_1$ rejected by the CAC at Edge$_1$ is relatively small in comparison with that at Edges$_{5,8}$. Because, after these flows traverse the core router$_1$, they have not any competitors on the EF reserved bandwidth along its path to their destination.

The third experiment was done in two parts. In the first part, we varied the number of EF flows generated by sources$_{1,2,3}$ from 50 to 1200 flows, and we fixed the values of D_{EF}, ϵ in the CAC at 100ms and 0.00001 respectively. Consequently, Fig. 4 shows that the number of rejected EF flows increases as long as the number of EF flows generated by sources$_{1,2,3}$ increases, because the CAC's works to guarantee the e2e delay bound of the accepted EF flows in the network. In the second part, during the simulation run time, precisely at 48s the EF source$_3$ becomes EF Dest$_3$ and EF Dest$_3$ becomes EF source$_3$. This transition was done to see how effectively the CAC's at edges$_{5,8}$ can exploit the available bandwidth to accept more EF flows. As a result, Fig. 5 shows that at the transition instant the CAC at Edge 5 exploited the available bandwidth to accept more EF flows belong to source 2. However, when the transition terminated the increasing rate of accepted flows generated by EF source$_2$ decreases to become less than that before the transition. Because, both EF sources$_{2,3}$ from 48s to 90s transmit in the same direction, so they compete on the same EF reserved bandwidth along the bottlenecks.

6   Conclusion

We presented a new call admission control to provide e2e statistical QoS in DiffServ network. The connection is accepted or rejected at the network ingress based on the in-band marking of resource probing packets. This makes the CAC scalable and simply to deploy on a DiffServ domain-by DiffServ domain basis within the Internet. It also ensures that no matter how much traffic arrives at the network ingresses, the resources assigned to EF flows are not affected. Furthermore, it exploits the available bandwidth as we can see in Fig. 5 to accept more EF flows, which it could be a key point in business. Simulation results show that the proposed CAC is efficient as we can conclude from Fig. 2, 3, and it can accurately control the admission region of different traffic generator models as it is shown in Fig. 2.

In the future, we would like expand this work to cover the synchronization problem between the CAC's at the different ingress routers in the DiffServ network shown in Fig. 1.

References

1. Jonathan Chao, H., Guo, X.: Quality of Service Control in High-Speed Networks. WILEY-Interscience, 2002.
2. Cetinkaya, C., Kanodia, V., Knightly, E.W.: Scalable Services via Egress Admission Control. IEEE Transactions on mutltimedia, vol. 3, no. 1, March 2001.
3. Zhang, Z., Duan, Z., Gao, L., Hou, Y.: Decoupling QoS Control from core routers: a novel bandwidth broker architecture for scalable support for guaranteed services. Proc. of ACM SIGCOMM, Stockholm, Sweden, August 2000.
4. Breslau, L., Knightly, E.W., Shenker, S., Stocia, I., Zhang, H.: Endpoint admission control: architectural issues and performance. Proc. of ACM SIGCOMM, Stockholm, Sweden, August 2000.
5. Knightly, E.W., Shroff, N.: Admission Control for statistical QoS: Theory and practice. IEEE Network, vol. 13, no. 2, March 1999, pp. 20-29.
6. Alshaer, H., Horlait, E.: Expedited Forwarding End to End Delay Jitter In the Differentiated Services networks. IEEE International Conference on High Speed Networks and Multimedia Communications-HSNMC'04, Lecture Notes in Computer Science (LNCS) 3079, Toulouse, France, June 2004.
7. Centinkaya, C., Knightly, E.W.: Egress Admission control. Proc. IEEE INFOCOM 2000, March 2000.
8. Ivars, I.M., Karlsson, G.: PBAC: Probe-Based Admission control. Proc. QoFIS 2001, pp. 97-109.
9. Cruz, R.: Quality of service guarantees in virtual circuit switched networks. IEEE Journal on selected Areas in communications, August 1995.
10. Braden, R., Zhang, L., Berson, S., Herzo, S., Jim, S.: Resource reservation protocol(RSVP) - version 1 functional specification. IETF RFC, Sept. 1997.
11. Kesidis, G., Konstantopoulos, T.: Worst-case performance of a buffer with independent shaped arrival processes.*IEEE Communication Letters.* 2000.
12. Chang, C.S., Song, W., Ming Chiu, Y.: On the performance of multiplexing independent regulated inputs. *Proc. of Sigmetrics 2001.* Massachusetts, USA, May 2001.
13. Papoulis, A.: Probability Random Variables and Stochastic Processes. 2nd ed., McGraw-Hill, 1983.
14. Knightly, W.E.: Second moment resource allocation in multi-service networks. In Proc. of ACM Sigmetrics'97. Pages 181-191, Seatle, June 1997.

Unbiased RIO: An Active Queue Management Scheme for DiffServ Networks*

Sergio Herrería-Alonso, Miguel Rodríguez-Pérez, Manuel Fernández-Veiga, Andrés Suárez-González, and Cándido López-García

Departamento de Enxeñería Telemática, Universidade de Vigo
Campus universitario, 36310 Vigo, Spain
sha@det.uvigo.es

Abstract. One of the more challenging research issues in the context of Assured Forwarding (AF) is the fair distribution of bandwidth among aggregates sharing the same AF class. Several studies have shown that the number of microflows in aggregates, the round trip time, the mean packet size and the TCP/UDP interaction are key factors in the throughput obtained by aggregates using this architecture. In this paper, we propose the Unbiased RIO (URIO) scheme, an enhanced RIO technique that improves fairness requirements among heterogeneous aggregates in AF-based networks. We validate URIO behaviour through simulation. Simulation results show how our proposal mitigates unfairness under many different circumstances.

1 Introduction

The Differentiated Services (DiffServ) architecture [1] is one of the most promising proposals to address Quality of Service (QoS) issues for data and multimedia applications in IP networks. The differentiated service is obtained through traffic conditioning and packet marking at the edge of the network combined with simple differentiated forwarding mechanisms at the core.

DiffServ defines a set of packet forwarding criteria called "Per Hop Behaviours" (PHBs). One of these packet-handling schemes standardised by the IETF is the Assured Forwarding (AF) PHB [2]. The basis of assured services is differentiated dropping of packets during congestion at a router. AF provides four classes of delivery for IP packets and three levels of drop precedence per class. Within each AF class, IP packets are marked based on conformance to their target throughputs. The Time Sliding Window Three Colour Marker (TSWTCM) [3] is one of the packet marking algorithms most used to work with AF. At the core of the network, the different drop probabilities can be achieved using the RIO (RED with In/Out) scheme [4], an active queue management technique that extends RED (Random Early Detection) gateways [5] to provide the service differentiation required.

---

* This work was supported by the project TIC2000-1126 of the "Plan Nacional de Investigación Científica, Desarrollo e Innovación Tecnológica" and by the "Secretaría Xeral de I+D da Xunta de Galicia" through the grant PGIDT01PX133202PN.

M. Freire et al. (Eds.): ECUMN 2004, LNCS 3262, pp. 60–69, 2004.
© Springer-Verlag Berlin Heidelberg 2004

Assured services should fulfil the two following requirements [6]:

1. Each aggregate should receive its subscribed target rate on average if there is enough bandwidth available (throughput assurance).
2. The extra unsubscribed bandwidth should be distributed among aggregates in a fair manner (fairness).

Unfortunately, as reported in [7,8,9], the fairness requirement of assured services cannot be met under some circumstances. Via simulation studies, these works confirm that the number of microflows in aggregates, the round trip time (RTT) and the mean packet size are critical factors for the fair distribution of bandwidth among aggregates belonging to the same AF class. Also, the interaction between responsive TCP traffic and unresponsive UDP traffic may impact the TCP traffic in an adverse manner.

Many smart packet marking mechanisms have been proposed to overcome these fairness issues. Adaptive Packet Marking (APM) [10] is one of these schemes able to provide soft bandwidth guarantees, but it has to be implemented inside the TCP code itself and thus, requires modifying all TCP agents. Intelligent traffic conditioners proposed in [11] handle some of these fairness problems using a simple TCP model when marking packets. However, they require external inputs and cooperation among markers for different aggregates. This feature complicates both implementation and deployment. Another marking algorithm based on a more complex TCP model is Equation-Based Marking (EBM) [12]. This scheme solves the fairness problems associated with heterogeneous TCP flows under diverse network conditions. EBM behaviour depends on the quality of the estimation of the current loss rate seen by TCP flows but, sadly, the calculation of this estimate is not an easy task and involves the deployment of the scheme awfully.

A different approach consists of addressing these problems by enhanced RIO queue management algorithms. One of the more interesting examples of these techniques is Dynamic RIO (DRIO) [6]. This scheme accomplishes per-flow monitoring at the core routers to share the available bandwidth among heterogeneous flows fairly. Since there can be thousands of active flows at the core of the network, DRIO needs to store and manage a great amount of state information and therefore, this renders it unscalable. In this paper, we first extend DRIO to work at the aggregate level. Applying DRIO to aggregated traffic instead of to individual flows reduces considerably the amount of stored information at the core improving scalability substantially. In addition, the mechanisms proposed for DiffServ are derived from a model that considers only aggregated traffic. Therefore, the application of DRIO to aggregates will be more natural and suitable for this kind of networks.

We then propose a new enhanced RIO queue management algorithm, namely Unbiased RIO (URIO), that uses proper level buffer usage policing to solve these fairness issues among heterogeneous aggregates. To validate the proposed schemes, we examine their behaviours under a variety of network conditions through simulation. Our simulation results show that URIO fulfils fairness requirement more satisfactorily than either RIO or DRIO.

The rest of the paper is organised as follows. Section 2 gives a brief overview of TSWTCM, RIO and DRIO schemes. In Section 3, we propose URIO, an enhanced version of RIO. Section 4 describes the simulation configuration used for the evaluation of RIO, DRIO and URIO schemes. In Section 5, we present the results obtained from the simulation experiments. We end the paper with some concluding remarks in Section 6.

2   Background

2.1   The TSWTCM Marker

The TSWTCM packet marking algorithm [3] defines two target rates: the Committed Information Rate (CIR) and the Peak Information Rate (PIR). Under TSWTCM, the aggregated traffic is monitored and when the measured traffic is below its CIR, packets are marked with the lowest drop precedence, AFx1 (green packets). If the measured traffic exceeds its CIR but falls below its PIR, packets are marked with a higher drop precedence, AFx2 (yellow packets). Finally, when traffic exceeds its PIR, packets are marked with the highest drop precedence, AFx3 (red packets).

2.2   The RIO Scheme

RIO [4] uses the same dropping mechanism as RED, but it is configured with three sets of parameters, one for each drop precedence marking (colour). These different RED parameters cause packets marked with higher drop precedence to be discarded more frequently during periods of congestion than packets marked with lower drop precedence. Incipient congestion is detected by computing a weighted average queue size, since a sustained long queue is an evidence of network congestion. There are several ways to compute the average queue length. For example, in the RIO-C mode, the average queue size for packets of different colours is calculated by adding its average queue size to the average queue sizes of colours with lower drop precedence. That is, for red packets, the average queue size will be calculated using red, yellow and green packets. For the configuration of RIO parameters, [13] recommends the staggered setting (Fig. 1).

2.3   The DRIO Scheme

DRIO [6] was originally proposed to improve fairness among individual flows for AF-based services. Upon arrival of a red packet, DRIO takes effect when the average queue length falls between min_{th} and max_{th} RIO thresholds. In DRIO, a history list of the last red packets seen is maintained. If the history list is not full, the red packet is added to it. Otherwise, two elements are randomly selected from the history list and compared with the red packet. A hit (miss) occurs when the red packet and the selected element (do not) belong to the same flow. Therefore, three outcomes are possible:

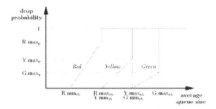

Fig. 1. Staggered setting for RIO

1. If a double hit occurs, the packet is dropped.
2. In the case of one hit and one miss, the packet is dropped with a certain probability. The value of this probability depends on the corresponding flow and it is adjusted accordingly to the number of hits.
3. Finally, if a double miss is obtained, the packet is enqueued and one of the two elements selected from the history list is replaced with it.

The number of hits is a good signal of whether a flow has sent excessive red packets. Thus, DRIO preferably discards packets from flows that have sent more red packets than their fair share. See [6] for a complete description of this scheme.

Albeit DRIO was originally proposed to be applied to individual flows, it can be adapted to work at the aggregate level with a minimum of variations. In fact, only an identification method for classifying packets from different aggregates is required. This method could be based on the value of a combination of several header fields, such as source and destination addresses, source and destination port numbers, protocol identifier, MPLS label, etc.

3   The Unbiased RIO Scheme

In this section, we present the Unbiased RIO (URIO) scheme, a modified version of RIO that improves fairness requirements in AF-based networks. As our extension of DRIO, URIO is also applied to aggregated traffic for scalability reasons. The technique employed by URIO to improve fairness requirement consists of discarding packets from aggregates that are using more than their fair share of the buffer. This technique is based on the one proposed in the Flow Random Early Drop (FRED) [14] scheme to isolate non-adaptive traffic.

3.1   URIO Operations

As in RIO, if the average queue length is less than min_{th}, no arriving packets are discarded. Likewise, when the average queue length is greater than max_{th}, all arriving packets are dropped. In fact, URIO only takes effect when the average queue length falls between min_{th} and max_{th} RIO thresholds. Therefore, URIO just acts when the router is likely to incur congestion.

```
/* Upon arrival of a packet pkt from aggregate i */
if (avg_queue_length > max_th) drop(pkt)
else if (avg_queue_length > min_th) {
    if (qbytes_i ≤ fairqbytes) {
        /* Non-greedy aggregate */
        qbytes_i ← qbytes_i + pkt_size
        enque(pkt)
    } else if ((pkt = UDP) || (rand() < p_TCP_i)) {
        drop(pkt)
    } else {
        qbytes_i ← qbytes_i + pkt_size
        enque(pkt)
    }
} else enque(pkt)
```

Fig. 2. URIO operations

To implement this scheme, it is necessary to record for each active aggregate the three following variables:

1. The aggregate identifier: as described in the previous section, a combination of several header fields can be used.
2. The number of bytes queued in the buffer by the aggregate ($qbytes$).
3. The timestamp with the last arrival of a packet from the aggregate.

We also need to estimate the fair share of the buffer that corresponds to each active aggregate. An easily computable estimation is the fair number of queued bytes per aggregate ($fairqbytes$). This value is calculated dividing the total number of bytes queued in the buffer by the current number of active aggregates (n_a):

$$fairqbytes = \frac{\sum_{i=1}^{n_a} qbytes_i}{n_a} \ . \tag{1}$$

The operations of URIO are shown in Fig. 2. Upon arrival of a packet belonging to an aggregate i, URIO reads the number of bytes already queued by this aggregate. If $qbytes_i$ exceeds $fairqbytes$, we consider that the aggregate is being greedy and, therefore, we should decide whether to accept the incoming packet. TCP packets are discarded with some probability p_{TCP} proportional to the greed degree of the corresponding aggregate. However, packets from UDP flows are always discarded. UDP traffic is treated in a worse manner than TCP traffic since UDP sources are unresponsive to the losses during periods of congestion. This selective dropping allows responsive traffic to send bursts of packets, but prevents greedy traffic from monopolising the buffer. When a packet is finally queued, the $qbytes$ variable of the corresponding aggregate must be incremented by the size of the packet. No extra operations must be accomplished when a packet leaves the queue.

In order to adapt to the current network conditions, periodically both the $qbytes$ value of all active aggregates is set to 0 and the outdated aggregates are

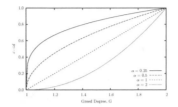

Fig. 3. Dropping probability vs. greed degree

purged based on their timestamp value. If the number of competing aggregates oscillates quickly, this reset timeout should be set to a small value. At the other extreme, if the number of aggregates is relatively constant over time, greater values can be used. The search for optimum values for the reset timeout is a topic for further research, as it does not affect the results presented in this paper.

3.2   Calculation of the p_{TCP} Dropping Probability

TCP packets from aggregates whose $qbytes$ does not exceed $fairqbytes$ are always queued in the buffer. Otherwise, the probability of dropping a TCP packet (p_{TCP}) should be proportional to the greed degree of the corresponding aggregate: the greater the greed degree of the aggregate, the higher the dropping probability. We define the greed degree G of an aggregate i as:

$$G_i = \frac{qbytes_i}{fairqbytes} \; . \tag{2}$$

Greedy aggregates verify that $G > 1$, so the p_{TCPi} probability can be obtained in the following manner:

$$p_{TCPi} = \min\left\{(G_i - 1)^{\alpha}, 1\right\} \; , \qquad \alpha > 0 \; , \tag{3}$$

where the α parameter controls the dropping probability in a inversely proportional manner. Figure 3 shows how the p_{TCP} probability varies with the greed degree for different values of α. The α parameter is a compromise: smaller values of α improve fairness at the expense of increasing packet loss ratios. Experimentally, we have found that the fairness condition is fulfilled satisfactorily with a moderate packet loss ratio when $\alpha = 0.5$. This result can be explained by the combined effect of the power-law dropping probability enforced by the router, coupled with the multiplicative-decrease reaction of TCP upon packet losses. For $\alpha < 1$, this leads to a sub-linear decrease of the sending rate only for the greedy sources that improves fairness.

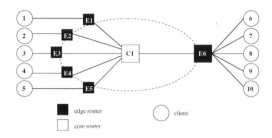

Fig. 4. Network topology. Each link has a 10 Mbps capacity and a 1 ms delay

4   Simulation Configuration

Figure 4 shows the network topology employed in simulations. We consider five competing aggregates sharing the same AF class: aggregate 1 runs between client 1 and client 6, aggregate 2 runs between client 2 and client 7, and so on. Therefore, all aggregates pass through a single bottleneck link between nodes C1 and E6. All TCP connections established are bulk data transfer and use the NewReno implementation. The packet size is set to 1000 bytes.

The marking scheme used in the edge routers is TSWTCM. Each edge node uses a single-level RED queue. The RED parameters $\{min_{th}, max_{th}, max_p, w_q\}$ used are $\{30,60,0.02,0.002\}$ and the queue size is set to 100 packets.

The core node implements RIO, DRIO and URIO mechanisms. Three sets of RED thresholds are maintained, one for each drop precedence (Table 1). The physical queue size is capped at 250 packets and w_q equals 0.002. The average core queue size is calculated as in the RIO-C mode.

Each aggregate has the same target: a 0.5 Mbps CIR and a 1.0 Mbps PIR. This subscription level implies a light load network condition. The fair bandwidth that should be allocated for each aggregate is given by.

$$\text{target rate} + \frac{\text{extra bandwidth}}{\text{number of aggregates}} \ . \tag{4}$$

Therefore, ideally all the aggregates will obtain a throughput of 2 Mbps.

All the simulations were performed with ns-2 [15]. We run the simulations for 100 seconds. We measure the average throughput achieved by each aggregate

Table 1. Core RED parameters

Precedence	min_{th} (pkts)	max_{th} (pkts)	max_p
$AFx1$ (green packets)	200	240	0.02
$AFx2$ (yellow packets)	160	200	0.06
$AFx3$ (red packets)	40	160	0.12

over the whole simulation period. Each experiment is repeated 10 times, and then an average and a confidence interval with 95% confidence level are taken over all runs.

5   Simulation Results

Five experiments have been simulated to study the performances of the proposed schemes. The fairness issues examined are the effect of differing number of microflows in an aggregate, the effect of differing RTTs, the effect of differing packet sizes, the impact of TCP/UDP interaction and the effect of different target rates.

5.1   Impact of Number of Microflows

Each microflow represents a single TCP connection. In this test, each aggregate contains a different number of TCP flows that varies from 5 to 25. Under RIO, the aggregate with a larger number of microflows obtains a greater share of the bandwidth. With both DRIO and URIO, each aggregate obtains an equal amount of bandwidth (Fig. 5(a)).

5.2   Impact of RTT

In this test, each aggregate comprises 10 TCP flows. In order to modify the RTT of the aggregates, we consider that the delay of each access link (the link that joins a client with its corresponding edge router) is different, varying from 1 to 100 ms. Under RIO, aggregates cannot achieve its proportional share of bandwidth. With both DRIO and URIO, this bias is fully overcome (Fig. 5(b)).

5.3   Impact of Packet Size

In this test, all the aggregates contain 10 TCP flows but each one has a different packet size, varying from 500 to 1500 bytes. Through RIO, the aggregate that is sending larger packets consumes more of the available bandwidth. DRIO cannot correct this bias either since it does not take into account that packets may have different sizes. Only URIO can make the sharing of bandwidth insensitive to the packet sizes of the aggregates (Fig. 5(c)).

5.4   TCP/UDP Interaction

In this test, there are two groups of aggregates. Aggregates 1, 2 and 3 comprise 10 TCP flows while aggregates 4 and 5 contain one UDP flow with a sending rate of 5 Mbps. With RIO, UDP traffic consumes more than its available throughput impacting TCP traffic negatively. Under DRIO, this bias is only slightly mitigated. As in the previous test, only URIO can completely isolate UDP traffic and share the extra bandwidth in a TCP-friendly manner (Fig. 5(d)).

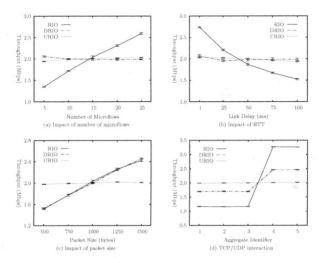

Fig. 5. Fairness evaluation

5.5   Impact of Target Rate

The previous experiments illustrate how URIO shares the extra bandwidth in a uniform manner among aggregates with equal traffic profiles. In a realistic environment, different users will contract different target rates and therefore, all aggregates will not have the same bandwidth expectations. In this test, each aggregate has different CIR and PIR values. The CIR value varies from 0.5 to 1.5 Mbps and the PIR value is set to the CIR value plus 0.5 Mbps. The expected fair share of bandwidth for each aggregate is obtained using Eq. (4). As shown in Table 2, RIO, DRIO and URIO schemes distribute bandwidth according to the subscribed profiles.

Table 2. Impact of target rate

Aggregates		Throughput (Mbps)			
Id	CIR	Expected	RIO	DRIO	URIO
1	0.50 Mbps	1.50	1.51 ± 0.01	1.50 ± 0.01	1.49 ± 0.01
2	0.75 Mbps	1.75	1.75 ± 0.01	1.74 ± 0.01	1.74 ± 0.01
3	1.00 Mbps	2.00	2.00 ± 0.01	2.00 ± 0.01	1.99 ± 0.01
4	1.25 Mbps	2.25	2.24 ± 0.01	2.24 ± 0.01	2.24 ± 0.01
5	1.50 Mbps	2.50	2.47 ± 0.01	2.49 ± 0.01	2.49 ± 0.01

6   Conclusions

In this paper, we have proposed an active queue management scheme called URIO that improves fairness requirements for assured services. Our proposal outperforms RIO and DRIO schemes using a selective discard mechanism to protect aggregates that are consuming less than their fair share. We have evaluated the performance of URIO under diverse network conditions through simulation. Simulation results confirm that URIO distributes the bandwidth among heterogeneous aggregates in a fair manner.

References

1. Blake, S., Black, D., Carlson, M., Davis, E., Wang, Z., Weiss, W.: An architecture for differentiated services. RFC 2475 (1998)
2. Heinanen, J., Baker, F., Weiss, W., Wroclawski, J.: Assured forwarding PHB group. RFC 2597 (1999)
3. Fang, W., Seddigh, N., Nandy, B.: A time sliding window three colour marker (TSWTCM). RFC 2859 (2000)
4. Clark, D.D., Fang, W.: Explicit allocation of best effort packet delivery. IEEE/ACM Transactions on Networking **6** (1998) 362–373
5. Floyd, S., Jacobson, V.: Random early detection gateways for congestion avoidance. IEEE/ACM Transactions on Networking **1** (1993) 397–413
6. Lin, W., Zheng, R., Hou, J.: How to make assured services more assured. In: Proc. ICNP. (1999)
7. Ibanez, J.A., Nichols, K.: Preliminary simulation evaluation of an assured service. IETF Draft (1998)
8. de Rezende, J.F.: Assured service evaluation. In: Proc. IEEE GLOBECOM. (1999)
9. Seddigh, N., Nandy, B., Pieda, P.: Bandwidth assurance issues for TCP flows in a differentiated services network. In: Proc. IEEE GLOBECOM. (1999)
10. Feng, W.C., Kandlur, D., Saha, D., Shin, K.: Adaptive packet marking for maintaining end-to-end throughput in a differentiated-services internet. IEEE/ACM Transactions on Networking **7** (1999) 685–697
11. Nandy, B., Seddigh, N., Pieda, P., Ethridge, J.: Intelligent traffic conditioners for assured forwarding based differentiated services networks. In: NetWorld+Interop 2000 Engineers Conference. (2000)
12. El-Gendy, M., Shin, K.: Equation-based packet marking for assured forwarding services. In: Proc. IEEE INFOCOM. (2002)
13. Makkar, R., Lambadaris, I., Salim, J.H., Seddigh, N., Nandy, B., Babiarz, J.: Empirical study of buffer management schemes for diffserv assured forwarding PHB. Technical report, Nortel Networks (2000)
14. Lin, D., Morris, R.: Dynamics of random early detection. In: Proc. ACM SIGCOMM. (1997)
15. NS: Network simulator, ns. http://www.isi.edu/nsnam/ns/ (2003)

Admission and GoS Control in Multiservice WCDMA System

Jean Marc Kelif[1] and Eitan Altman[2,*]

[1] France Telecom R&D, 38-40 Rue du General Leclerc, 92794 Issy les Moulineaux
Cedex 9, France
[2] INRIA, BP 93, 06902 Sophia Antipolis Cedex, France

Abstract. We consider a WCDMA system with both real time (RT) calls that have dedicated resources, and data non-real time (NRT) calls that use a time-shared channel. We assume that NRT traffic is assigned the resources left over from the RT traffic. The grade of service (GoS) of RT traffic is also controlled in order to allow for handling more RT call but at the cost of degraded transmission rates during congestion periods. We consider both the downlink (with and without macrodiversity) as well as the uplink and study the blocking probabilities of RT traffic as well as expected sojourn time of NRT traffic. We further study the conditional expected sojourn time of a data call given its size and the state of the system. We extend our framework to handover calls.

Keywords: WCDMA, call admission control, HSDPA, HDR, handover.

1   Introduction

An important performance measure of call admission policies is the probability of rejection of calls of different classes. In order to be able to compute these probabilities and to design the call admission control (CAC) policies, a dynamic stochastic approach should be used. A classical approach for CAC is based on adaptively deciding how many channels (resources) to allocate to calls of a given service class [2,7,8]. Then one evaluates the performance as a function of some parameters (thresholds) that characterize the admission policy. Where as this approach is natural to adapt to TDMA or FDMA systems in which the notion of "channel" and of "capacity" is clear, this is not the case any more with CDMA systems in which the capacity is much more complex to define. For the uplink case in CDMA, the capacity required by a call has been defined in the context of CAC see [12,5,3].

We focus on two type of calls, real-time (RT) and non-real time (NRT) data transfers. Where as all calls use CDMA, we assume that NRT calls are further time-multiplexed[1]. We propose a simple model that allows us to define in the

---

* Supported by a CRE research contract with France Telecom R&D.
[1] Time multiplexing over CDMA is typical for down link data channels, as the High Data Rate (HDR) [1] and the High Speed Downlink Packet Access (HSDPA) [11]

M. Freire et al. (Eds.): ECUMN 2004, LNCS 3262, pp. 70–80, 2004.

downlink case (also in presence of macrodiversity) the capacity required by a call when it uses a given GoS (transmission rate)[2]. We then propose a control policy that combines admission control together with a control of the GoS of real-time traffic. Key performance measures are then computed by modeling the CDMA system as a quasi birth and death (QBD) process. We obtain the call blocking probabilities and expected transfer times (see [3] for the uplink case). We further obtain the expected transfer time of a file conditioned on its size. We study the influence of the control parameters on the performances.

2   The Downlink

We use the model similar to [4]. Let there be S base stations. The minimum power received at a mobile k from its base station l call is determined by a condition concerning the signal to interference ratio, which should be larger than some constant

$$(C/I)_k = \frac{E_s}{N_o} \frac{R_s}{W} \Gamma, \tag{1}$$

where E_s is the energy per transmitted bit of type s, N_o is the thermal noise density, W is the WCDMA modulation bandwidth, R_s is the transmission rate of the type s call, and Γ is a constant is related to the shadow fading and imperfect power control [3]. Let $P_{k,l}$ be the power received at mobile k from the base station l. Assume that there are M mobiles in a cell l: the base station of that cell transmits at a total power $P_{tot,l}$ given by $P_{tot,l} = \sum_{j=1}^{M} P_{j,l} + P_{CCH}$ where P_{CCH} corresponds to the power transmitted for the orthogonal Common Channels. Note that these two terms are not power controlled and are assumed not to depend on l. Due to the multipath propagation, a fraction α of the received own cell power is experienced as intracell interference. Let $g_{k,l}$ be the attenuation between base station l and mobile k. Denoting by $I_{k,inter}$ and $I_{k,intra}$ the intercell and intracell interferences, respectively, we have

$$\left.\frac{C}{I}\right|_k = \frac{P_{k,l}/g_{k,l}}{I_{k,inter} + I_{k,intra} + N}$$

where N is the receiver noise floor (assumed not to depend on k), $I_{k,intra} = \alpha(P_{CCH} + \sum_{j \neq k} P_{j,l})/g_{k,l}$ and $I_{k,inter} = \sum_{j=1, j \neq l}^{S} P_{tot,j}/g_{k,j}$. Define $F_{k,l} = \frac{\sum_{j=1, j \neq l}^{S} P_{tot,j}/g_{k,j}}{P_{tot,l}/g_{k,l}}$. It then follows that

$$\beta_k = \frac{P_{k,l}/g_{k,l}}{(F_{k,l} + \alpha)P_{tot,l}/g_{k,l} + N} \qquad \text{where } \beta_k = \frac{(C/I)_k}{1 + \alpha(C/I)_k}. \tag{2}$$

We next consider two service classes (that will correspond to real time (RT) and non-real time (NRT) traffic, respectively). Let $(C/I)_s$ be the target SIR

---

[2] UMTS uses the Adaptive Multi-Rate (AMR) codec that offers eight different transmission rates of voice varying between 4.75 kbps to 12.2 kbps.

ratio for mobiles of service class s and let β_s be the corresponding value in (2). Let there be in a given cell M_s mobiles of class s. We shall use the following approximations. First we replace $F_{k,l}$ by a constant (e.g. its average value, as in [4]; this is a standard approximation, see [6]). Secondly, we approximate $g_{k,l}$ by their averages. More precisely we define G_s to be the average of $g_{l,k}$ over all mobiles k belonging to class s, $s = 1, 2$. With these approximations (2) gives the following value for $P_{tot,l}$ (we shall omit the index l):

$$P_{tot} = \frac{P_{CCH} + N \sum_s \beta_s M_s G_s}{1 - (\alpha + F) \sum_s \beta_s M_s}. \tag{3}$$

Further assuming that the system is designed so as to have $P_{CCH} = \psi P_{tot}$ and defining the downlink load as $Y_{DL} = \sum_s \beta_s M_s$, this gives

$$P_{tot} = \frac{N \sum_s \beta_s M_s G_s}{Z_2} \text{ where } Z_2 = (1 - \psi) - (\alpha + F)Y_{DL}. \tag{4}$$

In practice, to avoid instabilities and due to power limitation of the base stations, one wishes to avoid that Z_2 becomes close to zero, thus one poses the constraint $Z_2 \geq \epsilon$ for some $\epsilon > 0$. Define the system's capacity as $\Theta_c = 1 - \psi - \epsilon$ and the capacity required by a call as

$$\Delta(s) := (\alpha + F)\beta_s. \tag{5}$$

We note that β_s will allow to depend on $M_\sigma, \sigma = 1, 2$. Combining this with (1) and with (2) we get

$$R_s = \frac{\Delta(s)}{\alpha + F - \alpha\Delta(s)} \times \frac{N_o W}{E_s \Gamma}. \tag{6}$$

3   Downlink with Macro Diversity

Our approach is inspired by [4] who considered the single service case. A mobile i in macrodiversity is connected to two base stations, b and l. b is defined to be the station with larger SIR. Following [4] we assume that the Maximum Ratio Combining is used and hence the power control tries to maintain $\gamma_i = \frac{C}{T}|_i = \frac{C}{T}|_{i,b} + \frac{C}{T}|_{i,l}$ where γ_k is given by the constant in (1). We have $\Omega_i \leq 1$ where $\Omega_i := \frac{C/I|_{i,l}}{C/I|_{i,b}}$. ($\Omega_i$ is defined to be 0 for a mobile i that is not in macrodiversity.) This gives for the combined C/I [4]:

$$\frac{C}{I}\bigg|_i = \frac{(1 + \Omega_i)P_{i,b}/g_{bi}}{\alpha_b(P_{tot,b} - P_{i,b})/g_{bi} + F_{i,b}P_{tot,b}/g_{bi} + N}$$

The transmission power becomes $P_{i,b} = \kappa_i(\alpha P_{tot,b} + F_{i,b}P_{tot,b} + g_{bi}N)$, where

$$\kappa_i = \frac{(C/I)_i}{1 + \Omega_i + \alpha(C/I)_i} \tag{7}$$

Let there be M mobiles in a cell b (we shall omit this index) of which a fraction μ is in macrodiversity. Then as in [4], $P_{tot} = \sum_{i=1}^{(1-\mu)M} P_i + \sum_{j=1}^{2\mu M} P_j + P_{CCH}$. We now consider two classes of services $s = 1,2$ corresponding to RT and NRT mobiles. We make the following approximations, similar to [4]: For a given service class $s = 1,2$ Ω_i is replaced by a constant Ω_s (its average over all mobiles of the same service as i): we also replace $F_{i,b}$ by one of four constants F_s^{NMD} and F_s^{MD}, $s = 1,2$, where F_s^{NMD} (resp. F_s^{MD}) corresponds to an average value of $F_{i,b}$ over mobiles in service s which are not in macrodiversity (and which are in macrodiversity, resp.). Finally, we replace g_{bi} by one of the four constants G_s^{NMD} and G_s^{MD}, $s = 1,2$, where G_s^{NMD} (resp. G_s^{MD}) corresponds to an average value of g_{ib} over mobiles in service s which are not in macrodiversity (and which are in macrodiversity, resp.). This gives the total power of a base station b: $P_{tot} = Z_1/Z_2$ as long as Z_2 is strictly positive, where $Z_1 := (1-\mu) \sum_{s=1,2} M_s \kappa_s G_s^{NMD} N + 2\mu \sum_{s=1,2} M_s \kappa_s G_s^{MD} N$, $Z_2 := (1-\psi) - (1-\mu) \sum_{s=1,2} M_s \kappa_s (\alpha + F_s^{NMD}) - 2\mu \sum_{s=1,2} M_s \kappa_s (\alpha + F_s^{MD})$. In practice one wishes to avoid that Z_2 becomes close to zero, thus we pose the constraint $Z_2 \geq \epsilon$ for some $\epsilon > 0$. Define the system's capacity as $\Theta_\epsilon = 1 - \psi - \epsilon$, and the capacity required by a call of type s as $\Delta(s) = \kappa_s \Big[(1-\mu)(\alpha + F_s^{NMD}) + 2\mu(\alpha + F_s^{MD})\Big]$. Combining this with (1) and (7), we get

$$R_s = \frac{\Delta(s)(1+\Omega_s)}{(1-\mu)(\alpha + F_s^{NMD}) + 2\mu(\alpha + F_s^{MD}) - \alpha\Delta(s)} \cdot \frac{N_o W}{E_s \Gamma}, \quad s = 1,2. \quad (8)$$

4   Uplink

We recall the capacity notions from the case of uplink from [3]. Define

$$\tilde{\Delta}_s = \frac{E_s}{N_o} \frac{R_s}{W} \Gamma, \text{ and } \Delta'(s) = \frac{\tilde{\Delta}(s)}{1 + \tilde{\Delta}(s)}, \quad s = 1,2. \quad (9)$$

The power that should be received at a base station originating from a type s mobile in order to meet the QoS constraints is Z_1/Z_2 [3] where $Z_1 = N\Delta'(s)$ and $Z_2 = 1 - (1+f) \sum_{s=1,2} M_s \Delta'(s)$ (N is the background noise power at the base station, f is a constant describing the average ratio between inter and intra cell interference, and M_s is the number of mobiles of type s in the cell). To avoid instability one requires that $Z_2 \geq \epsilon$ for some $\epsilon > 0$. Define the system's capacity as $\Theta_\epsilon = 1 - \epsilon$, and the capacity required by a type s call as $\Delta(s) = (1+f)\Delta'(s)$. Combining this with (9) we get

$$R_s = \frac{\Delta(s)}{1 + f - \Delta(s)} \times \frac{N_o W}{E_s \Gamma}, \quad s = 1,2. \quad (10)$$

5   Admission and Rate Control

We assume that either the uplink or the downlink are the bottleneck in terms of capacity, so we can focus only on the more restrictive direction when accepting calls. All the notations will be understood to relate to that direction.

Capacity reservation. We assume that there exists a capacity L_{NRT} reserved for NRT traffic. RT traffic can use up to a capacity of $L_{RT} := \Theta_\epsilon - L_{NRT}$.

GoS control of RT traffic. UMTS will use the Adaptive Multi-Rate (AMR) codec that offers eight different transmission rates of voice that vary between 4.75 kb/s to 12.2 kb/s, and that can be dynamically changed every 20 ms. The lower the rate is, the larger the amount of compression is, and we say that the GoS is lower. For simplicity we shall assume that the set of available transmission rates of RT traffic has the form $[R^{\min}, R^{\max}]$. We note that $\Delta(RT)$ is increasing with the transmission rate. Hence the achievable capacity set per RT mobile has the form $[\Delta^{\min}, \Delta^{\max}]$. Note that the maximum number of RT calls that can be accepted is $M_{RT}^{\max} = \lfloor \Theta_\epsilon / \Delta^{\min} \rfloor$. We assign full rate $R^{\max}$ (and thus the maximum capacity $\Delta^{\max}$) for each RT mobile as long as $M_{RT} \leq N_{RT}$ where $N_{RT} = \lfloor L_{RT} / \Delta_{RT}^{\max} \rfloor$. For $N_{RT} < M_{RT} \leq M_{RT}^{\max}$ the capacity of each present RT call is reduced to $\Delta_{MR} = L_{RT} / M_{RT}$ and the rate is reduced accordingly (e.g. by combining (1), (2) and (5) for the case of downlink).

Rate control for NRT traffic. The capacity $C(M_{RT})$ unused by the RT traffic is fully assigned to one NRT mobile, and the mobile to which it is assigned is time multiplexed rapidly so that the throughput is shared equally between the present NRT mobiles. The available capacity for NRT mobiles is

$$C(M_{RT}) = \begin{cases} \Theta_\epsilon - M_{RT}\Delta^{\max} & \text{if } M_{RT} \leq N_{RT}, \\ L_{NRT} & \text{otherwise.} \end{cases}$$

The total transmission rate R_{NRT}^{tot} of NRT traffic for the downlink and uplink is then given by

$$R_{NRT}^{tot}(M_{RT}) = \frac{C(M_{RT})}{Z_3} \times \frac{N_o W}{E_s \Gamma},$$

where $Z_3 = \alpha + F - \alpha C(M_{RT})$ for the downlink and $Z_3 = 1 + f - C(M_{RT})$ for the uplink. The expression for downlink with macrodiversity is derived similarly.

6   Statistical Model and the QBD Approach

Model. RT and NRT calls arrive according to independent Poisson processes with rates λ_{RT} and λ_{NRT}, respectively. The duration of a RT call is exponentially distributed with parameter μ_{RT}. The size of a NRT file is exponentially distributed with parameter μ_{NRT}. Interarrival times, RT call durations and NRT file sizes are all independent. The departure rate of NRT calls depends on the current number of RT calls: $\nu(M_{RT}) = \mu_{NRT} R_{NRT}^{tot}(M_{RT})$

QBD approach. The number of active sessions in all three models (downlink with and without macrodiversity and uplink) can be described as a QBD process, and we denote by Q its generator. We shall assume that the system is stable. The stationary distribution of this system, π, is calculated by solving: $\pi Q = 0$, with the normalization condition $\pi e = 1$ where e is a vector of ones of proper dimension. π represents the steady-state probability of the two-dimensional process lexicographically. We may thus partition π as $[\pi(0), \pi(1), \dots]$ with the vector $\pi(i)$ for level i, where the levels correspond to the number of NRT calls in the system. We may further partition each level into the number of RT calls, $\pi(i) = [\pi(i,0), \pi(i,1), \dots, \pi(i, M_{RT}^{\max})]$, for $i \geq 0$. Q is given by

$$Q = \begin{bmatrix} B & A_0 & 0 & 0 & \cdots \\ A_2 & A_1 & A_0 & 0 & \cdots \\ 0 & A_2 & A_1 & A_0 & \cdots \\ & & & & \\ 0 & 0 & \ddots & \ddots & \ddots \end{bmatrix} \qquad (11)$$

where the matrices B, A_0, A_1, and A_2 are square matrices of size $(M_{RT}^{\max} + 1)$. The matrix A_0 corresponds to a NRT connection arrival, given by $A_0 = \mathrm{diag}(\lambda_{NRT})$. The matrix A_2 corresponds to a departure of a NRT call and is given by $A_2 = \mathrm{diag}(\nu(i); 0 \leq i \leq M_{NRT}^{\max})$. The matrix A_1 corresponds to the arrival and departure processes of RT calls. A_1 is tri-diagonal as follows: $A_1[i, i+1] = \lambda_{RT}, A_1[i, i-1] = i\mu_{RT}, A_1[i,i] = -\lambda_{RT} - i\mu_{RT} - \lambda_{NRT} - \nu(i)$. We also have $B = A_1 + A_2$. π is given by [3,9] $\pi(i) = \pi(0)\mathbf{R}^i$ where $\mathbf{R}$ is the minimal non-negative solution to the equation: $A_0 + \mathbf{R}A_1 + \mathbf{R}^2 A_2 = 0$. $\pi(0)$ is obtained by solving $\pi(0)(I - \mathbf{R})^{-1}e = 1$ [3]. Note that the evolution of number of RT calls is not affected by the process of NRT calls and the Erlang formula can be used to compute their steady state probability. The blocking probability of a RT call is:

$$P_B^{RT} = \frac{(\rho_{RT})^{M_{RT}^{\max}} / M_{MRT}^{\max}!}{\sum_{i=1}^{M_{RT}^{\max}} (\rho_{RT})^i / i!}$$

where $\rho_{RT} = \lambda_{RT}/\mu_{RT}$. This is the main performance measure for the RT traffic. For NRT calls the important performance measure is expected sojourn time which is given by Little's law as $T_{NRT} = E[M_{NRT}]/\lambda_{NRT}$.

Conditional expected sojourn times. The performance measures so far are similar to those already obtained in the uplink case in [3]. We wish however to present more refined performance measures concerning NRT calls: the expected sojourn times conditioned on the file size and the state upon the arrival of the call. We follow [10] and introduce a non-homogeneous QBD process with the following generator Q^* and the corresponding steady state probabilities π^*:

$$Q^* = \begin{bmatrix} B & A_0 & 0 & 0 & \cdots \\ (1/2)A_2 & A_1^* & A_0 & 0 & 0 & \cdots \\ 0 & (2/3)A_2 & A_1^* & A_0 & 0 & \cdots \\ 0 & 0 & (3/4)A_2 & A_1^* & A_0 & \cdots \\ & & & & \\ 0 & 0 & & \ddots & \ddots & \ddots \end{bmatrix} \qquad (12)$$

where the matrices A_0, A_2, B are the same as introduced before, and A_1^* is the same as A_1 defined before except that the diagonal element is chosen to be minus the sum of the off-diagonal elements of Q^*, i.e. $A_1^*[k, k] = -\lambda_{RT} - k\mu_{RT} - \lambda_{NRT} - \frac{k-1}{k}\nu(k)$. The conditional expected sojourn time of a NRT mobile given that its size is v, that there are i RT mobiles and $k-1$ NRT mobiles upon it's arrival, is obtained from [10]:

$$T_{k,i}(v) = \frac{v/R_{NRT}^{tot}(0)}{R^* - \rho^*} + \overline{1_{k,i}}\left[I - \exp\left(\frac{v}{R_{NRT}(0)}\mathcal{R}^{-1}Q^*\right)\right]\overline{w} \quad \text{where} \quad (13)$$

$$R^* := \sum_{k,i} \pi^*(k,i)\frac{R_{NRT}^{tot}(i)}{R_{NRT}^{tot}(0)}, \quad \mathcal{R} = diag\left[\frac{1}{k}\frac{R_{NRT}^{tot}(i)}{R_{NRT}^{tot}(0)}\right], \quad \rho^* := \frac{\lambda_{NRT}}{\mu_{NRT}R_{NRT}^{tot}(0)},$$

$\overline{1_{k,i}}$ is a vector whose entries are all zero except for the (k,i)th entry whose value is 1, and $\overline{w}$ is the solution [10] of $Q^*\overline{w} = \frac{1}{R^*-\rho^*}\mathcal{R}\cdot\overline{1_{k,i}} - \overline{1_{k,i}}$.

Remark 1. Suppose that the number of RT sessions stays fixed throughout the time (this can be used as an approximation when the average duration of RT sessions is very large). Then with $R^* = R_{NRT}^{tot}(i)/R_{NRT}^{tot}(0)$, (13) becomes

$$T_k(v) = \frac{v/R_{NRT}^{tot}(0)}{R^* - \rho^*} + \frac{1}{\mu_{NRT}R_{NRT}^{tot}(i) - \lambda_{NRT}} \times \left(k - \frac{1}{R^* - \rho^*}\right)$$
$$\times \left(1 - \exp\left(-\frac{v}{R_{NRT}^{tot}(0)}(\mu_{NRT}R_{NRT}^{tot}(0) - \lambda_{NRT})\right)\right) \quad (14)$$

In Fig. 1 we use (14) to compute the maximum number k of NRT calls present at the arrival instant of an NRT call (we include in this number the arriving call) such that the expected sojourn time of the connection, conditional on its size in kb and on k, is below 1 sec. For example, if the size of the file is 100 kb then its conditional expected sojourn time will be smaller than 1 sec as long as the number of mobiles upon arrival (including itself) does not exceed 12. This figure and the next one are obtained with $R_{NRT}^{tot}(0) = 1000kbps$, $\lambda_{NRT} = 1$, $\mu_{NRT}^{-1} = 160$ kbits so that $\rho^* = 0.16$ (We took no RT calls, i.e. $i = 0$).

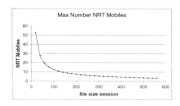
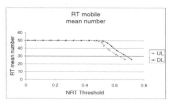

Fig. 1. Max number of NRT calls upon arrival s.t. the conditional expected sojourn time is below 1 sec

Fig. 2. Mean number of RT calls in a cell as a function of the reservation level for RT traffic

7   Numerical Results

We consider the following setting. Unless stated otherwize, the data are for both down and uplink. Transmission rates of RT mobiles: maximum rate is 12.2kbps, minimum rate 4.75kbps. NRT uplink data rate: 144 kbps. NRT downlink data rate: 384 kbps. E_{RT}/N_o (at 12.2 kbps) is 7.9 dB (UL) and 11.0 dB (DL). E_{NRT}/N_o is 4.5 dB (for 144 kbps UL) and 4.8 dB (for 384 kbps DL). Total bandwidth $W = 3.84MHz$. Mean NRT session size 160 kbits, arrival rate of NRT calls: $\lambda_{NRT} = 0.4$. Average duration of RT call: 125 s, arrival rate of RT calls: $\lambda_{RT} = 0.4$. Uplink interference factor is 0.73. Downlink interference factor is 0.55. $\alpha = 0.64$, $\psi = 0.2$, $\epsilon = 10^{-5}$. Γ is computed so as to guarantee that the probability of exceeding the target C/I ratio is 0.99. It corresponds to a standard deviation constant $\sigma = 0.5$ (see [3]).

Influence of NRT reservation on RT traffic. In Fig. 2 we depict the average cell capacity in terms of the average number of RT mobiles for both uplink and downlink as a function of the reservation threshold for NRT traffic. We see that it remains almost constant (50 mobiles per cell) for up to 50% of the load. In Fig. 3 we present the blocking rate of RT traffic. At a reservation L_{NRT} of 50% of the maximum load, the dropping rate is still lower than 1%.

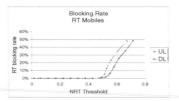

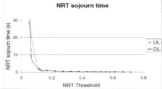

Fig. 3. Blocking rate for RT calls as a function of the reservation level for NRT traffic

Fig. 4. Expected sojourn times of NRT traffic as a function of the NRT reservation

Influence of NRT reservation on NRT traffic. Fig. 4 shows the impact of the reservation threshold L_{NRT} on the expected sojourn time of NRT calls both on uplink and downlink. We see that the expected sojourn times becomes very large as we decrease L_{NRT} below 0.15% of the load. This demonstrates the need for the reservation. In the whole region of loads between 0.16 to 0.5 the NRT expected sojourn time is low and at the same time, as we saw before, the rejection rate of RT calls is very small. This is thus a good operating region for both RT and NRT traffic.

Conditional expected sojourn time. Below, the reservation limit L_{NRT} is 0.27. In Fig. 5 we depict the expected sojourn time conditioned on the number of NRT and RT calls found upon the arrival of the call both being k and on the file being of the size of 100 kbits. k is varied in this figure.

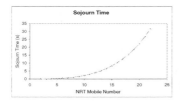

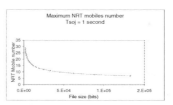

Fig. 5. Conditional expected Sojourn time of an NRT mobile as a function of the number of mobiles in the cell

Fig. 6. Max number of NRT calls upon arrival such that the conditional expected sojourn time is below 1 sec

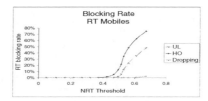

Fig. 7. RT dropping probabilities

Fig. 6 depicts for various file sizes, the maximum number k such that the conditional expected sojourn time of that file with the given size is below 1 sec. k is defined to be the total number of RT calls as well as the total number of NRT calls (including the call we consider) in the cell. We thus assume (as in the previous example) that the number of NRT and of RT calls is the same, and seek for the largest such number satisfying the limit on the expected sojourn time.

8   Extension to Handover Calls

Till now we did not differentiate between arrivals of new and of handover calls. We now wish to differentiate between these calls. We assume that RT new calls (resp. NRT new calls) arrive with a rate of λ_{RT}^{New} (resp. λ_{NRT}^{New}) where as the handover calls arrive at rate h_{RT} (resp. h_{NRT}). We assume that RT calls remain at a cell during an exponentially distributed duration with parameter μ_{RT}. Avoiding blocking of handover calls is considered more important than avoiding blocking of new ones. So we define a new threshold $\overline{M}_{RT}^{New} < M_{RT}^{\max}$. Now new RT calls are accepted as long as $M_{RT} \leq \overline{M}_{RT}^{New}$ where as handover RT calls are accepted as long as $M_{RT} \leq \overline{M}_{RT}^{New}$. The behavior of NRT calls is as before. Define

$\rho_{RT} = \lambda_{RT}/\mu_{RT}$ and $\rho_{RT}^{HO} = h_{RT}/\mu_{RT}$. Let $p_{RT}(i)$ denote the number of RT mobiles in steady state. It is given by

$$p_{RT}(i) = \begin{cases} \dfrac{(\rho_{RT})^i}{i!} p_{RT}(0) & \text{if } 0 \le i \le \overline{M}_{RT}^{New} \\[2ex] \dfrac{(\rho_{RT})^{\overline{M}_{RT}^{New}} (\rho_{RT}^{HO})^{i-\overline{M}_{RT}^{New}}}{i!} p_{RT}(0) & \text{if } \overline{M}_{RT}^{New} \le i \le M_{RT}^{\max} \end{cases}$$

where $\quad p_{RT}(0) = \left(\displaystyle\sum_{i=0}^{\overline{M}_{RT}^{New}} \dfrac{(\rho_{RT})^i}{i!} + \sum_{i=\overline{M}_{RT}^{New}}^{M_{RT}^{\max}} \dfrac{(\rho_{RT})^{\overline{M}_{RT}^{New}} (\rho_{RT}^{HO})^{i-\overline{M}_{RT}^{New}}}{i!} \right)^{-1}$

The QBD approach can be directly applied again to compute the joint distribution of RT and NRT calls and performance measures.

A numerical example. The data we consider are as before, except that now a fraction of 30% of arriving RT calls are due to handovers. In Fig. 7 we present the impact of the choice of the NRT threshold on the blocking rate of RT mobiles. We also illustrate the impact of the differentiation between New and Handover calls. The middle UL curve is obtained with no differentiation. The total dropping rate of the model with handover (the HO curve) is larger, but among the rate of dropping rate of calls already in the system (that arrive through a handover) is drastically diminished (the curve called "Dropping").

9 Conclusions

We have analyzed the performance of a CAC combined with a GoS control in a WCDMA environment used by RT and NRT traffic. RT traffic has dedicated resources where as NRT traffic obtains through time sharing the unused capacity leftover by RT traffic. We illustrated the importance of adding reserved capacity L_{NRT} for NRT traffic only and demonstrated that this can be done in a way not to harm RT traffic.

References

1. P. Bender et al, "CDMA/HDR: A bandwidth-efficient high-speed wireless data service for nomadic users", *IEEE Communications Magazine*, 70–77, July 2000.
2. Y. Fang and Y. Zhang, "Call admission control schemes and performance analysis in wireless mobile networks", *IEEE Trans. Vehicular Tech.* **51**(2), 371-382, 2002.
3. N. Hegde and E. Altman, "Capacity of multiservice WCDMA Networks with variable GoS", Proc. of IEEE WCNC, New Orleans, Louisiana, USA, March 2003.
4. K. Hiltunen and R. De Bernardi, "WCDMA downlink capacity estimation", VTC'2000, 992-996, 2000.
5. I. Koo, J. Ahn, H. A. Lee, K. Kim, "Analysis of Erlang capacity for the multimedia DS-CDMA systems", IEICE Trans. Fundamentals, Vol E82-A No. 5, 849-855, 1999.
6. J. Laiho and A. Wacker, "Radio network planning process and methods for WCDMA", *Ann. Telecommun.*, **56**, No. 5-6, 2001.

7. C. W. Leong and W. Zhuang, "Call admission control for voice and data traffic in wireless communications", Computer Communications 25 No. 10, (2002) 972-979.
8. B. Li, L. Li, B. Li, X.-R. Cao "On handoff performance for an integrated voice/data cellular system" Wireless Networks, Vol. 9, Issue 4 pp. 393 – 402, July 2003.
9. M. F. Neuts. *Matrix-geometric solutions in stochastic models: an algorithmic approach.* The John Hopkins University Press, 1981.
10. R. Núñez Queija and O. J. Boxma. Analysis of a multi-server queueing model of ABR. *J. Appl. Math. Stoch. Anal.*, 11(3), 1998.
11. S. Parkvall, E. Dahlman, P. Frenger, P. Beming and M. Persson, "The high speed packet data evolution of WCDMA", *Proc. of the 12th IEEE PIMRC*, 2001.
12. X. Tang and A. Goldsmith "Admission control and adaptive CDMA for integrated voice and data systems", available on http://citeseer.ist.psu.edu/336427.html

A QoS Based Service Reservation: Supporting Multiple Service in Optical Access Networks

NamUk Kim, HyunHo Yun, Yonggyu Lee, and Minho Kang

Optical Internet Research Center, Information and Communication University, 119,
Munjiro,Yuseong, Daejon, (305-714), Korea(ROK)
{niceguy, exhyho, yonggyu, mhkang}@icu.ac.kr

Abstract. In this paper, the service work reservation method for the efficient service differentiation and the bandwidth arbitration for the fair service allocation among access nodes are fully studied. In the polling based generalized TDM networks, QoS (Quality of Service) based multiple service is achieved by the reservation based service scheduling as well as the class based packet scheduling. These are tightly coupled with each other in every protocol step and determine the final performances of multiple service. For the seamless fair service, we propose the service quality pre-engagement (SQP) scheme and the equal weighted fair bandwidth arbitration (EWF-BA) mechanism. These make the thread based fair service allocation among access nodes possible while the class based service differentiation is maintained. The analytic and simulation results reveal that proposed mechanisms efficiently guarantee the QoS based multiple service with high network performances and the service fairness.

1 Introduction

The dynamic bandwidth allocation over the polling based generalized TDM networks is applied to many access networks such as Ethernet-PON (Passive Optical Networks) and WDM optical networks for upstream data transmission control[1-3]. Contrast to the simple broadcast and select mechanism of downstream[1], the upstream transmission scheduling is complicate because the channel must be efficiently and fairly shared by all ONU(Optical Network Unit)s. At the same time, the moderate QoS level must be guaranteed for supporting multiple service. In the generalized TDM network, the data channel consists of two kinds of time unit, the service slot and the frame cycle. The service slot represents for single ONU's allocated bandwidth in time domain during which ONU transmits queued data to OLT(Optical Line Termination). The frame cycle, generally expressed as the frame in circuit switched networks, is total sum of service slots of every ONUs. Hence, the efficiency of the service reservation for multiple services and the service work coordination for high utilization are the performance critical. These performances are tightly related with each other. Up to now, several protocols are contributed to this issue[1,3]. Among these, the simplest one is the static bandwidth allocation (SBA) which fixes the service slot and the frame

M. Freire et al. (Eds.): ECUMN 2004, LNCS 3262, pp. 81–90, 2004.

cycle to the constant value. Hence it guarantees the stable service to each ONU and the service isolation among ONUs is possible. But it inherits traditional TDM's problems like the low channel utilization and the weakness to burst traffic. DBA (Dynamic Bandwidth Allocation) can solve these problems with the three step reservation based transmission control[1]. But DBA easily causes the service monopolization and the fairness problems by bursty ONUs if there is no arbitration scheme like IPACT[1] in OLT side. It supports the static service fairness. Based on these service coordination mechanisms, general class based packet schedulings can be applied to the upstream port of ONU for multiple service[3]. But the performance of packet scheduling is tightly correlated with the service coordination(service work scheduling) because arbitration determines the service bandwidth within which packet scheduler sends data packets.

In this paper, we propose the noble SQP-DBA(Service Quality Preengagement with DBA) that is the QoS bandwidth reservation mechanism under the strict priority(SP) queuing packet service. Then, we also introduce a new service work scheduling mechanism, EWF-BA(Equal Weighted Fair - Bandwidth Arbitration), to achieve the high network performances as well as the real time service fairness in inter ONU domain. Although proposed mechanisms are introduced to the Ethernet-PON networks for analysis, our schemes can be applied to all kinds of generalized polling based TDMA networks independent from the physical media and topological differences.

2   Supporting Multiple Service in DBA Based Networks

Like other switch networks, general packet scheduling methods can be applied to ONU's upstream data transmission for supporting multiple service[2]. But due to the DBA's discrete operation, WFQ and the proportional delay differentiation (PDD) are hard to guarantee the tight delay bound and the efficient service differentiation in spite of high complexity. In this paper, we consider SP for the service differentiation. SP guarantees the absolutely prioritized service to higher priority traffic. But it easily causes the bandwidth monopolization and resultant service starvation of low class. This can be more serious when SP is applied with DBA. These problems mainly comes from the fact that the packet scheduling is separated from the channel reservation. The DBA mechanism for the service work scheduling is done with three steps. In next chapter, we briefly overview the DBA service reservation. For easy expressions, following notations are used when ONU_j, and m_{th} cycle are assumed. All notations are expressed for the class i of j_{th} ONU of total N ONUs networks.

- $Q_{i,j}^m(t)$: Size of all queued data frames at arbitrary time t
- $D_{i,j}^m$: Reported queue size at service slot ending instance for $m + 1_{th}$ cycle
- $A_{i,j}^m$:Allocated bandwidth by service work arbitration
- $S_{i,j}^m$:Total serviced data frames(service work) • R: Upstream link speed
- t_m :Arrival instance of m_{th} gate message   • λ_i : Traffic arrival rate

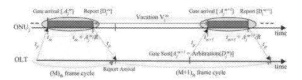

Fig. 1. Service reservation procedures of DBA.

We also assume that traffic is grouped into K classes of service which are ordered, such that class i is better than class i+1 from class 0 to class K-1, in terms of the queuing delay and the packet loss. Hence, the notation X_j^m for ONU_j (X can be Q, D, A, S) is equal to the sum of all classes, $\sum_{n=0}^{K-1} X_{i,j}^m$.

2.1   Service Work Reservation Mechanisms of DBA

Under DBA, the data transmission is done with three consecutive steps as shown in Fig.1. First, ONU reserves the next cycle's bandwidth(service slot) by sending the Report message in which D_j^m is set to $Q_j^m(t_m + A_j^m/R)$ at the end of the service slot. Secondly, when OLT receives Report message, it arbitrates the request service work of ONU_j with those of other ONUs. Then, it notifies ONU_j of the size of the following cycle's service slot, A_j^{m+1}, with the Gate message which will be sent with considering the reference time difference between OLT and ONU. Generally, that is little larger than the sum of the guard band time and two times of t_p. Along with this service work scheduling, OLT also must check the variation of t_p as well as physical device's performance parameters to avoid data collisions. Finally, when Gate message arrives, ONU_j transmits data frames to OLT within A_j^{m+1}. Hence, the real service work(S_j^{m+1}) is always no larger than A_j^{m+1}. Hence, major properties of DBA operation are the backward reservation and the discrete service with the long service vacation[1]. With this reservation based DBA, the pipelined service is possible with the high utilization and the robustness to burst traffic.

2.2   SQP-DBA for QoS Based Multiple Service Reservation

When SP is applied, the service work reservation under DBA is done as follows.

$$D_{i,j}^m = Q_j^m(t_m + \sum_{i=0}^{K-1} S_{i,j}^m/R), S_{i,j}^m \leq A_{i,j}^m \tag{1}$$

$$S_{i+1,j}^m = min(Q_{i+1,j}^m(t_m), A_j^m - \sum_{n=0}^{i} Q_{n,j}^m(t_m)), i \leq K - 2 \tag{2}$$

Therefore, the prioritized service for the high class is easily guaranteed. But as the traffic load increases or burst traffic arrives in short epoch, the large service reservation gap(SRG), which represents the reservation mismatch between the previous request bandwidth($D_{i,j}^m$) and the serviced work($S_{i,j}^{m+1}$) by SP, also increases due to $D_{i,j}^m = A_{i,j}^m < Q_{i,j}^m$. Hence SRG directly affects the performance of service differentiation. This SRG, $\triangle S$, is accumulated from higher classes by (2) and if every class except for the lowest priority K-1 class is fully serviced,

$$\sum_{i=0}^{K-2}(S_{i,j}^{m+1} - D_{i,j}^m) = \sum_{i=0}^{K-2}\triangle S_{K-1,j}^{m+1} = -\triangle S_{K-1,j}^{m+1} = D_{K-1,j}^m - S_{K-1,j}^{m+1} \quad (3)$$

Therefore it is impossible for the low class traffic to receive the reserved service due to the bandwidth shortage, $D_{i,j}^m >> A_{i,j}^{m \mid 1} = (A_{i,j}^{m+1} - \sum_{n=0}^{i-2}Q_{n,j}^{m+1})$. When burst traffic happens, this problem can continue to several cycles[3]. Therefore, increased SRG enlarges the average buffer size and arouses the long transfer delay of lower classes. As shown in (3), $\triangle S$ is inevitable due to the backward reservation property of DBA and is independently determined by higher priority class traffic's arrival. Therefore, for the efficient multiple service that guarantees each class the expectable service level even under the absolute SP, it is important to minimize the unpredictable SRG. Proposed SQP-DBA satisfies this by the QoS based service reservation. In SQP-DBA, ONU reserves the next service slot as the sum of total queued frame size and the QoS reservation bandwidth, D_{qos} determined by the service policy. Simply, we choose D_{qos} as the current cycle's service work of quality guaranteed p classes. Therefore, the bandwidth reservation of ONU_j is done

$$D_{i,j}^m = Q_{queue} + D_{qos} = Q_{i,j}^m(t_m + \sum_{i=0}^{K-1}A_{i,j}^m/R) + S_{i,j}^m, i \leq p \quad (4)$$

$$D_j^m = \sum_{i=0}^{K-1}Q_{i,j}^m(t_m + \sum_{i=0}^{K-1}A_{i,j}^m/R) + \sum_{i=0}^{p}S_{i,j}^m \quad (5)$$

By this, if there is no abrupt change of traffic arrival in the service vacation, $\triangle S$ can be reduced to $\sum_{i=p+1}^{K-2}S_{i,j}^m - D_{i,j}^m$. Hence, the lower priority traffic($i > p$) does not experience the serious bandwidth starvation. But, there can be two sorts of reservation mismatches as follows.

i. $D_{qos}^m < \sum_{i=0}^{p}\lambda_i V^m \approx \sum_{i=0}^{p}Q_{i,j}(t_{m+1})$ case : This happens when burst traffic arrives abruptly. By the SRG mismatch, lower priority classes from p+1 to K-1 lose their reserved bandwidth with amount of $\sum_{i=0}^{p}Q_{i,j}(t_{m+1}) - D_{qos}$ and backlogged. But this mismatch is rapidly reflected to the next cycle reservation by QoS bandwidth D_{qos}^{m+1}. and can be serviced by DBA in the following cycle.

ii. $D_{qos}^m > \sum_{i=0}^{p}\lambda_i V^m \approx \sum_{i=0}^{p}Q_{i,j}(t_{m+1})$ case :In this case, high class traffic rapidly diminishes after reserving D_{QoS} due to the change of the number of service applications or customers. But this is not a problem because there always are queued packets in low class buffers by the backward reservation property of DBA. Hence, the service work of lower classes increases with the amount of SRG as

$$\sum_{i=p+1}^{K-1}S_{i,j}^{m+1} = D_{qos}^m - \sum_{i=0}^{p}Q_{i,j}(t_m)(=\sum_{i=0}^{p}\lambda_i V^m) + \sum_{i=p+1}^{K-1}D_{i,j}^m \quad (6)$$

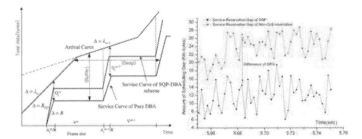

Fig. 2. (a)Service curve of SQP and SP under DBA (b)simulation result of SRG.

Therefore the delay of lower class traffic becomes short while there is no performance degradation of the channel utilization or throughput.

In any cases, it is possible that the lower prioritized non-real time traffic can be serviced by the original DBA's reservation mechanism while the objective QoS guaranteed p classes traffic is serviced with the minimum delay and the absolutely guaranteed throughput. Moreover, because D_{qos} is set to the current cycle's serviced work of p classes, the change of service pattern is rapidly reflected to the whole service reservation mechanism. This rapid service adaptation of SQP-DBA is easily explained by surveying the service tracking speed which means how fast the service scheduler follows and affects the traffic arrival pattern(demand) change to service scheduling. Fig. 2 shows the service curve and the simulation result of SP and SQP under DBA. The traffic arrival of each cycle is characterized by λ_m. Packet service is done during service slot with the service rate R which is same to the gradient(Δ) of the service curve. In service curve, the delay is examined as the horizontal deviation between the arrival and the service curve and the queue size is the vertical deviation. Hence, the delay and queue increases in the vacation and the service scheduler tracks the arrival curve by the service reservation. Hence, the performance critical factor is the service tracking speed. When m_{th} vacation time(V^m) is enough larger than the service slot Q_j^m/R), which is general, and is assumed as upper bounded by value V with OLT's service arbitration scheme, the service rate of SQP+DBA is approximately,

$$R_{SQP} \approx R[\sum_{i=0}^{K-1} Q_{i,j}^m(t_m + S_j^m/R) + \sum_{i=0}^{p} S_{i,j}^m]/\sum_{i=0}^{K-1} Q_{i,j}^m(t_m + S_j^m/R) \quad (7)$$

$$= R[1 + \sum_{i=0}^{p} S_{i,j}^m/(Q_j^m - S_j^m)] > R, 0 \leq p \leq K - 1 \quad (8)$$

The gradient of SQR-DBA service curve, R_{SQR}, is higher than SP+DBA case in every service slot. The access delay by the increased QoS bandwidth is negligible compared with the vacation period. Therefore, SQP-DBA can track the change of arrival traffic pattern more effectively in the average as well as short epoch. Fig.2.(b) shows that the SRG of SQP-DBA is the half of SP-DBA, which is the same amount of the D_{qos}. Therefore, the efficient differentiated service is more achievable even SP is applied. The numerical analysis result also shows the tracking performance enhancement. With same approaches of [4], when $\rho_p = \sum_{i=0}^{p} \rho_i$ and $\rho_{low} = \sum_{i=p+1}^{K-1} \rho_i$, the asymptotical average queue size of lower priority classes is driven as[4]

$$\overline{Q_{low}} = \frac{\lambda_{low}\overline{L}^2}{2(1-\rho_{low})} + \xi(\rho_p)\frac{\rho_p}{1-\rho_p}E[V] - \rho_{low}F[V] + \rho_{low}\frac{E[V]^2}{2E[V]} \qquad (9)$$

The $\xi(\rho_p)$ is used to intuitionally reflect the effect of higher priority classes. It proportionally increases according to ρ_p because large ρ_p brings about large backlogged queue of lower classes. In case of SP+DBA, p is equal to K-2. Hence, the lowest priority class is affected by $\rho_{SP} = \sum_{i=0}^{K-2}\rho_i$ and the large queue and delay are invoked by $\xi(\rho_p)\rho_{SP}/(1-\rho_{SP})$. But in SQP-DBA, D_{qos} is already reserved for higher p classes. Therefore, the effective load is changed to $\rho_{SQP} = \rho - \sum_{i=0}^{p}\rho_i = \sum_{i=p+1}^{K-2}\rho_i$ and this meets $\xi(\rho_{SQP})\rho_{SQP}/(1-\rho_{SQP}) << \xi(\rho_p)\rho_{SP}/(1-\rho_{SP})$. By this service load separation, low priority classes($i > p$) contend for service among only K-1-P classes. It minimizes the performance degradation of non-real time lower priority traffic although SP is strictly applied for the real time traffic.

This guaranteed and isolated service differentiation for the real multiple service is the objective of SQP-DBA. By SQP-DBA's QoS bandwidth pre-reservation scheme, the real traffic can be serviced always within the minimum queuing delay with the minimum interference to lower priority classes. This makes the service quality pre-engagement of each service class with the service policy possible. It can also solve the light load penalty problem of the one stage SP scheduler system[3]. This happens when SRG is same to the high classes' total traffic intensity of service vacation. This causes the low class' long queuing delay even in the light load condition. Especially, when higher class traffic is periodic, the light load penalty more easily happens. This problem can be resolved by SQP when D_{qos} is set to total sum of periodic traffic. Hence, the important issue of SQP is the choice of p value. In the best case, p can be K-1 but it causes the low channel utilization by the large SRG mismatch because it is not the backward reservation mechanism any more. Along with this, there needs two times of the original link bandwidth for supporting doubled request. Hence, generally, it is reasonable to set p as the total number of periodic real time traffic classes. By this, multiple service of SQR and the high utilization of DBA are guaranteed simultaneously. But the SQP-DBA client has no control authority for service allocation. Hence, there must be the optimized service work scheduling in OLT, defined as the arbitration, for the service fairness and high performances in network domain. In next section, we introduce the equal weighted fair-bandwidth arbitration (EWF-BA) mechanism to efficiently support these demands.

3   EWF-BA for Supporting Fair Service Allocation

In well arbitrated networks, the service work is fairly and efficiently distributed to all ONUs in the short epoch. Under DBA, the service fairness is directly determined by two elements, V_j^m and A_j^m. First, V_j^m affects the the queuing delay and zitter. A_j^m is a crucial factor of the service thread which reveals how stably and consistently the service is maintained with the consistent pattern. Proposed EWF-BA is a noble service work scheduling scheme which arbitrates the service slot with consideration of the service demand as well as the service thread of each ONU. V_c is assumed as the optimized frame cycle by which the QoS guaranteed service is possible to all N ONUs. The static arbitration that is based on the max-min theory allocates the service work by $A_j^{m+1} = \min(D_j^m, V_cR/N)$. Hence, it guarantees the static service fairness which every S_j^m satisfies fair conditions of $\mid S_j^{m+1} - S_j^m \mid \leq RT_c/N$ and $\mid S_j^m - S_k^m \mid \leq RT_c/N, j \neq k$. Therefore the service interval is not more than T_c in any cases. But it is impossible for burst ONUs that reports D_{burst}^m to efficiently use the unused bandwidth of under-flow cycle in which $D_{burst}^m >> V_cR/N$ happens. so, burst ONU must wait for several cycles if total network load is low. This is not fair if demand based best service is considered. EWF-BA targets for real time fair service arbitration that keeps the thread of each ONU. In general DBA which does not fix the frame cycle, V_m varies by the arbitration.

The objective of EWF-BA is to maintain the equal service fairness among ONUs in the short epoch as well as the long time period by maintaining the service ratio equal to all ONUs as $\mid S_j^{m+1} - S_j^m \mid = \mid S_k^{m+1} - S_k^m \mid, k \neq j$. This keeps the service fairness under any arbitration result. When the policy based weighted fair services is considered, it is also possible to apply the different service ratio μ_i to each class for different service weight as

$$\mid S_j^{m+1} - S_j^m \mid / \mid S_k^{m+1} - S_k^m \mid = \mu_j/\mu_k, \sum_{i=0}^{K-1} \mu_i = \mu, 0 \leq k \leq K-1 \quad (10)$$

In this paper, we target for only the equal weighted fair case in which the service ratio is same to all ONUs but the different weight is also applicable with same approach. For the thread based fair service, we introduce two parameters. The one is the bandwidth occupation ratio C_m. It shows the service ratio of each ONU compared with total service work of N ONUs in the previous cycle at the OLT's arbitration instance. C_m is expressed as following equation when $t_{j,R}^m$ is the time instance at which ONUj's m_{th} report frame arrives at OLT, $C_j^m = A_j^m / \sum_{n=0}^{N-1} A_n^m = A_j^m/(t_{j,R}^{m+1} - t_{j,R}^m)R$. It can be calculated by the simple time calculation as shown in Fig.3. By C_m, it is possible to reflect the service thread of each ONU to the arbitration. The other is the overflow ratio, $R_j^m = [\sum_{n=0,n\neq j}^{N-1} (A_j^m + D_j^m) - V_cR]/V_cR$, which is the excessively requested bandwidth ratio normalized by V_cR. It is meaningful only when $\sum_{n=0,n\neq j}^{N-1} (A_j^m + D_j^m)$ is greater than V_cR. These parameters are proportionally reflected to the final reduction weight value, W_j^{m+1}, to keep the thread as well as the service fairness.

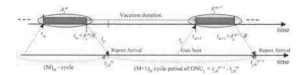

Fig. 3. Service reservation procedures of DBA.

In stable service condition, $(W_0^m D_0^m + W_1^m D_1^m + ... + W_{N-1}^m D_{N-1}^m) < V_c R$ is maintained. Hence, OLT calculates the reduction weight as $W_0^m =$ (normalized $C_j^m) R_j^m = A_j^m N[\sum_{n=0,n\neq j}^{N-1} (A_j^m + D_j^m) - V_c R]/V_c R \sum_{n=0}^{N-1} A_n^m$. Based on this, if $\sum_{n=0,n\neq j}^{N-1} (A_j^m + D_j^m) > V_c R$, OLT arbitrates service slot by following (11).

$$A_j^{m+1} = D_j^m[1 - A_j^m N[\sum_{n=0,n\neq j}^{N-1} (A_j^m + D_j^m) - V_c R]/V_c R(\sum_{n=0}^{N-1} A_n^m + A_j^m)]\,(11)$$

In other cases, A_j^m is same to the requested bandwidth, D_j^m. Hence, the unused bandwidth of under-flow frame cycle is efficiently reallocated to bursty ONUs and the heavy load of over-flow frame cycle is distributed to all ONUs by the service ratio and the occupation ratio. Additionally, the service slot of each ONU does not rapidly diminish and the service thread can gradually converge to the fair quality without the abrupt change. Although Vm can be larger that V_c in some frames in EWF-BA, it will not be long by this dynamic load distribution.

4   Simulation Results and Performance Evaluation

For effective simulations, we assume that all incoming data frames are classified into three service classes in ONU as class0 for the CBR and real time, class1 for non-real time and class2 for best effort service. The ON/OFF model is applied to Poisson and burst traffic sources. Traffic Load is uniformly distributed over three classes. The frame size distribution follows the real estimated traffic model in Poisson and burst traffic[5]. In burst traffic environment, we applied Pareto distributed(Hurst parameter 0.8) traffic generation scenario.

Fig.4.(a) shows DBA guarantees high utilization and the lower delay in low traffic load. But the delay gradually increases in high traffic load due to the increased cycle frame size. Fig.4.(b) shows the average delay of SQP-DBA based on EWF-BA is lower than SQP-DBA only case in every classes. This result comes from the fact that EWF-BA efficiently distributes the heavy traffic load to all ONUs with the equal weight until network goes into the stable service state. Moreover, Fig.5.(a) shows that the maximum delay of EWF-BA is smaller than SQP-DBA only case and total numbers of longer delayed data frames over targeted SLA value is relatively small. Finally, the congestion resolution period, time gap to overcome the reservation mismatch, is just several cycles and this

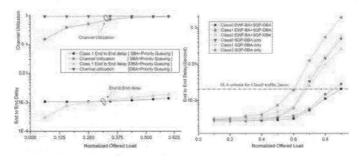

Fig. 4. (a) Average end to end delay and utilization when Poisson traffic applied. (b) Average end to end delay when burst traffic applied.

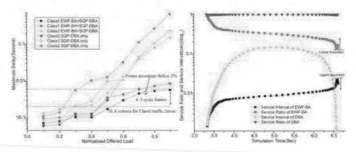

Fig. 5. (a) Maximum end to end delay when burst traffic applied.(b) Service ratio and interval when burst traffic applied(load=0.8,H=0.9).

means that EWF-BA can efficiently service some burst ONUs while the service fairness is maintained. This is more obvious in Fig.5.(b) This is derived from the same simulation scenario of above one but more burst load. In DBA case, the service ratio that is the service work normalized by the original request work is always one. But the service interval is too large to efficiently support the QoS guaranteed service. Moreover, the serious congestion duration is unpredictable and uncontrollable. Oppositely, EWF-BA guarantees the relatively short service interval and its value shows upper bounded pattern. This is possible by the load distribution and the service arbitration of EWF-BA. The service ratio is generally lower than DBA but is also lower bounded. Therefore, the differentiated service for QoS can be guaranteed within minimum delay by SQP and under stable service pattern and fairness by EWF-BA. In this simulation, the average queue size of EWF-BA is generally lower than DBA but the maximum queue size in short epoch is about 1.3 times larger than DBA case. This is the endurable performance degradation but there must be more performance improvement.

5   Conclusion

In this paper, we propose the SQP-DBA and EWF-BA for supporting QoS guaranteed fair service under the service work reservation and arbitration. The SQP-DBA supports the prioritized service for the QoS guaranteed high priority traffic while the service intervention among classes is effectively minimized. When EWF-BA scheme is applied, the network performances, especially the delay and the service fairness among access nodes, are enhanced even if burst traffic happens in some ONUs. The bandwidth starvation by some heavy loaded ONUs can be also resolved within relatively short time epoch by the EWF-BA's thread based load distribution scheme although more buffer size is needed.

Acknowledgement. This work was supported in part by the Korea Science and Engineering Foundation (KOSEF) through OIRC project and ETRI.

References

1. Kramer, G., Mukherjee, B., Pesavento, G.: IPACT-a dynamic protocol for an Ethernet PON (EPON). IEEE Communication Magazine, Vol. 40, (2002), pp. 74-80.
2. Rege, K., Dravida, S., Nanda, S., Narayan, S., Strombosky, J., Tandon, M., Gupta, D.: QoS Management in Trunk-and-Branch Switched Ethernet Network. IEEE Comm. Mag, Vol. 40 (2002) pp. 30-36.
3. Kramer, G., Mukherjee, B., Dixit, S., Ye, Y., Hirth, R.: Supporting differentiated classes of service in Ethernet passive optical networks. Journal of Optical Networking, vol. 1, (2002), pp. 280-298.
4. Leung, K., Disenberg, M.: A Single-Server Queue with Vacations and Gated Time-Limited Service. IEEE Transaction on Communication, Vol. 38, No. 9 (1990).
5 Leland, W., Taggu, M., Willinger, W., Wilson, D.: On the Self-Similar Nature of Ethernet Traffic. IEEE/ACM Transactions on Networking, Vol. 2, No. 1, (1994) pp. 1-15.

Strategies for the Routing and Wavelength Assignment of Scheduled and Random Lightpath Demands

Mohamed Koubàa, Nicolas Puech, and Maurice Gagnaire

Telecom Paris - LTCI - UMR 5141 CNRS,
46, rue Barrault F-75634 Paris-France,
{mohamed.koubaa|nicolas.puech|maurice.gagnaire}@enst.fr

Abstract. We propose three routing strategies for the routing and wave-length assignment (RWA) of scheduled and random lightpath demands in a wavelength-switching mesh network without wavelength conversion functionality. Scheduled lightpath demands (SLDs) are connection demands for which the set-up and tear-down times are known in advance as opposed to random lightpath demands (RLDs) which are dynamically established and released in response to a random pattern of requests. The routing strategies are studied and compared through rejection ratio.

1 Introduction

Wavelength Division Multiplexed (WDM) optical networks are emerging as promising candidates for the infrastructure of the next-generation Internet. Such networks are envisaged for spanning local, metropolitan and wide geographical areas. WDM optical networks go beyond technologies such as Synchronous Optical Network (SONET) and Asynchronous Transfer Mode (ATM) in realizing the full potential of the optical medium. Standardization bodies such as the International Telecommunications Union (ITU) have developed a set of standards that define the architecture of WDM optical transport networks (OTN) [1]. These networks are expected to be reconfigurable so that connections can be set-up, maintained, and torn down using either a management plane [2] or a control plane [3]. The decision on whether to use a management or a control plane to instantiate the lightpaths depends on factors such as the characteristics of the traffic demands (e.g., if the demands are static or scheduled, a management plane is enough, however if the demands are random, a control plane would be necessary) and the approach of the operator to control its network (e.g., an historical telco carrier is probably more familiar with a management plane. Conversely, an Internet service provider may prefer a multi-protocol label switching (MPLS)-like control plane). Moreover, both a management and a control plane are likely to coexist in some optical transport networks.

A service that could be provided by an OTN is the optical virtual private network (OVPN). An OVPN client company may request a set of static lightpaths to satisfy its minimal connectivity and capacity requirements from the

M. Freire et al. (Eds.): ECUMN 2004, LNCS 3262, pp. 91–103, 2004.

provider. The client may also require some scheduled lightpaths to increase the capacity of its network at specific times or certain links, for example, between headquarters and production centers during office hours or between data centers during the night, when backup databases is performed. Finally the unexpected peaks of traffic demands (random) in the client's network could be born by using dynamically established lightpaths. This example of OVPN shows that the services offered by the optical network operator can lead to three types of lightpath demands: static, scheduled, and random.

In this paper, we deal with the routing and wavelength assignment (RWA) problem in WDM all-optical transport networks. We consider two types of traffic demands: scheduled demands and random demands. Static demands are not considered here since, once established, these demands remain in the network for a long time. This can be seen as a reduction in the number of available wavelengths on some network links. We propose three routing strategies: the first strategy aims at establishing random lightpath demands (RLDs) dynamically, provided that the RWA for the scheduled lightpath demands (SLDs) has already been calculated. The second strategy indiscriminately computes the RWA for both the SLDs and the RLDs at the arrival time of each demand. The third routing strategy deals with the SLDs and the RLDs in the same way the second strategy does except that for any lightpath demand, all requested lightpaths have to be routed on the same path as opposed to the preceding strategies where the requested lightpaths may follow several paths between the source and the destination nodes. We compute the rejection ratio and discuss the advantages for each strategy.

The next section sets the framework of the study. In Section 3, we describe the mathematical model and the algorithms of the studied RWA strategies. We then (Section 4) propose some simulation results obtained with them and compare our strategies in terms of rejection ratio. Finally, in Section 5, we draw some conclusions and set directions for future work.

2   Description of the Problem

We distinguish the static RWA problem from the dynamic RWA problem. In the static RWA problem, the set of lightpath connections is known in advance, the problem is to set-up lightpaths for the connections while minimizing network resources such as the number of WDM channels or the number of wavelengths [4]. The set of established lightpaths remain in the network for a long period of time. Dynamic RWA deals with connections that arrive dynamically (randomly) [5]. The performance of dynamic RWA is often measured through the rejection ratio [6,7].

In this study, we consider two types of lightpath traffic demands referred to as scheduled lightpath demands and random lightpath demands. A scheduled lightpath demand (SLD) [8] is a connection demand represented by a tuple $(s, d, n, \alpha, \omega)$, where s and d are the source and destination nodes of the demand, n is the number of requested lightpaths, and α, ω are respectively the set-up and

tear-down dates of the demand. The SLD model is deterministic because it is
known in advance and is dynamic because it takes into account the evolution
of the traffic load in the network over time. A random lightpath demand (RLD)
corresponds to a connection request that arrives randomly and is dealt with on
the fly. We use the same tuple notation to describe an RLD. To the best of
our knowledge, this is the first time that both deterministic and dynamic traffic
demands are considered simultaneously to address the routing and wavelength
assignment problem in WDM all-optical transport networks.

In this study, we call a span the physical pipe connecting two adjacent nodes
u and v in the network. Fibers laid down in a span may have opposite directions.
We assume here that a span (u, v) is made of two opposite unidirectional fibers.
As opposed to a span, a link or a fiber-link refers to a single unidirectional fiber
connecting nodes u and v. The bandwidth of each optical fiber is wavelength-
division demultiplexed into a set of χ wavelengths, $\Lambda = \{\lambda_1, \lambda_2, \dots, \lambda_\chi\}$. When
a wavelength is used by a lightpath, it is occupied (in one direction). A lightpath
connecting a node s to a node d is defined by a physical route in the network
(a path) connecting s to d and a wavelength λ such that λ is available on every
fiber-link of this route. A path-free wavelength is a wavelength which is not used
by any lightpath on any optical fiber of the path. A lightpath demand is rejected
(blocked) when there are not enough available network resources to satisfy it.
The rejection ratio (Rr) is the ratio of the number of rejected demands to the
total number of lightpath demands.

3   The RWA Algorithms

In this section, we describe our algorithms used for the routing and wavelength
assignment of SLDs and RLDs.

3.1   Mathematical Model

We present the notations used to describe a lightpath demand (LD), be it scheduled or random.

- $G = (V, E, \vartheta)$ is an arc-weighted symmetrical directed graph with vertex
  set $V = \{v_1, v_2, \dots, v_N\}$, arc set $E = \{e_1, e_2, \dots, e_L\}$ and weight function
  $\vartheta : E \to \mathbb{R}_+$ mapping the physical length (or any other cost of the links set
  by the network operator for example).
- $N = |V|, L = |E|$ are respectively, the number of nodes and links in the
  network.
- D denotes the total number of SLDs and RLDs arriving at the network.
- The LD number i to be established is defined by a tuple $(s_i, d_i, n_i, \alpha_i, \omega_i)$.
  $s_i \in V$, $d_i \in V$ are source and destination nodes of the demand, n_i is the
  number of requested lightpaths, and α_i and ω_i are respectively, the set-up
  and tear-down dates of the demand.

- $P_{k,i}, 1 \leq k \leq K, 1 \leq i \leq D$, represents the kth alternate shortest path in G from node s_i to d_i of LD number i. We compute K alternate shortest paths for each source-destination pair according to the algorithm described in [9] (if so many paths exist, otherwise we only consider the available ones).
- $\kappa_{k,i,t} = (\gamma_{1,k}^{i,t}, \gamma_{2,k}^{i,t}, \ldots, \gamma_{\chi,k}^{i,t})$ is a χ-dimensional binary vector. $\gamma_{j,k}^{i,t} = 1$, $1 \leq j \leq \chi$, if λ_j is a path-free wavelength along the kth route, $P_{k,i}$, from s_i to d_i at time t. Otherwise $\gamma_{j,k}^{i,t} = 0$.
- $\sigma_{k,i,t} = \sum_{j=1}^{\chi} \gamma_{j,k}^{i,t}$ is the number of path-free wavelengths along $P_{k,i}$ at time t.

δ will denote an SLD whereas τ will denote an RLD. We also use n_i^δ, $P_{k,i}^\delta$ and $\kappa_{k,i,t}^\delta$ (respectively n_i^τ, $P_{k,i}^\tau$ and $\kappa_{k,i,t}^\tau$) for the parameters representing an SLD (respectively an RLD) when it is necessary to make a clear distinction between scheduled and random demands.

3.2 Separate Routing of Scheduled and Random Lightpath Demands

The first routing algorithm called RWASPR [7] (Routing and Wavelength Assignment of SLDs Prior to RLDs), deals with the SLDs and the RLDs separately. The first phase computes the RWA for the SLDs and aims at minimizing the number of blocked SLDs. Taking the assignment of the SLDs into account, the second phase computes the RWA for the RLDs. For each lightpath demand, one computes K alternate shortest paths between the source and the destination of the demand. We begin with routing all requested lightpaths on the shortest path if this is possible (i.e. if there are as many available path-free wavelengths along the shortest route as the requested number of lightpaths), otherwise several paths are used. Shortest paths are preferred since they require less WDM channels (we call a WDM channel the use of a wavelength on a link).

Routing and wavelength assignment of scheduled lightpath demands. Given a set of SLDs and a physical network with a fixed number of wavelengths per link, we want to determine a routing and a wavelength assignment that minimize the rejection ratio. We define the following additional notations:

- $\Delta = \{\delta_1, \delta_2, \ldots, \delta_M\}$ is the set of SLDs to be established.
- (G, Δ) is a pair representing an instance of the SLD routing problem.
- $\rho_\Delta = ((\rho_{1,1}, \rho_{2,1}, \ldots, \rho_{K,1}), (\rho_{1,2}, \rho_{2,2}, \ldots, \rho_{K,2}), \ldots, (\rho_{1,M}, \rho_{2,M}, \ldots, \rho_{K,M}))$ is called an admissible routing solution for Δ if $\sum_{k=1}^{K} \rho_{k,i} = n_i^\delta$, $1 \leq i \leq M$. The element $\rho_{k,i}$ indicates the number of lightpaths to be routed along $P_{k,i}^\delta$, the kth alternate route, of SLD δ_i.
- π_Δ is the set of all admissible routing solutions for Δ.
- $\mathcal{C} : \pi_\Delta \to \mathbb{N}$ is the function that counts the number of blocked SLDs. The combinatorial optimization problem to solve is:

$$\text{Minimize} \quad \mathcal{C}(\rho_\Delta)$$

$$\text{subject to} \quad \rho_\Delta \in \pi_\Delta$$

Table 1. A set of 3 SLDs

SLD	s_i	d_i	n_i^τ	α_i	ω_i	$P_{1,i}^\delta$	$P_{2,i}^\delta$
δ_1	1	14	2	122	457	1-4-9-14	1-3-6-13-14
δ_2	11	9	2	198	342	11-10-12-9	11-10-14-9
δ_3	2	12	2	203	403	2-8-10-12	2-1-4-9-12

We used a Random Search (RS) algorithm to find an approximate minimum of the function $\mathcal{C}$. The wavelengths are selected according to a first-fit scheme [10].

Let us consider the network of Figure 1. We assume that three wavelengths are available on each fiber-link. We compute $K = 2$ shortest paths for each SLD described in Table 1.

An admissible solution (among others) $\rho = ((2,0),(2,0),(2,0))$ is generated arbitrarily. Let us evaluate its cost. We assume that no SLD has already been routed, hence all the wavelengths are available. SLD δ_1 requires 2 lightpaths. All lightpaths are routed on $P_{1,1}^\delta$ as described by vector ρ. Indeed, we check that there are at least 2 path-free wavelengths on $P_{1,1}^\delta$ at time 122. $\kappa_{1,1,122}^\delta = (1,1,1)$ indicates that all wavelengths are available on $P_{1,1}^\delta$ and hence λ_1 and λ_2 are chosen. SLD δ_2 is to be set up whereas δ_1 is still active. Again, two lightpaths have to be established. Both lightpaths are routed on path $P_{1,2}^\delta$. $\kappa_{1,2,198}^\delta = (1,1,1)$. SLD δ_2 is serviced using wavelengths λ_1 and λ_2. When SLD δ_3 arrives, one has to evaluate $\kappa_{1,3,203}^\delta$. Two path-free wavelengths are required on $P_{1,3}^\delta$. $\kappa_{1,3,203}^\delta = (0,0,1)$ shows that both wavelengths λ_1 and λ_2 are already used whereas λ_3 is still available (λ_1 and λ_2 are assigned to SLD δ_2 on $P_{1,2}^\delta$). According to the current value of vector ρ, δ_3 is rejected due to resource shortage. The cost of this solution measured in terms of rejected SLDs is $\mathcal{C}(\rho_\Delta) = 1$ ($Rr = 1/3$). This process is repeated $\mathcal{I}$ times. At the end of a number $\mathcal{I}$ of iterations the solution with the minimal cost $\mathcal{C}$ is selected as the best solution for the RWA for the SLDs. Similarly, one can easily see that when vector ρ equals to $((2,0),(2,0),(1,1))$, we get $\mathcal{C}(\rho_\Delta) = 0$ ($Rr = 0$).

Routing and wavelength assignment of random lightpath demands. Once the RWA for the SLDs have been calculated, we establish the RLDs sequentially, that is demand by demand at arrival dates. When a new RLD arrives, K alternate shortest paths are computed between its source and its destination nodes. We look for as many path-free wavelengths along the K shortest paths as the number of required lightpaths. Again, the wavelengths are assigned according to a first-fit scheme. The following pseudo-code (Algorithm 1) shows the way we compute the RWA for the RLDs.

To illustrate the algorithm, let us go back to the previous example. We assume that $\rho = ((2,0),(2,0),(1,1))$ ($\mathcal{C}(\rho_\Delta) = 0$). Hence the established lightpaths are $(P_{1,1}^\delta, \lambda_1)$, $(P_{1,1}^\delta, \lambda_2)$, $(P_{1,2}^\delta, \lambda_1)$, $(P_{1,2}^\delta, \lambda_2)$, $(P_{1,3}^\delta, \lambda_3)$, and $(P_{2,3}^\delta, \lambda_3)$. In addition to the SLDs defined in Table 1, we want to instantiate the RLDs described in Table 2. We compute $K = 2$ alternate shortest paths for each RLD.

Algorithm 1 RWA algorithm for the RLDs

---

  for RLD i arriving at the network **do**
    boolean $\leftarrow 0$; $k \leftarrow 1$
    while $(k \leq K)$ and *boolean* $= 0$ **do**
      if $\sum_{j=1}^{k} \sigma_{j,i,t}^{\tau} \geq n_i^{\tau}$ **then**
        boolean $\leftarrow 1$ *the RLD can be established*
      end if
      $k \leftarrow k + 1$
    end while
    if *boolean* $= 1$ **then**
      instantiate the lightpaths
    else
      the RLD is rejected due to resource shortage
    end if
  end for

Table 2. A set of 2 RLDs

RLD	s_i	d_i	n_i^{τ}	α_i	ω_i	$P_{1,i}^{\tau}$	$P_{2,i}^{\tau}$
τ_1	4	9	2	153	478	4-9	4-5-7-8-10-12-9
τ_2	10	12	3	450	734	10-12	10-14-9-12

When RLD τ_1 has to be set up, we first evaluate the vector $\kappa_{1,1,153}^{\tau}$. We compute $\kappa_{1,1,153}^{\tau} = (0,0,0)$. Wavelengths λ_1 and λ_2 are already used by SLD δ_1, whereas λ_3 is to be used at time $t = 203$ on $P_{2,3}^{\delta}$ by SLD δ_3. Let us now evaluate $\kappa_{2,1,153}^{\tau}$. $\kappa_{2,1,153}^{\tau} = (0,0,0)$. All available wavelengths will be used on $P_{1,2}^{\delta}$ and $P_{1,3}^{\delta}$ by δ_2 and δ_3 at times $t = 198$ and $t = 203$ respectively. No more path-free wavelengths are available on the computed shortest paths connecting nodes 4 and 9. RLD τ_1 is thus rejected. When RLD τ_2 has to be set up, SLDs δ_2 and δ_3 have already been released. $\kappa_{1,2,450}^{\tau} = (1,1,1)$ and RLD τ_2 is satisfied using λ_1, λ_2, and λ_3 on $P_{1,2}^{\tau}$. We notice that the rejection ratio for the total set of demands (SLDs+RLDs) is equal to $Rr = 1/5$.

3.3   Sequential Routing and Wavelength Assignment Algorithm

This section presents the sequential Routing and Wavelength Assignment algorithm (sRWA) [7]. The algorithm described in subsection 3.2 (Algorithm 1) is used to compute the RWA for both the SLDs and the RLDs. If an SLD and an RLD have to be set up at the same time, the SLD is serviced first. Let us take the example of the preceding subsection to describe the process of the sRWA algorithm. SLD δ_1 arrives when all wavelengths are available. $\kappa_{1,1,122}^{\delta} = (1,1,1)$ so that λ_1, and λ_2 are chosen on $P_{1,1}^{\delta}$ to satisfy the demand. At time $t = 153$, we want to set up RLD τ_1. We determine the number of free wavelengths along $P_{1,1}^{\tau}$. $\kappa_{1,1,153}^{\tau} = (0,0,1)$ shows that wavelength λ_3 is still free whereas λ_1 and λ_2 have already been used. $\kappa_{2,1,153}^{\tau} = (1,1,1)$ and RLD τ_1 is serviced using wavelengths

λ_3 and λ_1 on $P_{1,1}^\tau$ and $P_{2,1}^\tau$ respectively. Later SLD δ_2 arrives. $\kappa_{1,2,198}^\delta = (0,1,1)$ since λ_1 is used by RLD τ_1 on link $(10,12)$. SLD δ_2 is set up using wavelengths λ_2 and λ_3 on $P_{1,2}^\delta$. At time $t = 203$ arrives SLD δ_3. We compute $\kappa_{1,3,203}^\delta = (0,0,0)$ and $\kappa_{2,3,203}^\delta = (0,0,0)$. SLD δ_3 is thus rejected due to resource leakage. Finally, RLD τ_2 must be set up when SLD δ_2 is already released. $\kappa_{1,2,450}^\tau = (0,1,1)$ and $\kappa_{2,2,450}^\tau = (1,1,1)$. RLD τ_2 is serviced using λ_2, and λ_3 on $P_{1,2}^\tau$ and λ_1 on $P_{2,2}^\tau$. We notice that the RWASPR and the sRWA algorithms have the same rejection ratio ($Rr = 1/5$), however, using the RWASPR algorithm, τ_1 is blocked whereas δ_3 is blocked using the sRWA algorithm.

3.4 Sequential Atomic Routing and Wavelength Assignment Algorithm

Our third routing algorithm we present here, called sARWA (sequential Atomic Routing and Wavelength Assignment), is similar to those described in subsections 3.2 and 3.3 except that for any LD, all required lightpaths have to be routed on the same path among the considered K alternate shortest paths. This is called atomic routing and is also known as non bifurcated routing [11]. Whenever there are not as many path-free wavelengths as the required number of lightpaths on any alternate shortest path, the LD is rejected, even if the cumulative number of path-free wavelengths among the K alternate shortest paths is larger than the number of required lightpaths. The following pseudo-code (Algorithm 2) shows the way we compute the RWA for the SLDs and the RLDs.

Algorithm 2 Sequential atomic RWA algorithm

  for LD i arriving at the network **do**
    $boolean \leftarrow 0$; $k \leftarrow 1$
    while $(k \leq K)$ and $boolean = 0$ **do**
      if $\sigma_{j,i,t} \geq n_i$ **then**
        $boolean \leftarrow 1$ *the LD can be established*
      end if
      $k \leftarrow k+1$
    end while
    if $boolean = 1$ **then**
      instantiate the lightpaths
    else
      the LD is rejected due to resource shortage
    end if
  end for

In order to describe the process of the sARWA algorithm, let us still consider the same example as before. At time $t = 122$, $\kappa_{1,1,122}^\delta = (1,1,1)$ and δ_1 is serviced using wavelengths λ_1, and λ_2 on $P_{1,1}^\delta$. At time $t = 153$, $\kappa_{1,1,153}^\tau = (0,0,1)$. RLD τ_1 cannot be established on $P_{1,1}^\tau$ since only λ_3 is available. $\kappa_{2,1,153}^\tau = (1,1,1)$ and λ_1 and λ_2 are used on $P_{2,1}^\tau$. SLD δ_2 is to be set up at time $t = 198$.

$\kappa^{\delta}_{1,2,198} = (0,0,1)$ and $\kappa^{\delta}_{2,2,198} = (1,1,1)$. SLD δ_2 is hence set up using λ_1 and λ_2 on $P^{\delta}_{2,2}$. The next LD to be serviced is δ_3. $\kappa^{\delta}_{1,3,203} = (0,0,1)$ and $\kappa^{\delta}_{2,3,203} = (0,0,1)$. SLD δ_3 is rejected because the number of free wavelengths on both considered shortest paths is lower than the number of requested lightpaths even if the cumulative number of path-free wavelengths along $P^{\delta}_{1,3}$ and $P^{\delta}_{2,3}$ equals 2. When RLD τ_2 has to be set up, SLD δ_1 and RLD τ_1 are still alive, all other lightpaths have been released. $\kappa^{\tau}_{1,2,450} = (0,0,1)$ and $\kappa^{\tau}_{2,2,450} = (1,1,1)$. RLD τ_2 is set up using wavelengths λ_1, λ_2, and λ_3 on $P^{\tau}_{2,2}$. We notice that compared to the sRWA algorithm, the lightpaths associated with RLDs τ_1 and τ_2 have not been split on the available shortest paths but follow the same paths $P^{\tau}_{2,1}$ and $P^{\tau}_{2,2}$ respectively. SLD δ_3 is rejected although the cumulative number of path-free wavelengths along the different shortest paths equals the number of required lightpaths. The physical paths followed by the established lightpaths are longer than those computed by the sRWA and the RWASPR algorithms. This may affect the number of rejected lightpath demands. The sARWA still computes the same rejection ratio as the sRWA and the RWASPR algorithms.

4   Experimental Results

In this section we experimentally evaluate the algorithms proposed in the previous sections. We used two network topologies: the former is the 14-node NSFNET network (Figure 1), the latter is the hypothetical US backbone network of 29 nodes and 44 spans (Figure 2).

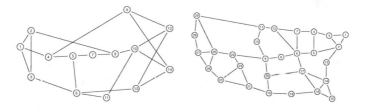

Fig. 1. 14-node NSFNET network   **Fig. 2.** 29-node network

For the sets Δ, the source/destination nodes are drawn according to a random uniform distribution in the interval $[1, 14]$ (respectively $[1, 29]$) for the 14-node network (respectively the 29-node network). We also used uniform random distributions over the intervals $[1, 5]$ and $[1, 1440]$ for the number of lightpaths and the set-up/tear-down dates of the SLDs respectively. We assume observation periods of about a day (1440 is the number of minutes in a day). Random connection requests (RLDs) arrive according to a Poisson process with an arrival rate $\nu = 1$

and if accepted, will hold the circuit for exponentially distributed times with mean $\mu = 250$ much larger than the cumulated round-trip time and the connection set-up delay. The number of lightpaths required by an RLD is drawn from a random uniform distribution in the interval $[1, 5]$. We call a scenario the set of demands be they scheduled or random that occur from start to finish of a day. We assume that we compute $K = 5$ alternate shortest paths between each source/destination pair and that there are $\chi = 24$ available wavelengths on each fiber-link in the network. We want to asses the gain obtained using the RWASPR algorithm compared to the sRWA and the sARWA algorithms.

We generated 25 test scenarios, run the three algorithms on them and computed rejection ratio averages for each algorithm. In the following pictures, each triplet of bars shows the average number of blocked demands computed using the RWASPR (left bar), the sRWA (middle bar) and the sARWA (right bar) algorithms respectively. Each bar is divided into two segments. The height of the black segment indicates the average number of rejected SLDs whereas the height of the white one shows the number of rejected RLDs. Each couple of figures show the same simulation results when computed using the 14-node network and the 29-node network respectively.

Figure 3(a) (respectively Figure 3(b)) shows the average number of rejected SLDs and RLDs computed for the 14-node network (respectively the 29-node network) when D, the total number of LDs varies from 100 to 1000 and for $\mu = 250$ (respectively $\mu = 500$). We notice that when $D = 100$ all lightpath demands are satisfied. The number of rejected demands increases when D increases. The figures show that the RWASPR computes a solution with a lower cost in terms of rejected SLDs while the global rejection ratio (total number of rejected SLDs and RLDs) is higher compared to the global rejection ratios computed by the sRWA and the sARWA algorithms. This can be explained by the fact that when the SLDs are accepted, it becomes more difficult to find sufficient available path-free wavelengths to route the RLDs when they have to be set up. We also notice that whenever the traffic load increases during the day, more and more lightpath demands are rejected due to resource shortage. The rejection ratio computed by the sRWA and the sARWA algorithms remain roughly constant.

In the following pictures, we present simulation results obtained with four different values of D (250, 500, 750, and 1000). Each figure shows four subfigures. The upper left-hand (respectively right-hand) subfigure shows results obtained when D has a value of 250 (respectively $D = 500$). The lower left-hand (respectively right-hand) subfigure presents simulation results obtained with $D = 750$ (respectively $D = 1000$).

We first study the influence of the factor K on the average number of rejected LDs. Figure 4(a) (respectively Figure 4(b)) shows the average number of rejected LDs computed using the 14-node network (respectively the 29-node network) when K, the number of alternate shortest routes, varies from 1 to 10 and $\mu = 250$. We notice that the average number of blocked demands falls when the value of K increases and becomes roughly constant when $K \geq 5$. This is due to the fact that when K increases, several links of the network are shared by the considered

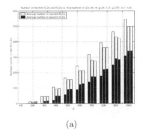
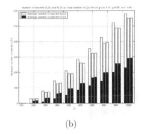

Fig. 3. Average number of rejected LDs computed with the RWASPR, the sRWA, and the sARWA algorithms w.r.t. the number of LDs, D

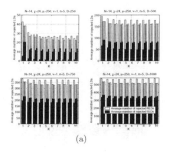
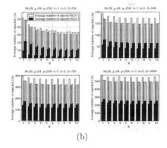

(a)   (b)

Fig. 4. Average number of rejected LDs computed with the RWASPR, the sRWA, and the sARWA algorithms w.r.t. K

K alternate shortest paths. Once one or several paths are assigned to an LD, the remaining paths sharing one or more links with the assigned paths can not be used by other LDs in the absence of path-free wavelengths.

Figure 5(a) (respectively Figure 5(b)) shows the average number of rejected LDs when the number of available wavelengths on each fiber-link varies from 4 to 32 using the 14-node network (respectively the 29-node network) and $\mu = 250$. We notice, as expected that the number of rejected LDs decreases when χ, the number of available wavelengths, increases. The RWASPR still computes the minimum number of blocked SLDs whereas the global rejection ratio is higher than the global rejection ratios computed by the sRWA and the sARWA algorithms.

In Figure 6(a) (respectively Figure 6(b)), we present the average number of rejected LDs as computed by our three RWA algorithms using the 14-node

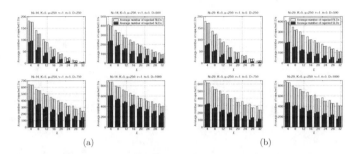

Fig. 5. Average number of rejected LDs computed with the RWASPR, the sRWA, and the sARWA algorithms w.r.t. χ

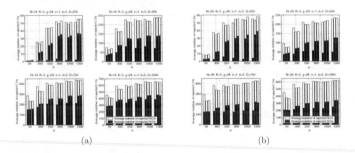

Fig. 6. Average number of rejected LDs computed with the RWASPR, the sRWA, and the sARWA algorithms w.r.t. μ

network (respectively the 29-node network) as a function of μ, the exponential duration of RLDs. We notice that the number of blocked demands increases when the duration of RLDs increases. This is due to the fact that whenever the RLDs' holding time is higher, when accepted, RLDs maintain network resources for longer periods and hence fewer demands (may they be SLDs or RLDs) can be satisfied. One can also notice a higher average number of rejected RLDs computed using the RWASPR compared to the sRWA and the sARWA algorithms.

Finally, we evaluate the CPU execution time required by each routing algorithm. Figure 7 shows the average CPU execution time measured using the RWASPR algorithm when considering the 14-node network (respectively the 29-node network) w.r.t. the number of LDs, D. We notice that the RWASPR requires long times to compute the RWA for SLDs and RLDs as opposed to the

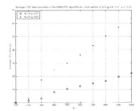

Fig. 7. Average CPU execution time of the RWASPR algorithm w.r.t. D

sRWA and the sARWA algorithms which compute the routing solution in very short times. We have not plotted here the CPU execution time required by the sRWA and the sARWA algorithms since their CPU execution times are negligible compared to the CPU execution time of the RWASPR algorithm. This is primarily due to the computing time necessary to the random search algorithm to determine the routing and wavelength assignment of SLDs.

5   Conclusions

We proposed and presented three algorithms for the routing and wavelength assignment of scheduled lightpath demands and random lightpath demands in an all-optical transport network. The RWASPR algorithm based on a random search algorithm, computes the RWA for SLDs prior to the RWA for RLDs. The sRWA algorithm consider the LDs sequentially as they arrive in the network. Both the RWASPR and the sRWA algorithms authorize an LD to use one or more alternate paths between the source and the destination of the LD as opposed to the third routing algorithm (sARWA) which requires that all lightpaths follow the same path. The LDs are considered in the same way they are considered with the sRWA algorithm. We computed the average number of rejected SLDs and RLDs for different simulation scenarios. The results show that the RWASPR performs better in terms of average number of rejected SLDs while the global rejection ratio is higher compared to the rejection ratios computed by the sRWA and the sARWA algorithms. Our simulation results also show that in the majority of the cases, one and only one path is used to satisfy an LD as the average number of rejected LDs computed by the sRWA and the sARWA algorithms is remaining roughly constant. The sRWA and the sARWA routing strategies are simpler to implement and do not require a long time to compute the routing solution as opposed to the RWASPR routing strategy.

Future works will focus on how decreasing the rejection ratio. Wavelength rerouting techniques may alleviate the wavelength continuity constraint imposed by the all-optical cross-connect switches.

References

1. ITU-T. Recommendation G.852: Management of the Transport Network. (1999)
2. ITU-T. Recommendation G.807: Requirements for the Automatic Switched Transport Network (ASTN). (2001)
3. IETF CCAMP Group: Generalized Multi-Protocol Label Switching (GMLPS) Architecture. (2002)
4. Banerjee D., Mukherjee B.: A Practical Aprroach for Routing and Wavelength Assignment in Large Wavelength-Routed Optical Networks. IEEE Journal on Selected Areas in Communications (1996) 14(5):903–908
5. Ramamurthy R., Mukherjee B.: Fixed-Alternate Routing and Wavelength Conversion in Wavelength-Routed Optical Networks. IEEE/ACM Transactions on Networking (2002) 10(3):351–367
6. Ramaswami R. Sivarajan K. N.: Routing and Wavelength Assignment in All-Optical Networks. IEEE/ACM Transactions on Networking (1995) 3(5):489–500
7. Koubàa M., Puech N., Gagnaire M.: Routing and Wavelength Assignment of Scheduled and Random Lightpath Demands. Proc. Wireless and Optical Communications Networks (2004) 7–9
8. Kuri J., Puech N. Gagnaire M., Dotaro E., Douville R.: Routing and Wavelength Assignment of Scheduled Lightpath Demands. IEEE JSAC Optical Communications and Networking Series (2003) 21(8):1231–1240
9. Eppstein D.: Finding the k Shortest Paths. SIAM Journal of Computing (1998) 28(2):652–673
10. Zang, Jue J. P., Mukherjee B.: A Review of Routing and Wavelength Assignment Approaches for Wavelength-Routed Optical WDM Networks. Optical Networks Magazine (2000) 1(1):47–60
11. Banerjee D., Mukherjee B.: Wavelength-Routed Optical Networks: Linear Formulation, Resource Budgeting Tradeoffs, and a Reconfiguration Study. IEEE/ACM Transactions on Networking (2000) 8(5):598–607

A Restoration Technique Incorporating Multi-objective Goals in WDM Optical Networks

Sung Woo Tak[1] and Wonjun Lee[2,*]

[1] Department of Computer Science and Engineering, Pusan National University, San-30, Jangjeon-dong, Geumjeong-gu, Busan, 609-735, Republic of Korea, Tel: +1 605 688-6519
swtak@pusan.ac.kr
[2] Department of Computer Science and Engineering, Korea University, 1, 5-ka, Anam-dong Sungbuk-ku, Seoul, 136-701, Republic of Korea
wlee@korea.ac.kr

Abstract. Restoration techniques available in literature have not addressed their performance in terms of significant, multiple objective goals. Some of these methods have shown good performance for a single objective function. However, restoration must consider a number of objective functions. In this paper, we evaluate existing models and their performance in an attempt to verify their performance and efficacy based on literature. Our research has found not only inefficiency in some of these methods of restoration, but a general incompatibility. Consequently, this paper proposes eight objective functions that yield objective goals significant to the optimal design of a WDM (Wavelength Division Multiplexing) optical network. Each objective function model is presented and is examined by experimentation. Four proposed restoration algorithms are evaluated: KSDPR (k-Shortest Disjoint Path Restoration based on multiple uphill moves and heuristic rule), DCROS (Deep Conjectural Reinforced Optimal Search), RWWA (Random Walk-based Wavelength Assignment), and PTCI (Physical Topology Connectivity Increase). Numerical results obtained by experimental evaluation of KSDPR, DCROS, RWWA, and PTCI algorithms confirm that MTWM (objective function of Minimizing Total Wavelengths with Multi-objective goals) based on the DCROS algorithm is a technique for efficient restoration in WDM optical networks.

1 Introduction

The goal of a restoration technique in WDM (Wavelength Division Multiplexing) optical networks is to find optimal survivable network planning with the best network performance and the smallest network costs. The best network performance is evaluated by the following factors: total number of wavelengths, wavelength link distance, even distribution of traffic flows, total and average restoration time of backup lightpaths, and escape from physical and virtual topology constraints. The smallest network costs are evaluated by wavelength mileage costs required for the design of a survivable WDM optical network. Unfortunately, even if some of

---

* Corresponding author

M. Freire et al. (Eds.): ECUMN 2004, LNCS 3262, pp. 104–114, 2004.
© Springer-Verlag Berlin Heidelberg 2004

restoration techniques available in literature have showed good performance for a single objective function, they have not addressed their performance in terms of significant, multiple objective goals [1-9]. Therefore, considerable objective functions, which can deal with the significant objective goals, should be proposed and evaluated extensively to verify the performance of restoration algorithms proposed in this paper. The objective goals measured in this paper are the network performance factors and costs, which are described the above. In this paper, eight objective functions that yield objective goals considerable for the optimal design of a survivable WDM optical network are developed. Each of those objective functions needs to be verified if it can simultaneously optimize the network performance and costs for the design of a survivable WDM optical network. This paper is organized as follows. Section 2 addresses eight objective functions. Section 3 describes the design of restoration algorithms. In Section 4, eight objective functions are evaluated in terms of significant, optimal objective goals. Section 5 concludes this paper.

2   Design of Objective Functions

Traditionally two topologies exist in WDM optical networks, which are a physical topology and a virtual topology. The physical topology is physically composed of optical fiber links and photonic nodes. The virtual topology consists of a graph representing all optical communication paths called lightpaths underlying the physical topology. G_p represents a physical topology consisting of N_p, $|P_{mn}|$, $k(G_p)$, and $un/uni/2$-uni/bi. N_p denotes the number of nodes in the physical topology. $|P_{mn}|$ denotes the number of optical fiber links between node m and node n. $k(G_p)$ denotes the node connectivity of the physical topology. $k(G_p)$ ensures that the minimum number of nodes separating node u from node v in G_p is equal to the maximum number of disjoint paths between u and v in G_p. Thus, the minimum value of $k(G_p)$ should be at least two to survive a node failure or a link failure. un denotes that the link type is undirected, uni denotes that the link type is unidirectional, and bi denotes that the link type is bi-directional. 2-uni denotes that two unidirectional links in opposite directions are established for a bi-directional link. G_v represents a virtual topology that consists of Δ_l, N_v, T_{sd}, Γ_{set}, Γ_{option}, W, S_H or M_H. Δ_l denotes the maximum degree of the virtual topology. N_v denotes the number of nodes in the virtual topology. T_{sd} denotes a $[N_v \times N_v]$ traffic flows matrix between source s and destination d. Γ_{set} denotes a set of fundamental constraints, where $\Gamma_{set} \subseteq \{\Gamma_{MAXDEG}, \Gamma_{FLOW}, \Gamma_{DCA}, \Gamma_{WA}, \Gamma_{LC}, \Gamma_{NC}\}$. Γ_{option} denotes a set of optional constraints, where $\Gamma_{option} \subseteq \{\Gamma_{CONT}, \Gamma_{UNIQ}, \Gamma_{DUP}, \Gamma_{SWA}, \Gamma_{DWA}, \Gamma_{MILE}, \Gamma_{PROP}, \Gamma_{HOP}, \Gamma_{SYM}\}$. W denotes the number of wavelengths required to construct the virtual topology, where $W \geq 1$. S_H is referred to that the virtual topology is referred to as a virtual single-hop WDM optical network topology. M_H is referred to as a virtual multi-hop WDM optical network topology. Γ_{MAXDEG} denotes the maximum degree constraint of the virtual topology, where $\Gamma_{MAXDEG} \geq \Delta_l$. Γ_{FLOW} denotes the flow conservation constraint. Γ_{DCA} denotes the distinct wavelength assignment constraint. Γ_{WA} denotes the fundamental wavelength assignment constraint. Γ_{LC} denotes the link capacity constraint. Γ_{NC} denotes the node capacity constraint. Γ_{CONT} denotes the wavelength continuity constraint. Γ_{UNIQ} denotes the unique wavelength assignment constraint. Γ_{DUP} denotes the duplicate wavelength assignment constraint. Γ_{SWA} denotes the same wavelength assignment constraint for a primary lightpath and its backup lightpath. Γ_{DWA} denotes no same wavelength assignment constraint for a primary lightpath and its backup lightpath. Γ_{MILE} denotes the lightpath mileage bound constraint. Γ_{PROP} denotes the lightpath propagation delay bound constraint. Γ_{HOP} denotes the hop bound constraint. Γ_{SYM} denotes the symmetric constraint.

One of main concerns in eight objective functions is to find one near optimal objective function that minimizes significant objective goals simultaneously. Eight objective functions are considered in terms of only virtual topology. The objective functions are formulated in a given physical topology $G_P (N_P, |P_{mn}| \geq 2, k(G_P) \geq 2, 2\text{-}uni)$.

First, MMBW (objective function of Minimizing Max Bound of Wavelengths) is developed to evaluate the proposed software framework in this paper. It minimizes the maximum offered wavelengths on any link. MMBW may reduce the cost of a WDM optical network by minimizing the maximum bound of wavelengths on any link. Second, MTWC (objective function of Minimizing Total Wavelength mileage Costs) is developed. Wavelength mileage can be an indicator of network resources. The wavelength mileage costs interconnecting node m and node n are the sum of the wavelength costs multiplied by the corresponding distance between node m and node n. Third, MTWL (objective function of Minimizing Total Wavelength Link distance) is developed. Wavelength link is defined as a wavelength between two adjacent fiber terminating equipment over the optical fiber interconnecting node m and node n. Fourth, MTLH (objective function of Minimizing Total Lightpath Hops) is developed. It minimizes the number of hops traversed by lightpaths. Fifth, MMSW (objective function of Minimizing Max bound of Spare Wavelengths) is developed. For the design of an efficient restoration technique, we need to consider the efficient utilization of spare wavelengths when backup lightpaths are assigned. Sixth, MSDT (objective function of Minimizing Standard Deviation of Traffic flows) is developed. It evenly distributes traffic demands. It can be directly formulated by minimizing the standard deviation of traffic flows over links. The minimal standard deviation of traffic flows in the WDM optical network represents that the number of wavelength differences between the lightpath flow, which will be assigned, and the average of traffic flows, which have already assigned, is minimal. Seventh, MTWM (objective function of Minimizing Total Wavelengths with Multi-objective goals) is developed. It minimizes total wavelengths while satisfying goals of other objective functions: MMBW, MTWC, and MTWL. MTWM finds a near optimal solution that minimizes total wavelengths subject to minimal wavelengths, wavelength link distance, and wavelength mileage costs in a WDM optical network. When the minimal number of hops traversed by lightpaths is considered, MTWL included in the goals of MTWM is replaced with MTLH. Eighth, MBLR (objective function of Minimizing Backup Lightpaths Restoration time) is developed. It minimizes the restoration time of backup lightpaths.

The idea of MMBW is used in [1-3]. [1-3] exploit the idea of MMBW for their objective functions. The objective functions considered are minimizing the maximum offered load. If the switching speed at nodes is limited, then minimizing the maximum offered load will be appropriate because it enables the traffic carried per wavelength to increase. In Section 4, we show that all traffic demands are evenly distributed over the WDM optical network by MMBW, and so the affected wavelengths or lightpaths can be reduced in the event of a node failure or a link failure. Objective functions used in [4] take advantage of MTWC. In [4], the WDM layer cost is defined as the total wavelength mileage (i.e., miles of working and protection wavelengths) required in the network to support a given set of traffic demands. The objective function considered in [5] obtains a virtual topology that maximizes total one-hop traffic demands in the network. The goal of [5] is similar to that of MTLH. The objective functions considered in [6-7] are related to MBLR. Objective functions considered in [8-9] are to minimize total wavelengths required in survivable WDM optical networks. Previous work available in literature does not consider survivable network planning in terms of significant, multiple objective goals.

3   Design of Restoration Algorithms

3.1   KSDPR (k-Shortest Disjoint Path Restoration Based on Multiple Uphill Moves and Heuristic Rule)

Four objective functions – i.e. MTWC, MTWL, MTWM, and MBLR – consider the optical fiber distance between node m and node n. As the value of optical fiber distance decreases, each solution generated by four objective functions may be expected to approach to a near optimal solution. The concept of shortest link and node disjoint paths is prevailed in WDM optical networks [10-16]. The following shows three reasons for the wide deployment of shortest disjoint paths in WDM optical networks. First, if longer hops lead to longer distances in a WDM optical network, a lightpath with longer hops is not right for the restoration strategy according to Lemma 1.

Lemma 1. Let G_p be a connected physical topology with at least two nodes. G_p contains at least two nodes that are not cut points. A lightpath with longer hops between two nodes is not right for the restoration strategy in WDM optical networks.

Proof. Assume two nodes s and t, where the distance between s and t is maximal. If t is a cut point, then $G_p - t$ will consist of two components. If a node w is not contained in the component where s belongs, any paths from s to w have to contain t. Therefore, the distance from s to t has to be at least distance between s and t plus 1. However, it is contradiction. Therefore, s and t cannot be cut points. Meanwhile, we can obtain $k(K_n) = n - 1$ and $k(G_p) \leq n - 2$ when G_p is not a complete physical topology and n is total number of nodes in G_p. Any paths with n hops contains precisely $n - 2$ cut points. Therefore, one lightpath with minimum hops may be more survivable against node failures or link failures than any other lightpaths with more than minimum hops.

Second, shortest link and node disjoint paths make restoration speed faster by reducing the lightpath propagation delay over optical fiber links. Third, shortest link and node disjoint paths can avoid additional 3R (Reshaping, Retiming, and Regeneration) optical amplifiers over links in a WDM optical network. The issue that needs to consider shortest disjoint paths and the even distribution of traffic flows at the same time belongs to a ranking paths problem and a multi-objective paths problem. In the ranking paths problem, it is intended to determine not only the shortest path but also k-shortest paths according to an objective function. In the multi-objective paths problem, a set of k-shortest paths has to be determined, being all of them considered as optimal. Thus, k-shortest disjoint paths may be favorable for the design of a survivable WDM optical network. In k-shortest disjoint paths, as k increases, a bounded candidate set of k-shortest paths grows more and a chance to approach a near optimal solution may become bigger.

The developed KSDPR (k-Shortest Disjoint Path Restoration based on multiple uphill moves and heuristic rule) algorithm based on k-shortest disjoint paths search consists of two procedures: a procedure of uphill moves with multiple probabilities and a procedure of disobedience of the strong and the weak optimal principles. It is important to find a set of optimal paths from a global point of view. Optimal path problems can be classified into unconstrained and constrained path problems. A set of optimal lightpath routing in a WDM optical network belongs to constrained path problems because of wavelength continuity and distinct wavelength assignment constraints. If it is NP-Hard to find an optimal path, near optimal solution should be considered to find a sequence of optimal subpaths. For optimal considerations, we need to examine the optimal principle in a network topology with nodes i, j, and k. The optimal principle states that if node j is on the optimal path from node i to node k, then the optimal path from node j to node k also falls along the same optimal path. The optimal principle can be classified into the strong and the weak optimal principles. The strong optimal principle means that every optimal path is formed by optimal subpaths. The weak optimal principle means that there is an optimal path formed by optimal subpaths. If some optimal path problems belong to the strong or the weak optimal principles, an optimal solution can be determined by finding successive optimal sub-solutions.

A k-shortest disjoint paths problem in a WDM optical network is finite if there is an optimal path that generates an optimal solution for a given objective function; otherwise, the problem is non-finite. Besides, the k-shortest disjoint paths problem is bounded if the optimal solution is a finite value. However, if the k-shortest disjoint paths problem in a WDM optical network is non-finite and unbounded, the multiple probabilities of accepting uphill moves are used to escape local minima and improve current solutions. It is possible to allow certain solutions to be acceptable even if the solutions are not minimal. A path is selected among a set of k-shortest paths by using the multiple probabilities of accepting uphill moves. If a path chosen in a set of k-shortest paths that satisfies the strong and the weak optimal principles cannot make traffic flows distribute evenly, the procedure of disobedience of the strong and the weak optimal principles may escape the bad effect of the optimal principle by distributing traffic flows evenly in a WDM optical network. The procedure of disobedience of the strong and the weak optimal principles is developed by adopting Heuristic Rule 1. Heuristic Rule 1 describes a tabu list of heavily congested links.

Heuristic Rule 1. A tabu list of heavily congested links: If the proposed KSDPR algorithm obeys the strong and the weak optimal principles, the usage of links included in optimal paths increases. Then the number of wavelengths required in a WDM optical network increases. The heavily congested links are stored in a tabu list. The tabu list will be used to enhance the proposed KSDPR algorithm by forbidding moves included in the tabu list. This rule is derived from MMBW that minimizes the maximum offered wavelengths on the links.

Remark: If the usage of some links increases in a WDM optical network, the links are heavily congested. The congested links need to pretend to be excluded in the WDM optical network to distribute traffic flows evenly. The efficient size of the tabu list turns out to be $\lceil k(G_p) \rceil$ through the experimentation of the KSDPR algorithm. Heuristic Rule 1 is employed in the procedure of disobedience of the strong and the weak optimal principles.

3.2   DCROS (Deep Conjectural Reinforced Optimal Search Algorithm)

The DCROS (Deep Conjectural Reinforced Optimal Search) algorithm improves the solutions generated by the KSDPR algorithm. The KSDPR algorithm generates near optimal solutions quickly but we found that the generated solutions could be annealed by an annealing algorithm called the DCROS algorithm. The requirements for the DCROS algorithm are classified into four categories: independence from objective functions, intense exploration, diverse exploration, and reinforced exploration. As the independence from objective functions, it yields near optimal solutions for any objective functions without incorporating specific heuristic rules. As the strong optimality acceptance ratio, it is successfully accepted for any problem domain related to the design of a survivable WDM optical network. By the intense exploration procedure, it intensifies a process of exploration in some neighborhoods because there are several local optima that lead the way to a global optimal solution. By the diverse exploration procedure, it spreads the search process effort over different regions because they may contain some acceptable solutions to escape local minima from the current region of problem domain. As the reinforced exploration procedure, it improves near optimal solutions generated by intense exploration and diverse exploration procedures. The following definitions are used to develop the DCROS algorithm.

Definition 1. Let Z_1 and Z_2 denote polynomial local optimal problems. A polynomial local search optimal problem reduction from Z_1 to Z_2 consists of two polynomial time computable functions f_1 and f_2 such that f_1 maps instances x of Z_1 to instances $f_1(x)$ of Z_2. f_2 maps a pair of $f_1(x)$ and x to solutions of x. For all instances x of Z_1, if i^* is a local optimum for instance $f_1(x)$ of Z_2, then $f_2(i^*, x)$ is a near optimal solution for x. If there is such a reduction, we say that Z_1 reduce the polynomial local search optimal problem of Z_2.

Definition 2. Let Z_1 and Z_2 denote polynomial local optimal problems and let (f_1, f_2) be a reduction from Z_1 to Z_2. We say that the reduction is tight if we can choose a subset ϑ of feasible solutions for the instance y of Z_2, where $y = f_1(x)$, so that the following properties are satisfied: ϑ contains all local optima of y, where a solution in ϑ is one of reasonable solutions of the instance y. For every solution p of x, we can construct a solution $q \in \vartheta$ of y such that $f_2(q, x) = p$.

Definition 3. If Z_1, Z_2, and Z_3 are polynomial local optimal problems such that Z_1 reduces the polynomial local search optimal problem of Z_2 and Z_2 reduces the polynomial local search optimal problem of Z_3, then Z_1 reduces the polynomial local search optimal problem of Z_3.

Definition 1 through Definition 3 applies to develop the DCROS algorithm. Let Π denote the NP-Hard optimization problem in a survivable WDM optical network, which needs to design and develop an efficient restoration software framework. Let Z denote the proposed DCROS algorithm. Let Z_1 denote the DS (Deep Search) procedure. Let Z_2 denote the CS (Conjectural Search) procedure. Let Z_3 denote the RS (Reinforced Search) procedure. The framework of Z can be described as follows to solve the NP-Hard optimization problem Π in a survivable WDM optical network:

$$\Pi : Z \Rightarrow \{ Z_1, Z_2, Z_3 \} \tag{1}$$

To apply equation (1) to the DCROS algorithm, minor modifications are needed to make it applicable to a given objective function as follows:

$$\Pi : DCROS \Rightarrow \left\{ \left[DS, \left(P_{CS} \cdot CS^{D_{conjectural}} \right) \right]^{D_{deep}}, [RS]^{D_{reinforced}} \right\}, \text{ where } D_{deep} \geq 1, D_{conjectural} \geq 1,$$

$$D_{reinforced} \geq 1, \text{ and } 0 \leq P_{CS} \leq 1 \tag{2}$$

In equation (2), the DCROS algorithm (Z) consists of the DS procedure (Z_1), the CS procedure (Z_2), and the RS procedure (Z_3). $D_{conjectural}$ represents the number of iteration depths for the CS procedure. If solutions estimated by the CS procedure are near optimal, the DCROS algorithm will keep them; otherwise, P_{CS} needs to be considered. P_{CS} is a penalty factor required to make the estimate of CS procedure accurate. If solutions estimated by the CS procedure are not near optimal and $P_{CS} > 1$, a bad overestimate somewhere along the true optimal solution may cause the DCROS algorithm to wander away from the optimal solution. If solutions estimated by the CS procedure are near optimal and $P_{CS} < 1$, an underestimate cause the right move to be overlooked. D_{deep} represents the number of iteration depths for the DS procedure. $D_{reinforced}$ represents the number of iteration depths for the RS procedure. If a part of the near optimal solution is replaced with another one through the RS procedure, the near optimal solution will be improved. If the DCROS algorithm that exploits Definition 1 through Definition 3 produces a near optimal solution for Π, it will turn out to be a successful restoration technique for the design of a survivable WDM optical network.

3.3   RWWA (Random Walk-Based Wavelength Assignment)

An optimal k-wavelength assignment algorithm has not been known so far. A well-known wavelength assignment strategy is a RFF (Random First-Fit wavelength assignment) strategy. We propose the following two wavelength assignment strategies: SHF (Shortest-Hop lightpath First wavelength assignment) and LHF (Longest-Hop lightpath First wavelength assignment) based on Heuristic Rules 2 and 3.

Heuristic Rule 2. Given a set of lightpaths in a WDM optical network with the distinct wavelength assignment constraint, the SHF strategy can generate the lower bound of wavelengths in the WDM optical network.
Remark: Heuristic Rule 2 implies that the shortest-hop lightpath among the whole of lightpaths needs to be first considered because the first wavelength assignment of the shortest-hop lightpath may increase the maximum reuse of wavelengths and decrease the lower bound of wavelengths.

Heuristic Rule 3. Given a set of lightpaths in a WDM optical network with the distinct wavelength assignment constraint, the LHF strategy can generate the minimum lower bound of wavelengths in the WDM optical network.

Remark: Heuristic Rule 3 implies that the longest-hop lightpath among the whole of lightpaths needs to be first considered because the last wavelength assignment of the longest-hop lightpath may interfere with optimal k-wavelength assignment and increase the lower bound of wavelengths.

In the RFF strategy, the best first-fit lightpath is randomly selected among a set of lightpaths and an available wavelength is assigned to it. In the SHF strategy, the shortest-hop lightpath among a set of lightpaths is selected and an available wavelength is assigned to it. In the LHF strategy, the longest-hop lightpath among a set of lightpaths is selected and an available wavelength is assigned to it. The key to three wavelength assignment strategies is to maximize the reuse of wavelengths that are already assigned to some lightpaths. To design the RWWA (Random Walk-based Wavelength Assignment) algorithm, Random Walk framework with a greedy uphill movement probability is first used and a probability of accepting uphill moves will be gradually varied to anneal the generated solution repeatedly. Hence, the randomized nature enables asymptotic convergence to an optimal solution under certain mild conditions.

4   Experiments

The original 14-node NSFNET physical topology G_p ($N_p = 14$, $|P_{mn}| = 2$, $k(G_p) = 2$, 2-*uni*) is used for the measurement of proposed algorithms. In the experimentation, we consider additional physical topology constraints, such as planar and nonplanar topologies, link and node connectivity, independent node sets. For example, if a physical topology is more than 5-connected, it becomes nonplanar topology [17]. The most important physical topology constraint in WDM optical networks is link and node connectivity. As additional links are added to the NSFNET physical topology, link and node connectivity, survivability for link and node failures, and the chance of finding optimal paths increase while independence node sets decrease. We develop the PTCI (Physical Topology Connectivity Increase) algorithm based on Heuristic Rule 4. The PTCI algorithm constructs a new k-connected NSFNET physical topology.

Heuristic Rule 4. If an additional link is required to increase the connectivity of a physical topology, the shortest link is first considered among a possible set of candidate links.

Remark: If a lightpath traverses over longer hops in WDM optical networks, potential node cuts increases. Therefore, a lightpath with longer hops is not right for the restoration strategy in WDM optical networks. To reduce the number of hops in the lightpath, an additional shortest link is required.

In the experimentation, one static symmetric uniform lightpath traffic demands are considered. So, traffic demands with one wavelength are fully meshed among all nodes. Additionally, the following experiment factors are considered: First, the path-based dedicated backup lightpath restoration method with 100 percent restoration guarantee against a link or a node failure is used. Second, the RWWA algorithm based on the LHF strategy is exploited. It can yield a virtual single-hop WDM optical network that requires the same number of wavelengths as a virtual multi-hop WDM optical network. A virtual single-hop WDM optical network is better than a virtual multi-hop WDM optical network for survivable network dimensioning [18]. Third, the wavelength mileage cost is set to uniform cost. Finally, 2 through 13-connected NSFNET physical topology G_P ($N_p = 14$, $|P_{mn}| = 2$, $2 \leq k(G_P) \leq 13$, 2-*uni*) and G_V (Δ_f, N_V = 14, T_{sd} = one static symmetric uniform lightpath traffic demands, $\Gamma_{set} = \{\Gamma_{MAXDEG}, \Gamma_{FLOW}, \Gamma_{DCA}, \Gamma_{WA}, \Gamma_{LC}, \Gamma_{NC}\}$, $\Gamma_{option} = \{\Gamma_{CONT}, \Gamma_{DUP}, \Gamma_{DWA}, \Gamma_{MILE}, \Gamma_{PROP}$, and $\Gamma_{HOP}\}$, W, S_H) apply to experiments. The hop bound constraint Γ_{HOP} is 13 hops for any lightpaths. 13 hops can prevent the partial cycling in a primary or a backup lightpath because there are 14 nodes in the NSFNET physical topology. In Figure 1 through Figure 5, we do not show the performance of some objective functions proposed in Section 2 because their performance seems worse than the performance of objective functions illustrated in the Figures. The performance of KSDPR, DCROS, RWWA and PTCI algorithms is compared under the node disjoint path constraint because the node disjoint path constraint is strong by considering node failures as well as link failures in a WDM optical network. Additionally, in the Figures, "Objective function A subject to Objective function B" means that if there are some candidates that generates the same optimal solution for A, B is considered to select one of the candidates. In the experimentation of MBLR, the following parameters are set: Message processing time at a node is 10µs, which is corresponding to the execution of 10000 instructions on a 1GHz CPU. The queueing delay of control messages at a node is assumed to be included in the message processing time. There is no good estimate of switching configuration time at this time. Thus, the impact of switching configuration time will be studied by assuming that it can take on a low value of 10µs, and a high value of 500µs. Figure 1 through Figure 5 shows that the performance of MTWM seems to be the best among eight objective functions in terms of wavelengths, wavelength link distance, wavelength mileage costs, even distribution of traffic flows, total and average restoration time of backup lightpaths, and escape from physical and virtual topology constraints. The DCROS algorithm improves solutions generated by the KSDPR algorithm.

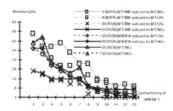

Fig. 1. Total wavelengths in 2 through 13-connected NSFNET physical topology

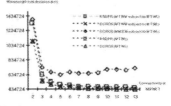

Fig. 2. Total wavelengths in 2 through 13-connected NSFNET physical topology

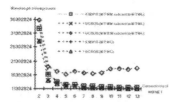 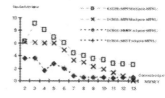

Fig. 3. Wavelength mileage costs in 2 through 13-connected NSFNET physical topology

Fig. 4. Standard deviation of traffic flows in 2 through 13-connected NSFNET physical topology

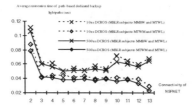

Fig. 5. Average restoration time of backup lightpaths in 2 through 13-connected NSFNET physical topology

5   Conclusion

We develop four restoration algorithms for a restoration software framework in a survivable WDM optical network as follows: KSDPR, DCROS, RWWA, and PTCI algorithms. Even if the KSDPR algorithm based on the concept of k-shortest disjoint paths produces near optimal solutions for given objective functions, the DCROS algorithm improves solutions generated by the KSDPR algorithm. The performance of MTWM seems to be the best among proposed objective functions by achieving the following significant objective goals simultaneously: minimal wavelengths, minimal wavelengths, minimal wavelength link distance, minimal wavelength mileage costs, even distribution of traffic flows, fast total and average restoration time of backup lightpaths, and escape from physical and virtual topology constraints. Additionally, in 2 through 13-connected NSFNET physical topology, MTWM produces a survivable virtual single-hop WDM optical network with the number of wavelengths required by a survivable virtual multi-hop WDM optical network.

References

1. Krishaswamy, R.M., Sivarajan, K.N.: Design of Logical Topologies. IEEE/ACM Transactions on Networking, Vol. 9, No. 2, (2001) 186-198
2. Baroni S., Bayvel, P.: Wavelength Requirements in Arbitrarily Connected Wavelength-Routed Optical Networks. IEEE Journal of Lightwave Technology, Vol. 15, No. 2, (1997) 242-251
3. Yener, B., Boult, T.E.: A Study of Upper and Lower Bounds for minimum Congestion Routing in Lightwave Networks. Proceedings of INFOCOM, Toronto, Canada, (1994) 138-148
4. Fumagalli, A., Valcarenghi, L.: IP Restoration vs. WDM Protection. IEEE Network, Vol. 14, (2000) 34-41
5. Banerjee, S., Yoo, J., Chen, C.: Design of Wavelength-Routed Optical Networks for Packet Switched Traffic. IEEE Journal of Lightwave Technology, Vol. 15, No. 9, (1997) 1636-1646
6. Clouqueur, M., Grover, W.D.: Availability Analysis of Span-Restorable Mesh Networks. IEEE Journal on Selected Areas in Communications, Vol. 20, No. 4, (2002) 810-821
7. Newport, K.T., Varshney, P.K.: Design of Survivable Communications Networks under Performance Constraints. IEEE Transaction on Reliability, Vol. 40, No. 4, (1991) 433-440
8. Xiong, Y., Xu, D., Qiao, C.: Achieving Fast and Bandwidth-Efficient Shared-Path Protection. IEEE Journal of Lightwave Technology, Vol. 21, No. 2, (2003) 1636-1646
9. Iraschko, R., Grover, W.D.: A Highly Efficient Path-Restoration Protocol for Management of Optical Network Transport Integrity. IEEE Journal on Selected Areas in Communications, Vol. 18, No. 5, (2000) 779-794
10. Dacomo, A., Patre, S., Maier, G., Pattavina, A., Martinelli, M.: Design of Static Resilient WDM Mesh Networks with Multiple Heuristic Criteria. Proceedings of IEEE INFOCOM, New York, (2002) 1793-1802
11. Modiano, E., Narula-Tam, A..: Survivable Lightpath Routing: A New Approach to the Design of WDM-Based Networks. IEEE Journal on Selected Areas in Communications, Vol. 20, No. 4, (2002) 800-809
12. Ho, P-H., Mouftah, H.: A Framework for Service-Guaranteed Shared Protection in WDM Mesh Networks. IEEE Communication Magazine, Vol. 40, No. 2, (2002) 97-103
13. Sengupta, S., Ramamurthy, R.: From Network Design to Dynamic Provisioning and Restoration in Optical Cross-Connect Mesh Networks: an Architectural and Algorithmic Overview. IEEE Network, Vol. 15, No. 4, (2001) 46-54
14. Lumetta, S., Medard, M., Tseng, Y-C.: Capacity Versus Robustness: a Tradeoff for Link Restoration in Mesh Networks. IEEE Journal of Lightwave Technology, Vol. 18, No. 12, (2000) 1765-1775
15. Dunn, D., Grover, W., Gregor, M.: Comparison of k-shortest paths and maximum flow routing for network facility restoration. IEEE Journal on Selected Areas in Communications, Vol. 12, No. 1, (1994) 88-89
16. Suurballe, J.: Disjoint paths in a network. Networks, Vol. 4, (1974) 125-145
17. Liebers, A.: Planarizing Graphs–A Survey and Annotated Bibliography. Journal of Graph Algorithms and Applications, Vol. 5, (2001) 1-74
18. Caenegem, B.V., Parys, W.V., Turck, F.D., Demeester, P.M.: Dimensioning of Survivable WDM Networks. IEEE Journal on Selected Areas in Communications, Vol. 16, No. 7, (1998) 1146-1157

Cost Effective Design of TDM ADM Embedded in WDM Ring Networks

Jibok Chung[1], Heesang Lee[2], and Hae-gu Song[3]

[1] Daejeon University, Daejeon, Korea
[2] Sungkyunkwan University, Suwon, Korea
[3] i2 Technologies, Seoul, Korea

Abstract. In this paper, we study a network design problem for TDM rings that are embedded in a WDM ring. We propose an optimization model based on the graph theory for the minimum use of TDM ADMs for these networks. We propose a branch-and-price algorithm to solve the suggested model to the optimality. By exploiting mathematical structure of ring networks, we develop a polynomial time column generation subroutine and a branching rule that conserves the mathematical structure for an efficient column generation. In computational experiments, the suggested model and algorithm can find not the approximation solutions but the optimal solutions for sufficient size networks within reasonable time. We also compare the solution qualities of the suggested algorithm with some known heuristics in literature.

1 Introduction

Wavelength Division Multiplexing (WDM) transmission systems are being deployed and ever increasing in its demand to support high speed telecommunication networks. WDM systems have specially been favorable with traditional Time Division Multiplexing (TDM) ring networks since TDM (SDH in Europe or SONET in USA) rings have been used for dense traffic areas for its cost effectiveness and self-healing robustness. Therefore WDM ring networks are being deployed by a growing number of telecommunication carriers to support TDM self-healing rings. This hybrid transmission architecture is called as TDM embedded in WDM rings, TDM over WDM rings, or TDM/WDM [1], [2].

Recently, two concerns have emerged in planning and operations of TDM/WDM ring networks. First, individual traffic streams that WDM networks will carry are more likely to have small bandwidth required comparing to the huge bandwidth available in a single WDM system[1] [3]. Second, for cost effective TDM/WDM ring networks, network design problem must consider the optimization of the main cost components of the networks.

In this paper, we study the second issue. In literature, many suggest that for TDM/WDM networks the cost of electronic components like TDM add-drop

---

[1] To answer this problem, *Traffic Grooming Problem*, which is defined as a problem to integrate effectively low speed traffic into high speed traffic, has been studied. In this paper we assume that traffic grooming problem has been solved.

M. Freire et al. (Eds.): ECUMN 2004, LNCS 3262, pp. 115–124, 2004.
© Springer-Verlag Berlin Heidelberg 2004

multiplexers (ADMs) is a more meaningful cost factor than the number of wave-lengths that has been considered as a main design parameter for WDM networks [1], [2]. Hence we study the minimum cost TDM /WDM ring problem, which was studied in [4], [5], and recently in [6]. The minimum cost TDM/WDM ring problem is an optimization problem to assign the wavelengths to the lightpaths so as to minimize the number of TDM ADMs while satisfying the wavelength-continuity constraints.[2]

The remainder of this paper is organized as follows. In Sect. 2, we present some related research works. In Sect. 3, we propose an integer linear program-ming (ILP) formulation based on the graph theory. In Sect. 4, we suggest an exact algorithm to solve the formulation proposed in Sect. 3. In Sect. 5, we show, by computational experiments, that the suggested algorithm can find optimal solu-tions for real-sized TDM rings embedded in WDM ring networks within reason-able time. We also do the performance comparisons with some generic heuristics. In Sect. 6, we give some concluding remarks and discuss further research topics.

2   Assumptions and Previous Research

In this paper, we assume that the routes of the lightpaths in the ring are pre-fixed. Hence we focus on the wavelength assignment problem and do not consider the routing problem. This assumption is true for many WDM ring architectures. In some other architectures, where the routes of the lightpaths in the ring are not pre-fixed, a routing problem need to be solved before wavelength assignment.

Many works have been conducted for the problems related with WDM ring networks. [4] showed that the number of wavelength and the number of TDM ADMs can not be minimized simultaneously. They proposed two kinds of heuris-tics and some kinds of post-assignment transformations such as merging, combin-ing, and splitting. They also analyzed the lower bound and a worst-case example for each heuristic. [5] showed that the minimum cost TDM/WDM ring problem belongs to NP-Complete and has a dual problem of the maximum ADM sharing problem. They proposed an ILP formulation which is based on the integer multi-commodity model. It, however, has too many variables and constraints even for small sized problem to solve ILP to the optimality. Recently, [6] proposed a flow-based ILP formulation that has exponentially many constraints with respect to the number of lightpaths such that it can find the optimal solution only for some small-size problems. They also suggested the least interference heuristic, which is a greedy type breadth first search algorithm to maximize the ADM shar-ing. [7], [8] considered the wavelength assignment problem in all-optical WDM rings that do not embed TDM ADMs. They proposed graph-theoretic ILP for-mulations and Branch and Price (B & P) algorithms. Through computational experiments, they demonstrated the effectiveness of their algorithms. [9] pro-posed a simple explicit formula for determining wavelength index in all-optical WDM rings when there exists a traffic demand for all node pairs.

---

[2] Wavelength-continuity constraints is defined as a technical requirement that two lightpaths with a common link must not use the same wavelength.

3   Problem Formulation

In this section, we present a new ILP formulation for the minimum cost TDM/WDM ring problem. Wavelength assignment for route-fixed lightpaths in a WDM network can be interpreted as the coloring problem of the "paths", where all paths going trough a "link" of the WDM network should have different colors. Hence each color represents a wavelength. When the physical topology of the WDM network is the ring, lightpaths around the ring can be viewed as a collection of arc paths on a circle [10]. Hence we can convert the wavelength assignment problem on lightpaths for a WDM ring network into a vertex coloring problem on a conflict graph that is constructed as the followings. (See [7] for details.)

For each lightpath of the original WDM ring network, we define a vertex in the conflict graph, and define an edge between two vertices of the conflict graph if the associated two lightpaths overlap at any link of the original WDM ADM ring network. This conversion technique can convert a path coloring problem not only for the ring network but also for an arbitrary topology network into a vertex coloring problem of the related conflict graph. When the original WDM network is a ring, the conflict graph is a special class of graphs so called circular arc graph (CAG).

Now we introduce the minimum cost TDM/WDM ring problem as the following ILP formulation. Let $G = (V, E)$ be an undirected conflict graph, where V is the set of vertices and E is the set of edges. The set of vertices such that there is no edge connecting any pair of vertices is defined as an independent set (IS) [12]. Note that the set of vertices with the same color of the graph can be found by finding an IS. Let S be the set of all ISs of G. We define a binary variable $x_s = 1$ if an IS, $s \in S$ will be given an unique color (wavelength), while $x_s = 0$, otherwise. Let w_s be the weight of IS s. In this paper, we define w_s as the number of TDM ADMs required to assign the wavelength for all the lightpaths in s. By using the above notation, we have:

(MCRP)

$$Minimize \sum_s w_s x_s \tag{1}$$

$$subject\ to \sum_{\{s:i \in s\}} x_s = 1,\ \forall\, i \in V, \tag{2}$$

$$x_s \in \{0,\ 1\},\ \forall\, s \in S. \tag{3}$$

This graph theoretic formulation has an advantage that it has reasonable number of constraints and has very strong LP relaxation without symmetry property.[3] The objective function (1) means to minimize the total number of TDM ADMs required to assign the wavelength for each lightpaths. The constraints (2) and (3) imply that one vertex of the conflict graph must be included

---

[3] Here the symmetry means that the variables for each color appear in exactly the same way for deciding number of colors.

in one IS, (i.e., The unique wavelength must be assigned to each lightpath). We claim that w_s can be expressed as $2(\sum_{i \in s} x_i - \sum_{i,j \in \bar{E}} x_i x_j)$, where $\bar{E}$ is the subset of E such that the ADMs sharing between the lightpaths is possible. Note that w_s is not an explicit function with respect to the lightpaths. This makes it some difficulty to solve (MCRP). In the following section, however, we will resolve this difficulty.

4   Branch-and-Price (B & P) Approach

Note that our ILP formulation has only one constraint for each vertex of the conflict graph. However, the number of decision variables is huge since the number of all ISs of a graph G can be exponentially large. Therefore generating all ISs of a graph to get the explicit formulation is intractable (except for very small size TDM rings embedded in WDM rings). However the technique, called column generation technique for linear program (LP), which is using only subset of the variables of LP and "generating" more variables and the related column of the LP when they are needed, is successfully used to solve the LP problems with many variables [7]. To solve (MCRP) we suggest a B & P approach that is using a column generation techniques for the LP relaxations of ILP within a branch-and-bound method. In many cases an ILP formulation based on a column generation technique has a stronger LP relaxation than a canonical compact ILP formulation, which is a very important advantage for avoiding long computation time of a usual branch-and-bound algorithm. See [10] for general expositions of the B & P approach for ILP formulations.

In our problem, the column generation procedure for the LP relaxation is described as follows. Begin with $\bar{S}$, a subset of S, the set of all ISs. Solve the restricted LP relaxation for all $s \in \bar{S}$. This gives a feasible solution for the restricted LP relaxation and a dual value π_i for each constraint i of the original (primal) LP relaxation. Now, determine if we need more columns (i.e., if we need more ISs) by solving the following subproblem.

(SP)

$$Maximize \sum_{i \in s} \pi_i - w_s, \forall\ s \in S \qquad (4)$$

or equivalently,

$$Maximize \sum_{i \in s} \pi_i x_i + \sum_{i,j \in \bar{E}} x_i x_j - 2 \sum_{i \in s} x_i, \forall\ s \in S \qquad (5)$$

Note that (SP) is to find the set of ISs that maximizes the function (4) or the function (5). The first term of the objective function (4) is the sum of LP dual variables associated with each IS and the second term is the weight of the IS. As noted in [11], when the weight of IS can be defined explicitly, Maximum Weighted Independent Set (MWIS) problem in CAG can be solved by computing MWIS problem in the IntervalGraph (IG) iteratively, where IG is a special case of CAG without forming a cycle [12]. However, since the weight in our problem is not explicitly defined (i.e., the objective is not a simple summation of lightpaths), it requires a modified algorithm to deal with (SP).

4.1   Algorithm for (SP) in IG

We suggest that (SP) in IG is transformed to the longest path problem in a modified network. Note that longest path problem belongs to NP-Hard in general network but the problem is solvable in polynomial time for acyclic network. Since our modified network for (SP) in IG has no directed cycle, we can find the longest path in polynomial time using the topological ordering algorithm [13]. Therefore, (SP) in IG can be computed within polynomial time bound. The detailed procedures are as the followings:

(SI-Algorithm: An algorithm for (SP) in IG)

1. Step 0: <u>Initialization</u>. Sort the lightpaths from the most left node.
2. Step 1: <u>Network Construction</u>. Construct the associated network as follows: For each pair of lightpaths i, j ($i < j$),
   if two lightpaths do not overlap, add edge(i, j).
   If two lightpaths have a common end point, set the edge weight as w_j+1.
   Else, set the edge weight as w_j.
3. Step 2: <u>Network Transformation</u>. Add artificial node s, t as follows:
   For all lightpaths j, add edge (s, j) with edge weight w_j and (j, t) with edge weight 0.
4. Step 3: <u>Solve</u>. Find the longest path from node s to node t.
5. Step 4: <u>Get Solution for (SP)</u>. The set of arcs included in the longest path is the solution for (SP) in IG.

4.2   Algorithm for (SP) in CAG

In this subsection we show that (SP) in CAG can be solvable in polynomial time by using the algorithm for (SP) in IG iteratively. When graph G is a CAG, the remaining graph after deleting all intersecting nodes with arbitrary node reduces to an IG. Since the number of nodes in G is finite, it is obvious that (SP) in CAG can be solved in polynomial time bound. We have:

(SC-Algorithm: An algorithm for (SP) in CAG)

1. Step 0: <u>Initialization</u>. Label all nodes as unmarked.
   Let wt_{max} be negative infinite and $wt(v)$ be the weight of node v.
   $N(v)$ is the set of adjacent node with node v.

2. Step 1: <u>Network Construction</u>. Select any unmarked node v and construct IG $\bar{G}$ by deleting the node v and $N(v)$ from G. Increase the weight of nodes which has a common end point with v by 1.
3. Step 2: <u>Solve</u>. Solve (SP) in corresponding $\bar{G}$ and find the solution wt^*.
4. Step 3: <u>Solution Update</u>. If $wt(v) + wt^* > wt_{max}$, then $wt_{max} = wt(v) + wt^*$.
5. Step 4: <u>Terminate</u>. If all nodes are marked, terminate. Else, let node v as marked and go to step 1.

Theorem 1. (SP) can be solved in $O(n^3)$ by the SC-Algorithm.

Proof. First, we show that the complexity of the SI-Algorithm is $O(n^2)$. In step 0 of the SI-Algorithm, $O(n\log n)$ time is needed to sort the lightpaths. In step 1 of the SI-Algorithm, we need to compare the endpoints of all lightpaths. So, $O(n^2)$ time is needed to construct the network. In step 2 of the SI-Algorithm, $O(n)$ time is needed to add artificial nodes. The topological ordering algorithm of [13] is $O(m)$, where m is the number of edges in transformed network and is smaller than n^2. Thus, overall complexity of the SI-Algorithm is $O(n^2)$. Since the SC-Algorithm iterates the SI-Algorithm n times, the overall complexity of the SC-Algorithm is $O(n^3)$.

4.3   Branching Rule

When the optimal solution of LP relaxations obtained through the above column generation procedure is not integral, we need to branch some fractional variables. For effective implementation of B & P approach, it is necessary to have a branching rule that conserves the structure of column generation structure. We can do it by the followings:

First, moving clockwise once around the ring from arbitrary starting point, we can index the vertices according to the order in which the counterclockwise endpoints of the corresponding lightpaths occur (tie can be broken arbitrarily). We select the pair of vertex which must be independent and most nearest with respect to the index. We call it as minimal distance branching rule.

The suggested branching rule is composed of two operations, called SAME(i,j) and DIFFER(i,j). SAME(i,j) is an operation to collapse two vertices i and j ($i < j$) into a single vertex. It means to merge two lightpaths corresponding to the vertices i and j into one lightpath by extending the counterclockwise endpoint of j to the clockwise endpoint of i. DIFFER(i,j) is an operation to add edge between the vertices i and j. It means to extend the clockwise endpoint of i such that it is larger than the counterclockwise endpoint of j but smaller than the counterclockwise endpoint of $j+1$.

We want to show that the suggested branching rule conserves the structure of the column generation procedure by using the quasi-circular 1's property. Let

$M(G)=(m_{ij})_{nn}$ be the augmented adjacent matrix for CAG G and U_i (V_i) be the set of consecutive 1's below (rightward) from the ith diagonal element of $M(G)$: (Consider the matrix as wrapped around a cylinder.) If the union of U_i and V_i covers all 1's in $M(G)$, then $M(G)$ satisfy the quasi-circular 1's property. Refer to [14] for the details of the quasi-circular 1's property.

Proposition 1. An undirected graph G is a CAG if and only if $M(G)$ has the quasi-circular 1's property. (See [14].)

Theorem 2. The suggested branching rule conserves the structure of column generation procedure.

Proof. We are going to show that the graph $\bar{G}$ after performing the branching operation for CAG G is also a CAG by showing the augmented adjacent matrix still satisfies the quasi-circular 1's property. Let $M(\bar{G}) = (m'_{ij})_{nn}$ be the augmented adjacent matrix for $\bar{G}$ and U'_i (V'_i) be the set of consecutive 1s below (rightward) from the ith diagonal element of $M(\bar{G})$. Let A_i be the set of n elements in $\{ 1, \dots , i \}$ at the ith column and B_i be the set of elements in $\{ i+1, \dots , n \}$ at the ith column. Without loss of generality, assume that the index of vertex i precedes that of vertex j and SAME(i,j) operation collapses two vertices i and j into i'. By definition, SAME(i,j) operation takes the value 1 if at least one of the row i and j has value 1. For the new 1's in B_i at the ith column of $M(\bar{G})$, it is easy to know that $U'_{i'}$ includes the all 1's which was in U_i and U_j. Furthermore, $U'_{i'}$ is consecutive because we select the vertex j which is the first 0's when we are moving below (rightward) from the ith diagonal element of $M(G)$. For the new 1's in A_i at the ith column of $M(\bar{G})$, they must be included U_j of $M(G)$. If it is not true, it contradicts the assumption that the G has the quasi-circular 1's property. Therefore, SAME(i,j) operation conserves the quasi-circular 1's property. By definition, DIFFER(i,j) operation changes the value of m'_{ij} (m'_{ji}) from 0 to 1. Because we select the vertex j which is first 0's when we are moving below (rightward) from the ith column of $M(\bar{G})$, it is easy to know that U'_i (U'_j) includes the 1s of m'_{ij} (m'_{ji}) and is consecutive. Therefore, DIFFER(i,j) operation conserves the quasi-circular 1's property.

5   Computational Experiments

The proposed algorithm for the suggested ILP formulation has been coded in C and experimented on an engineering workstation using an ILP optimization software. By the experiment, we want to show that the suggested algorithm is computationally feasible to implement in real-sized TDM rings embedded in WDM rings.

We experiment four classes of problem instances that have 5, 10, 15, and 20 WDM ring nodes. To know the effect of demand on the ring to the performance, we divide each class of node sizes into four demand sets by setting the demand density of 0.3, 0.5, 0.7, and 0.9, which is defined as the probability that requires one lightpath for a pair of ring node. Input parameters of twenty sets are summarized in Table 1. In Table 1, $G(n, d)$ represents that n is the number of ring

Table 1. Problem Instance (Average of 5 Instances)

set	$G(n,d)$	VERT	EDGE	DEGREE
1	$G(5,0.3)$	6	11.2	4
2	$G(5,0.5)$	10	34.8	6.8
3	$G(5,0.7)$	14	65.2	7.8
4	$G(5,0.9)$	18	113.6	9.8
5	$G(10,0.3)$	27	282.6	16.6
6	$G(10,0.5)$	45	799.6	26
7	$G(10,0.7)$	63	1564.8	34.4
8	$G(10,0.9)$	81	2615	43.2
9	$G(15,0.3)$	63	1594.8	36.4
10	$G(15,0.5)$	105	4325.6	56.8
11	$G(15,0.7)$	147	8736.2	77.8
12	$G(15,0.9)$	188	14226.4	96.8
13	$G(20,0.3)$	114	5307.2	62.8
14	$G(20,0.5)$	190	14677.8	102.8
15	$G(20,0.7)$	266	28681.4	139.4
16	$G(20,0.9)$	342	47771.2	175.6

nodes from 5 to 20, and the d is the demand density from 0.3 to 0.9. "VERT", "EDGE" and "DEGREE" in Table 1 denotes, respectively, the number of vertices, the number of edges, and the average degree in the conflict graph. Note that "VERT" is the same in a given set but the "EDGE" and "Degree" can be different according to the overlapping conditions for each instance. The largest size of the tested conflict graphs has 342 nodes and 47,886 edges.

Average performances of five instances for each set are displayed in Table 2. In Table 2, "Heuri" denotes the number of TDM ADMs obtained by the heuristic of [4]. This heuristic is also used as an initial integer solution of the our B & P algorithm. "COLs" denotes the average number of columns generated and "BnBs" denotes the average number of branch and bound tree nodes to get the final integer optimal solution. "LP" denotes the average of the optimal objective values of the LP relaxations and "OPT" denotes the average of the minimum numbers of TDM ADMs obtained by the suggested algorithm. As we can see in Table 2, our B & P algorithm can solve any instance of 100 instances in less than 523 seconds. Many instances can be solved within a few seconds. Note that the suggested B & P algorithm does not require generating huge number of columns to get the optimal solution for the LP relaxations. Moreover, after getting an LP optimal solution on the root node of branch and bound tree, the algorithm can be terminated without traversing many branch and bound tree nodes. This can be possible since our graph theoretic formulation (MCRP) has very strong LP relaxations as we can see in Table 2. Note that the value of LP is very near the value of OPT.

We also want to compare the suggested optimization algorithm with the generic heuristics developed by [4]. Our B & P algorithm can find an exact op-

Table 2. Average Performance of B & P Approach

set	Heuri	Cols	LP	BnBs	OPT	Time
1	10	2.2	9.6	0	9.6	0.01
2	15.4	3.6	14.8	0	14.8	0.02
3	19.8	9	17	0	17	0.04
4	25.4	17.2	19.8	0	19.8	0.09
5	43.2	21.2	39.2	0	39.2	0.15
6	70.6	52.2	59.4	0	59.4	0.60
7	94.6	101.4	77	0	77	1.75
8	118.6	177.6	88.2	0	88.2	4.48
9	100.6	74.8	91.2	0.8	91.2	1.20
10	158.6	216.8	138.6	0	138.6	7.12
11	224	431.4	176.2	0.4	176.2	27.02
12	272.8	769.6	201.8	0.2	201.8	75.14
13	178	197.8	161.8	0.2	161.8	8.07
14	290.6	475	247.2	0.2	247.2	41.64
15	398.8	1206	316.6	2.6	316.6	219.29
16	498.2	1763	367	0	367	522.64

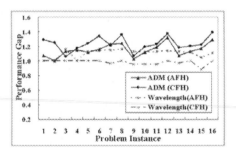

Fig. 1. Performance Comparison with heuristics

timal solution, but the heuristics may find a near optimal solution. Note that
in our problem "optimality" is defined as the minimum number of used TDM
ADMs. As we can see in Fig. 1, our B & P algorithm outperforms the existing
heuristics AFH (Assign First Heuristic) and CFH (Cut First Heuristic) about
15%, 24%, respectively in terms of the number of used TDM ADMs. Furthermore,
the performance gap increases as the traffic load becomes heavy. In terms of numbers of used wavelengths, which is not a main cost component of TDM/WDM
networks, our B & P algorithm requires 10% less than AFH and only 3% more
than CFH as we see in Fig. 1.

6   Conclusions

In this paper, we have shown that the minimum cost TDM/WDM ring problem can be formulated as an ILP problem, and the problem can be exactly solvable up to 20 ADM nodes that are enough large to meet fast growing telecommunication traffic demands. To accomplish this, we exploited the graph-theoretic properties of ring networks and suggested a B & P algorithm to get the exact optimal solutions. The computer simulation shows that the suggested algorithm could find the optimal solution within reasonable time and better performance than the known heuristic methods. In this paper, we do not consider the routing and traffic grooming of TDM traffic. Combining the results of this paper to the routing or traffic grooming optimization is one of the further research topics for TDM rings embedded in WDM ring networks.

References

1. T. E. Stern and K. Bala, "Multiwavelength Optical Networks: A layered approach", Addison-Wesley, 1999.
2. O. Gerstel et al. , Combined WDM and SONET Network Design, IEEE Infocom'99, 1999.
3. R. Dutta, and G. Rouskas, On Optimal Traffic Grooming in WDM Rings, IEEE JSAC, v20, n1, pp. 110-121, 2002.
4. O. Gerstel et al. , Wavelength Assignment in a WDM Ring to Minimize Cost of Embedded SONET Rings, IEEE Infocom'98, 1998.
5. L. Liu, X. Li, P. Wan, and O. Frieder, Wavelength assignment in WDM rings to minimize SONET ADMs, Proc. IEEE Infocom'00, vol. 2, pp. 1020-1025, 2000.
6. X. Yuan, and A. Fulay, Wavelength Assignment to Minimize the Number of SONET ADMs in WDM Rings IEEE ICC'02, 2002.
7. H. Lee and J. Chung, Wavelength Assignment Optimization in Uni-Directional WDM ring, IE interface, v13, n4, 2000.
8. T. Lee et al., Optimal Routing and Wavelength Assignment in WDM Ring Networks, IEEE JSAC, v18, n10, pp. 2146-2154, 2000.
9. X. Lu and S. He, Wavelength assignment for WDM ring, Electronics Letters, v39, n19, 2003.
10. A. Mehrotra and M. A. Trick, A Column Generation Approach for Graph Coloring, INFORMS Journal on Computing, 8, pp. 344-354, 1996.
11. J. Y. Hsiao, C. Y. Tang, and R. S. Chang, An efficient algorithms for finding a maximum weighted 2 independent set on Interval graph, Information Processing Letters, n43, pp. 229-235, 1992.
12. M. C. Golumbic, Algorithmic Graph Theory and Perfect Graphs, Academic Press, 1980.
13. R. K. Ahuja, T. L. Magnanti and J. B. Orlin, Network flows: theory, algorithms and applications, Prentice Hall Inc, 1992.
14. A. Tucker, Matrix Characterization of Circular-Arc Graph, Pacific J. of Mathematics, v39, n2, pp. 535-545, 1971.

An Efficient Mobility Management Scheme in Hierarchical Mobile IPv6

Taehyoun Kim, Bongjun Choi, Hyunho Lee, Hyosoon Park, and Jaiyong Lee

Department of Electrical & Electronic Engineering, Yonsei University,
134 Shinchon-dong Seodaemun-gu Seoul, Korea
{tuskkim, jyl}@nasla.yonsei.ac.kr

Abstract. In the mobile Internet, the wireless link has far less available bandwidth resources and limited scalability compared to the wired network link. Therefore, the signaling overhead associated with mobility management has a severe effect on the wireless link. Moreover, each cell becomes smaller and this increases handoff rate yielding more signaling overhead in the wireless link. In this paper, we propose the IP-Grouping scheme. In the proposed scheme, Access Routers (ARs) with a large rate of handoff are allocated into a Group Zone. The signaling cost in the wireless link can be greatly reduced as the current Care-of Address (CoA) of Mobile Nodes (MNs) is not changed whenever the handoffs occur between ARs within the same Group Zone. The performance of the proposed scheme is compared with the Hierarchical Mobile IPv6 (HMIPv6). We present the simulation results under various condition for IP-Grouping scheme and HMIPv6.

1 Introduction

The Mobile Node(MN) requires special support to maintain connectivity as they change their point-of-attachment. In Mobile IPv6 [5], when a MN moves from a network coverage cell to another cell, it gets a Care-of-Address(CoA) from the visited network. After receiving the CoA, the MN registers the association between the CoA and the home address by sending a binding update message to its Home Agent(HA) or Correspondent Node(CN). HA and CN keeps a binding cache that is updated when new binding update arrives. The signaling load for the binding update may become very significant as the number of MNs increases. To overcome this problem, Hierarchical Mobile IPv6 (HMIPv6) [1] [2] uses a local anchor point called Mobility Anchor Point (MAP) to allow the MN to send binding update messages only up to the MAP when it moves within the same MAP domain. This reduces additional signaling cost in the wired network link between the MAP and the CN that exists in MIPv6. But, HMIPv6 cannot reduce the binding update messages in the wireless link. In addition, IETF proposed the fast handoff over HMIPv6 [3] [4] that integrates HMIPv6 and the fast handoff mechanism to reduce the handoff latency by address pre-configuration. Since the fast handoff over HMIPv6 inherits the basic signaling structure of HMIPv6, the signaling cost in the wireless link is unchanged from HMIPv6.

M. Freire et al. (Eds.): ECUMN 2004, LNCS 3262, pp. 125–134, 2004.
© Springer-Verlag Berlin Heidelberg 2004

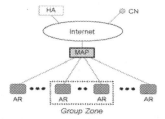

Fig. 1. Network architecture of IP-Grouping scheme

In the mobile Internet, the wireless link has far less available bandwidth resources and limited scalability compared to the wired network link [3]. Therefore, the signaling overhead associated with mobility management has a severe effect on the wireless link. Moreover, each cell becomes smaller [6][8] and this increases handoff rates yielding more signaling overhead in the wireless link.

In this paper, we propose the IP-Grouping scheme to reduce the wireless signaling cost in areas with a large rate of handoff. IP-Grouping scheme has benefits since MNs within the Group Zone does not need the binding update procedure. First of all, wireless network resource is saved. Second, power consumption of the MN is reduced. Third, interferences in the wireless link are reduced and a better communication quality can be achieved.

The rest of the paper is organized as follows. Section 2 provides the overview of IP-Grouping scheme and Section 3 presents the detailed description of IP-Grouping scheme. Section 4 presents the simulation results. Finally, we summarize the paper in Section 5.

2   The Overview of IP-Grouping Scheme

2.1   The Reference Architecture

The reference architecture of IP-Grouping scheme as shown in Fig. 1 is based on the HMIPv6 [1] [2] consisting of a two level architecture where global mobility and local mobility are separated. The reference architecture consists of followings.

- **MAP** : It commands to corresponding ARs to create a Group Zone and routes the packets of MN to the new AR.
- **AR** : It monitors the Movement Status(MS) that has a trace of handoff history. And it count the number of handoffs to its neighboring ARs. When it detects that the measured handoff rate exceeds the threshold value or drops below the threshold value, it sends its status and IP address of its neighboring AR to MAP. According to the command of MAP, it sends the group network prefix or the original network prefix to the MNs.

Group Zone

Fig. 2. Definition of Group Zone

- **GCoA and LCoA** : The GCoA is a Group CoA configured on the MN based on the group network prefix advertised by AR. And LCoA is an On-link CoA configured on the MN based on the original network prefix advertised by AR.
- **RCoA** : The RCoA is a Regional CoA configured by a MN when it receives the MAP option.
- **Movement Update** : When the MN moves under a new AR within the Group Zone, old AR sends a Movement Update to the MAP in order to establish binding among RCoA, GCoA and new AR IP address.
- **Local Binding Update** : MN sends a Local Binding Update to the MAP in order to establish a binding between RCoA and LCoA, or between RCoA and GCoA

2.2 The Definition of Group Zone

AR monitors the Movement Status(MS) that has a trace of handoff history. ARs count the number of handoffs to its neighboring ARs. In this framework, ARs create the Group Zone by using the measured handoff rate derived from the history of the handoff in ARs. For example, we assume the linear AR topology as shown in Fig. 2 and if the condition in (1) is met, AR-i send a request to the MAP to create Group Zone with AR-j, and the MAP commands AR-i and AR-j to create a Group Zone. Therefore, a Group Zone is formed at AR-i and AR-j.

$$rateHO(i, j) \geq THi(G) \tag{1}$$

where *rateHO(i, j)* is the number of handoff from AR-i to AR-j during a certain period of time, and *THi(G)* is the threshold value of handoff for Grouping in AR-i.

Also, if the condition in (2) is met, AR-i sends a request to the MAP to release a Group Zone with AR-j, and the MAP commands AR-i and AR-j to release a Group Zone. Therefore, a Group Zone is released at AR-i and AR-j.

$$rateHO(i,j) \leq THi(UG) \tag{2}$$

where *THi(UG)* is the threshold value of handoff for Ungrouping in AR-i.

Here, *THi(G)* to create and *THi(UG)* to release a Group Zone have different value from some hysteresis (i.e., ensure that the trigger condition for the creation of a Group Zone is sufficiently different from the trigger condition for the release of a Group Zone) to avoid oscillation in stable condition.

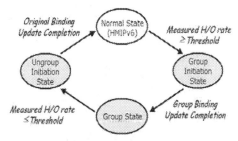

Fig. 3. State Diagram of IP-Grouping scheme

3 The Description of IP-Grouping Scheme

As shown in Fig.3, IP-Grouping scheme operates in four states. First, the Normal State operates as the HMIPv6 until the measured handoff rate of the AR (calculated from the rate of handoffs to its neighboring ARs) exceeds the threshold value and it switches to the Group Initiate State when the measured handoff rate of the AR exceeds the threshold value. Second, in the Group Initiate State, the ARs send the group network prefixes to MNs in their area and switches to Group State. Third, in the Group State, Group Zones are created with ARs with the same group network prefix. When the MN moves to a new AR within the same Group Zone, the handoff occurs through L2 Source Trigger without the current CoA being changed. As a result, local binding updates are not generated and it greatly reduces the signaling cost in the wireless link. Also, when the measured handoff rate of the AR within the Group Zone drops below the threshold value, the Group State switches to the Ungroup Initiation State. Finally, the Ungroup Initiation State switches to the Normal State by sending different original network prefixes in each AR of the Group Zone.

More detailed description of operations of each state is as follows. Note that for simplicity, the paper explains IP-Grouping scheme of only one out of many MNs under each AR and also Group Zone adapted just two ARs out of many ARs. There are actually many MNs and ARs operating simultaneously by IP-Grouping scheme.

3.1 Normal State (HMIPv6)

In the Normal state, it operates as HMIPv6 proposed by IETF. An MN entering a MAP domain will receive Router Advertisements containing information on one or more local MAPs. The MN can bind its current CoA(LCoA) with an CoA on the subnet of the MAP(RCoA). Acting as a local HA, the MAP receives all packets on behalf of the MN. And then, the MAP encapsulates and forwards them directly to the

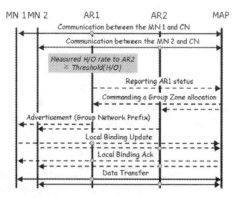

Fig. 4. Message flow in Group Initiation State

LCoA of the MN. If the MN changes its current address within a local MAP domain, it only need to register the new address with the MAP. Hence, only the RCoA needs to be registered with the CNs and the HA. The RCoA does not change as long as the MN moves within the same MAP domain. This makes the MN mobility transparent to the CNs it is communicating with. The boundaries of the MAP domain are defined by means of the ARs advertising the MAP information to the attached MNs.

3.2 Group Initiation State

Fig. 4 shows the message flow in the Group Initiation State. MN1 and MN2 are communicating with AR1 and AR2 respectively in the Normal State (binding cache of MN1 - Regional CoA1(RCoA1) : On-link CoA1(LCoA1), binding cache of MN2 - RCoA2 : LCoA2, binding cache of MAP - RCoA1 : LCoA1, RCoA2 : LCoA2). When AR1 detects that the measured handoff rate exceeds the threshold value of handoff, AR1 sends its status and IP address of AR2 to the MAP. Then, the MAP commands to AR1 and AR2 to create a Group Zone.

Following the procedure, AR1 and AR2 send a group network prefix using Router Advertisement instead of the original network prefix. MN1 and MN2 receive this network prefix and compare to its original network prefix. They each recognize an arrival of a new network prefix and generate new Group CoA1(GCoA1) and GCoA2 by auto-configuration. This mechanism causes MN1 and MN2 to perceive as if they are separately being handed over to a new AR, and causes them to change their current CoA. MN1 and MN2 register newly acquired GCoA1 and GCoA2 to the MAP. The MAP updates its binding cache (as MN1 - RCoA1 : GCoA1 : AR1, MN2 - RCoA2 : GCoA2 : AR2). Through this procedure, Group Zone is performed on AR1 and AR2 and the state then switches to the Group State.

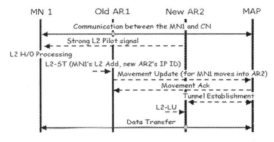

Fig. 5. Message flow of handoff in Group State

3.3 Group State

In the Group State, the packet data sent from CN to MN1 and MN2 are encapsulated with the new GCoA1 and GCoA2 by the MAP and forwarded to MN1 and MN2 respectively. Fig. 5 shows the message flow of handoff in the Group State. MN1 is communicating with AR1(binding cache of MN1 and MAP - RCoA1: GCoA1: AR1). When MN1 approaches new AR2, it receives a strong L2 pilot signal and performs L2 handoff. Here, old AR1 determines the IP address of new AR2 using L2 Source Trigger(L2-ST) [4]. L2-ST includes the information such as the L2 ID of the MN1 and the IP ID of new AR2 (it is transferred using the L2 message of L2 handoff. Therefore, it is not the newly generated message from the MN). Through this procedure, old AR1 detects MN1 moving towards new AR2 and sends a Movement Update message to the MAP. Then, the MAP updates its binding cache(as RCoA1: GCoA1: AR2), and establishes a tunnel to new AR2. And new AR2 sets up a host route for GCoA1 of MN1. After that, MN1 moves to new AR2 and sets up a L2 link after completing a L2 handoff. At the same time, since MN1 receives same group network prefix as old AR1 from new AR2, it does not perform a binding update. From this point on, the packet data sent from CN to MN1 is encapsulated with GCoA1 and forward to MN1 through new AR2 by MAP. Hence, even if MN1 is handed over to new AR2 in Group Zone, binding update requests and acknowledges need not to be sent over the wireless link. Therefore, the signaling cost in the wireless link is greatly reduced. If a MN in the Group Zone moves to AR out of the Group Zone or if a MN in outside the Group Zone moves into AR in the Group Zone, in both cases, a MN receives a different network prefix.

3.4 Ungroup Initiation State

MN1 and MN2 are communicating with AR1 and AR2 respectively in the Group State (binding cache of MN1 - RCoA1 : GCoA1, binding cache of MN2 - RCoA2 : GCoA2, binding cache of MAP - RCoA1 : GCoA1 : AR1, RCoA2 : GCoA2 : AR2).

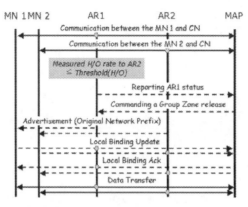

Fig. 6. Message flow in Ungroup Initiation State

Fig. 6 shows the message flow in the Ungroup Initiation State. When AR1 detects that the measured handoff rate drops below the threshold value of handoff, AR1 sends its status and the IP address of the AR2 to the MAP. Then the MAP commands AR1 and AR2 to release a Group Zone. Following the procedure, AR1 and AR2 independently send different original network prefix instead of the group network prefix. MN1 and MN2 each receive this network prefix and compare to its group network prefix. And MN1 and MN2 separately recognize the arrival of a new network prefix and generate new LCoA1 and LCoA2 by auto-configuration. MN1 and MN2 register newly acquired LCoA1 and LCoA2 to the MAP. The MAP updates its Binding cache (as RCoA1 : LCoA1, RCoA2 : LCoA2). Through this procedure, Ungroup Initiation State is finished at AR1 and AR2 and the state switches to the Normal State.

4 Simulation

4.1 Simulation Environment

We study the performance of IP-Grouping scheme using the ns-2.1b7a [7]. Constant bit rate(CBR) sources are used as traffic sources. A source agent is attached to the CN and sink agents are attached at MNs. The duration of each simulation experiment is 180s. The total number of cells in the MAP area is 49 and the Group Zone size is varied between 4, 9, 16, 25, and 36 cells. And the movement of MN is generated randomly. Also we investigate the impact of the continuous movement and discrete movement of MNs in the Group Zone and MAP Area respectively.

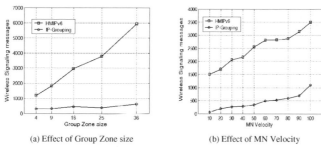

(a) Effect of Group Zone size   (b) Effect of MN Velocity

Fig. 7. Effect of continuous movement (within Group Zone case)

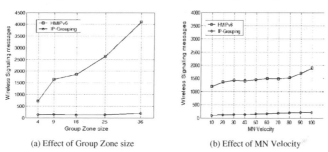

(a) Effect of Group Zone size   (b) Effect of MN Velocity

Fig. 8. Effect of discrete movement (within Group Zone case)

4.2 Simulation Results

4.2.1 Wireless Signaling Cost in the Group Zone

In this section, we investigate the effect of IP-Grouping scheme within Group Zone. All MNs operate with a pause time of 0, supporting continuous movement. This movement pattern has similarity to the fluid flow model in which all MNs constantly move. The simulation results for continuous movement are shown in Fig.7. Simulation results imply that the performance of the IP-Grouping scheme is closely related to the size of the Group Zone and the velocity of the MNs. We also investigate the impact of the discrete movement of mobile nodes. All MNs are now set to operate with a pause time of 30, supporting discrete movement. In contrast to the continuous movement discussed above, MNs move to a destination, stay there for certain period of time and then move again.

The results shown in Fig.8 show that IP-Grouping scheme can reduce the wireless signaling cost for MNs that move infrequently. Besides, the wireless signaling cost

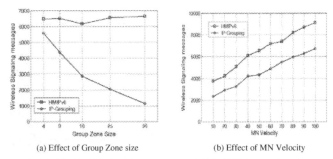

(a) Effect of Group Zone size   (b) Effect of MN Velocity

Fig. 9. Effect of continuous movement (within MAP Area case)

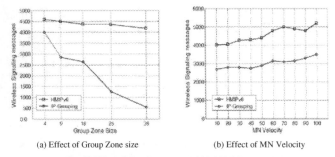

(a) Effect of Group Zone size   (b) Effect of MN Velocity

Fig. 10. Effect of discrete moment (within MAP Area case)

savings are little less than the continuous movement. This is because the number of cell boundary crossings is reduced as the MNs move less frequently.

4.2.2 Wireless Signaling Cost in the MAP Area

In this section, we investigate the effect of IP-Grouping within MAP area with 49 ARs. Fig.9(a) shows the effect of Group Zone size in case of continuous movement. In this figure, we can observe that the total wireless signaling cost of HMIPv6 maintains constantly in MAP area regardless of the change in the Group Zone size. In contrast, the total wireless cost gradually decrease within the MAP area as the Group Zone size increases in IP-Grouping since binding updates do not occur within the Group Zone. In addition, Fig.10(a) shows the effect of Group Zone size in case of discrete movement. We can see that it follows the similar trends shown in Fig.9(a). Also, Fig.9(b) and Fig.10(b) shows that wireless signaling cost of IP-Grouping is much lower than that of HMIPv6 in the MAP area as the velocity increases. As a result, IP-Grouping also reduces wireless signaling cost within MAP area.

5 Conclusion

In this paper, we proposed the IP-Grouping scheme that can reduce the wireless signaling cost in the wireless link. The proposed scheme establishes the Group Zone by using the measured handoff rate derived from the history of the handoff in ARs. The simulation results show that the wireless signaling cost in IP-Grouping scheme is much lower than that of the HMIPv6 in both the Group Zone and MAP area. Furthermore, if the Group Zone persists for several minutes, it will reduce several tens of thousands of binding update message in the wireless link than the HMIPv6. As IP-Grouping scheme is deployed to reduce the wireless signaling cost in the with a large rate of handoff, we expect that mobile users will be offered with more reliable service in the mobile Internet.

References

1. C. Castelluccia, and L. Bellier : Hierarchical Mobile IPv6. Internet Draft, draft-ietf-mobileip-hmipv6-08.txt, work in progress, (2003)
2. Sangheon. Park and Yanghee Choi : Performance Analysis of Hierarchical Mobile IPv6 in IP-based Cellular networks. IPCN, Conference, (2003)
3. Robert Hsieh, Zhe Guang Zhou, Aruna Seneviratne : SMIP : A Seamless Handoff Architecture for Mobile IP. Infocom, (2003)
4. Rajeev Koodli : Fast Handoffs for Mobile IPv6. Internet Draft, draft-ieft-mobileip-fast-mipv6-06.txt, work in progress, (2003)
5. D. Johnson, and C. Perkins : Mobility Support in IPv6. Internet Draft, draft-ietf-mobileip-ipv6-24.txt, work in progress, (2003)
6. R. Berezdivin, R. Breinig, and R. Topp.: Next-Generation Wireless Communication Concepts and Technologies. IEEE Communication Magazine, (2002) 108-116
7. ns2 simulator, version 2.1b7a : http://www.isi.edu/nanam/ns
8. G. Evans, K. Baughan : Visions of 4G. IEEE, Electronics & Communications, Vol.12, No.6. (2000), 293-303

Scalable Local Mobility Support for the IPv6-Based Wireless Network Using Improved Host Based Routing

Seung-Jin Baek

Department of Information and Communication, Kyungpook National University
1370, Sankyuk-Dong, Buk-Gu, Taegu, Korea 702-701
sjbaek@ee.knu.ac.kr

Abstract. There have been active research activities to solve the problems of frequent handoff and encapsulation overhead in the 3G or 4G wireless IP networks. The host based routing scheme has been proposed to provide the micro-mobility management. One of the drawbacks of the host based routing scheme is the limitation in scalability. Accordingly, the current paper presents a new host-based routing scheme that improves the scalability, while providing a fast handoff and various efficient forwarding schemes. Furthermore, it is compatible with the basic Mobile IPv6 protocol. When applied to the HAWAII protocol in an IPv6 network, the new scheme eliminates the limitations of the legacy micro-mobility protocols by providing both a seamless and transparent service and a fast and smooth handoff. A performance evaluation is done to show the scalability improvement.

Keywords: Micro-mobility, Mobile IP, Host Based Routing, Mobility Management

1   Introduction

An internet protocol (IP) is used in almost all internet-based application software and hardware, and as the number of hosts connected to the Internet continues to increase, the use of IPv6 in the real world is not too far off. In particular, Mobile IPv6 and mobility management are essential components for mobile terminals and users.

Much research is currently focusing on the provision of mobility to mobile terminals, including a Mobile IP. Local mobility management is one such approach and is needed to support frequent handoffs and minimize packet losses in 3G or 4G wireless access networks, where handoffs occur frequently. Corresponding to this need, micro-mobility protocols, such as Hawaii, Cellular IP, and Hierarchical Mobile IP, have already been proposed by the IETF working groups [1], [2].

The above protocols can be classified in two kinds: one group is HBR (host-based routing) protocols that create and manage routing entries for each host, while the other group is tunneling-based protocols that use tunnels to communicate between the access network gateway, home agent, and host. According to previous studies, HBR-based schemes show a better performance than tunnel-based schemes [3], however, both types of scheme still have critical problems as regards scalability when the number of mobile nodes (MNs) increases, which also increases the number of routing entries or tunnels. In a tunnel-based scheme, each tunnel requires a kind of agent as

M. Freire et al. (Eds.): ECUMN 2004, LNCS 3262, pp. 135–144, 2004.

the tunnel end-point, which creates overheads in capsulation and decapsulation and disobeys the basic architecture of Mobile IPv6 that only has MNs, CNs, and HAs. In particular, a tunnel-based scheme is very weak in the case of real-time traffic. In particular, since Cellular IP uses its proprietary protocol architecture, it requires pure Mobile IP clients to install new protocol stacks or applications.

Accordingly, the current paper proposes a new local mobility support architecture and Improved HBR scheme to solve the abovementioned problems of other micro-mobility protocols. As such, the proposed Improved HBR (IHBR) suppresses the routing entry population and produces a better handoff performance by modifying HBR, which has already shown a good performance. Meanwhile, the proposed architecture also allows an MN and CN to create a direct communication channel, thereby eliminating the need for an agent, like a tunnel end-point or foreign gateway. In addition, the proposed architecture minimizes the overheads that can appear in routing or capsulation, and follows the basic communication model of Mobile IPv6. Plus, since it only changes the routing scheme among the routers and BSs in the access network, MNs visiting the access network do not need any other protocols except for the basic Mobile IPv6.

The remainder of this paper is organized as follows: Section 2 describes the existing micro-mobility protocols, including Hawaii, and their limitations, then section 3 presents the Improved HBR and new local mobility management architecture. A performance evaluation is conducted in Section 4 and a final summary given in section 5.

2   Previous Work

2.1   Micro-Mobility Protocols

Mobile IP, the *de facto* standard in mobility research, has been currently developed up to version 6. As such, Mobile IPv6 (MIPv6) solves the triangular routing problem in version 4, and enables an MN and CN to communicate directly [4]. The handoff in MIPv6 is an essential issue supporting the mobility of MNs, yet it has a problem related to a high handoff latency. Therefore, in the case of a 3G or 4G network, where the size of the subnet or network is smaller, handoffs occur more frequently, which results in an increased latency.

Various solutions have already been proposed, including omitting the Mobile IP registration and updating processes in the case of an intra-domain handoff, referred to as a micro-mobility protocol [2]. In other words, the movement of an MN within a specific domain does not create any Mobile IP messages between the CN and MN. Yet, when the MN moves out from the specific domain to another domain, the MN should facilitate the Mobile IP handoff process.

Figure 1 shows the differences between macro-mobility and micro-mobility. The two circles in the lower part of Figure 1 represent wireless access networks. Thus, an intra-domain handoff is when an MN moves within one circle and a micro-mobility protocol is responsible for the mobility management, whereas an inter-domain handoff is when an MN moves between the circles and a macro-mobility protocol, like MIPv6, becomes responsible for the mobility management. Consequently, macro-

and micro-mobility protocols have a coexisting relationship, where the micro-mobility protocol can compensate for the weak points of the macro-mobility protocol.

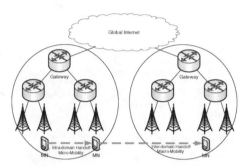

Fig. 1. Macro-mobility and Micro-mobility

There are many micro mobility protocols, such as Hierarchical Mobile IP (HMIP), Cellular IP (CIP), Hawaii, MER-TORA, and so on. First, HMIP uses tunnels to provide local mobility and a MAP (Mobile Anchor Point) for communication between an MN and CN, where the MAP is a kind of agent [5]. However, if the number of MNs is increased, this also increases the number of tunnels and overheads due to the heavy load in capsulation. In contrast, Cellular IP supports local mobility using HBR and is simple and easy to use. However, CIP also has a routing entry population problem and uses a proprietary protocol stack, which means that CIP can only support mobility for CIP clients. While MER-TORA is somewhat similar to the proposed architecture, it does not differentiate between the backbone network and the access network, and is designed to be operated in an ad-hoc network [6].

2.2   HAWAII

HAWAII (Handoff Aware Wireless Internet Infrastructure) also uses HBR, like CIP. As such, all routers in the HAWAII domain have routing entries for all nodes. HAWAII uses specific messages to update the routing information among intra-domain routers, while CIP uses packet snooping [1], [2].

HAWAII exchanges its own specific messages between routers, thus mobile nodes do not need any other protocol stacks, except the basic Mobile IP, for mobility support. In other words, HAWAII can support mobility for basic Mobile IP clients. In addition, HAWAII provides various forwarding schemes for a smooth handoff, such as a non-forwarding scheme and forwarding scheme, to minimize packet losses that can occur during a handoff procedure. However, HAWAII still suffers from a routing entry population problem.

3   Efficient Local Mobility Support for IPv6-Based Wireless Network

3.1   Improved HBR

The current paper proposes a new Improved HBR scheme to overcome the limitations of the existing micro-mobility protocols mentioned in Section 2. As such, the proposed scheme offers a better performance and scalability in a wireless access network that requires micro-mobility support. The new scheme mixes HBR and prefix-based routing, which is already being used in the Internet. Although this may sound similar to MER-TORA, the proposed scheme restricts the applicable area to a wireless access network, rather than the entire Internet. Plus, the topology of IHBR is restricted to a tree, whereas MER-TORA uses a mesh.

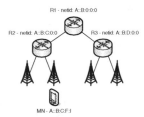

R1 - netid: A::B:0:0:0

R1 - netid: A::B:0:0:0/80	
Destination	Next Hop
A::B:C:0:0/96	R2
A::B:D:0:0/96	R3

R2 - netid: A::B:C:0:0   R3 - netid: A::B:D:0:0

R2 - notid: A::B:C:0:0/96	
Destination	Next Hop
A::B:C:E:0/112	Interface a
A::B:C:F:0/112	Interface b

R3 - netid: A::B:D:0:0/96	
Destination	Next Hop
A::B:D:G:0/112	Interface a
A::B:D:H:0/112	Interface b

MN - A::B:C:F:I

Fig. 2. MN addition in IHBR domain

The routers in an HBR-based access network have routing entries for all nodes or hosts. Thus, if a new node joins or a handoff occurs, the routing entries need to be updated from the root router's table to the attached routers' tables. Therefore, to remove this overhead, the IHBR scheme assigns an IP address to the newly joined nodes according to specific rules, which follow the routing and subnetting rules in generic IP addressing. The details are presented below:

- Each router has its own network or subnet ID (prefix) to distinguish other networks in the access network and perform prefix-based routing.
- All routers in the access network build prefix-based IP forwarding tables according to the assigned prefix, and the table is identical with the case of the Internet.
- Each subnetwork assigns IP addresses (CoA) to newly added nodes based on its own ID and mask.
- The BN (Base Network) is where nodes are assigned their addresses for the first time. The NN (Neighbor Network) refers to any subnetwork in the access network, except for the BN.

Figure 2 shows the routing table construction and address assignment for a new MN in the sample network. When handoffs occur between subnets, the detailed handoff procedures are as follows:

- The NN does not assign a new address to the moved MN, as the MN uses the address assigned by the BN.
- The routers related to the moved MN add or remove its routing entry from each IP forwarding table.
  - When an MN moves from the BN to an NN, a host-specific routing entry is added to the routing tables from the crossover router of the BN and NN to the BN and NN.
  - When an MN moves from an NN to another NN, the host-specific routing entry is removed from the routing tables of the routers from the crossover router of the BN and old NN to the BN and old NN. The entry is then added to the routing tables from the crossover router of the BN and new NN to the BN and new NN.
  - When an MN moves from an NN to the BN, the host-specific routing entry is removed from the routing tables from the crossover router of the BN and NN to the BN and NN.
- Since IHBR only changes the information from the attached routers to the crossover router, this creates a simple signal process and minimizes the size of the IP forwarding table and route caches.
- Plus, since the CoA is not changed, regardless of the occurrence of a handoff in the access network, the CN and MN can communicate transparently.

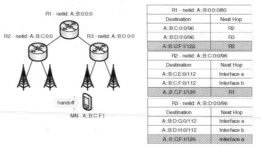

Fig. 3. Handoff in IHBR domain

Figure 3 shows the inter-domain handoff process and CoA unchangeability. The difference between IHBR and HBR is in the routing table operation or management and routing within the domain . In HBR, the downstream is always forwarded to the parent node and the root router makes decisions whether to send the traffic to the global Internet or inside the network. However, in IHBR, each router has its own forwarding table and decides whether to send to the parent node, child nodes, or MNs, which is extremely efficient when the MN and CN exist within the same access network.

The proposed IHBR scheme is also a generic routing algorithm, making it applicable to other HBR-based micro-mobility protocols with certain modifications.

3.2   Local Mobility Support with HAWAII Extension for Mobile IPv6

3.2.1   Design Requirements

This section presents IHBR-based HAWAII in a Mobile IPv6 network. In the recent IETF WG's draft, there are 12 requirements for micro-mobility protocols, and the proposed architecture satisfies all these requirements as follows [7]:

Intra-domain Mobility. The most important factor in LMM (Local Mobility Management) is intra-domain mobility. In the proposed architecture, an MN only maintains one CoA until it leaves the access network. The CN only knows one address, regardless of any handoffs. No signaling procedure is needed between the CN and an MN when an intra-domain handoff occurs.

Security. The proposed scheme is related to the routing mechanism, while other factors simply follow IPv6. The proposed architecture also uses the security mechanisms provided by IPv6.

Induced LMM Functional Requirement. The proposed architecture provides local mobility by only modifying HBR. There is no need to modify MIPv6. In other words, the routing mechanism between MNs and the routers in the access network is changed, yet an MIPv6-based CN, MN, and HA can be applied without any modification.

Scalability, Reliability, and Performance. The IETF draft requires that the LMM complexity is linearly proportional to the size of the local domain. While other HBR-based protocols have the same number of routing entries as number of mobile nodes, the proposed architecture only has the same number as other protocols in the worst case, when all mobile nodes are moved from their own BN, which has a very low probability. As such, the proposed architecture has a lower routing entry population rate. Detailed proof will be shown in Section 4.

Mobility Management Support. IHBR-based local mobility management can facilitate a fast handoff and minimize latency and packet loss. No modification or extension to an MIPv6-based CN, HA, and MN is needed.

Auto-configuration Capabilities for LMM Constituents. In the proposed scheme, there is no need for manual configuration after topology construction or routing table building. Plus, since there is no LMM agent, no agent configuration is needed either. The CoA assignment uses the mechanism of the IPv6 auto-configuration or DHCP (Dynamic Host Configuration Protocol) to minimize the manual configuration.

LMM Inter-working with IP Routing Infrastructure Requirement. As IHBR defines the routing rules within a local access network, there is no corruption or modification of the global Internet, in contrast to MER-TORA. In other words, IHBR does not disturb the core IP routing mechanism.

Sparse Routing Element Population Requirement. The routing entry population rate is lower than an HBR-based system, because the proposed system only adds a routing entry if an MN disobeys the basic IHBR routing rules.

Support for Mobile IPv4 or Mobile IPv6 Handover. The proposed scheme defines routing rules for the efficient operation of MIP, and supports the interwork with an MIP handoff.

Simple Network Design Requirement. Generic IP subnetting and CIDR are used, and they allow a simple access network design, such as a tree model.

Stability. As the proposed LMM uses a tree model and makes immediate table updates according to the movement of the MNs, it does not make any forwarding loops.

Quality of Service requirements. As a security mechanism, the proposed LMM uses the QoS mechanism provided by IPv6.

3.2.2 Operation
- Path Setup

Figure 4 shows the process of the path setup in the IHBR-based HAWAII protocol. An MN that visits this access network tries to access the BS and acquire a CoA. Thereafter, the MN sends a BU (binding update) message to the HA and CN to announce its new CoA. This procedure is identical to MIPv6, plus the CN can communicate with the MN without any agent.
- Handoff

Figure 5 presents the intra-domain handoff process. The handoff procedure can be divided into three phases, 1) handoff phase, and 2) data transmission before and 3) after handoff completion.

First, in the handoff phase, the MN detects layer 2 signals and requests connection to the new BS. Since a BU message is only sent when the CoA of an MN is changed, the MN does not send a BU message. The routers in the access network perform the following processes: The new router that receives the request message from the new BS delivers the message to the crossover router. When the ID or prefix of the new network is different from the host ID of the MN, this means that the MN is disobeying basic routing rules, so the routers between the new router and the crossover router add a routing entry for the MN. If the old network is the BN, a routing entry is also added to the router between the old router and the crossover router for data forwarding and intra-domain routing. When the MN moves from an NN, the routing entry for the MN is removed from all routers between the crossover router and the old router. The routing entry is then updated in the routers from the crossover router of the NN and BN to the attached router.

This procedure is for an intra-domain handoff, and Mobile IP supports the mobility when an inter-domain handoff occurs.

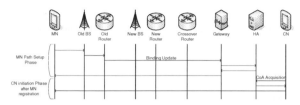

Fig. 4. Signal flow for path setup

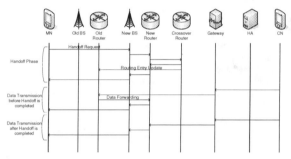

Fig. 5. Signal flow for local handoff

4   Evaluation

This section presents a performance evaluation of the proposed scheme. As mentioned in the IETF's draft [7], scalability is one of important factors in measuring the performance of micro-mobility protocols, as represented by the routing entry population. If there are n nodes in the access network, the number of routing entries, E, can be as follows.

- With the normal HBR scheme

$$E_{HBR} = n \text{, where } n \text{ is the number of nodes} \tag{1}$$

If p is the probability of the occurrence of a handoff within a time unit and l is the number of group routing entries, the number of routing entries in the IHBR domain is:

- If there is no handoff-node in the IHBR domain

$$E_{IHBR_0} = l \text{, where } l \text{ is relatively very small to } n \tag{2}$$

- When handoffs occur in the IHBR domain

If there are k subnets in the access network, the number of entries after a time unit t can be:

$$E_t = \frac{k-2}{k-1}E_{t-1} + p\left(n - E_{t-1}\right) + l = \sum_{i=1}^{t}\left(\frac{k-2}{k-1} - p\right)^{i-1}pn + l \qquad (3)$$

Equation 3 can be generalized as follows:

$$E_{IHBR} = \lim_{t\to\infty} E_t = \left\{ p \middle/ 1 - \left(\frac{k-2}{k-1} - p\right)\right\}n + l = Kn + l,$$

$$\text{where } K = p\middle/\left\{1 - \left(\frac{k-2}{k-1} - p\right)\right\} \qquad (4)$$

In this case, E means the number of hosts that make a handoff or the sum of the routing entries in the routers at the same tree depth. Suppose there is a binary tree model, the root router only has two child nodes. Thus, when $k = 2$ at the root router and K becomes smaller than 0.5, the root router only has half of n routing entries, while the root router with the HBR scheme has n routing entries.

From equations 1, 2, 3, and 4, the proposed scheme had a lower population rate than the HBR-based schemes when the probability p was a small value. Plus, when n in equation 1 was replaced with the number of tunnels, the tunnel population rate of tunneling-based schemes was easily estimated.

The proposed scheme was simulated using the network model shown in Figure 2. The computer simulation was conducted with n mobile nodes, a handoff probability p, and time unit t. The mobile nodes were designed to freely move to any subnet after unit time with the probability p. The routing entry population was measured by counting the number of moved nodes after time t. Figure 6(a) shows the routing entry population rate according to the number of mobile nodes and probability. In Figure 6(b), the dotted line is from equation 4, while the solid line is from the simulation. 1000 mobile nodes were assumed and only the root router considered. The simulation results exhibited a similar shape to the calculated result. When the probability approached one, the number of entries converged to $500(= n/2)$. Since a probability of 1 is unfeasible, the number of entries was a smaller value ($n*0.2$ at $p = 0.1$, $n*0.3$ at $p = 0.2$).

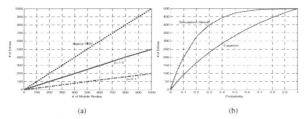

(a)   (b)

Fig. 6. Comparison of routing entry population

5   Conclusion

This paper proposed a new local mobility management architecture and IHBR scheme based on the HBR scheme. The proposed scheme overcomes the limitations of other micro-mobility protocols, satisfies the requirements proposed by the IETF WG, and produced a better scalability based on simulation results showing a lower routing entry population rate. In addition, no modification is needed to the existing Mobile IP protocol, while transparent and seamless services are provided.

References

1. A. T. Campbell, Gomez, J., Kim, S., Turanyi, Z., Wan, C-Y. and A, Valko, "Comparison of IP Micro-Mobility Protocols," IEEE Wireless Communications Magazine, Vol. 9, No. 1, Feb. 2002.
2. A. T. Campbell and J. Gomez, "IP Micro-Mobility Protocols," ACM SIGMOBILE Mobile Computer and Communication Review (MC2R), Vol. 4, No. 4, pp 45-54, Oct. 2001.
3. Sudhir Dixit, Ramjee Prasad, "Wireless IP and Building the Mobile Internet," Artech House Publishers, 2003.
4. D. Johnson, C. Perkins, J. Arkko, "Mobility Support in IPv6," Internet-Draft, Dec. 2003.
5. Hesham Soliman, Claude Castelluccia, Karim El-Malki, Ludovic Bellier, "Hierarchical Mobile IPv6 mobility management (HMIPv6)," Internet-Draft, Jun. 2003.
6. A W O'Neill and G Tsirtsis, "Edge mobility architecture – routing and hand-off," BT Technology Journal, Vol. 19, No.1, Jan. 2001.
7. Carl Williams, "Localized Mobility Management Requirements," Internet-Draft, Oct. 2003.

Fuzzy Mobility Management in Cellular Communication Systems

José J. Astrain and Jesús Villadangos

Dpt. Matemática e Informática
Universidad Pública de Navarra
Campus de Arrosadía
31006 Pamplona (Spain)
{josej.astrain,jesusv}@unavarra.es

Abstract. There is growing need to provide location mechanisms to determine the movement of a mobile user in order to grant quality of service (QoS) in cellular communication systems. Predicting in advance the movement of a mobile user, the system can tune the operation point: blocking and interruption probabilities in circuit switched networks, and flow weights in packet commutation networks. The use of fuzzy symbols allows to better manage imprecise (fuzzy) location information due to attenuation and multi-path propagation. A fuzzy automaton deals with the strings of fuzzy symbols that represent the mobile user movement across the geography of the cellular system, and grants a high recognition rate of the path followed by the mobile terminal. This fuzzy automaton can select the blocking and interrupting probabilities and then ensure QoS for different scenarios of the cellular system.

1 Introduction

The scenario of this work is a mobile cellular communication system, where mobility of terminals (mobile stations, MS), users and services must be granted and managed.

Mobility management requires the location of the mobile stations (MSs) and handoff control. When some intelligent services, as location services, are provided, some knowledge concerning the direction followed by MSs during their movement across a cell is also required. MSs cross different cells when traveling across a cellular system: the process that governs those transitions between contiguous cells are handoff processes. At this step, resources located at the next arrival cell must be allocated in order to ensure the transition between cells without lost of information or communication interruption. If any mobility prediction systems is provided, the system has a certain time to determine the best resource allocation for a given scenario. As soon the prediction system determines the path followed by a certain MS, the system can estimate the interval time when the MS is going to traverse from the current cell to the destination one. Then, the system can maximize the occupation of the communication channels without decreasing the QoS of the users.

M. Freire et al. (Eds.): ECUMN 2004, LNCS 3262, pp. 145–154, 2004.
© Springer-Verlag Berlin Heidelberg 2004

In the literature a great variety of algorithms to manage the mobility of users or terminals has been described, the main part of them considering predictive algorithms because in the new communication environments, both resources and services present mobility. A lot of concepts of mobility were first introduced in cellular mobile networks for voice, now data and multimedia is also included. Mobility models classification (used in computing and communication wireless networks) is provided in [2]. Mobility models (either macroscopic or microscopic) are fundamental to design strategies to update location and paging, to manage resources and to plan the whole network. The mobility model influence will increase as the number of service subscribers grows and the size of the network units (cells) decreases.

The progress observed on wireless communications makes it possible to combine a predictive mobility management with auxiliary storage (dependent on the location, caching, and prefetching methods). Mobile computing in wireless communication systems opens an interesting number of applications including quality of service management (QoS) and final user services. Some examples of predictive applications are: a new architecture to browse web pages [4], where the predictive algorithm is based on a learning automaton that attributes the percentage of cache to the adjacent cells to minimize the connection time between servers and stations; the adaptation of the transport protocol TCP to wireless links in handoff situations [5], that uses artificial intelligence techniques (learning automaton) to carry out a trajectory prediction algorithm which can relocate datagrams and store them in the caches of the adjacent cells; an estimation and prediction method for the ATM architecture in the wireless domain [8], where a method improving the reliability on the connection and the efficient use of the bandwidth is introduced using matching of patterns with Kalman filters (stability problems); and so on.

The solution here presented follows a dynamic strategy where the MSs calculate periodically their situation by triangulating the seven signal proceeding from the base stations (BSs) of the current cell, and the six adjacent cells. By this way, the MS can encode its movement in terms of fuzzy symbols that represents the membership degree of each cell. Different types of codification can be used, but all of them provides strings of fuzzy symbols that represent the path followed by MSs in their movement. Each fuzzy string is dynamically built, and dynamically compared with a dictionary of strings containing the most frequent followed paths for a given cell. The degree of resemblance between two patterns is measured in terms of a fuzzy similarity [3] by way of a fuzzy automaton. Is a classical imperfect string matching problem where a fuzzy automaton is used to compute the similarity between the path followed by the MS and the paths contained in the dictionary. Sensitivity to deviations from the real path followed by a MS is controlled by means of editing operations [6], where the insertion of a fuzzy symbol models the effects of multi-path propagation, the deletion of a fuzzy symbol models the effects of an attenuation, and the change of a fuzzy symbol models both effects. Fuzzy techniques perform better than statistical ones for this problem because in cellular systems, the signal power used to determine the location of a MS are affected by fading, attenuation and multi-path propagation problems. So, location uncertainties due to the signal triangulation are better described and managed using fuzzy techniques. An advantage of this predictive fuzzy method is that it deals with a non limited number of signal propagation errors.

The knowledge of the time interval when the MS is going to perform the handoff process allows to increase the occupation of the channel. GPRS and UMTS networks allows to transmit data flows where the weights of the WFQ algorithm can be syntonized since the path followed by the user can be predicted in advance. In GSM, a new channel allocation is needed to ensure the communication without interruption when the user moves from one cell to another. This allocation can be performed maximizing the occupation of the communication system without decreasing the QoS of the user communication. This work is devoted to the GSM scheme.

The rest of the paper is organized as follows: section 2 is devoted to present the mobility management system model used; section 3 introduces the predictive capabilities of the mobility management system model previously presented; section 4 evaluates the results obtained for this technique; finally conclusions and references end the paper.

2 Mobility Model

As introduced in [7], mobility modelling, location tracking and trajectory prediction are suitable in order to improve the QoS of those systems. By predicting future locations and speeds of MSs, it is possible to reduce the number of communication dropped and as a consequence, a higher QoS degree. Since it is very difficult to know exactly the mobile location, an approximate pattern matching is suitable.

Mobility prediction is easier when user movements are regulars. So, we need to detect the most frequent followed paths of each cell that will be stored by each BS as the cellular dictionary. In the same way, the user dictionary is built taking into account the regular movements followed by a certain user across a specific cell. The user dictionary is stored in the MS of the user. Both, user and cell dictionaries, constitute a hybrid dictionary that is used by an imperfect string matching system to analyze, by terms of edit operations, the similarity between the path followed by a MS and all the paths included in that dictionary. When the MS follows a non regular movement, the resulting path is not included into its user dictionary, so non exact coincidence with any paths of the dictionary occurs. As the hybrid dictionary also contents the most frequent paths followed by all the users of this cell, the only case whenever the system cannot determine the trajectory followed by the MS is when its movement is not usually followed neither the user, neither the rest of users. At this point, no prediction must be performed, but in order to avoid this circumstance later, the string corresponding to that movement is added to the user dictionary. Such as, next time the movement could be predicted.

MSs can measure seven different pilot signals powers, since each cell of a cellular system is enclosed by six cells (see figure 1). So, a fuzzy symbol including the ownership degree for each cell is built every time that the MS measures that power. During its movement across the network, a MS builds a string of fuzzy symbols (the known locations of the MS) by concatenating those symbols. The strings obtained are matched with the strings contained in the hybrid dictionary, obtaining a similarity measure calculated by a fuzzy automaton [1,3]. Then, a macroscopic mobility model is obtained, but a microscopic one is also desired. Then, the same concept is applied to a single cell, but now, the cell is split into different divisions. Three different codification schemes to describe the paths are considered. First one considers cell division in non-overlapped rectangular zones, each one uniquely identified by an alphabetic symbol. Second scheme considers sectorial division of a cell, and third one considers a k-connectivity scheme, where eight alphabetic characters represent all the possible movements that the MS can perform (see figure 2). Now, the ownership degrees managed correspond to the different divisions performed for a cell.

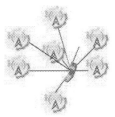

Fig. 1. Signal triangulation.

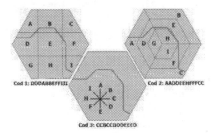

Fig. 2. Path codification schemes

3   Fuzzy Automata for the Trajectory Prediction

Trajectory prediction is performed by the MS (it can also be performed by the BS if desired), so dictionaries resides in the own terminal. When the MS is going to access a new cell, the BS of the incoming cell transmits the cell profile (cell dictionary) to the MS. Since the user dictionary is stored in the MS, and the cell dictionary has already been recovered, the MS can perform the calculus of its trajectory itself. So, the MS stores the dictionaries (user and cell profiles), measures the signal power in order to obtain the fuzzy symbol, and calculates the similarity between the string containing the path followed by the MS and the possible paths contained in the hybrid dictionary. Each time that the MS calculates the fuzzy symbol $\tilde{\alpha} = \{S_1\ S_2\ S_3\ S_4\ S_5\ S_6\ S_7\}$ by measuring the pilot signal power for each one of the adjacent cells, it obtains the existing similarity between the string of fuzzy symbols built by concatenation of the observed symbols and the pattern strings contained in the dictionary. Signal triangulation allows the MS to estimate the proximity degree for each cell (S_i, $\forall\ i = 1, 2\ldots 7$). Since the proximity degree is certainly a fuzzy concept, a fuzzy automaton [3] is proposed to deal with the required imperfect string of fuzzy symbols $\tilde{\alpha}$ comparison.

The finite fuzzy automata with empty string transitions AFF_ε used is defined by the tuple $(Q, \Sigma, \mu, \mu_\varepsilon, \sigma, \eta)$, where:

Q : is a finite and non-empty set of states.

Σ : is a finite and non-empty set of symbols.

μ : is a ternary fuzzy relationship over $(Q \times Q \times \Sigma)$; $\mu : Q \times Q \times \Sigma \to [0, 1]$. The value $\mu(q, p, x) \in [0, 1]$ determines the transition degree from state q to state p by the symbol x.

μ_ε : is a ternary fuzzy relationship over $Q \times Q$; $\mu_\varepsilon : Q \times Q \to [0, 1]$. The value $\mu_\varepsilon(q, p) \in [0, 1]$ determines the transition degree from state q to state p without spending any input symbol.

σ : is the initial set of fuzzy states: $\sigma \in \mathcal{F}(Q)$; $\sigma : Q \to [0, 1]$.

η : is the final set of fuzzy states: $\eta \in \mathcal{F}(Q)$; $\eta : Q \to [0, 1]$.

The AFF_ε operation over a string $\alpha \in \Sigma^*$, is defined by $(AFF_\varepsilon, T, \hat{\mu}, \hat{\mu}_\varepsilon, \mu^*)$, where:

(i) $AFF_\varepsilon \equiv (Q, \Sigma, \mu, \mu_\varepsilon, \sigma, \eta)$ is finite fuzzy automaton with transitions by empty string.

(ii) T is a t-norm $T : [0, 1]^2 \to [0, 1]$.

(iii) $\hat{\mu} : \mathcal{F}(Q) \times \Sigma \to \mathcal{F}(Q)$ is the fuzzy state transition function. Given a fuzzy state $\tilde{Q} \in \mathcal{F}(Q)$ and a symbol $x \in \Sigma$, $\hat{\mu}(\tilde{Q}, x)$ represents the next reachable fuzzy state. $\hat{\mu}(\tilde{Q}, x) = \tilde{Q} \circ_T \mu[x]$ where $\mu[x]$ is the fuzzy binary relation over Q obtained from μ by the projection over the value $x \in \Sigma$. Then, $\forall\ p \in Q : \hat{\mu}(\tilde{Q}, x)(p) = max_{\forall q \in Q}\{\mu_{\tilde{Q}}(q) \otimes^T \mu(q, p, x)\}$.

(iv) $\hat{\mu}_\varepsilon : \mathcal{F}(Q) \to \mathcal{F}(Q)$, is the fuzzy states transition by empty string. Given a fuzzy state $\tilde{Q} \in \mathcal{F}(Q)$, $\hat{\mu}_\varepsilon(\tilde{Q})$ represents next reached fuzzy state without consuming an input symbol. $\hat{\mu}_\varepsilon(\tilde{Q}) = \tilde{Q} \circ_T \hat{\mu}_\varepsilon^T$ where $\hat{\mu}_\varepsilon^T$ is the T-transitive

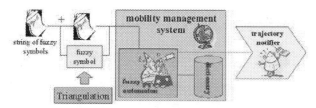

Fig. 3. Mobility management system architecture.

closure of the fuzzy binary relationship $\mu_\varepsilon \circ_T \hat{\mu}_\varepsilon^T = \mu_\varepsilon^{(n-1)T}$ if Q has cardinality n. So, $\forall\, p \in Q : \hat{\mu}_\varepsilon(\tilde{Q})(p) = max_{\forall q \in Q}\{\mu_{\tilde{Q}}(q) \otimes^T \hat{\mu}_\varepsilon^T(q,p)\}$.

(v) $\mu^* : \mathcal{F}(Q) \times \Sigma^* \to \mathcal{F}(Q)$, is the main transition function for a given string $\alpha \in \Sigma^*$ and it is defined by:

   a) $\mu^*(\tilde{Q}, \varepsilon) = \hat{\mu}_\varepsilon(\tilde{Q}) = \tilde{Q} \circ_T \hat{\mu}_\varepsilon^T,\ \forall\, \tilde{Q} \in \mathcal{F}(\mathcal{Q})$.

   b) $\mu^*(\tilde{Q}, \alpha x) = \hat{\mu}_\varepsilon(\hat{\mu}(\mu^*(\tilde{Q}, \alpha), x)) - (\mu^*(\tilde{Q}, \alpha) \circ_T \mu|x|) \circ_T \hat{\mu}_\varepsilon^T,\ \forall\, \alpha \in \Sigma^*,$
   $\forall\, x \in \Sigma,\ \forall\, \tilde{Q} \in \mathcal{F}(\mathcal{Q})$.

As figure 3 shows, the fuzzy automaton is used to compare each string of fuzzy symbols with all the known paths for the current cell stored in the dictionary. The automaton provides as output the ownership degree of the string of fuzzy symbols to the paths contained in the dictionary. $\frac{D_1}{D_{2...k}} > 10^{dd}$ where k is the number of strings (paths) contained in the dictionary, and dd is the decision degree. dd can be selected in order to increase or reduce the recognition rate, taken into account that a low value for dd can increase the number of false recognitions. A false recognition is the fact to decide that a MS follows a certain path when it does not follow this path.

Each 480 milliseconds, the MS makes the triangulation and obtains the new fuzzy symbol that is concatenated to the prefix representing the followed path. As soon as the MS identifies the movement pattern followed, it notifies the predicted trajectory to the BS. Then, the local BS contacts the BS placed in the destination cell and negotiates the allocation of the resources needed to grant the service in the handoff process. Such as, BS can manage adequately available resources in order to grant QoS. In the same way, once known the path followed, the MS can receive information about some interesting places located in its trajectory. It can be seen as a way of sending/receiving selective and dynamic publicity or/and advertisements or location services. The detection of the trajectory followed does not stop the continuous calculus performed by the MS, because the user can decide to change its movement, and then, prediction must be reformulated. However, the MS only contacts the BS when prediction is performed. Absence of information makes the BS suppose that no prediction has obtained; or if a previous one was formulated, that it has not been modified. Then, the number of signaling messages is considerably reduced.

4   QoS Guarantee

QoS in cellular communication systems can be measured in terms of blocking and interruption probabilities, P_b and P_i respectively. In order to ensure the correct transition of a MS among cells, without losses and communication interruption, the system must provide the demanded resources at the destination cell. This section is devoted to analyze the influence of the trajectory prediction in P_b and P_i, an then, in the QoS guarantee.

The handoff process in circuit switched networks requires the allocation of available communication channels at the destination cell. When all the resources of this cell are busy, the communication will be interrupted. In order to avoid this undesirable situation, the system can allocate a certain number of handoff channels. Those channels ensure the communication continuity. In this sense, some experiments concerning the number of allocation channel of the destination cell and the resource allocation strategy are presented.

Three different strategies referring the number of handoff channels are considered: ensure a minimum number of them, limit the maximum number of them and no reservation of them. Figures 4 and 5 illustrate respectively the blocking and interruption probabilities for the minimum granted strategy. Some different calls per second rates (cps) are considered for the incoming calls. The total number of communication channels considered is 20, and 8 of them are used as handoff channels. When using the minimum granted strategy, as minimum 8 of the 20 channels are reserved for the handoff process, and can be reserved by the origin cell when the prediction system estimates the arrival of the MS to the destination cell. When using the maximum strategy, no more than 8 channels are reserved for the handoff process but a lower number of channels can be also considered. And when no reservation is made, the allocation can be performed only if any channel is available at the allocation instant. Results obtained shows that values obtained for the three strategies only differ in a 1%.

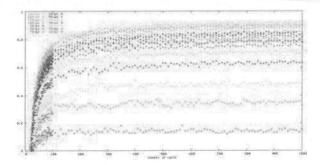

Fig. 4. Blocking probability for the minimum granted strategy.

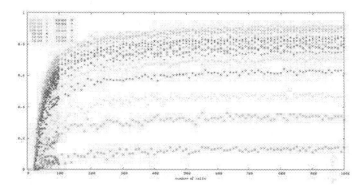

Fig. 5. Interruption probability for the minimum granted strategy.

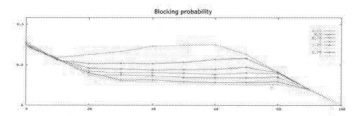

Fig. 6. Blocking probability (up) and interruption probability (down).

The average duration of the calls is 180 seconds, having an average number of handoffs per call of 3. Considering figures 4 and 5, the number of calls considered is 300. Results obtained when the incoming call rate is 20 cps, are presented in figure 6. In the same way, figure 8 shows the results obtained when the incoming call rate considered is 80 cps. The rate indicated in both figures represents the relationship between P_b and P_i, and the X axis indicates the handoff rate in %. Using the prediction system, we can select the desired values of P_i and P_b (the operation point) for a given allocation strategy. So, QoS can be improved since the system can know in advance the path followed by each MS in a cell, and then, taken into account the QoS selected for each user, perform the resource allocations (handoff channels) at the destination cell.

In [9] QoS is measured as $QoS = P_b + \alpha P_i$, although $QoS = \alpha P_b + (1-\alpha) P_i$ is also considered in the literature. In both cases, α allows to modulate the influence of P_i and P_b. Selecting the value of α we obtain the QoS value, and then, the

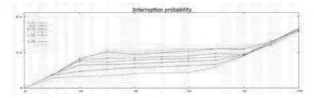

Fig. 7. Interruption probability.

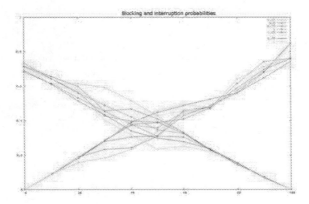

Fig. 8. Blocking and interruption probabilities.

instant when the system must allocate the resources at the destination cell. The prediction system allows an intelligent allocation of resources (handoff channels) that ensures a certain QoS that can be selected by the system for each user.

5   Conclusions

Mobility management increases benefits of any mobile environment. The mobility model used is a 2D, intra-cell microscopic model which can be extended to inter-cell movements and the user can fix its degree of randomness considering the constitutional characteristics.

We propose a fuzzy automaton as a trajectory predictor in order to anticipate the movement of a mobile station in a cellular system. The fuzzy automaton allows to manage users mobility in cellular communication systems improving their properties (QoS), and growing the number and quality of services offered by them. The management of imperfect or imprecise location (fuzzy) information obtained in the power triangulation process due to fading, signal attenuation and multi-path propagation as fuzzy symbols, makes possible to perform a good location prediction. Fuzzy techniques deals with the existing uncertainties very well, being a really good alternative to other classical statistical techniques.

Some studies concerning the incoming call rates are performed in order to evaluate the relationship existing between blocking and interruption probabilities when different channel allocation strategies are considered. Results obtained show that the system can reserve a certain number of handoff channels in order to ensure QoS without decreasing the occupation and availability of the system.

References

1. J. J. Astrain, J. R. Garitagoitia, J. Villadangos, F. Fariña, A. Córdoba and J. R. González de Mendívil, *An Imperfect String matching Experience Using Deformed Fuzzy Automata*, Frontiers in Artificial Intelligence and Applications, Soft Computing Systems, vol. 87, pp 115-123, IOS Press, The Nederlands, Sept. 2002.
2. C. Bettsetter, "Mobility modeling in wireless networks: categorization, smooth movement, and border effects", *ACM Mobile Computing and Communications Review*, vol. 5, no. 3, pp. 55-67, July 2001.
3. J. R. Garitagoitia, J. R. González de Mendívil, J. Echanobe, J. J. Astrain, and F. Fariña, "Deformed Fuzzy Automata for Correcting Imperfect Strings of Fuzzy Symbols", *IEEE Transactions on Fuzzy Systems*, vol. 11, no. 3, pp. 299-310, 2003.
4. S. Hadjefthymiades and L. Merakos, "ESW4: Enhanced Scheme for WWW computing in Wireless communication environments", *ACM SIGCOMM Computer Communication Review*, vol. 29, no. 4, 1999.
5. S. Hadjefthymiades, S. Papayiannis and L. Merakos, "Using path prediction to improve TCP performace in wireless/mobile communications", *IEEE Communications Magazine*, vol. 40, no. 8, pp. 54-61, Aug. 2002.
6. V. I. Levenshtein, *Binary codes capable of correcting deletions, insertions, and reversals*, Sov. Phys. Dokl., vol. 10, no. 8, pp. 707-710, 1966.
7. G. Liu and G. Q. Maguire Jr., "A Predictive Mobility Management Scheme for Supporting Wireless Mobile Computing", *Technical Report*, ITR 95-04, Royal Institute of Technology, Sweden, Jan. 1995.
8. T. Liu, P. Bahl and I. Chlamtac, "Mobility Modeling, Location Tracking, and Trajectory Prediction in Wireless Networks", *IEEE JSAC*, vol. 16, no. 6, pp. 922-936, Aug. 1998.
9. M. Oliver i Riera, *Mecanismos de gestion de recursos basados en criterios de calidad de servicio para una red de comunicaciones moviles de tercera generacion*, Doctoral thesis, Universitat Politecnica de Catalunya, Barcelona (Spain), Dec. 1998.

Mobility Prediction Handover Using User Mobility Pattern and Guard Channel Assignment Scheme

Jae-il Jung[1], Jaeyeol Kim[2], and Younggap You[2]

[1] School of Electrical and Computer Engineering Hanyang University,
17 Haengdang-dong, Sungdong-gu, Seoul, 133-791, Korea
jijung@hanyang.ac.kr
[2] School of Electrical and Computer Engineering, Chungbuk Nat'l University, San
48, Gaesin-dong, Heungdug-ku, Cheongju City, Chungchungbuk-do, Korea

Abstract. In the next generation of mobile communication networks, the radius of the cell will become smaller, causing more frequent handovers and disconnection of existing handover calls if the channel capacity for handover is insufficient. This paper proposes a mobility prediction handover scheme to prevent these phenomena. If the next cell into which a mobile user will move could be predicted, then the necessary bandwidth could be reserved to maintain the user's connection. The proposed scheme predicts the next cell to which a mobile user will move, based on the user's mobility pattern history and current mobility trace. In addition, this paper suggests a Mobility Prediction Guard Channel Reservation Scheme (MPGCRS) that adjusts guard channel size in accordance with the handover prediction ratio.

1 Introduction

Future mobile communication systems will be required to support broadband multimedia services with diverse quality of service (QoS) requirements [1]. As a mobile user may change cells a number of times during the lifetime of a connection, the avail-ability of wireless network resources cannot be guaranteed. Thus, users may experience performance degradations due to movement. This problem will be amplified in future micro/pico-cellular networks, where handover events may occur more frequently than in today's macro-cellular networks [2]. From a user's perspective, a forced termination in the middle of a call due to handover failure is far more annoying than the blockage of a new call. Therefore, the service availability of handover calls must be guaranteed at the same time that high utilization of wireless channels is permitted.

If the next cell that a mobile user will enter could be predicted, the necessary bandwidth could be reserved to maintain his connection. To this end, this paper proposes a mobility prediction handover scheme that considers the fact that most users have a constant movement pattern. This scheme predicts the next cell to which a mobile user will move, based on the user's mobility pattern history as

M. Freire et al. (Eds.): ECUMN 2004, LNCS 3262, pp. 155–164, 2004.
© Springer-Verlag Berlin Heidelberg 2004

well as the user's current mobility trace. The mobile terminal captures the user's current movement and stores the movement pattern in an autonomous manner.

There have been some research efforts to predict user mobility. In [3], the next cell to which a mobile user will move is predicted by movement probability, which is the mobile user's visited frequency. In [4], the future movement of a mobile user is predicted based on the mobile user's recent and past movement history as well as his movement pattern. In [5], the mobility prediction scheme is performed based on a detailed history of the mobile user's movement pattern. A general problem with using movement patterns is that a heavy and complex database is required. Therefore, the first part of this paper proposes not only an efficient method of managing a movement pattern history database, but also a method of mobility prediction that is simpler and more accurate than existing schemes.

The second part of this paper suggests a channel assignment scheme called Mobility Prediction Guard Channel Reservation Scheme (MPGCRS), which can be adapted to the above mobility prediction algorithm. Existing channel assignment schemes do not allow for a mobility-predicted handover. Therefore, the proposed scheme modifies existing channel assignment schemes in accordance with the handover prediction ratio.

The rest of the paper is organized as follows. Section 2 presents our mobility prediction handover scheme using a user's mobility pattern. Section 3 describes an efficient channel assignment scheme for mobility prediction handover. Section 4 shows some results of performance evaluation and Section 5 gives some conclusions.

2   Mobility Prediction Handover Using User Mobility Pattern

2.1   Data Structures

Generally, mobile users maintain constant movement patterns on a daily, weekly or monthly basis. In our scheme, the data structure for the user's mobility history is divided into the CMC (Current Movement Cache) and MPH (Movement Pattern His-tory). The CMC is a cache that traces the mobile user's current location, and the MPH is a database that stores and manages the user's mobility pattern history. These data structures are managed by the mobile terminal itself in order to reduce the network signaling overhead for mobility prediction. In general, a mobile terminal's location information is managed by a VLR (Visitor Location Register) and HLR (Home Location Register) [6,7,8]. Implementing CMC and MPH in VLR and HLR, the networks including the serving and surrounding base stations, requires a relatively heavy signaling procedure to retrieve and store data. The VLR and HLR would store and manage these data structures on a large scale as well. To avoid these problems, these data structures are managed by the mobile terminal itself.

2.1.1   CMC (Current Movement Cache). Table 1 describes the CMC fields. The Cell ID shows the cell location number, the enter time field indicates the time at which the user enters the cell, and the dwell time field shows the time a user stays in the cell before moving to another cell.

Table 1. CMC (Current Movement Cache)

Cell ID	Enter Time	Dwell Time (min.)
1	9:00 p.m.	605
2	7:05 a.m.	3
8	7:08 a.m.	4

When a mobile user enters a new cell, the CMC_generation function executes. If the dwell time is greater than the fixed period of time T, the cell should be the starting or ending cell of one movement pattern. If there is no entry in the CMC, the current cell could be the starting cell. Thus, the storage of a new CMC is begun. Otherwise, the cell may be an ending cell. The CMC storing operation is complete, and the MPH_generation function executes. If the dwell time is not greater than T, the cell could be a moving cell; in this instance, one CMC entry is added to the CMC.

2.1.2   MPH (Movement Pattern History). The MPH (Movement Pattern History) stores and manages a user's mobility pattern history. The fields of the MPH shown in Table 2 are based on the CMC. The day of week field represents weekday. The time field is classified into morning, afternoon, and evening. In the path sequence field, a path sequence consists of cell number, average dwell time, minimum dwell time, where the average dwell time is the average time of staying in the cell, and the minimum dwell time is the minimum time. We define a fixed time duration, T. Thus, an integer k in the dwell time fields means that the staying time in the cell is slightly longer than $k \cdot T$.

Table 2. MPH (Movement Pattern History)

Day of week	Time	Path sequence = (cell number, average dwell time T_{ave_dwell}, minimum dwell time T_{min_dwell})	Count
M,T, W,T, F	morning	(1,T180,T60),(2,T3,T2),(8,T4,T2),(21,T2,T2),(20,T120,T80) (home→institute, health club)	20
M,T W,T, F	morning	(institute, health club→school, company)	20
M,T,F	morning	(home→school, company)	4

The count field manages the core or random pattern. When a new MPH entry is generated, the new MPH entry's count field is set to one. When the MPH entry is referenced, this count is increased by one. On the other hand, the count value is periodically (e.g., once a month) decreased by one. This field indicates the frequency of visits and describes the user's core pattern.

Once a CMC operation is complete, the MPH_generation function executes. First, we compare a new CMC's path sequence field with all MPH entries. If there is a matched entry, the count value is increased by one. If there is no matched entry, a new MPH entry is generated.

2.2 Mobility Prediction Handover Algorithm

The determination of a matched path is the first step in the prediction algorithm. Here, the current cell numbers stored in the CMC are compared with those of path sequences in each MPH entry. We search the matched path for the following information, listed in order of priority:
1. day of week + time + path sequence
2. day of week + path sequence
3. time + path sequence
4. path sequence.

When the mobile terminal begins to enter a new cell, the path-matching function executes. If there is a matched path, we then determine whether the current cell is a starting, ending, or moving cell. If there is no matched path, a normal (unpredicted) handover executes.

If the current cell is a starting or moving cell, we expect that the mobile terminal will move into the next cell indicated in the MPH's path sequence immediately or after time $k \cdot T$, respectively. Thus, the handover reservation algorithm executes. If the current cell is an ending cell, we terminate a series of reservations for a specific hand-over call. The determination of T and the signaling procedure for handover reservation will be described in further studies.

3   Mobility Prediction Guard Channel Reservation Scheme (MPGCRS)

3.1   Guard Channel Reservation Scheme (GCS)

GCS gives higher priority to handover calls than to new calls. In GCS, guard channels are exclusively reserved for handover calls, and the remaining channels (called normal channels) can be shared equally by handover and new calls (see Figure 1).

Thus, whenever the channel occupancy exceeds a certain threshold T, GCS rejects new calls until channel occupancy goes below the threshold. In contrast, handover calls are accepted until channel occupancy exceeds the total number of channels in a cell. This offers a generic means of decreasing the probability that handover calls will drop, but it causes reduction in the total amount of

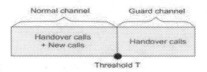

Fig. 1. Channel allocation of GCS

traffic carried. The reason why total carried traffic is reduced is that few channels (except the guard channels) are granted to new calls. The demerits of this system become more serious when handover requests are rare. In the end, this situation may bring about inefficient spectrum utilization and increased blocking probability, as only a few handover calls are able to use the reserved channels exclusively. The use of guard channels requires careful determination of the optimum number of channels, knowledge of the traffic pattern of relevant areas, and estimation of the channel occupancy time distribution [9-11].

3.2 Mobility Prediction Guard Channel Reservation Scheme (MPGCRS)

If we adapt mobility prediction handover to GCS, channel reservation for the predicted handover can be made on a normal channel and a guard channel. In this paper, we propose the MPGCRS as an efficient channel assignment scheme for mobility prediction handover. Our MPGCRS adjusts the size of the guard channel in accordance with the prediction ratio of handover calls. Figure 2 shows the channel allocation of MPGCRS.

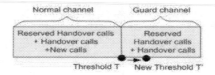

Fig. 2. Channel allocation of MPGCRS

Eq. 1 shows a formula to calculate the modified threshold in the MPGCRS

$$T' = T + (C - T) \cdot \beta \cdot \text{k} \tag{1}$$

where T is the guard channel threshold in GCS and C is the number of total channels in the cell. T' is the modified guard channel threshold. T' can move between T and C according to the handover prediction ratio β. β ($0 < \beta < 1$) is defined as the ratio of the predicted handover call arrival rate to the total

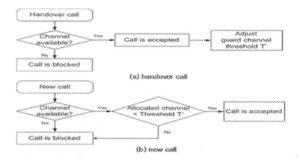

Fig. 3. Call processing flow diagram for MPGCRS

handover call arrival rate. k is a safety factor that properly adjusts modified threshold T'. In Section 4, k will be tuned by computer simulation to achieve a better balance between guaranteeing handover dropping probability and maximizing resource utilization.

Figure 3 shows the call processing flow diagram for new and handover calls in MPGCRS. In the case of handover calls, if there are available channels, the call is accepted, and the guard channel threshold is readjusted by the handover prediction ratio. In the case of new calls, MPGCRS determines whether to block a new call according to T'.

4 Performance Evaluation

We compare the performances of the proposed MPGCRS with the existing scheme by computer simulation. We assume that the number of mobile users is much larger than the total number of channels in the base station so that call arrivals may approximate to a Poisson process [10,12].

The input parameters in our simulation are the following:

- λ : call arrival rate. Call arrives according to a Poisson process of rate λ.

- μ : mean call completion rate. Call holding time is assumed to be exponentially distributed with a mean of $1/\mu$.

- η : portable mobility. User residual time is assumed to be exponentially distributed with a mean of $1/\eta$.

In our simulation, the mean call holding time $1/\mu$ is 6 min, and the mean user residual time $1/\eta$ is 3 min. In addition, we assume that the total number of channels in a base station is 60, and that all wireless channels have the same fixed capacity [9]. We start from the mean call arrival time of 3.2 sec (low call load) and gradually decrease the mean call arrival time by 0.05 sec. The total simulation time is 7,200 sec.

The performance metrics of the proposed MPGCRS are as follows:

- P_b: blocking probability for a new call
- P_d: dropping probability for a handover call
- U_c: channel utilization of base station.

We compare MPGCRS with GCS in terms of dropping probability (P_d), blocking probability (P_b), and channel utilization (U_c). We assume that handover call arrivals are 50% of total call arrivals, and that predicted handover call arrivals are 25% and 75% of total handover call arrivals. In the MPGCRS, we analyze performance metrics with varying safety factors, k (0 <k < 1).

4.1   Dropping Probability of Handover Calls

Figure 4 (a) shows the dropping probability (P_d) of handover calls when the handover prediction rate is 25%. On average, the P_d of MPGCRS represents a 20% reduction of the P_d of GCS. Due to the readjustment (i.e., the increment) of T ', the guard channels are reserved preferentially for more handover calls. In other words, the decreased number of new calls in the guard channel allows more predicted handover calls to be reserved. When the handover prediction rate is 75%, the P_d of MPGCRS is 40% less than that of GCS. Figure 4 (b) shows the effect of the safety factor, k, on the P_d in the MPGCRS. As k decreases, the P_d in MPGCRS decreases. This relationship is attributable to the fact that more channels in the guard channel can be reserved for predicted handover calls.

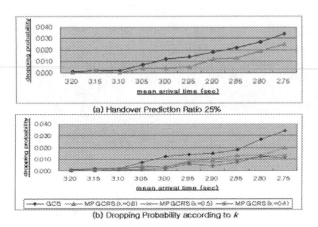

(a) Handover Prediction Ratio 25%

(b) Dropping Probability according to k

Fig. 4. Dropping Probability of Handover Calls -P_d

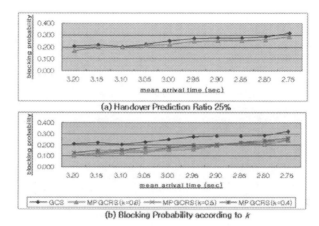

Fig. 5. Blocking Probability of New Calls - P_b

4.2   Blocking Probability of New Calls

Figure 5 (a) shows the blocking probability (P_b) of new calls. When the handover prediction rate is 25%, the P_b of the MPGCRS is an average of 10% less than that of the GCS. When the handover prediction rate is 75%, the P_b of MPGCRS is approximately 33% less than that of the GCS. In the both cases, the P_b of the MPGCRS is much lower than that of the GCS. The effect becomes greater as the handover prediction ratio increases. This is because MPGCRS decreases the guard channel size for handover calls only, which then increases the normal channel for new calls.

Figure 5 (b) shows P_b according to the safety factor, k, in the MPGCRS, where k is the value that properly adjusts modified threshold T ′. As k increases, the P_b of the MPGCRS decreases. A larger k means that the smaller guard channel is used only for handover calls while the larger, normal channel is used for new calls and handover calls.

4.3   Channel Utilization

Figure 6 (a) shows channel utilization (U_c). When the handover prediction rate is 25%, U_c of MPGCRS is 0.8% higher than that of GCS. When the handover prediction rate is 75%, U_c of MPGCRS is 2.0% higher than that of GCS. The effect becomes greater as the handover prediction ratio increases. These results show that U_c is maximized as the guard channel for handover calls is decreased by the handover prediction ratio in the MPGCRS.

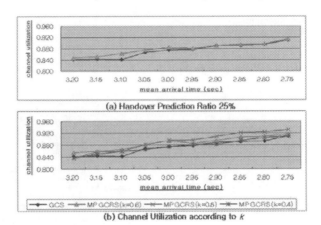

Fig. 6. Channel Utilization - U_c

Figure 6 (b) shows U_c according to the safety factor, k, in the MPGCRS. When the k is 0.5, U_c is maximized.

5   Conclusion

In this paper, we have proposed a mobility prediction handover algorithm using the user's mobility pattern and effective channel assignment schemes for mobility prediction. The proposed mobility prediction handover algorithm predicts the next cell a user will enter. Here, the channel for a handover call is reserved in a predicted cell. Therefore, it is possible to maintain the QoS of the handover call, even when handovers occur frequently, as in future micro/pico-cellular networks.

We suggest the MPGCRS for a mobility prediction handover algorithm. In the proposed scheme, we adjust the guard channel size in accordance with the handover prediction ratio. As the predicted handover call is pre-assigned with a channel in the mobility prediction handover, the dropping probability of the handover call can be decreased far below the required level, while the blocking probability of a new call can be decreased and the overall channel utilization can be increased. The computer simulation results show that the effect improves as the call load and handover prediction ratio increase.

Future work should refine our mobility prediction scheme, in terms of signaling protocol and proper parameter values. A refined scheme should also consider the channel assignment scheme concerning the QoS requirements of various handover calls: real-time calls, non-real-time calls, etc.

164   J.-i. Jung, J. Kim, and Y. You

Acknowledgements. This work was supported in part by the Center of Innovative Design Optimization Technology (ERC of Korea Science and Engineering Foundation) and by the ITRC program of the Ministry of Information and Communication, Korea.

References

1. M. Zeng, A. Annamalai, and Vijay K. Bhargava, "Recent Advances in Cellular Wireless Communications", IEEE Communications Magazine, vol. 37, No. 9, pp. 128-138, September 1999.
2. W.C.Y. Lee, "Smaller cells for greater performance," Communications Magazine, vol. 29, pp. 19-23, November 1991.
3. Hoon-ki Kim, Jae-il Jung, "A Mobility Prediction Handover Algorithm for Effective Channel Assignment in Wireless ATM," IEEE GLOBECOM, vol 6, pp. 3673-3680, November 2001.
4. Ing-Ray Chen, Naresh Verma, "Simulation Study of a Class of Autonomous Host-Centric Mobility Prediction Algorithms for Wireless Cellular and Ad Hoc Networks," IEEE COMPUTER SOCIETY, March 2003.
5. Fei Yu, Victor Leung, "Mobility-based predictive call admission control and bandwidth reservation in wireless cellular networks," Computer Networks, vol. 38, pp. 577-589, 2002.
6. Clinet Smith, Daniel Collins, "3G Wireless Networks," McGraw-Hill.
7. Abbas Jamalipour, "The Wireless Mobile Internet," Willy.
8. Yuguang Fang, "General Modeling and Performance Analysis for Location Management in Wireless Mobile Networks," IEEE Transactions on computers, vol. 51, No. 10, pp. 1169-1180, Oct. 2002.
9. Daehyoung Hong, Stephen S. Rappaport, "Traffic Model and Performance Analysis for Cellular Mobile Radio Telephone Systems with Prioritized and Non-Prioritized Handoff Procedures," IEEE Trans. Vehic. Tech., vol. VT-35, no. 3, pp. 77-92, Aug. 1986.
10. S. Tekinay and B. Jabbari, "A Measurement-Based Prioritization Scheme for Handovers in Mobile Cellular Networks," IEEE JSAC, pp. 1343-50, vol. 10, no. 8, October 1992.
11. Yi-Bing Lin, Seshdri Mohan, Anthony Noerpel, "PCS Channel Assignment Strategies for Handoff and Initial Access," IEEE Personal Communications, pp. 47-56, Third Quarter, 1994.
12. Young Chon Kim, Dong Eun Lee, Bong Ju Lee, Young Sun Kim, "Dynamic Channel Reservation Based on Mobility in Wireless ATM Networks," IEEE Communication Magazine, pp. 47-51, November, 1999.

Lost Retransmission Detection for TCP Part 1: TCP Reno and NewReno

Beomjoon Kim, Yong-Hoon Choi, and Jaesung Park

LG Electronics Inc., LG R&D Complex, 533, Hogye-1dong, Dongan-gu, Anyang-shi,
Kyongki-do, 431-749, Korea, {beom,dearyonghoon,4better}@lge.com

Abstract. As a well-known issue, the performance of transmission control protocol (TCP) is affected by its loss recovery mechanism working based on two algorithms of fast retransmit and fast recovery. In particular, the fast recovery algorithm has been modified for improvement and become a basis on which each TCP implementation is differentiated such as TCP Tahoe, Reno, NewReno, and selective acknowledgement (SACK) option. Despite the recent improvement achieved in TCP NewReno and SACK option, TCP still has a problem that it cannot avoid retransmission timeout (RTO) when a retransmitted packet is lost. Therefore, in this paper, we introduce a simple algorithm for detecting a lost retransmission that can be applied to TCP implementations that do not use SACK option such as TCP Reno and NewReno. Using *ns* simulations, we show the microscopic behaviors of the lost retransmission detection algorithm when it works with each TCP. The numerical results based on existing analytic models are also shown.

1 Introduction

So far, fast recovery [1], [2] has been modified continuously for better performance of transmission control protocol (TCP). The problem for multiple packet losses of the original fast recovery algorithm [3] was improved in later TCP implementations such as TCP NewReno [4] and selective acknowledgement (SACK) option [5]. Limited Transmit [7] was proposed to avoid the lack of duplicate acknowledgement (ACK) and subsequent retransmission timeout (RTO).

In recent, a simple algorithm has been proposed to make it possible for TCP sender to detect a lost retransmission that can be applied to each TCP implementation [7], [8]. However, Duplicate Acknowledgement Counting (DAC) presented in [7] is not evaluated from the aspect of the additive-increase-multiplicative-decrease (AIMD) [1] principle to keep the 'conservation of packets' [2] rule of conventional TCP congestion control. Therefore, DAC is revisited in this paper in order to show its microscopic behavior as in [8], [9].

The remainder of this paper is organized as follows. Section 2 provides a brief presentation of DAC and its operations obtained from simulations for two scenarios. In Section 3 we derive the loss recovery probability of each TCP considered in this paper. Section 4 contains the numerical results and their discussion. Finally, some conclusions are summarized in Section 5.

M. Freire et al. (Eds.): ECUMN 2004, LNCS 3262, pp. 165–174, 2004.
© Springer-Verlag Berlin Heidelberg 2004

2   Duplicate Acknowledgement Counting

DAC detects a lost retransmission on the basis of the number of duplicate ACKs. In the rest of this paper, each TCP using DAC is indicated by a plus sign such as TCP Reno+ and NewReno+.

2.1   Description

The principle of DAC is very simple: if a packet is retransmitted successfully, a packet transmitted after the retransmitted packet would not deliver a duplicate ACK for it [7].

Two variables are maintained in TCP sender for DAC. One variable is to measure the expected number of duplicate ACKs for each retransmitted packet, and denoted by D_i for the ith retransmitted packet in a window. If more than D_i duplicate ACKs are received, the ith retransmitted packet can be decided to be lost.[1] When a sender performs fast retransmit, it is not aware of how many packets are lost in a window but only knows that at least one packet is lost. Therefore, the sender needs to store the size of 'congestion window' (cwnd) [1] just before fast retransmit in another variable, which is denoted by W_D. Then, D_1 is always equal to $W_D - 1$.

DAC can also be applied for multiple packet losses. The only difference is that D_j for $j \geq 2$ is determined when a partial ACK for the jth lost packet is received. In this case, D_j is equal to the number of new packets transmitted after the $(j - 1)$th packet loss.

2.2   Consistency with Conventional TCP Congestion Control

From the aspect of respecting the 'conservation of packets' rule [2], DAC is perfectly consistent with AIMD scheme specified in [1]. As described above, a lost retransmission is detected based on the packets transmitted after the retransmission. It means that, if congestion is so heavy that no packets should be transmitted, TCP sender to which DAC is applied does not transmit additive packets by detecting a lost retransmission. It can be justified by the fact that, in such a congested situation, the packets after the retransmission would be likely to be lost as well. However, since a lost retransmission detected means that there was a serious congestion, the proposed algorithm decreases cwnd and 'slow-start-threshold' (ssthresh) twice after recovery of a lost retransmission [1].

2.3   Simulations

We implemented DAC using ns simulations. In the following two scenarios, several specific packets are forced to be dropped as in [8], [9].

When a packet and its retransmission are lost, the loss recovery behavior of TCP NewReno is shown in Fig. 1. At about 0.7 second, packets 7–14 are

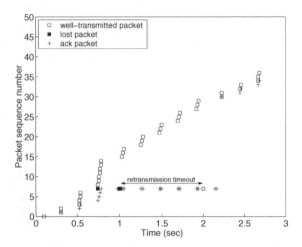

Fig. 1. Behaviors of TCP NewReno for a single packet loss and lost retransmission during fast recovery (RTT=100msec).

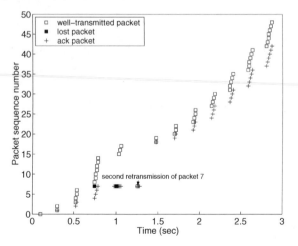

Fig. 2. Behaviors of TCP NewReno+ for a single packet loss and lost retransmission during fast recovery (RTT=100msec).

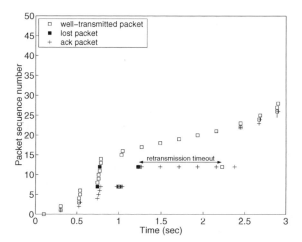

Fig. 3. Behaviors of TCP NewReno for two packet losses and lost retransmission during fast recovery (RTT=100msec).

transmitted with cwnd of 8 and packet 7 is dropped. When the sender receives three duplicate ACKs at about 0.9 second, the sender retransmits packet 7, and it is lost again. The last duplicate ACK for packet 7 inflates the usable window [9] to $11(= \lfloor 8/2 \rfloor + 7)$ so that newly included packets 15–17 are transmitted. Because the sender cannot perceive that the retransmitted packet 7 has been lost, it transmits three new packets per every round-trip time (RTT) till eventual RTO occurs. It can be seen that the sender restarts in slow start at about 2 second.

Using DAC shown in Fig. 2, the sender sets D_1 to $7(= 8 - 1)$ when it retransmits packet 7. When the sender receives the eighth duplicate ACK corresponding to the receiver's receipt of packet 15 at about 1.3 second, since it receives more duplicate ACKs than D_1, it can be informed that the retransmission of packet 7 is lost again. It can be seen that the sender transmits the second retransmission of packet 7 at about 1.3 second. After the retransmission, the sender halves cwnd and ssthresh again to be $2(= \lfloor 4/2 \rfloor)$. Two more duplicate ACKs by packet 16 and 17 inflate the window to be 4, but no packet can be transmitted because all packets in the window are outstanding. At about 1.5 second, an ACK that acknowledges packets up to packet 17 brings the sender out of fast recovery and congestion avoidance starts with cwnd of 2.

---

[1] Note that *delayed acknowledgement* [15] does not affect the operation of DAC because the receiver generates an ACK immediately if it is duplicate [1].

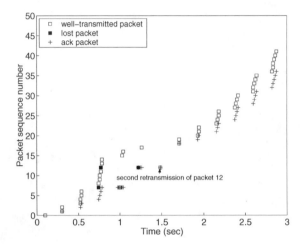

Fig. 4. Behaviors of TCP NewReno+ for two packet losses and lost retransmission during fast recovery (RTT=100msec).

We perform the same simulations for two packet losses and the results are shown in Figs. 3–4. In these simulations, packet 7 and 12 are lost and the retransmission of packet 12 is lost again. TCP NewReno's behavior shown in Fig. 3 can be explained in the same way as Fig. 1: RTO occurs at about 2.3 second.

For using DAC, when the sender receives a partial ACK for packet 12, D_1 set to 7 is cleared because the partial ACK assures that the retransmission of packet 7 has been well received. Note that only 6 duplicate ACKs have been received for packet 7 at this time. After retransmitting packet 12, the sender sets $D_2 = 2$ since two packets 15–16 have been transmitted after the retransmission of packet 7. Because packet 17 delivers the third duplicate ACK, the sender has received more duplicate ACKs than expected so that it transmits packet 12 again at about 1.5 second. When an ACK that acknowledges up to packet 17 is received, congestion avoidance follows with cwnd of 2.

3   Probabilistic Analysis

Similarly as the approach presented in [7], [8], the benefit of DAC when it is applied to TCP Reno can be derived for random packet loss p as following way. We mainly follow the procedures presented in [10], [11] under the same assumptions such as fixed packet size, no ACK loss, and infinite packet transmission.

We also follow several notations such as W_{max} for receiver's advertised window and K for ssthresh.

If we denote the recovery probability of TCP Reno by $R_R(u)$ for a window of u packets,[2] the recovery probability of TCP Reno+ is given by

$$R_{R+}(u) = R_R(u) + p \cdot \Delta_R(u) \tag{1}$$

where $\Delta_R(u)$ is the probability that a single lost retransmitted packet may be recovered by DAC. Consequently, the term $p \cdot \Delta_R(u)$ means the probability that a retransmission is lost and recovered. Recalling that D_1 is always equal to $u-1$, the condition for successful lost retransmission detection can be derived by

$$(u - 1) + (\lfloor u/2 \rfloor - 1) \geq u$$
$$\lfloor u/2 \rfloor \geq 2. \tag{2}$$

Its probability is given by

$$\Delta_R^{(1)}(u) = p \cdot (1 - p)^{(u-1)+(\lfloor u/2 \rfloor - 1)+1}. \tag{3}$$

Unlike TCP Reno, when DAC is applied to TCP NewReno, a single retransmission loss can be always recovered regardless of the number of packet losses. Whether a retransmitted packet may be recovered by DAC or not depends on the position and the sequence of each packet loss. The final loss recovery probability of TCP NewReno+ is given by

$$R_{N+}(u) = R_N(u) + p(1-p)^u \left\{ \begin{array}{l} \sum_{n=1}^{\lfloor u/4 \rfloor} \binom{u-1}{n-1} p^{n-1} + \sum_{n=2}^{\lfloor u/2 \rfloor} \binom{\lfloor u/2 \rfloor - 1}{n-1} p^{n-1} \\ + \sum_{n=3}^{u-K} \sum_{h=3}^{n} \binom{u-b}{h-2} \binom{b-1}{n-h+1} p^{n-1} \end{array} \right\} \tag{4}$$

where $R_N(u)$ is the loss recovery probability of TCP NewReno [7], [10]. Detailed derivation of (4) can be found in [7].

4   Numerical Results

We calculate the loss recovery probability normalized to the stationary distribution of Ω.[3] The x-axis of each graph indicates packet loss probability, which corresponds to p. We assume that K is always three.

In Figs. 5–6, we compare the loss recovery probability of TCP Reno, Reno+, NewReno, and NewReno+ for two different values of W_{max}. The loss recovery probability of all TCP starts to drop rapidly when p exceeds 10^{-2}. The drop of the loss recovery probability for packet loss probabilities from 10^{-2} to 10^{-1} can

---

[2] In [12], [13], we have proposed a model and derived the condition for successful fast retransmit of TCP Reno in terms of the number of packet losses in a window. Finally, the loss recovery performance of TCP Reno can be evaluated by fast retransmit probability.

[3] The procedures of Markov Chain in order to obtain the distribution of Ω in steady-state can be found in [10].

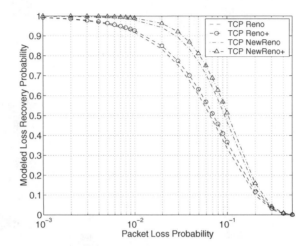

Fig. 5. Comparison of the loss recovery probability predicted by the model of Reno, Reno+, NewReno, and NewReno+ ($W_{max} = 8$).

be explained by the fact that fast retransmit for l_1 cannot be triggered well due to lack of duplicate ACKs. All TCP show a poor performance identically for a large p exceeding 10^{-1}.

There is a considerable improvement (over 10% in maximum) in the loss recovery probability of TCP NewReno. However, there is a little difference in loss recovery probability (at most 5% in maximum) between TCP NewReno and NewReno+: it comes from lost retransmissions. As p increases, more packet losses may occur and their retransmissions also tend to be lost again. On the other hand, Ω cannot keep its size large enough due to frequent loss recovery processes that multiple packet losses are not likely to be included in Ω. That is, the probability itself that a retransmitted packet is lost again is quite low, which is the reason for the insignificant difference in the loss recovery probability between TCP NewReno and NewReno+.

As shown in Fig. 6, further increment of W_{max} to 32 makes no significant differences to the loss recovery probability. For small value of p, it is unlikely that there are more than two losses in Ω. Note that a single packet loss can be retransmitted if only three duplicate ACKs are received, in other words, the size of Ω is greater than or equal to four. Even if multiple packets are lost with a large p, it only matters whether the first fast retransmit is successful or not. Consequently, even if a large W_{max} may improve TCP throughput under low packet loss rate, it offers little benefit to the loss recovery probability.

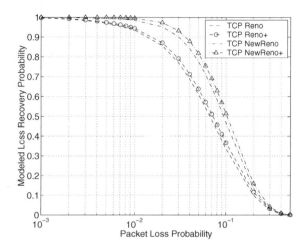

Fig. 6. Comparison of the loss recovery probability predicted by the model of Reno, Reno+, NewReno, and NewReno+ ($W_{max} = 32$).

We evaluate our analysis based on the results obtained from ns simulations. A sender and a receiver are connected with a link of 1 Mbps and 100 msec where packets are dropped in random. Using FTP, the sender transmits data consist of 10^5 packets whose size is 1 kbytes, and cwnd can grow up to W_{max}. The loss recovery probability is defined as the ratio of the number of the packets recovered by retransmissions to the total number of packet losses.

In Figs. 7–8, we compare the loss recovery probability predicted by our model to the results obtained from the simulations. The loss recovery probability of TCP Reno presented in [11] is underestimated because of the assumption that once a packet in a given round is lost, all remaining packets in that round are always lost as well. On the other hand, our assumption that packet losses are independent even if they are transmitted within a round gives more chances for successful fast retransmit under the same value of p.

There are significant differences between our model and [10] in the loss recovery probability of TCP NewReno. In [10], the author does not consider retransmission losses but assumes that its loss recovery always succeeds if only three duplicate ACKs can be received for l_1. It corresponds to an ideal case of TCP loss recovery and provides an upper bound of its performance without Limited Transmit [6]. It can be seen that its loss probability is rather higher than TCP NewReno+. The slight difference between the ideal case and NewReno+ reflects that DAC cannot recover every lost retransmission. Note that DAC can detect

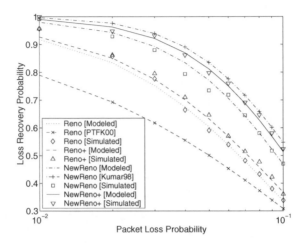

Fig. 7. Verification of the derived recovery probability for each TCP ($W_{max} = 8$).

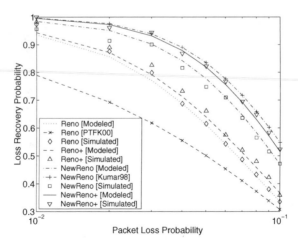

Fig. 8. Verification of the derived recovery probability for each TCP ($W_{max} = 32$).

a lost retransmission if there is at least a well-transmitted packet after the re-transmitted packet. It can be shown that the overall predictions of our model provides a good fit with the simulated results.

5   Conclusions

In this paper, we have evaluated an algorithm called DAC that enables a TCP sender to detect and recover a lost retransmission without RTO. Through sim-ulations, the microscopic behaviors of fast recovery can be investigated when DAC is applied to TCP NewReno.

Although DAC works only for a limited situation, it contributes to make TCP more robust in that DAC enables TCP sender to avoid unnecessary RTO caused by a lost retransmission by requiring only simple and feasible modifications. Additionally, we have justified that DAC does not violate the conservative rule of conventional TCP congestion control.

References

1. M. Allman, V. Paxson, and W. Stevens: TCP Congestion Control. RFC 2581, (1999)
2. V. Jacobson: Congestion Control and Avoidance. ACM SIGCOMM'88, (1988)
3. V. Jacobson: Modified TCP Congestion Avoidance Algorithm. note sent to end2end-interest mailing list, (1990) 314–329
4. Janey C. Hoe: Improving the Start-Up Behavior of a Congestion Control Scheme for TCP. ACM SIGCOMM'96, (1996)
5. E. Blanton, M. Allman, K. Fall, and L. Wang: A Conservative Selective Acknowl-edgment (SACK)-based Loss Recovery Algorithm for TCP. RFC 3517, (2003)
6. M. Allman, H. Balakrishnan, and S. Floyd: Enhancing TCP's Loss Recovery Using Limited Transmit. RFC 3042, (2001)
7. B. Kim and J. Lee: Retransmission Loss Recovery by Duplicate Acknowledgement Counting. IEEE Communications Letters, 8 (1) (2004) 69–71
8. B. Kim, et al.: Lost Retransmission Detection for TCP Part 2: TCP using SACK option. Springer LNCS 3042, (2004) 88–99
9. K. Fall and S. Floyd: Simulation-based Comparisons of Tahoe, Reno, and SACK TCP. ACM Computer Communication Review, 26 (3) (1996) 5–21
10. A. Kumar: Comparative Performance Analysis of Versions of TCP in a Local Net-work with a Lossy Link. IEEE/ACM Trans. Networking, 6 (4) (1998) 485–498
11. J. Padhye, V. Firoiu, D. F. Towsley, and J. F. Kurose: Modeling TCP Reno Per-formance: A Simple Model and Its Empirical Validation. IEEE/ACM Trans. Net-working, 8 (2) (2000) 133–145
12. B. Kim and J. Lee: Analytic Models of Loss Recovery of TCP Reno with Random Losses. Springer LNCS 2662, (2003) 938–947
13. B. Kim and J. Lee: A Simple Model for TCP Loss Recovery Performance over Wire-less Networks. Int'l Journal of Communications and Networks (JCN) published by Korean Institute of Communication and Science (KICS), 6 (3) (2004)
14. B. Kim, D. Kim, and J. Lee: Lost Retransmission Detection for TCP SACK. ac-cepted for publication in IEEE Communications Letters, (2004)
15. R. Braden: Requirements for Internet Hosts. RFC 1122, (1989)

A TCP Protocol Booster for Wireless Networks

Jeroen Hoebeke[1], Tom Van Leeuwen[2], Ingrid Moerman, Bart Dhoedt, and
Piet Demeester

Department of Information Technology (INTEC),
Ghent University - IMEC,
Sint-Pietersnieuwstraat 41,
9000 Ghent, Belgium
{Jeroen.Hoebeke, Tom.VanLeeuwen, Ingrid.Moerman, Bart.Dhoedt,
Piet.Demeester}@intec.UGent.be
http://www.intec.ugent.be

Abstract. Standard TCP has been developed for wired networks and stationary
hosts. In wireless networks, consecutive handoffs between access points introduce packet loss that TCP will consider as an indication of congestion, which is
followed by a reduction of its congestion window. This will result in a considerable degradation of the performance in terms of throughput, although sufficient bandwidth is available between handoffs. In this paper we present the design of a TCP protocol booster, which combines retransmissions based on
negative selective acknowledgements, delayed duplicate acknowledgements
and detection of congestion in order to protect the sender in a transparent way
from the handoff packet losses. A test-bed implementation demonstrates that
the booster improves the TCP performance considerably in a wireless network.

1 Introduction

Wireless LAN hotspots have become part of everyday life, while workgroup 11
within the IEEE 802 LAN/MAN standard committee [1] is still trying to push the
limits of the 802.11 MAC Wireless LAN protocol. Several task groups are addressing
some unresolved issues like Quality of Service (802.11e), higher throughput
(802.11n), security (802.11i) and also inter-access point handoff (802.11f). The problem with current handoff between 802.11 access points is that the connection between
a mobile host and its previous access point has to be terminated before the association
to the next access point can be established. This process can be a source of considerable packet loss, which is detrimental for the performance of the end-to-end Transmission Control Protocol (TCP).

---

[1] Jeroen Hoebeke is Research Assistant of the Fund for Scientific Research - Flanders
(F.W.O.-V., Belgium)
[2] Research funded by a Ph.D. grant for Tom Van Leeuwen of the Institute for the Promotion
of Innovation through Science and Technology in Flanders (IWT-Vlaanderen)

M. Freire et al. (Eds.): ECUMN 2004, LNCS 3262, pp. 175–184, 2004.
© Springer-Verlag Berlin Heidelberg 2004

Standard TCP has been developed for traditional networks with wired links and stationary hosts and not for modern wireless networks. TCP delivers a reliable service over the unreliable IP network and, to this end, it provides flow control, which takes into account the limited buffer size at the receiver side, and congestion control, which prevents buffer overruns in the intermediate routers of the IP network [2]. When the network gets congested, packet losses will occur and the TCP sender will react by reducing its send speed. This mechanism works fine in networks where congestion is the only cause of packet loss. However in wireless networks characterized by frequent handoffs, TCP wrongly interprets packet loss due to handoff as buffer exhaustion. As a result, an excessive amount of time will be spent waiting for acknowledgements that do not arrive, and, although these losses are not congestion-related, the TCP sender will reduce its send speed in order to protect the network from traffic overload.

Several solutions have been proposed to improve the end-to-end performance of TCP over wireless networks. For instance, the HAWAII handoff protocol deals with handoff packet loss by forwarding the packets that were not delivered to the mobile host from the previous access router to the next access router after the mobile host has re-established its connection to the network. Unfortunately, this is an IP layer approach which cannot be used between link layer access points and it also introduces the reception of out-of-order packets at the mobile host, triggering duplicate acknowledgements and hereby reducing TCP throughput at the sender side [3]. Other approaches handle packet loss at the link level [4, 5], such as TCP-aware link-layer protocols that provide local reliability (e.g. SNOOP [6]), end-to-end protocols, where loss recovery is performed by the sender (e.g. TCP SACK [7]), split-connection protocols that break the end-to-end connection into several parts (e.g. I-TCP [8]). These solutions however either require modifications to the TCP implementation, work only when just one wireless link is involved or either damage the layering of the protocol stack.

In this paper we present a novel approach to improve the TCP performance in a wireless network based on the concept of protocol boosters. A protocol booster is a software module, which improves in a transparent way (i.e. without modifying end-to-end protocol messages) the performance of the original protocol [9]. The booster can be inserted at any place in the network (e.g. at the access router and the mobile host) and can add, remove or delay protocol messages. The main advantages of the presented protocol booster are its easy deployment, as there is no need to modify the standard TCP protocol, its flexibility, as the booster can be easily adapted to the specific network environment, and its robustness, as transmission will still continue if the booster fails, though with reduced performance.

2 Design

When being used in a wireless network with several access points and moving users, TCP wrongly interprets handoff packet loss caused as congestion related loss which results in congestion mechanisms being invoked at the source. This causes a severe performance degradation. A TCP protocol booster can offer a solution if it is able to hide handoff losses from the TCP sender. More specifically, TCP timeouts and fast retransmissions not caused by congestion should be avoided because they unnecessarily reduce the congestion window.

The development of the TCP protocol booster is based on the following design considerations:

- A booster element at the access router temporary buffers packets that are sent to the mobile host via its access point. Packets, which are lost due to handoff, are retransmitted. We assume a sufficiently fast wireless network, making it possible to do a retransmission before a timeout occurs. In order to detect lost packets, a second booster element is required at the mobile host, which notifies the access router of packet loss by means of a negative selective acknowledgement strategy. The retransmission of lost packets should be done before the end-to-end TCP retransmission timeout (RTO) timer expires. For a detailed description of the mechanism for retransmissions based on negative selective acknowledgements, we refer to section 2.1.
- Fast retransmissions at the TCP sender, caused by duplicate acknowledgements that were sent for new data packets that successfully arrive after the handoff process is completed, have to be prevented because they unnecessarily reduce the congestion window and interfere with the local retransmissions from the TCP protocol booster. To this end, the mechanism of delayed duplicate acknowledgements is introduced (see section 2.2).
- Finally, the protocol booster must be aware that congestion in the wired network can still occur. When this happens, normal TCP congestion control is necessary. Therefore, a mechanism to distinguish between packet losses due to congestion and due to handoff is required (see section 2.3).

Based on the above design considerations, the functional layout, presented in Fig. 1, is proposed. The complete protocol booster consists of two elements, one at the access router and one at the mobile host. We remark that the protocol booster is inserted between TCP and IP in the OSI protocol stack.

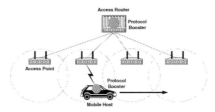

Fig. 1. Protocol booster network architecture

2.1 Retransmissions Based on Selective Negative Acknowledgements

When TCP data packets arrive at the access router, a copy is stored in a circular FIFO buffer before being transmitted to the mobile host's access point. Between handoffs the wireless MAC protocol ensures the correct and in-sequence delivery of data packets to the mobile host. The TCP protocol booster can also be used in case the MAC protocol is unable handle packet loss on the wireless link, e.g. in case of deep signal fading and high bit error rate. In this case we refer to an alternative compatible TCP booster in our earlier work [10].

At handoff between access points, the mobile host is no longer connected to its access point and TCP packets are dropped. When the mobile host has re-established its connection to the network and new TCP packets arrive, the TCP protocol booster at the mobile host will send a selective negative acknowledgement (SNACK) back to the booster at the access router hereby indicating the range of TCP sequence numbers that were lost due to handoff. Please note that this SNACK is not a TCP message but a booster message, which has only meaning between boosters. Upon arrival of the SNACK in the booster of the access router, the missing packets are immediately retransmitted (see Fig.2.a).

It is however possible that during handoff a whole (or the last packets of a) TCP send window is lost. After this window, the TCP sender will stall until the transmitted data packets, which never arrived at the mobile host, are acknowledged. When handoff has finished, the TCP protocol booster at the mobile host will receive no more packets and is therefore not able to send back a SNACK. The TCP sender can eventually time-out, which seriously degrades throughput. The solution to this problem is to use link layer triggers (e.g. LINK_UP) that indicate when connection to the network has been re-established after handoff. The 'triggered booster' requires that the IP layer forwards this MAC layer trigger to the booster layer such that upon reception the booster can send a negative acknowledgement (NACK) back to the booster at the access router with the TCP sequence number of the last data packet it received correctly. When this booster receives this NACK, all packets in the buffer with higher TCP sequence number are retransmitted, thus triggering acknowledgements at the mobile TCP receiver and preventing timeout of the TCP sender (Fig.2.b).

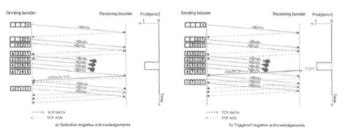

Fig. 2. Retransmissions based on selective negative acknowledgements (a) and on negative acknowledgements using link layer triggers (b)

2.2 Delayed Duplicate Acknowledgements

Although the booster takes care of rapid retransmission of TCP data packets after handoff, duplicate acknowledgements (dupACKs) can be generated by the TCP receiver if new TCP packets arrive before the booster retransmissions. This may give rise to a fast retransmission at the TCP sender after the third duplicate acknowledgement, resulting in a reduced congestion window. To prevent unnecessary fast retransmissions at the TCP sender, the protocol booster at the access router will store every third and next dupACK in a buffer (the dupACK buffer) and delay the dupACK by δ. In the meantime the booster has time to retransmit the lost packets. If a retransmitted packet arrives in the TCP receiver, the latter generates an acknowledgement. If this acknowledgement arrives in the access router booster within the delay time δ, the delayed dupACKs will be stopped. If the acknowledgement does not arrive within the delay time δ, the dupACKs are sent to the TCP sender, causing a fast retransmission. If δ is too small, the acknowledgement after retransmission by the booster might arrive too late, resulting in a useless fast retransmission at the TCP receiver. If δ is too large, any fast retransmission, also those needed to react to congestion in the wired part of the network, is suppressed. Ideally δ is a few times the round trip time of the wireless link [10].

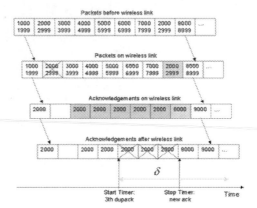

Fig. 3. Delayed duplicate acknowledgement scheme

Suppose packet 2000:2999 is lost in the wireless network due to handoff (Fig. 3). After the handoff process has completed, the following five data packets get through to the receiver before a retransmission of the lost packet by the protocol booster occurs, e.g. at the access router in the case of a fixed TCP sender in the wired network. All third and following duplicate acknowledgements are withheld for a time span δ until the accumulated acknowledgment with number 8000, caused by the booster retransmission of packet 2000:2999, arrives at the access router booster. The old dupACKs are deleted and the new accumulated acknowledgement is passed to the TCP sender. In this way an unnecessary reduction in congestion window size and throughput has been avoided.

2.3 Detection of Congestion

Congestion related losses cause duplicate acknowledgements which must not be delayed because they need to trigger a fast retransmission at the sender side as soon as possible. We have devised an algorithm to distinguish congestion losses from other packet losses. We consider congestion in the wired network before the wireless network and distinguish two cases. Case 1 deals with a fixed TCP sender in the wired network and case 2 considers a mobile TCP sender.

Case 1: The access router booster will check the incoming TCP data packets and will store the TCP sequence number SEQ of the next expected packet in a special buffer (SEQ buffer). When the next packet arrives, its sequence number s is compared with SEQ. If s is larger than SEQ, then congestion occurred and a dupACK is to be expected from the mobile host, which should be forwarded immediately to the TCP sender without delay. Therefore the delayed dupACK booster mechanism will be notified that this dupACK must not be delayed.

Case 2: In case the mobile host is the TCP sender and congestion occurs in the wired network, the TCP receiver will send back duplicate acknowledgements that pass through the access router booster. If no special action is taken, the booster at the mobile host will delay these dupACKs, as it considers them as caused by handoff packet loss. In this case the booster at the access router will mark these unexpected and unpredicted dupACKs with a booster flag in the options field of their IP header. Upon reception of these marked acknowledgements the booster dupACK mechanism at the mobile host will forward them immediately to the TCP sender, without delay (Fig. 4).

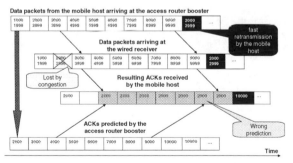

Fig. 4. Congestion detection when the mobile host is the TCP sender

Suppose packet 2000:2999 is lost due to congestion on the wired network after the wireless link. As can be seen in figure 4, a number of acknowledgements arrive at the access router booster that are not predicted. These acknowledgements are marked and passed without delay in the mobile host booster, triggering a fast retransmission at the mobile TCP sender. After retransmission by the TCP sender, the predictions stabilize and the delayed duplicate acknowledgement scheme can be applied again.

3   Click Router Implementation Results

We evaluated the TCP protocol booster on a Click Router test-bed [11] (Fig. 5). A web server was installed on the server PC and a web browser on the client PC for downloading a 20MB file from the server. Three click router PC's connect the server with the client: the access point and the mobile host are running the TCP protocol booster software, while the central PC runs an error model emulating periodic packet loss due to consecutive handoffs. The client and server PC run the Linux 2.4.x operating system which uses the Linux TCP NEWRENO/SACK implementation by default [7].

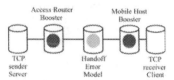

Fig. 5. Click router test-bed

3.1   Cumulative Throughput Versus Time Without Wired Network Congestion

Fig. 6 shows the cumulative throughput of a TCP connection for standard TCP without error model, i.e. the maximum achievable throughput in the test-bed taking into account the click router overhead: 285 kB/s. Also the throughput for standard Linux TCP is depicted with a handoff duration of 100ms and with time between handoffs of 10s, i.e. comparable to a mobile host moving at a speed of 36 km/h with cell sizes of 100m. With the help of the TCP protocol booster a considerable increase (up to 4 times) in throughput is achieved under the same handoff conditions.

3.2   Booster Throughput for Different Handoff Conditions

In Fig. 7 TCP throughput for different handoff conditions and with a wired network delay of 100 ms is depicted. Again, a considerable increase in throughput is obtained using the TCP protocol booster, even for short handoff periodicities, i.e. the case where the mobile host is moving very fast. If the duration of the handoff takes too

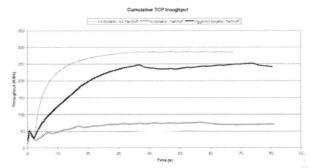

Fig. 6. Cumulative Linux standard TCP throughput vs. TCP booster throughput without wired network congestion

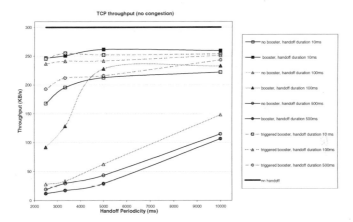

Fig. 7. TCP booster throughput under different handoff conditions

long in comparison with the TCP round trip time, the TCP sender will timeout before the booster had the opportunity to retransmit the lost data packets. As a consequence the TCP sender retransmissions as well as the booster retransmissions will arrive at the TCP receiver causing unexpected duplicate acknowledgements and preventing the TCP send window to increase. In this case the booster will perform worse than standard Linux TCP, while the 'triggered booster' is able to react faster to the handoff and gives much better performance.

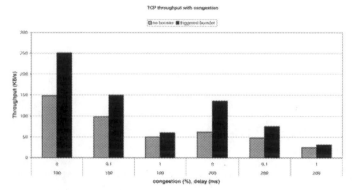

Fig. 8. Cumulative TCP throughput

3.3 Booster Throughput with Wired Network Congestion

Figure 8 shows the TCP throughput for three values of wired network congestion and two values of wired network delay. The handoff duration is 100ms and handoff periodicity 10s. Increasing congestion in the wired network before the wireless link to the mobile host severely limits the TCP throughput, however the protocol booster with congestion detection is able to provide slightly better performance, even under difficult congestion conditions.

4 Conclusions

We have developed a novel TCP protocol booster, which combines retransmission based on negative selective acknowledgements, delayed duplicate acknowledgements and detection of congestion. The TCP protocol booster prevents improper action by TCP in case of handoff packet losses in a wireless network. In addition to this, the congestion detection mechanism guarantees the excellent performance of the booster when there is congestion in the wired network. The performance of the protocol booster has been validated by Click router implementations in a test network.

Acknowledgement. Part of this research is funded by the Belgian Science Policy Office through the IAP (phase V) Contract No. P5/11, by the Flemish IWT through the GBOU Contract 20152 'End- to-end Qos in an IP Based Mobile Network' and by the Fund for Scientific Research - Flanders (F.W.O.-V.).

References

1. IEEE Workgroup for Wireless LANs: http://grouper.ieee.org/groups/802/11/
2. W.R. Stevens, "TCP/IP illustrated 1: The protocols", Addison,-Wesley, Reading, Massachusetts (1994)
3. A. Campbell, J. Gomez, S. Kim, C. Wan, Z. Turanyi, A. Valko, "Comparison of IP Micromobility Protocols", IEEE Wireless Communications, February 2002, pp. 72-82.
4. H. Balakrishnan, V.N. Padmanabhan, S. Seshan, and R.H. Katz, "A comparison of mechanisms for improving TCP performance over wireless links," IEEE/ACM Transactions on Networking, December 1997.
5. G. Xylomenos, G.C. Polyzos, "TCP and UDP performance over a wireless LAN", IEEE INFOCOM, 1999, 439-446.
6. The Berkeley Snoop Protocol: http://nms.lcs.mit.edu/~hari/papers/snoop.html
7. K. Fall, S. Floyd., "Comparisons of Tahoe, Reno, and Sack TCP", ACM Computer Communication Review, 1996, 26 (3), 5-21
8. A. Bakre, B.R. Badrinath, I-TCP: Indirect TCP for Mobile Hosts , Proceeding of the 15th International Conference on Distributed Computing system (IDCS), 1995, 136-143
9. D. C. Feldmeier, A. J. McAuley, J. M. Smith, D. S. Bakin, W. S. Marcus, and T. M. Raleigh, "Protocol Boosters", IEEE Journal on Selected Areas in Communications, April 1999, 16(3): 437-444.9.
10. J. Hoebeke, T. Van Leeuwen, L. Peters, K. Cooreman, I. Moerman, B. Dhoedt, P. Demeester, "Development of a TCP protocol booster over a wireless link", SCVT 2002 Louvain-La-Neuve, 17/10/2002, p.27-34.
11. Click Router website: http://www.pdos.lcs.mit.edu/click/

Fairness Property and TCP-Level Performance of Unified Scheduling Algorithm in HSDPA Networks

Yoshiaki Ohta, Masato Tsuru, and Yuji Oie

Dept. of Computer Science and Electronics, Kyushu Institute of Technology
Kawazu 680–4, Iizuka, Fukuoka 820–8502, Japan
yoshiaki@infonet.cse.kyutech.ac.jp, {tsuru,oie}@cse.kyutech.ac.jp

Abstract. Channel-state-aware scheduling strategies on wireless links play an essential role for enhancing throughput performance of elastic data traffic by exploiting channel fluctuations. Especially in HSDPA (High-Speed Downlink Packet Access), while various channel-state-aware schedulers have been proposed, each of them has pros and cons, and thus, none of them can be the best solution for all situations and policies in the base station management. To respond to the variety of the base station management, in our previous paper, a unified scheduler that can flexibly vary its characteristics by adjusting its parameters was proposed, but only the basic packet-level performance was evaluated. In this paper, therefore, the detailed performance characteristics of the proposed unified scheduler are investigated. First, it is shown that the unified scheduler meets a general fairness criterion—(p, α)-proportional fairness—in terms of the asymptotic behavior of the long-term averaged throughput. Second, the TCP-level throughput performance of the unified scheduler with some link-layer technologies actually employed in HSDPA is evaluated by network simulation. Lastly, two scenarios to set the parameters of the scheduler—the semi-proportional fair scheduling and the semi-greedy scheduling—are examined, which demonstrate the flexibility and effectiveness of the unified scheduler.

Keywords: Wireless Communications, HSDPA, Channel-State-Aware Scheduling, (p, α)-Proportional Fairness, TCP.

1 Introduction

In the next-generation wireless networks, the extended wireless access method of 3G W-CDMA systems with high-speed downlink shared channels is referred to as HSDPA (High-Speed Downlink Packet Access), which can provide a peak data rate of about 10 Mb/s [1]. In such high-speed wireless networks, channel-state-aware resource allocation and scheduling strategies play an important role for getting high throughput performance. In particular, the delay-tolerant elastic traffic raises the possibility of efficient communication by exploiting channel fluctuations. Thus, various channel-state-aware schedulers have been proposed and evaluated [2,3,4,5,6,7,8,9]. Among them, the Proportional Fair (PF) scheduler [3,7] is known as one that takes both the throughput fairness among users and the total throughput to all users into account, and approximately meets a fairness criterion—proportional fairness [10]. However, the proportional fairness is not always

M. Freire et al. (Eds.): ECUMN 2004, LNCS 3262, pp. 185–195, 2004.
© Springer-Verlag Berlin Heidelberg 2004

the best solution, depending on how different channel states of users are and what kind of policy is employed in terms of fairness and throughput performance to be achieved in managing the base station. Thus, operators of the base station may want to selectively use different schedulers, instead of using one typical scheduler. In order to respond to the variety of situations and policies in the base station management, we have previously proposed a unified scheduler that can flexibly vary its characteristics by adjusting its parameters [11]. The most important benefit of the proposed scheduler is the inclusiveness of typical few existing schedulers, e.g., the Maximum Carrier-to-interference power Ratio (Max CIR) scheduler, the PF scheduler, and the Round Robin (RR) scheduler, as well as the capability of the scheduling with compound characteristics. This opens up the possibility of flexible base station management, i.e., operators can flexibly adjust the characteristics according to their requirements. In our previous paper, however, we have only evaluated the basic packet-level performance of the unified scheduler, assuming infinite backlogs as well as neglecting the aspects of link layer and transport layer [11].

The objective of this paper is, therefore, to clarify the detailed performance of the proposed unified scheduler. First, in Sec. 3, we will show that the unified scheduler can theoretically meet a general fairness criterion—$(\mathbf{p}, \alpha)$-*proportional fairness* [12]—in terms of the asymptotic behavior of the long-term averaged throughput. The fairness parameter $(\mathbf{p}, \alpha)$ depends on the long-term averaged channel-states of individual users and the adjustable parameter (p, q) of the unified scheduler. Second, in Sec. 5.1–5.2, by carrying out network simulations, we will investigate the TCP-level end-to-end throughput performance of the unified scheduler with some underlying link-layer technologies actually employed in HSDPA, such as adaptive modulation and cording schemes (MCS), link-layer ARQ (Automatic Repeat reQuest), and in-sequence packet delivery [13, 14, 15]. In fact, these link-layer technologies considerably affect the TCP-level performance as we have previously shown [16, 17]. Note that, since the unified scheduler includes the existing schedulers, our investigation of the TCP-level performance also exposes the characteristics of typical existing schedulers such as the PF scheduler. Finally, in Sec. 5.3, to demonstrate the flexibility and effectiveness of the unified scheduler, we will show two example scenarios to set the parameters of the scheduler; one is the semi-proportional fair scheduler and the other is the semi-greedy scheduler.

2   Packet Delivery in HSDPA Networks

For error-free packet delivery in HSDPA, various link-layer control technologies are employed. The RLC (Radio Link Control) protocol [14] and MAC (Media Access Control) protocol [15] are standardized by 3GPP as link-layer protocols. In the RLC layer, a SDU (Service Data Unit) corresponding to an IP datagram received from upper layers is segmented into a number of RLC PDUs (RLC Protocol Data Unit) of a fixed size, which are submitted to the lower layer. The size of RLC PDUs is considerably short compared with that in transport layer, which makes a fine-grained transmission control possible. The RLC layer provides error-free packet delivery by using the link-layer ARQ with in-sequence packet transmission. With the link-layer ARQ, if a SDU overtakes the preceding SDUs suffering from bit error and still being retransmitted, which is referred to as out-of-order SDU delivery, the TCP layer recognizes it as packet losses due to congestion and hence reduces the transmission rate. The detailed description of the RLC protocol is given in [14]. The MAC layers provide fast and adaptive packet control function, including adaptive MCS selection, adaptive transport block size (the size of a data unit in the physical layer) based on a selected MCS, and fast scheduling. The TTI (Transmission Time Interval) or interleaving period is short compared with the 3G W-CDMA TTI, and thus enables a short round-trip delay between a base station and receivers. The detailed description of the MAC protocol is given in [15]. Recently, several approaches with transport layer modification have been proposed for enhancing throughput performance and alleviating negative effects caused by a high bit-error rate, e.g., [18, 19]. However, some of them are difficult to be implemented, and some others violate the end-to-end communication model of the TCP connections. Thus, in this paper, we will only investigate the impact of the above existing mechanisms in link layers on the TCP performance over HSDPA environment, without any additional modifications of transport layers.

3   Description of Scheduler and Fairness

We assume that M users share a high-speed downlink channel. A scheduler serves one of those M users during every TTI. We call TTI as slot. The metric of user selection is denoted by $M_m(n)$ for user m in the nth slot, where $m = 1, 2, \ldots, M$. User $m = U(n)$ is selected when $M_m(n)$ takes the maximum value among users, i.e., $U(n) = \arg \max_{m=1,2,\ldots,M} M_m(n)$. If more than one user take the maximum of metric $M_m(n)$ in the nth slot, the scheduler randomly selects one of the users taking the maximum metric.

3.1   Algorithm Description

We first briefly show the algorithm of the PF scheduler [3, 7], which was originally proposed in CDMA 1x EV-DO systems. In the nth slot, $T_m(n)$ denotes the (exponentially) averaged throughput of user m, i.e., $T_m(n + 1) = (1 - 1/t_c) T_m(n) + 1/t_c R_m(n)\delta_{m,U(n)}$, where $t_c := \min \{n, c\}$; c is a predetermined constant and δ is a Kronecker delta. $R_m(n)$ denotes feasible rate, i.e., instantaneous peak rate in the nth slot. For the PF scheduler, $M_m(n)$ is defined by $M_m(n) = R_m(n)/T_m(n)$.

In addition to using $T_m(n)$ and $R_m(n)$, we have previously proposed a unified scheduler that uses the averaged feasible rate of each user m denoted by $Q_m(n + 1) := (1 - 1/t_d) Q_m(n) + 1/t_d R_m(n + 1)$, where $t_d := \min \{n, d\}$; d is a predetermined constant, which must be a large value. For the proposed scheduler,

$$M_m(n) = \frac{Q_m(n)^{1-q} R_m(n)^q}{T_m(n)^p}. \qquad (1)$$

where p and q ($0 \le p, q \le 1$) are parameters to control the impact of the user throughput and the channel condition on the metric, respectively. By letting (p, q) be $(0,1)$, $(1,1)$, or $(1,0)$, the scheduler can act as the existing Max CIR scheduler, PF scheduler, or approximate RR scheduler, respectively, which is the reason why we call it the unified scheduler. Furthermore, by letting p and/or q be intermediate values, the scheduler can show compound characteristics. The details are given in [11].

3.2   Fairness Property of Proposed Scheduler

Fair allocation of throughput among users has been discussed in various forms so far. The most common fairness criterion is the max-min fairness [20], which focuses only on the throughput balance among users and does not consider the total throughput. An alternative fairness criterion is the proportional fairness [10], which takes both the throughput balance among users and the total throughput to all users into account. The $(\mathbf{p}, \alpha)$-*proportional fairness* [12] can unify the max-min fairness, the proportional fairness, and the worst case fairness (total-throughput maximization).

By focusing on the long-term averaged throughput achieved by each scheduler, we can apply the above fairness criterion to scheduling algorithms. In the following, superscript S means that a variable is one in a scheduler denoted by S, and N is a time duration for averaging the throughput. Here, we explain the $(\mathbf{p}, \alpha)$-proportional fairness. Let $T_{m,N}^{S} := \sum_{n=1}^{N} V_{m}^{S}(n)/N$, where $V_{m}^{S}(n) := R_{m}(n)\delta_{m,U^{S}(n)}$, $U^{S}(n)$ denotes a user selected by S in the nth slot. We suppose that the long-term averaged throughput exists, i.e., $T_{m}^{S} := \lim_{N \to \infty} T_{m,N}^{S}$. For $\mathbf{p} = (p_1, p_2, \dots, p_M)$ and $\alpha > 0$, scheduler S^* is called $(\mathbf{p}, \alpha)$-proportionally fair if the following inequality holds for any other scheduler S,

$$\sum_{m=1}^{M} p_m \frac{T_m^S - T_m^{S^*}}{(T_m^{S^*})^{\alpha}} \leq 0.$$

In particular, S^* is proportionally fair when $p_1 = p_2 = \cdots = p_M$ and $\alpha = 1$. The PF scheduler is proportionally fair under some assumptions [7].

The proposed unified scheduler satisfies $(\mathbf{p}, \alpha)$-proportional fairness in the following sense. Let $\overline{T}_{m,N}^{S} := \sum_{n=1}^{N} T_m^S(n)/N$ and $\overline{Q}_{m,N} := \sum_{n=1}^{N} Q_m(n)/N$. Here, we denote the proposed scheduler as S^*, where predetermined constants c and d take large values. In addition, we assume a condition of stability in which, for a sufficiently large n and N, the difference of $T_m^{S^*}(n)$ and $\overline{T}_{m,N}^{S^*}$, and that of $Q_m(n)$ and $\overline{Q}_{m,N}$ are so small that the allocation of time slots depends only on the rate fluctuations. In such a condition, the user selection metric (1) of scheduler S^* can be replaced with

$$M_m(n) = \frac{\overline{Q}_{m,N}^{1-q} R_m(n)^q}{\left(\overline{T}_{m,N}^{S^*}\right)^p}. \tag{2}$$

Let $\overline{p}_{m,N} := \overline{Q}_{m,N}^{\frac{1-q}{q}}$, and $\overline{p}_m := \lim_{N \to \infty} \overline{p}_{m,N}$.

Theorem. *For* $\mathbf{p} = (\overline{p}_1, \overline{p}_2, \dots, \overline{p}_M)$ *and* $\alpha = \frac{p}{q}$, *the proposed scheduler satisfies* $(\mathbf{p}, \alpha)$-*proportional fairness.*

Proof. Due to space limitation, we give a sketch of proof in Appendix.   □

In addition to the above-mentioned theoretical property on the asymptotic behavior of the long-term averaged throughput, we will further evaluate practical performance of the unified scheduler in the next section, that is, TCP-level throughput performance for end-to-end connections through wired networks and a HSDPA network.

4   Simulation Model

Figure 1 shows the simulation model for evaluating the TCP-level performance of the unified scheduler. We used Network Simulator Version 2 (ns-2.27) [21], developed by VINT project with some modifications, including link-layer ARQ, in-sequence packet delivery, and the scheduler itself.

In our simulation, the major simulation parameters are shown in Tab. 1. The wired access links are 100 Mb/s in bandwidth and 1–5 ms in propagation delay to avoid TCP synchronization. The wired core link is 45 Mb/s (T3 link speed) in bandwidth and 10 ms in propagation delay. The condition of wireless links varies over time, which are 2.4 Mb/s–7.2 Mb/s in bandwidth according to predetermined allowable MCS sets shown in Tab. 2 and 30 ms in propagation delay including the processing delay in a base station (BS) and receivers. The radio link parameters are based on [13], excluding shadowing effect to focus on evaluating the impact of the path loss and Rayleigh fading on the throughput performance. The physical layer encodes PDUs in a TTI and thus bit errors happen per TTI. We assume the selective repeat ARQ protocol is employed in which the ACK and NACK (Negative ACK) in link layers are error free. As stated in Sec. 2, ARQ may cause the out-of-order SDU transmission. Thus, to preserve SDU integrity, we employ a reorder buffer with a timeout timer at each receiver. When an out-of-order delivery happens, the timer is started and the receiver keeps the correctly received SDUs waiting in the reorder buffer until the overtaken SDU arrives during the timeout interval. If the timer expires before the overtaken SDU arrives, the SDUs in the reorder buffer are forwarded to the upper layer, and thus the recovery is conducted by TCP layer. The timeout value is denoted by WT (Waiting Timeout). The size of a reorder buffer here is infinite to prevent SDU losses at receivers, but the queue length is upper-bounded because the stored packets are flushed out immediately after the timer expires.

The greedy file transfer is employed as the TCP traffic. The TCP variant used here is NewReno [22], which is most widely used in the current Internet. The BS segments an incoming IP packet into RLC PDUs, which are stored in the dedicated BS buffers. The buffer size is denoted as B, where B is the number of packets (SDUs) instead of PDUs.

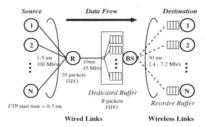

Fig. 1. Simulation model

Table 1. Simulation parameters

Cell Layout	3 sectored 19 cells
# of Multipaths	1, 2
Doppler Frequency	5.55 Hz
Path Loss Factor	3.76
Max. # of ARQ	10
TCP Algorithm	NewReno
TCP Packet Size	1500 bytes
L2 PDU Header/Payload Size	2/40 bytes

Table 2. Allowable MCS sets

Chip Rate		3.84 Mc/s
Spreading Factor		16
Multicode		10
Mod.	Code Rate	$R_m(n)$ [Mb/s]
QPSK	1/2	2.4
QPSK	3/4	3.6
16QAM	1/2	4.8
16QAM	3/4	7.2

Note that the above-mentioned model of link-layer mechanisms may not fully reflect the standard specifications and/or the actual implementations in detail. However, we believe this can be adopted as a useful and suitable model for investigating the TCP-level performance characteristics of the scheduler.

5   Simulation Results and Discussions

In the simulation, ten users are randomly selected in total from three sectors in a cell so that the normalized distance between the BS and User 1/User 2 is 0.1, that between the BS and User 3/User 4 is 0.3, ..., and that between the BS and User 9/User 10 is 0.9, respectively. A user with a smaller ID stays nearer to the BS, and thus, is in an inherently better channel condition, which is expected to result in high throughput performance. Each user conducts one TCP connection. We set c to 1000, which is the recommended value in the PF scheduler, and then d to 1000. We carry out the simulation for 200 s, and examine the time-averaged throughput performance of each TCP connection and the total throughput to all users.

5.1   Impact of B and WT on Total Throughput

We investigate the impact of WT (timeout duration for the packet reorder) and B (size of per-flow buffer at BS) on the total throughput. Since a large WT prevents packets from the out-of-order delivery in the case of the patient link-layer ARQ and a large B can help packets not to be dropped at the BS even in the case of bad link conditions, the throughput may increase as WT or B increases. However, too large WT and B may cause too long packet delays, which make an agile response to packet losses in wired links difficult and damage the overall congestion control. Thus, we examine proper values of WT and B to get adequate throughput. Figure 2 shows the impact of B on the total throughput when $WT = 240$. "PF" represents the case when (p, q) is set to (1,1); "Proposed" represents the case when (p, q) is set to (0.4,0.1), which is one of the suggested settings reasoned later. From this figure, we can say that $B = 20$ is sufficient to get large throughput despite of the number of multipaths and the type of schedulers.

Figure 3 shows the impact of WT on the total throughput when $B = 20$ and 100. From this figure, the total throughput gradually increases with the increase of WT, and then it saturates. This shows that the throughput degradation due to the out-of-order

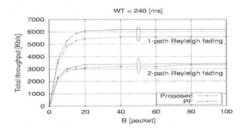

Fig. 2. Total throughput as a function of B

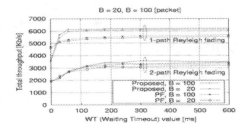

Fig. 3. Total throughput as a function of WT

delivery can be eliminated when WT is a certain value, e.g., $WT = 240$, despite of the number of multipaths and the type of schedulers. In addition, the increase of B does not give large contribution to the increase of the total throughput for all of WT.

From the above results, we set $WT = 240$ and $B = 20$, and then examine the detailed performance of the unified scheduler in the following subsections.

5.2 Impact of p and q on Total Throughput

We investigate the impact of the parameter (p, q) on the total throughput in 2-path Rayleigh fading environment. Figure 4 shows the impact of (p, q) when $WT = 240$ and $B = 20$. From this figure, the total throughput roughly increases as p decreases or q increases. When q is equal to 1, as p decreases from 1 to 0, the scheduler gradually varies from the PF scheduler to the Max CIR scheduler, and we can thus get large total throughput. Furthermore, when p is equal to 1, as q decreases from 1 to 0, the scheduling algorithm gradually changes from the PF scheme to the approximate RR scheme, and so the throughput fairness among users increases while the total throughput is sacrificed. Interestingly, Fig. 4 shows that, for each small fixed q (e.g., 0.1), the value of p exhibiting the largest total throughput is not 0 but some small intermediate value (e.g., 0.2).

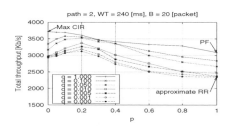

Fig. 4. Total throughput as a function of p

From these results, the total throughput is increasing with the increase of q and the decrease of the value of p to a certain point, e.g. 0.2. This is a pivotal characteristic for setting the parameters of the scheduler.

5.3 Examples of Scheduling Strategies

We now demonstrate the flexibility and effectiveness of the unified scheduler by examining two scenarios to determine the parameters of the scheduler.

First, we consider a scheduler that can get larger total throughput than that of the PF scheduler, with slightly sacrificing fairness among users compared with the PF scheduler. We call this scheduler Semi-Fair scheduler according to the following guidelines:

Guideline 1. The total throughput outperforms that achieved by the PF scheduler.
Guideline 2. Users in relatively good channel conditions get larger throughput than that achieved by the PF scheduler; while users in relatively bad channel conditions get throughput not less than that achieved by the RR scheduler.

To agree with these guidelines, we found (0.4,0.1) as a desired pair in a trial-and-error manner. The total throughput achieved by the Semi-Fair scheduler is then 3354.95 Kb/s; while that of the PF scheduler is 3086.44 Kb/s, and the gain of which is 10.8%, which indicates Guideline 1 is satisfied. This throughput is equivalent to 90.2% of the Max CIR scheduler (3716.71 Kb/s). Figure 5 shows the throughput of each user in the Semi-Fair scheduler. The throughput of User 1–4 outperforms that achieved by the PF scheduler, and the throughput of User 7–10 is nearly equivalent to that by the RR scheduler, which indicates that Guideline 2 is satisfied.

Next, we consider a scheduler that can get throughput performance similar to the Max CIR scheduler, with slightly improving fairness among users compared with the Max CIR scheduler. We call this scheduler Semi-Greedy scheduler according to the following guidelines:

Guideline 1. The total throughput is as similar to that achieved by the Max CIR scheduler as possible.

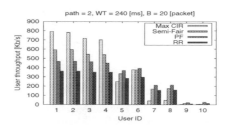

Fig. 5. User throughput (Semi-Fair settings)

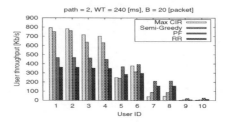

Fig. 6. User throughput (Semi-Greedy settings)

Guideline 2. Users in relatively good channel conditions get throughput similar to that
achieved by the Max CIR scheduler; while users in relatively bad channel conditions
get throughput equal to the medium throughput between the RR scheduler and the
Max CIR scheduler.

To agree with Guideline 1, we let p be 0.2 for the total throughput becomes close to that
achieved by the Max CIR scheduler. Then, in a trial-and-error manner, we let q be 0.05
as a desired setting to agree with Guideline 2. Figure 6 shows the user throughput in the
case that (p, q) is equal to (0.2,0.05). The total throughput of the Semi-Greedy scheduler
is 3527.93 Kb/s, which is equivalent to 94.9% of the total throughput achieved by the
Max CIR scheduler.

6   Concluding Remarks

In this paper, we have focused on investigating the throughput performance of a unified
scheduler for HSDPA downlink, which we previously proposed. It has been mathematically shown that the scheduler has $(\mathbf{p}, \alpha)$-*proportional fairness* property. The TCP-level

end-to-end throughput performance of the scheduler with some underlying link-layer technologies has also been shown by network simulation, which indicated that the unified scheduler could flexibly vary its characteristics by adjusting its parameters. Two example scenarios determining the parameters of the unified scheduler have been examined. One is the semi-proportional fair scheduling, by which the total throughput can outperform that by the PF scheduler, while slightly sacrificing fairness among users compared with the PF scheduler. The other is the semi-greedy scheduling, by which the total throughput is close to the best possible value by the Max CIR scheduler, while slightly improving fairness among users compared with the Max CIR scheduler. In future work, we hope to further investigate the performance characteristics of the unified scheduler with various types of traffic in HSDPA networks.

References

1. 3GPP: High Speed Downlink Packet Access (HSDPA); overall description. Technical Specification 25.308, Version 5.4.0, Release 5, 3GPP, Sophia Antipolis, France (2003)
2. Lu, S., Bharghavan, V., Srikant, R.: Fair scheduling in wireless packet networks. IEEE/ACM Transactions on Networking 7 (1999) 473–489
3. Jalali, A., Padovani, R., Pankaj, R.: Data throughput of CDMA-HDR a high efficiency-high data rate personal communication wireless system. In: IEEE Vichuliar Technology Conference 2000 (Spring). (2000) 1854–1858
4. Bender, P., et al.: CDMA/HDR: A bandwidth-efficient high-speed wireless data service for nomadic users. IEEE Communications Magazine 38 (2000) 70–77
5. Liu, X., Chong, K., Shroff, N.: Opportunistic transmission scheduling with resource sharing constraints in wireless networks. IEEE Journal on Selected Areas in Communications 19 (2001) 2053–2064
6. Borst, S., Whiting, P.: Dynamic rate control algorithms for HDR throughput optimization. In: IEEE INFOCOM 2001. (2001) 976–985
7. Viswanath, P., Tse, D., Laroia, R.: Opportunistic beamforming using dumb antennas. IEEE Transactions on Information Theory 48 (2002) 1277–1294
8. Liu, Y., Gruhl, S., Knighly, E.: WCFQ: an opprutnistic wireless scheduler with statistical fainress bounds. IEEE Transactions on Wireless Communications 2 (2003)
9. Panigrahi, D., Khaleghi, F.: Enabling trade-offs between system throughput and fairness in wireless data scheduling techniques. In: World Wireless Congress (3G Wireless). (2003)
10. Kelly, F.: Charging and rate control for elastic traffic. European Transactions on Telecommunications 8 (1997) 33–37
11. Ohta, Y., Tsuru, M., Oie, Y.: Framework for fair scheduling schemes in the next generation high-speed wireless links. In: the 8th International Conference on Cellular and Intelligent Communications (CIC 2003). (2003) 411
12. Mo, J., Walrand, J.: Fair end to end window-based congestion control. IEEE/ACM Transactions on Networking 8 (2000) 556–567
13. 3GPP: Pyisical layer aspects of UTRA high speed downlink packet access. Technical Report 25.848, Version 4.0.0, Release 4, 3GPP, Sophia Antipolis, France (2001)
14. 3GPP: Radio link control (RLC) protocol specification. Technical Specification 25.322, Version 5.4.0, Release 5, 3GPP, Sophia Antipolis, France (2003)
15. 3GPP: Media access control (MAC) protocol specification. Technical Specification 25.321, Version 5.4.0, Release 5, 3GPP, Sophia Antipolis, France (2003)

16. Koga, H., Ikenaga, T., Hori, Y., Oie, Y.: Out-of-sequence in packet arrivals due to layer 2 ARQ and its impact on TCP performance in W-CDMA networks. In: Symposium on Applications and the Internet 2003 (SAINT 2003). (2003) 398–401
17. Koga, H., Kawahara, K., Oie, Y.: TCP flow control using link layer information in mobile networks. In: SPIE Conference of International Performance and Control of Network Systems III. (2002) 305–315
18. Bakre, A., Badrinath, B.: I-TCP: Indirect TCP for mobile hosts. In: International Conference on Distributed Computing Systems. (1995) 136–143
19. Balakrishnan, H., Seshan, S., Katz, R.H.: Improving reliable transport and handoff performance in cellular wireless networks. ACM Wireless Networks **1** (1995) 469–481
20. Bertsekas, D., Gallager, R.: 6. In: Data Networks. Second edn. Prentice-Hall, Englewood Cliffs, New Jersey 07632 (1991)
21. VINT Project Network Simulator (ns-2): http://www.isi.edu/nsnam/ns/
22. Floyd, S., Henderson, T.: The NewReno modification to TCP's fast recovery algorithm. RFC2582 (1999)

Appendix: Sketch for the Proof of Theorem

Let S^* denote the proposed scheduler and S denote any other scheduler. Let J be the number of all possible vectors $(R_1(n), R_2(n), \ldots, R_M(n))$ of feasible rates. Thus, the whole set $\{1, 2, \ldots, N\}$ of time-slot indexes can be divided into $I_{i,j,k}$ ($i = 1, 2, \ldots, M$, $j = 1, 2, \ldots, J$, $k = 1, 2, \ldots, M$), where $I_{i,j,k}$ indicates the set of time-slot indexes such that user i is selected by scheduler S^* in the slot, the feasible rate vector of the slot takes the jth value, and user k is selected by scheduler S in the slot. Note that, if there in no slot satisfying the above condition, we define $I_{i,j,k}$ as $\emptyset$. Let $R_m^{(i,j)} := R_m(n)$ for $n \in \bigcup_k I_{i,j,k}$ and $r_m^{(i,j)} := |I_{i,j,m}|/N$. Therefore, $T_{i,N}^{S^*}$ and $T_{m,N}^{S}$ are $T_{i,N}^{S^*} = \sum_{j=1}^{J} \sum_{m=1}^{M} r_m^{(i,j)} R_i^{(i,j)}$ and $T_{m,N}^{S} = \sum_{i=1}^{M} \sum_{j=1}^{J} r_m^{(i,j)} R_m^{(i,j)}$, respectively.

As we now use the metric for user selection defined by (2), for any i, j, m, the following inequality holds: $\overline{p}_{m,N} R_m^{(i,j)} / \left(\overline{T}_{m,N}^{S^*} \right)^\alpha \le \overline{p}_{i,N} R_i^{(i,j)} / \left(\overline{T}_{i,N}^{S^*} \right)^\alpha$. Thus,

$$
T_{m,N}^{S} = \sum_{i=1}^{M} \sum_{j=1}^{J} r_m^{(i,j)} R_m^{(i,j)} \le \sum_{i=1}^{M} \sum_{j=1}^{J} r_m^{(i,j)} \frac{\left(\overline{T}_{m,N}^{S^*} \right)^\alpha \overline{p}_{i,N}}{\overline{p}_{m,N} \left(\overline{T}_{i,N}^{S^*} \right)^\alpha} R_i^{(i,j)}.
$$

Hence, we can show

$$
\sum_{m=1}^{M} \frac{\overline{p}_{m,N}}{\left(\overline{T}_{m,N}^{S^*} \right)^\alpha} T_{m,N}^{S} \le \sum_{m=1}^{M} \sum_{i=1}^{M} \sum_{j=1}^{J} r_m^{(i,j)} \frac{\overline{p}_{i,N}}{\left(\overline{T}_{i,N}^{S^*} \right)^\alpha} R_i^{(i,j)} = \sum_{i=1}^{M} \frac{\overline{p}_{i,N}}{\left(\overline{T}_{i,N}^{S^*} \right)^\alpha} T_{i,N}^{S^*}.
$$

With easy computation, when $N \to \infty$, $\overline{T}_{m,N}^{S} \to T_m^{S}$ holds. Thus, when $N \to \infty$, we can eventually obtain

$$
\sum_{m=1}^{M} \overline{p}_m \frac{T_m^{S} - T_m^{S^*}}{(T_m^{S^*})^\alpha} \le 0,
$$

which completes the proof of Theorem.   □

A Generic Approach for the Transmission of Real Time Data in Multicast Networks*

Miguel Rodríguez-Pérez, Sergio Herrería-Alonso, Manuel Fernández-Veiga,
Andrés Suárez-González, and Cándido López-García

E.T.S.E. Telecomunicación
Campus universitario s/n
36310 Vigo, Spain
Miguel.Rodriguez@det.uvigo.es

Abstract. We present a new single-rate multicast congestion control protocol aimed at the transmission of real time multimedia data. The protocol extends previous work based on the use of a representative host (leader) decentralising its election. We also introduce an open-loop RTT estimation algorithm that avoids the need to send regular feedback to the sender, and that lets receivers decide to claim for leadership only when needed. Our solution is able to compete against TCP traffic in a fair manner so it can be safely deployed in the Internet. Moreover, we have also made it possible to change the rate adjustment algorithm both at the sender and at the receivers, so that it can be elected according to the application needs or network characteristics.

1 Introduction

Multimedia streaming across the Internet to a large and disperse number of receivers has still to improve before being safely deployed. At the network level, the essence of the problem lies in devising a bandwidth-efficient transport protocol which does not compromise the stability of the current Internet. The last years have witnessed important advances in this area, resulting in two broad approaches: the multiple-layer proposals and the single-rate approach. While it is better to use a multiple-layer approach if best-quality delivery to a very different set of receivers is the concern [1], single-rate protocols fit better for keeping the sender and the encoding of the data simpler [2]. In addition, single-rate protocols are generally easier to understand and analyze because of their similarity with unicast protocols. Hence, in this paper we focus on a single-rate algorithm.

There exist two main strategies aimed to avoid network congestion suffered with the single-rate multicast streaming protocols, both borrowed from unicast streaming solutions and based on the robust congestion control mechanisms embedded in TCP. One of the options is to rely on an equation-based algorithm

---

* This work was supported by the project "TIC2000-1126 of the Plan Nacional de Investigación Científica, Desarrollo e Innovación Tecnológica" and by the grant PGIDT01PX132202PN of the "Secretaría Xeral de I+D da Xunta de Galicia."

M. Freire et al. (Eds.): ECUMN 2004, LNCS 3262, pp. 196–205, 2004.
© Springer-Verlag Berlin Heidelberg 2004

that, by using an approximate throughput formula for TCP as a non-linear control rule, tries to track the same transmission rate that a real TCP connection would have between the two endpoints. However, since the TCP mechanism produces rapidly varying rates, the equation-based protocols react too conservatively to congestion and are a far-from-ideal solution for streaming applications [3]. The alternative pursued by TCP emulators is to reproduce directly the behaviour of TCP. In this case, and considering a multicast transmission paradigm, the main problem is getting enough feedback information (acknowledgements) from all receivers without flooding the sender.

In this paper we address the design of a new general TCP emulator that can be configured to use different kinds of rate estimation algorithms. Our solution uses a representative host like pgmcc [4], but moves the election mechanism to the receivers. This reduces the information that receivers need to transmit to the sender and avoids the use of complicate mechanisms in the network for aggregating this feedback information. Receivers need to calculate their fair share of bandwidth, but must refrain to ask for leadership until the actual sending rate exceeds their fair rate, in order to avoid the feedback implosion problem [2].

We also propose a novel approach for the calculation of the Round Trip Time (RTT) at the receivers, so that it can be carried in a completely open-loop fashion. This approach is key for the protocol to be able to operate in a decentralised way and delegate the election of the representative host on the receivers. Our proposed method is able to recover for multiple consecutive packet losses avoiding drifts in the RTT estimation.

A second goal is to obtain a smooth throughput suitable for the transmission of multimedia content. To this end, the Generic Multicast Congestion Control (GMCC) supports the use of different rate adjustment algorithms. By using a rate adjustment algorithm compatible with TCP, GMCC is able to achieve a fair transmission rate while obeying the requisites of multimedia transmission. We advocate the use of either binomial algorithms [5] or a variation of TCP-Vegas [6, 7] adapted for a multicast environment.

The rest of this paper is organized as follows. In Section 2 we present a protocol description. Section 3 describes our new RTT estimation algorithm. Then, in Section 4, we show some experimental results. We conclude in Section 5.

2   Protocol Description

Throughout this section we will describe the main elements conforming the GMCC family of congestion control protocols. We will proceed by showing first the main architecture of the protocols.

2.1   Functional Description

Our proposed solution for designing real time multicast congestion control protocols distributes the tasks among all the nodes participating in the transmission.

Like in other single-layer protocols as [2,4,7] the sender simultaneously transmits the information to a single multicast group where all the receivers are listening. To be able to regulate the transmission rate of the sender node, one of the receivers is selected as a representative of all them and it emits regular feedback to the sender. We call the special receiver the session leader. The session leader must be the host experiencing the worst network conditions among all the receivers.

The rest of the receivers in the transmission, called followers, just have to help the sender to elect an appropriate leader. For this, every follower just calculates the transmission rate they would experiment if they were to be chosen as the session leader, and, whenever they detect this rate is lower than the current one, they warn the sender asking to become the new leader. The sender decides if it acknowledges the warning follower, appointing it in the new leader, or it ignores the petition if obeying to it would severely degrade the transmission quality. In this case, the warning follower should give up the transmission as it cannot cope with the minimum exigible rate.

2.2 The Sender

The sender in GMCC is very simple. While most congestion control protocols put a severe burden in the sender, in GMCC we have decided to make it rather simple, so that it can devote resources to other tasks, such as media codification or the transmission of several independent sessions.

With the help of the feedback it receives from the session leader, the sender calculates an appropriate sending rate. The algorithm used to calculate this rate must exhibit small variations in its throughput to be usable for multimedia applications [8,9,2,10]. We have experimented with two window based algorithms: binomial algorithms [5] and a modification of the TCP-Vegas algorithm [6,7] that we have called VLM, mostly because they are TCP-friendly algorithms with smooth variations in their throughput. Because changes in the session's leadership are treated by the sender as variations in the available network resources, we discarded equation-based algorithms [2] since under these dynamic conditions they perform sub-optimally [3].

Finally the sender implements a shaper that smoothes the instantaneous transmission rate. This shaper imposes a maximum length to the bursts of packets that the sender may transmit. We have chosen to impose a delay of T^{-1} every time two consecutive packets are transmitted, where T is the current transmission rate (Thus, $T = W/rtt$ for window based algorithms, where W is the congestion window size and rtt es the estimated RTT).

2.3 The Followers

The followers are nodes who are experiencing better network conditions than the Leader. So, their job is to monitor network conditions and, provided such conditions worsen, warn the sender requesting to become the new leader.

The tricky part is, of course, monitoring the network conditions: notwithstanding, the followers cannot send any regular feedback as that would render the protocol unscalable. With the information provided by detected packet losses and RTT measurements, the followers calculate the expected throughput a TCP-friendly connection to the sender would achieve under the same circumstances. For this calculation the same algorithm used by the sender is employed.

The sender adds information about the current sending rate to every data packet. Followers use this information to calculate a weighted moving average of the difference between their calculated expected throughput, and the rate announced by the sender. Every time this difference exceeds a certain threshold an alert in sent by the affected follower back to the sender, asking the sender to elect the alerting follower as the new session leader. In our design, we send an alert every time the average difference between the announced rate and the expected throughput is larger than the equivalent of increasing W in the follower with N_{extra} packets; that is, if

$$sender\_rate > \frac{W + N_{\text{extra}}}{\widehat{rtt}} \qquad (1)$$

The N_{extra} value is a compromise between having a transmission that is completely fair on all the network links with competing flows, but with a high frequency of leader changes, and one that may take more bandwidth in some links, but is much more stable. The amount of extra bandwidth allocated is bounded by N_{extra}, so GMCC can be adjusted to be as much fair as needed.

For the above to work receivers need to have an estimation of the RTT value from their point of view. As followers must avoid sending regular feedback to the sender, the RTT estimation must be carried in an open-loop fashion. Although having all the nodes in the transmission synchronised via GPS receivers or the NTP [11] protocol could be a solution to the problem, we believe it is not very practical in all circumstances, so we present a better solution in Section 3.

2.4   Connection Establishment

We have not addresses so far how the sender starts transmitting data when the connection is first created, as there is no designated leader at that time.

As we have seen in Section 3, nodes need to send an initial packet to the server to obtain an initial rtt estimation. The sender uses this packet to obtain information about the receivers present in the session. When the sender has data to send it can choose as the initial leader the last receiver that challenged it to obtain the rtt.

2.5   Leadership Loss

GMCC needs the presence of a host responsible of acknowledging every packet sent by the sender. If this leader host is missing, the sender must manage somehow to find a substitute. In case the protocol did not provide a mechanism for selecting a new leader, the congestion window size at the sender would drop to 1, and packets would only be sent due to timeouts. Moreover, as probably every host in the transmission will be able to admit higher throughput, no challenges will be produced, and the no-leader situation would perpetuate itself. To avoid this deadlock, the protocol needs to resolve two problems, the first is detecting a leadership loss, and the second is being able to recover from it without causing feedback-implosion.

There is obviously no reliable way to know that the leader host is down or that it has dropped its connection, because in the first case it may not be able to warn the sender. For this reason GMCC uses an heuristic to detect leadership loss. The sender assumes that the leader is not responsible when more than a whole window of packets is sent without getting back a single acknowledgement.

Once the leader loss has been detected the sender has to elect a new leader. For this, it ignores the fact that it is not getting acknowledgments and keeps on sending packets at the previous sending rate, so that the connection throughput does not degrade unnecessarily. As it is a fact that all the nodes were not experiencing worse network conditions than the leader, the sender increments this rate each RTT to force an alert from some host in the network. Once this alert arises, the sender can choose the challenger host as the new leader. For the window based algorithms we have used, we increment the window size by one packet per RTT.

We have developed a simple simulation to help to understand the behavior of the leadership recovery algorithm. In Fig. 1(a) we show the network topology used in the simulation. There is a single source and four receivers under two different bottlenecks. At the beginning of the simulation all hosts are subscribed to the appropriate multicast group. The host under the 1 Mbps link is selected as the leader, but after five seconds it leaves the multicast group. In Fig. 1(b) the reaction of the sender is shown. Before second five the window size is kept at a constant value, but once the sender discovers that the leader is not acknowledging the data packets any more, it starts increasing the window size. Around the second 6.6 of the simulation a new leader is finally elected and the window size converges to the value needed to fill the 5 Mbps available in the bottleneck.

3   The RTT Open-Loop Estimation Algorithm

For the followers to be able to calculate an admissible receiving rate, they need an estimation of the RTT. For this GMCC followers do an initial estimation of rtt when they first join the session sending a special timestamped packet to the sender, that must always be answered.

After this initial estimation, the follower adjusts rtt according to

$$rtt \leftarrow rtt + \Delta_{rtt} \tag{2}$$

every time two consecutive packets are received. With this rtt value, the estimation $\widehat{rtt}$ can be computed using a weighted moving average. But how is Δ_{rtt}

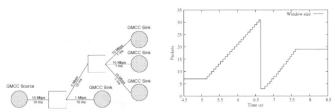

(a) Network used to show the behaviour of the leadership recovery algorithm.

(b) Window size variation at the sender when the session leader is lost.

Fig. 1. Leadership recovery algorithm behaviour.

calculated? The sender has to provide the needed information in every packet by inserting the time since the previous packet was sent: `drift_prev`. So

$$\Delta_{rtt} = \text{now} - (\text{lastArrival} + \text{drift\_prev}) \qquad (3)$$

This method works very well when there are few packet losses, but can cause the RTT estimation to drift if the number of packet losses raises.

This is not a severe problem most of the time, because usually the RTT does not suffer sharp variations over short periods of time, so single packet losses are not normally significant in practice, but some scenarios, as the one depicted in Fig. 2 can happen that make this RTT measurement unreliable.

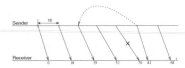

Fig. 2. Scenario to illustrate rtt measurement drift at the receiver.

In Fig. 2, packets are being sent every 16 time units. In the bottom line the time the packets are actually received at a chosen follower node is shown. During the first five packets, the inter-arrival time increases, for instance because a router's queue is getting filled, but the 6th packet sees the same network congestion as the first one. In Table 1 the measurements of the RTT done by the receiver are shown. We use 0 as the initial *rtt* value for simplicity, without any loss of generality in our argument.

Table 2 shows the same measures when the packet marked with a cross in the figure is lost. When the 6th packet reaches the receiver it cannot know how much did the *rtt* vary since the previous packet, because it never reached the

Table 1. RTT and RTT variation measures done by the receiver in Fig. 2 when no packets are lost.

Packet	1	2	3	4	5	6	7
Δrtt	?	+2	+1	+1	+1	−4	+1
rtt	0	2	3	4	5	1	2

receiver. So, it assumes it has not changed and keeps the same rtt value, hoping the best. After the last packet is received the rtt value is 7 time units instead of 2. This overestimation of the RTT causes the receiver to underestimate its available bandwidth and may fie a superfluous challenge packet to be sent.

Table 2. RTT and RTT variation measures done by the receiver in Fig. 2 when the 5 th packet is lost.

Packet	1	2	3	4	5	6	7
Δrtt	?	+2	+1	+1	+1	?	+1
rtt	0	2	3	4	5	5	7

One way to overcome this problem is having not only the information about the offset to the previous packet, but also the information about the offset to previously received packets. We have reserved an optional header field in the data packets to carry this information, whenever the sender decides to send it. Thus all the receiver has to do is to record the time the last packets were received and the rtt value at that time.

For instance, if the sender had sent this information in the 6 th packet about the offset to the 3 rd packet, the receiver could have corrected the RTT measurement. In the figure, we have depicted this with a dotted line. In the Table 3 we can see the effect of this information. When the 6 th packet arrives the receiver uses this additional information to decide that rtt should be decreased in 2 units, taking as a reference the value it had after the 3 rd packed was processed.

Table 3. RTT and RTT variation measures done by the receiver in Fig. 2 when the 5 th packet is loss, but the 6 th carries enough information to correct the error.

Packet	1	2	3	4	5	6	7
Δrtt	?	+2	+1	+1	+1	−2	+1
rtt	0	2	3	4	5	1	2

The last problem to solve was deciding when the sender should send this information. In our implementation we decided to send information about the offset with a packet sent W packets before in every packet, where W is the

Fig. 3. Illustration of information carried by *Data* packets to avoid drifts in the RTT measurement at the receivers. In the figure, packets from 33 to 36 have been lost, so as $W = 4$ there will be error in the RTT estimation.

congestion window size. This is depicted in Fig. 3. In the figure $W = 4$ and it can be easily observed why a whole window of consecutive packet losses is needed, for the receiver to err in the RTT measurement.

4   Experimental Results

We have run several simulation with the help of ns-2 [12] to compare the performance obtained by the GMCC protocol when configured with different rate estimators. In particular we have tested the protocol with the three more important binomial algorithms (AIMD, IIAD and SQRT) and our modification of the Vegas algorithm (VLM) configured with the default parameters that their TCP counterparts have in ns-2. We present here the most significant results.

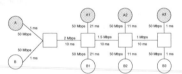

Fig. 4. Network configuration used for the simulations. All routers use simple FIFO queues.

For the first simulation he have configured the network depicted in Fig. 4 with three TCP flows, all originating from node A and with destinations A1, A2 and A3 respectively. At the same time, a GMCC session is established with node B as the source, and B1, B2 and B3 acting as destinations. We have represented in Fig. 5(a) the results obtained for nodes A2 and B2 (the rest of the nodes have similar results). For each rate estimator we have run a simulation ten times for 100 seconds and represented the averaged throughput obtained by GMCC (left) and TCP (right). We can see how all estimators, except VLM (the one bases in TCP-Vegas), are able to share the network fairly with TCP. VLM get less than its fair share because the routers are configured with quite big queues and TCP connections monopolise them. In this situation delay-based algorithms get less throughput than the fair share [13]. For a Vegas-like algorithm to perform optimally it should be used in isolation, for instance by using different QoS classes.

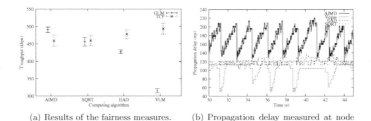

(a) Results of the fairness measures.   (b) Propagation delay measured at node B3.

Fig. 5. Basic properties of the different congestion control protocols implemented in GMCC.

In Fig. 5(b) we show the jitter that is produced by each rate estimation algorithm when confronted to itself. For this we have replaced the three TCP flows with a second GMCC session from node A to A1, A2 and A3 and then measured the propagation delay for each of the rate estimators. We can observe here that it is the delay based estimator the one that has the lowest average propagation delay, but, at the same time, both SQRT and IIAD exhibit a very good behaviour. However, AIMD, because of its highly oscillatory nature exhibits a very high jitter that is not appropriate for real time applications.

5   Conclusions

We have presented a novel multicast congestion control protocol suitable for the transmission of multicast multimedia data. Our work has several innovations. It uses a completely decentralised model for the election of the representative receiver that frees the server of some load. This decentralisation also reduces the feedback that receivers have to send back to the sender, up to the point that there is no need to regulate it to avoid feedback-implosion.

We have also shown an open-loop algorithm for the estimation of the RTT at the receivers, that is a key part of the decentralisation process. Finally GMCC permits the use of different rate adjustment algorithms to tailor the characteristics of the transmission to our needs. Using binomial algorithms when competing with greedy TCP traffic or a Vegas-like algorithm when we have bandwidth reservation for our multimedia traffic.

Finally we have tested that the protocol is fair against competing TCP-traffic and thus is safe for deployment in the Internet.

References

1. McCanne, S., Jacobson, V., Vetterli, M.: Receiver-driven layered multicast. In: ACM SIGCOMM. Volume 26,4., New York, ACM Press (1996) 117–130
2. Widmer, B., Handley, M.: Extending equation-based congestion control to multicast applications. In: Proceedings of the 2001 conference on applications, technologies, architectures, and protocols for computer communications, ACM Press (2001) 275–285
3. Bansal, D., Balakrishnan, H., Floyd, S., Shenker, S.: Dynamic behavior of slowly-responsive congestion control algorithms. In: Proceedings of the 2001 conference on Applications, technologies, architectures, and protocols for computer communications, ACM Press (2001) 263–274
4. Rizzo, L.: pgmcc: a TCP-friendly single-rate multicast congestion control scheme. In: Proceedings of the conference on Applications, Technologies, Architectures, and Protocols for Computer Communication, ACM Press (2000) 17–28
5. Bansal, D., Balakrishnan, H.: TCP-friendly congestion control for real-time streaming applications. MIT Technical Report. MIT-LCS-TR-806 (2000)
6. Brakmo, L.S., O'Malley, S.W., Peterson, L.L.: TCP Vegas: New techniques for congestion detection and avoidance. ACM SIGCOMM Computer Communication Review **24** (1994) 24–35
7. Rodríguez Pérez, M., Fernández Veiga, M., Herrería Alonso, S., Suárez González, A., López García, C.: A receiver based single-layer multicast congestion control protocol for multimedia streaming. In: Proceedings of the third International IFIP-TC6 Networking Conference. Number 3042 in LNCS, Athens, Greece, Springer-Verlag (2004) 550–561
8. Tan, W., Zakhor, A.: Real-time Internet video using error resilient scalable compression and TCP-friendly transport protocol. IEEE Transactions on Multimedia **1** (1999) 172–186
9. Rejaie, R., Handley, M., Estrin, D.: RAP: An end-to-end rate-based congestion mechanism for realtime streams in the internet. In: Proceedings of the IEEE INFOCOM. Volume 3. (1999) 1337–1345
10. Feamster, N., Bansal, D., Balakrishnan, H.: On the interactions between layered quality adaptation and congestion control for streaming video. In: 11th International Packet Video Workshop. (2001)
11. Mills, D.L.: Network time protocol (version 3) specification, implementation (1992)
12. NS: ns Network Simulator (2003) http://www.isi.edu/nsman/ns/.
13. Martin, J., Nilsson, A., Rhee, I.: Delay-based congestion avoidance for tcp. IEEE/ACM Trans. Netw. **11** (2003) 356–369

A Multi-channel Route Relay Protocol for Path Breakage and Inter-channel Interference in Mobile Ad Hoc Networks

Kyung-jun Kim, Jin-nyun Kim, and Ki-jun Han*

Department of Computer Engineering, Kyungpook National University
{Kjkim,duritz}@netopia.knu.ac.kr, kjhan@bh.knu.ac.kr

Abstract. Quality-of-Service (QoS) routing and bandwidth reallocation is one of the most important issues to support multimedia service in wireless multi-hop networks. The goal of QoS routing is to find a satisfactory path that can support the end-to-end QoS requirements of the multimedia flow. In this paper, we propose a multi-channel route relay MAC protocol to alleviate the adverse effects by inter-channel interference and path breakage caused by mobility in multi-channel mobile ad hoc networks. Our protocol relays an alternate channel via the neighbor which finds the released node when a communicating node moves out of transmission range. Simulation results demonstrated that our protocol may reduce communication overheads associated with exchanging control messages for re-establishing a path.

1 Introduction

The next-generation mobile/wireless networks are envisioned to constitute a variety of infrastructure substrates, including fixed single and multi-hop wireless/mobile networks. A multi-hop mobile radio network, also called mobile ad hoc network is a self-organizing and rapidly deployable network in which neither a wired backbone nor a centralized control exists [8]. A mobile ad hoc network consists of a number of geographically distributed. The network nodes communicate with one another over scarce wireless channels in a multi-hop fashion. Mobile ad hoc networks are a new paradigm of wireless wearable devices enabling instantaneous person-to-person, person-to-machine, or machine-to-person communications immediately and easily.

The work on multi-channel MAC protocol includes two issues: 1) Channel assignment is to decide which channels are to be used by which hosts. 2) Medium access is to resolve the contention/collision problem when using a particular channel. There are many related works that study the benefits of using multiple channels [1], [8], [12].

Wu et al. [10], [12] has proposed a protocol called Dynamic Channel Assignment (DCA) that assigns channels dynamically, in an on-demand style. This protocol maintains one dedicated channel for control messages, and other channels for data.

---

* Correspondent author

M. Freire et al. (Eds.): ECUMN 2004, LNCS 3262, pp. 206–215, 2004.
© Springer-Verlag Berlin Heidelberg 2004

When the number of channels is small, one channel dedicated for control messages can be costly [6]. Chang et al. [7] proposed a protocol that used a similar scheme as [9] in having one control channel and two or three data channels. This scheme assigns channels to hosts dynamically.

In this paper, we propose a multi-channel MAC protocol with route relay capability. To this end, in order to support QoS sensitive multimedia applications, both increasing the overall system capacity by efficiently reallocating channel as well as when there is inter-channel interference or path breakage due to mobility in mobile ad hoc networks. Our protocol dynamically negotiates channels to allow multiple communications in the same region simultaneously, each in a different channel without any interference problem. Our protocol offers a higher throughput, and a more enhanced QoS capability as well as less transmission delay than the conventional ones.

The remainder of this paper is organized as follows: In Section 2, we propose a multi-channel route relay protocol for minimum path breakage and inter-channel interference reduction in mobile multimedia ad hoc networks with dynamic route relay capability. In Section 3, we present an analysis and simulation model for performance evaluation. Finally, Section 4 contains conclusion and future work, reveals that our protocol improves the capacity and effectively copes with the path breakage and inter-channel interference of communication of ad hoc networks.

2 Dynamic Channel Routes Relaying Protocol

2.1 Preliminary Study

As a host moves into the communication range of another host pair, there will be a path breakage and communication interference with each other. In mobile ad hoc networks, most MAC protocols are designed for sharing a single channel to resolve potential contention and collision [12]. In this reason, we propose our protocol. Several assumptions have been made to reduce complexity of protocol descriptions.

1) All channels have the same bandwidth. None of the channels overlap, so the packets transmitted on different channels do not interfere with each other. Hosts have prior knowledge on how many channels are available through control messages.

2) Each host is equipped with two half-duplex transceivers. So, a host can both transmit and listen simultaneously. When listening to one channel, it can senses carrier on the other channels.

3) The transceiver is capable of switching its channel dynamically, and channel switching requires a negligible overhead.

4) We assume a TDMA-based system on a multi-channel shared by all hosts. To avoid collisions, for example, consider the currently hot CDMA technology, CDMA can be overlaid on the top of the TDMA system. This may mean that a mobile host can utilize multiple codes simultaneously, or dynamically switch from one code to another as needed.

The IEEE 802.11 standard is based on the single channel model although it has multiple channels available for use, for example, the IEEE 802.11b and 802.11a physical layers (PHY) have 14 and 12 channels, respectively [1], [6]. This is because each IEEE 802.11 host is equipped with one half-duplex transceiver, so it can only transmit or listen on one channel at a time. When a host is listening on a particular channel, it cannot hear communications taking place on a different channel [6]. When a single-channel MAC protocol is applied in a multi-channel environment wherein each node may dynamically switch channels, the network performance is inevitably degraded unless additional precautions are taken to manage dynamic channel selection [12].

However, consider, for example, the Code Division Multiple Access (CDMA) technology; this may mean that a mobile host can utilize multiple codes simultaneously, or dynamically switch from one code to another as needed [8]. A multi-channel MAC protocol can be defined as one with such capability.

2.2 Dynamic Channel Routes Relaying Protocol

Consider the early stage of network topology shown in Fig. 1(a), where a circle node represents a host, and the number on a link specifies the channel that is occupied by the pair of hosts connected by the link. A ticker-connecting link between host pairs D-A, F-C, M-L-K and E-H indicates that they are within the communicative range in the Fig. 1. Host D, however, is located within radio range of host A, host B and host G. Host D and host F are not within reception range of each other. Four connections have already been established, first between host D and host A, second between host F, host C and host M, third between host L and host K and forth between host E and host H. A channel that is to be used by one pair of hosts can be reused by another pair of hosts, only if their communication ranges do not overlap. Host D and host A are in the communication state and they use *ch#1* for communication.

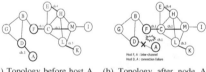

(a) Topology before host A moves   (b) Topology after node A moves out of transmission range

Fig. 1. Network Topology

When a host is listening on a particular channel, it cannot hear communications taking place on a different channel. Due to this, as observed in [6], if a single channel MAC protocol is applied in a multi channel environment wherein each node may dynamically switch channels, performance degradation may occur (unless additional precautions are taken to manage dynamic channel selection.) So we assume, as refer to our analysis, that the four channels, *ch#1, ch#2, ch#3* and *ch#4* are provided by the system.

Consider that the destination host gradually moves away from the source host, as shown in the Fig. 1(b). For example, host A moves towards host F. Then there will be a communication breakage between the source host D and the destination host A since the destination host moves out of communication range of the source host.

As shown in Table 1, each mobile node maintains a communications states table (CST) that includes source/destination node and the channel used by source node or not. When one of the communication pairs detects an interference or a link failure, the source host broadcasts a Node Search Message (NSM) to its neighbours as depicted in Fig. 3. Every host will be flooded with the NSM until the released node is discovered.

Table 1. Communications state table

Source	Destination	Busy	Idle
D	A	$ch\#1$	$ch\#2, ch\#3, ch\#4$
F	M	$ch\#1$	$ch\#2, ch\#3, ch\#4$
L	K	$ch\#3$	$ch\#1, ch\#2, ch\#4$
E	H	$ch\#4$	$ch\#1, ch\#2, ch\#3$

But, when the released host A receives the NSM, consider host A, which hopes to communicate with another host. It checks the CST and selects a proper channel (for example, the smallest channel). If host A determines to use $ch\#2$ for communication, the host pairs will change their communication channel to avoid co-channel interference with specific F-C host. The host pair simultaneously compare the received message with the information stored in their CST. And, then it replies with a Network Search Message Acknowledgement (NSM_ACK) to the finder directly. If the released node, however, experiences interference with someone before receiving the NSM, then it can immediately initiate the route relay process by sending NSM_ACK to the interfering host. The finder will forward the NSM_ACK to the source host to report where the released host is finally found. The NSM_ACK contains the information about idle channels available for route relay.

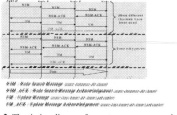

Fig. 2. The timing diagram for route recovery procedure

For example, if host F will find the released host A in Fig. 1(b), as the communicating host A gradually moves toward each other, there will be an inter-channel interference. Then, the source D will know that its released node is currently located near host F. At this time, if the source host and the finder have been using different channels for communicating before the breakage, then the finder may establish a relaying route from the source to the destination via itself.

In Fig 2, host F may establish a relaying route to D from host A. As soon as a new communication link has been built up, a update message should be transmitted to those neighbors that are currently in the communication state to maintain up-to-date communication information.

By contrast, if they have been using the same channel for communication before the breakage, there will be channel interference between the destination and the finder since they are sharing a common channel within transmission range of each other.

In this case, the finder and the source should choose another channel, for example one with the lowest channel number, if they agree to communicate via the finder. Fig. 3 plots the communication state graph for route recovery.

Fig. 3. Topology after a relay route is established

Finally, as shown in Fig. 3, the *A-D* path use *ch#2* for communication. After relaying the route, the source host *D* creates an update message to notify their neighbors of changed communication channels.

However, consider the case where the *B-F* link of the *A-D* path could use *ch#3* instead of *ch#2*, freeing such a channel in node *B* for proper usage. On receiving the update message, neighbors of hosts *D*, *A*, *F* and *B* update their CST's and the route relaying procedure is completed.

Fig. 4. An example for executing channel relay in case of congestion.

As shown in Table 2, the following functions are defined to formally describe our relay protocol.

Table 2. Notation for route relay algorithm

FUNCTION	DESCRIPTIONS
OneHop(x)	*OneHop(x)* defined to be the set of hosts located in communicative range of host x
FreeChannel(x)	*FreeChannel(x)* is a function that seeks a free (idle) channel to establish a communication link.
Neighbor(OneHop(x))	*Neighbor(OneHop(x))* computes the set of hosts located in communicative range of *OneHop(x)*.
Connect(x)	*Connect(x)* computes the number of channels which have been connected.

Consider the example in Fig. 4, host *L*, which hopes to communicate with another host *H*. Host *L* perform *FreeChannel(L)={ch#1, ch#2}* function and determine a idle channel for channel assignment. Host *L* selects a smallest idle channel *{ch#1}* and

ALGORITHM FOR SATURATED CHANNEL ASSIGNED STATE

DESCRIPTION
Host x which hopes to communicate with another host y.
At this moment, host H performs the following operations which
calculate the number of assigned channel.

ALGORITHMS :
repeat for ever
    When host x wants to communicate with host y.
    if FreeChannel(x) not NULL   /* Host x process free channels */
        send CRM message to host y :
        if CRM message received   /* Host y idle channel check */
            CHECK CST table;
        if FreeChannel(y) not NULL && /* Host y */
                Connect(y) > maximum number of channels {
            send CRM_ACK to host x;
            call OneHop(y);
        }
        else {
                call Channel negotiation process; /* see Fig. 3 */
                exchange host y and host x; /* their information */
                end loop;
        }
        if CRM_ACK message received   call OneHop(x); /* Host x */
    else end loop;
    if Neighbor(OneHop(y) ∩ OneHop(x)) is available /* Host x */
    {
        if (FreeChannel(y) ∩ FreeChannel(x)) /* All channel are used */
            discard CRM message request;
        else   call Channel negotiation process;
    }
    broadcast update message; /* update CST table */
    end loop;

Fig. 5. Pseudocode for route relay procedure (See Table 2 for notions)

then host L send CRM message to host H. On receiving the CRM message, host H check a number of channel in the communication states table.

If $FreeChannel(H)=\{ch\#1\}$ function is available and $Connect(H)=\{ch\#4\}$ function is larger than four, then host H sends a CRM_ACK to host L and perform $OneHop(H)=\{E,C,L,M\}$. If all channel are busy or a number of channel connected less than four, calling the channel negotiation process. Host L and H exchange their information. On receiving CRM_ACK message, at that moment, host L execute $OneHop(L)=\{C,M,H,K\}$ function. If $Neighbor(OneHop(H) \cap OneHop(L))=\{C,E\}$ function is not available, this request will discard.

If not, calling the channel negotiation process. Hosts update their CST table.

3 Analysis and Simulations

A simulation was conducted in order to evaluate the proposed protocol. For simulation in an MANET, it is assumed that a network consists of 20 through 100 nodes in a physical area the size of 1000m x 1000m, and each node has a transmission radius of 200m. The node travels for 1000 seconds. The location of the node is updated once per second. When the node arrives at its randomly chosen destination, it sets a pose time of zero. Each run is executed for 1000 seconds of simulation time.

The success rate of channel assignment is generally proportional to the number of channels provided. The proposed protocol always selects the smallest channel to increase the success rate of channel route relay. Suppose that we are given a fixed channel bandwidth. Let the bandwidth of the control channel be B_c, the number of data channels be n, and that of each data channels B_d.

Let the lengths of control packet and data packet be denoted by L_c and L_d, respectively. Based on the assumption, the number of data channels should be limited by the scheduling capability of the control channel. Since the control channel can schedule a data packet by sending at least *3* control packets, the maximum number of data channels should be limited by

$$n \leq \frac{L_d/B_d}{3L_c/B_c} \qquad (1)$$

Similarly, the success ratio s_b is given by

$$s_b \leq \frac{n * B_d}{n * B_d + B_c} \qquad (2)$$

We can see that the success ratio is a function of the lengths of control and data packets. Thus, decreasing the length of control packets or increasing the length of data packets will improve the success ratio. Hence, the success ratio can be rewritten by

$$s_b = \frac{L_d}{3 * L_c + L_d} = \frac{n * B_d}{n * B_d * B_c} = \frac{B_c}{n * B_d} \qquad (3)$$

Therefore, the total success ratio S_t can be obtained by

$$S_t = \frac{3L_c}{L_d} \qquad (4)$$

Fig. 6 shows the success ratio versus the number of nodes. The curves labeled as random refers to performance by the conventional channel assignment scheme [7], and the curves labeled as *'channels 2'*, *'channels 3'*, and *'channels 4'* indicate performance by our protocol with *2, 3*, and *4 channels* available, respectively. Fig. 6 shows that our protocol offers a higher success ratio than the conventional one [7, 11] that randomly selects a channel from the set of available channels.

This is because it exploits channel reuse opportunity before assigning a channel. As previously stated, our protocol dynamically assigns channels to the nodes, based on the topological and channels status. It can be seen that the success ratio of channel negotiation is generally proportional to the number of channels provided.

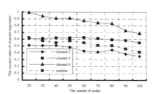

Fig. 6. Success ratio of channel negotiation process when there is inter-channel interference between two communicating pair.

Fig. 7 shows the control message overhead associated with the route relay. The conventional channel assignment scheme generally creates more control messages than our scheme. The conventional scheme [7] leads to inappropriate switch in the new channel. In general, the improper channel reassignment causes co-channel interference over the ad hoc networks. It may again introduce new inter-channel

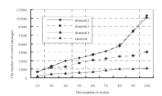

Fig. 7. The number of control messages versus the number of communication hosts.

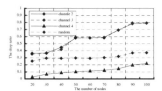

Fig. 8. Communication drop ratio

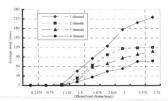

Fig. 9. Average frame delay versus the offered loads.

interference between communication host pairs. Therefore, it should extend the range of inter-channel interference. Our scheme not only exploits channel reuse opportunities but also eliminates the adverse effects by inter-channel interference, and thus may reduce the possibility of route relay failure and decrease the number of control messages.

In Fig. 8, it is shown that the communication drop ratio increases slowly as the number of available channels provided by the system increases. When two pairs of communicating hosts gradually move toward each other, they will undergo channel interference. When the number of communicating hosts increases, the success ratio of the route relay will decrease.

Fig. 9 compares the average frame delay of our MAC protocol against wireless protocol with single channel under different offered loads. The advantage of our multi-channel protocol is that as the load becomes heavier, the level of contention will not increase proportionally because the channel change cost is low.

However, when there is co-channel interference, the communication will fail if a number of hosts have been distributed in sparsely. This is because the mobile node exchanges route relay messages when there is co-channel interference or failure. Since the messages are propagated hop by hop, they cannot reach all nodes immediately. Also, it requires each node to maintain a updated knowledge of (part at least) of the overall topology.

This is very costly in an ad-hoc setting, where power saving is a major concern, and the maintenance of the topology knowledge requires the sending and receiving of many control messages which are propagated with the very costly technique of flooding.

4 Conclusions and Future Work

We proposed an efficient dynamic channel negotiation protocol for minimizing the effects by path breakage and inter-channel interference in mobile multimedia ad hoc networks. Our protocol with route relay capability may alleviate some problems due to inter-channel interference and path breakage. Simulation studies show that our protocol improves the capacity of ad hoc networks by effectively coping with the breakage of communication in a mobile ad hoc network. We are now investigating underlying performance limit and applicability extension of our protocol to the sensor-based wireless network.

Acknowledgement. University Fundamental Research Program supported by Ministry of Information & Communication in Republic of Korea

Reference

[1] IEEE 802.11a Working Group, "Wireless LAN Medium Access Control (MAC) and Physical Layer (PHY) specifications -Amendmant 1: High-speed Physical Layer in the 5 GHz band," 1999.
[2] R.Jurdar, C.V.Lopes and P.Baldi. "A Survey, Classification and Comparative Analysis of Medium Access Control Protocols for Ad Hoc Networks," IEEE Commu. Survey & Tutorial, Vol. 6. No.1, 2004, pp 2-16.
[3] Z. Tang and J. J. Garcia-Luna-Aceves, "Hop-Reservation Multiple Access (HRMA) for Ad-Hoc Networks," in IEEE INFOCOM, 1999.
[4] A. Muqattash and M. Krunz, "Power Controlled Dual Channel(PCDC) Medium Access Protocol for Wirelesss Ad Hoc Networks," in IEEE INFOCOM, 2003
[5] W. Navidi and T. Camp, "Stationary Distributions for the Random Waypoint Mobility Model," Technical Report MCS-03-04, The Colorado School of Mines, 2003.
[6] J.M. So and N.H. Vaidya, "A Multi-channel MAC Protocol for Ad Hoc Wireless Networks," Technical Report, University of Illinois at Urbana Champaign, 2003.
[7] C.Y. Chang, P.C. Huang, C.T. Chang, and Y.S. Chen, "Dynamic Channel Assignment and Reassignment for Exploiting Channel Reuse Opportunities in Ad Hoc Wireless Networks," in IEICE TRANS. COMMUN., Vol.E86-B, No.4, April 2003.

[8] G. Aggelou, "On the Performance Analysis of the Minimum-Blocking and Bandwidth-Reallocation Channel-Assignment (MBCA/BRCA) Methods for Quality-fo-Service Routing Support in Mobile Multimedia Ad Hoc," *in IEEE Transactions on Vehicular Technology Vol.53 No. 3*, May. 2004.

[9] N. Jain and S. Das, "A Multi-channel CSMA MAC Protocol with Receiver-Based Channel Selection for Multihop Wireless Networks," in Proceedings of the *9th Int. Conf. on Computer Communications and Networks (IC3N)*, October 2001.

[10] S.L. Wu, C.-Y. Lin, Y.-C. Tseng, and J.-P. Sheu, "A New Multi-Channel MAC Protocol with On-Demand Channel Assignment for Mobile Ad Hoc Networks", *Int'l Symposium on Parallel Architectures, Algorithms and Networks(I-SPAN)*, 2000, pp. 232-237.

[11] A. Nasipuri, J. Zhuang and S. R. Das, "A Multichannel CSMA MAC Protocol for Multihop Wireless Networks," in *IEEE Wireless Communications and Networking Conference (WCNC)*, September 1999.

[12] S.L.Wu, S.-Y. Ni, Y.-C. Tseng, and J.-P. Sheu, "Route Maintenance in a Wireless Mobile Ad Hoc Network," *Telecommunication Systems*, vol. 18, issue 1/3, September 2001, pp. 61-84.

An Adjustable Scheduling Algorithm in Wireless Ad Hoc Networks

Marc Gilg and Pascal Lorenz

University of Haute-Alsace, 34 rue du Grillenbreit, 68 008 Colmar, France
Tel: + 33 3 89 33 63 48
M.Gilg@uha.fr, P.Lorenz@uha.fr

Abstract. In a half duplex mobile Ad-Hoc networks, fair allocation of bandwidth and maximisation of channel utilisation are two antagonistic parameters. In this paper, we propose to use a maximal clique based scheduling algorithm with one control parameter. This algorithm permits to adjust the balance spot between fairness and throughput. A theoretical proof of the algorithm's behaviour will be done. We will apply this algorithm to an Ad-Hoc wireless network in some relevant cases.

1 Introduction

In recent research, fair resource distribution of bandwidth and maximisation in mobile ad hoc networks (MANETs) have been identified as two important goals [2]. The fact that fair allocation bandwidth and maximisation of channel utilisation are two antagonistic parameters are crucial. This is the consequence of the existence of some bottleneck in the network due to the hight contention of some nodes. Consideration of bottelnecks can be used in some scheduling algorithm [9] to find a balance spot between the two seemingly incompatible design goals: fairness and maximal throughput. Another approach is to allocate a minimum channel utilisation to each flow [3]. An algorithm based on bounding unfairness between flows is given in [6]. This algorithm assure some maximum bandwidth utilisation using constraint on the maximal clique numbers of the flows. This algorithms give a fixed compromise between fairness and maximal throughput.

The algorithm will use a parameter calles $TpsW$ to adjust the balance spot between fairness and throughput. This parameter controls the transmission of each flow between fairness and maximal bandwith utilisation. It is based on maximal clique numbers to assure fairness. One part is distributed, and another one is centralised to compute throughput, which is a global parameter to synchronise the transmissions.

The first section of this paper will give a theoretical background on flow contention graph, maximal clique and throughput. The second section describes our algorithm. The third section is a theoretical analysis of our algorithm, and establishs some convergence. The last section presents some simulations with an example of contention graph. We conclude the paper with some considerations of future development of this algorithm.

M. Freire et al. (Eds.): ECUMN 2004, LNCS 3262, pp. 216–226, 2004.

2   Generalities and Basic Definitions

2.1   Network Model

The considered mobile ad hoc network is a set of mobile nodes who form a packet-switched multihop wireless network. The nodes are working in half-duplex mode. This means that they can't transmit or receive packets at the same time. When a node is in broadcasting mode, only the nodes in some transmission mode will receive the sending packets. A stream of packets transmitted from one node to another is called single-hop flow. Each multihop flow can be broken in single-hop flow, and each single-hop flow is handle locally by its source and destination nodes. In this paper, each flow will be refered to a single-hop flow.

To modelise the wireless network we do the following assumptions:

- time is divided into discrete slots
- packets have all the same size and each packet can be transmitted in a slot
- the wireless channel is noise free and unsuccessful receptions are generated by collisions
- a single physical channel is available for wireless transmissions

2.2   Flow Contention Graph

Definition 1. A network topology graph is an undirected graph $G = (V, E)$, where V represents all nodes in the network and the set E represents all flows.

Definition 2. Two flows are **contending flows** if they are within a two hops distance.

Definition 3. A **flow contending graph** of a graph G is an undirected graph $\bar{G} = (E. \bar{E})$, where $\bar{E}$ is the set of edge $(f_i, f_j) \in E \times E$ such that the flow f_i, f_j are contending.

Example 1. This is an example coming from [6] that represent a network topology graph and its contending graph.

2.3   Maximal Clique

Definition 4. A maximal clique of a graph G is a maximal complete subgraph.

Remark 1. In a flow graph, the vertices in a maximal clique represent a maximal set of flows that contend with each other. At most one flow in the clique can transmit at any time.

Definition 5. The **maximal clique number** of a flow f is the number of maximal cliques which contents f. If the maximal clique number of the flow f is equal to 1, then f is a **single-clique flow**, else it is a **multi-clique flow**.

Example 2. For the contention graph describe in the figure 1 we have three cliques: $C_1 = \{f_1, f_2, f_3, f_4\}$, $C_2 = \{f_3, f_4, f_5, f_6\}$ and $C_3 = \{f_4, f_5, f_6, f_7\}$.

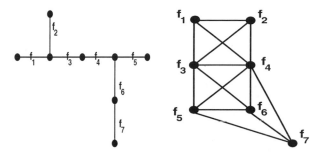

Fig. 1. Network topology graph and its flow contention graph

2.4   The Average Throughput

Definition 6. The average throughput noted *Tps* is defined as the ratio of the number of packets transmitted in the network and the number of slots used during the transmission.

Remark 2. The maximisation of bandwidth utilisation is equal to the maximun of Tps.

Definition 7. The maximum of Tps without any fairness consideration, noted Tps_{max} is given by the cardinality of the maximum independent set of the graph G, if all flows have the same weight.

Definition 8. The maximun Tps subject to fairness constraint Tps_{fair} is given by $Tps_{fair} = \sum_i W_i/X_w(G)$, where $X_w(G)$ is the weighted chromatic number of the graph G.

3   A Maximal Clique Based Packet Scheduling Model

The introduced algorithm try that Tps will go to the limit define by a pre-selected value $TpsW$, where

$$Tps_{fair} \leq TpsW \leq Tps_{max}$$

The choice of $TpsW$ have an incidence to the algorithm fairness. For each flow f_j, we define a **refute coefficient** e_j such as the maximal clique number of flow f_j increased with the number of activation. The flow with the lower refute coefficient will have the greater priority to transmit.

4   Algorithm Description

4.1   Initialisation

We suppose that each node knows the maximal cliques. This is done by a control message broadcasting and by using the Bierstone algorithm given in [1]. Each node will then initialise its refute coefficient e_j to the maximal clique number. Each node has an acknowledge array noted a_k from its neighbours noted f_k. A node is becoming the network master and sent a Master Ready message to all nodes.

4.2   Node Message Proceding

Each node f_j gets control messages. For each message, it will do the following task:

  − Get the message Master Ready:
    If $P_j = True$ then it send data and it do a Permission Computation,
    else it do a Permission Computation.
  − Get the message Master Init:
    It reset e_j to maximal clique number and it do a Permission Computation
  − Get the P_k and e_k messages from the neighbours f_k:
    it send Ack message to f_k.

  − Get the Ack message from f_k:
    It set $a_k = 1$
    If get Ack message from all neighbours then

      • it look for $\tilde{f}_k$ neighbots of f_j, f_j included, such that $P_k = 1$ and $\tilde{f}_k$ has less transmission.
      • it look for $\tilde{f}_h$ with h Minimum.
      • If $h \neq j$ then $P_j = 0$ and $e_j = e_j - 1$.
      • it send a Read message to master

4.3   Permission Computation

Each node f_j will determine its sending permission. In this case the algorithm can be describe as follow:

  − We set P_j to $False$.
  − If neighbours parameters P_k, e_k are not known then we read them.
  − If for all neighbours flows f_k we have $P_k = False$ and $e_j \leq e_k$ then we set $P_j = True$ and $e_j = e_j + 1$. We set a_k to 0 for all neighbours f_k and we send the parameters P_j, e_j to the neighbours
    else we send Ready message with P_j, e_j to the master.
  − We wait for messages

4.4 Master Message Management

If the master get the Ready message from all nodes then it will compute

$$Tps = \frac{\text{number of True Permission} - \text{number of False}}{\text{timestamp}}$$

If $Tps < TpsW$ then it send a Master Init message to all nodes else it send a Master Ready message to all nodes.

5 Influence of the Parameter $TpsW$

In this section, we will give a theoretical approach to the influence of the $TpsW$ parameter. A first theorem will decompose the contention graph into subgraph which will help us to prove the behaviour of the algorithm.

5.1 Subgraphs

We will define the subgraph $K_n(E_n, V_n), n \subset \mathbb{N}$ of the contending graph $G(E, V)$ by:

$$\left\{ C_f(E_f, V_f), \text{ clique of } G \setminus \bigcup_{i=1}^{n-1} K_i / \exists f \in E_n \text{ with } e_f = min\left\{ e_{\bar{f}}, \bar{f} \in G \setminus \bigcup_{i=1}^{n-1} K_i \right\} \right\}$$

Theorem 1. Let $f_1, \ldots, f_p$ be the transmissions flows with the coefficient e_f. then $\{f_1, \ldots, f_p\}$ is a maximal independent set of some K_n.

Proof. Let $f_1^1, f_2^1, \ldots, f_{q_1}^1, f_1^2, \ldots, f_{q_2}^2, \ldots$ be the transmission flows of G such that e_r is the coefficient of flows $f_1^r, \ldots, f_{q_r}^r$. We order the coefficient such that $e_1 < e_2 < \ldots < e_r < \ldots < e_n$.

By the selection of the algorithm, the transmission flow f_i^1 will have the minimum coefficient e_1 of its cliques $C_{f_i^1}$. Then we have $C_{f_i^1} \subset K_1$ by the definition of K_1. We notice that K_1 is not empty, because it exists at least a transmission flow f_1^1 with $e_1 = \min\{e_f, f \in E\}$.

As $f_1^1, \ldots f_p^1$ are the transmission flows, it is an independent set of K_1. Suppose it is maximal. If not, it exists a flow transmission flow f^1 of G such that $e_{f^1} = e_1$ and $f^1 \notin E_1$. As f^1 has a minimum coefficient of a cliques C_{f^1} then $f^1 \in K_1$, we have a contradiction.

By induction, we suppose that for $r < n$, $f_1^r, \ldots k_{q_r}^r$ is a maximal Independent set of K_r.

Let $f_1^n, \ldots, f_q^n$ be the transmission flows of G with coefficient e_n. Let $C_{f_1^n}, \ldots, C_{f_q^n}$ the cliques where f_j^n is a vertex. We will prove that $C_{f_1^n}, \ldots, C_{f_q^n} \subset K_n$.

If $f_s^n \in E_n$, then we have $C_{f_s^n} \subset K_n$ as f_s^n has the minimal coefficient of its clique.

If $f_s^n \notin E_n$ then $f_s^n \in E_r$ for some $r < n$ because $K_n \subset G \backslash K_1 \cup K_2 \cup \ldots \cup K_{n-1}$. This implies that f_s^n has a neighbour f_s^r with $e_r < e_n$. This is a contradiction with the fact that f_s^n is a transmission flow.

This prove that $f_1^n, \ldots, f_q^n$ is an independent set of K_n. We will suppose that this set is not maximal. Then it exists a flow $f^n \notin E_n$ of G which is a transmission flow and $e_{f^n} = e_n$. It exists a clique C_{f^n} such that f^n has minimum coefficient. As $f^n \notin E_n$ he have $C_{f^n} \subset K_r$ for some $r < n$. The definition of K_r implies that $e_{f^n} = e_r < e_n$. This is a contradiction with the definition of f^n.

We have proved that $f_1^n, \ldots, f_q^n$ i s a maximal independent set of K_n. By induction we prove that for every $n \in \mathbb{N}$.

Corollary 1. If we have $\forall f \in E$ $e_f =$ maximal clique number of f then $Tps_n = Tps_{max}$.

Proof. Let $f_1^1, \ldots, f_{q_1}^1, f_1^2, \ldots$ be the transmissions flows of G such that f_i^r has a coefficient e_r. Let $F_p = \{f_1^p, \ldots, f_{q_p}^p\}$ be the maximal independent set of K_p given by theorem 1.

I claim that $F = \cup_p F_p$ is a maximal independent set of G. The set is independent because all flows are in transmission mode. We will prove that it is maximal.

As the coefficients e_i are equal to the maximal clique number of the flows f_i, the set of transmission flows F_1 has the lowest maximal clique number of G. This is due the the fact the the algorithm select the lowest refute coefficient. Hence the graph $G \setminus K_1$ has the maximal number of transmission flow.

For each F_p, we chose a maximal set of transmission flow. This way we construct a maximal independent set of G.

5.2   Convergence to Tps_{max}

Theorem 2. If $TpsW \geq Tps_{max}$ then we have:

$$\lim_{n \to +\infty} Tps_n = Tps_{max}$$

Proof. If $TpsW \geq Tps_{max}$ then $Tps_n \leq TpsW$. The coefficients e_i will be always initialised to the maximal clique number of f_i by the algorithm. By using the corollary 1 we have $Tps_n = Tps_{max}$.

5.3   Convergence to Tps_{fair}

Lemma 1. Suppose that at tim e n_0 $e_i = c_{n_0}, \forall i$ then it exists $n_1 \geq n_0$ such that

$$Tps_{n_1} = Tps_{fair} \text{ and } e_i = c_{n_0} + 1 = c_{n_1}$$

Proof. At the time n_0 the transmission flows in F_1 will be a maximal independent set of $K_1 = G$. At the date $n_0 + 1$ the transmission flows $f_1^1, \ldots f_{q_1}^1$ will have

$e_i = c_{n_0} + 1$. As the other flows have a coefficient c_{n_0} then the transmission flows will be chosen in this set. It will be a maximal independent set of $E \setminus F_{n_0}$.

By induction, we have $F_p \subseteq G \setminus F_1 \cup F_2 \cup \ldots \cup F_{p-1}$. There exists n_1 such that $G \setminus F_1 \cup F_2 \cup \ldots \cup F_{n_1-1} = \emptyset$.

At this step, each flow transmited one packet, and $e_i = c_{n_0} + 1$.

Theorem 3. Let Tps_{min} be the minimum value of Tps_n. If $TpsW < Tps_{min}$ then

$$\lim_{n \to +\infty} Tps_n = Tps_{fair}$$

Proof. As $TpsW < Tps_{min}$ the coefficients e_i will never be initialised by the algorithm. Each time f_i is transmitting, its coefficient e_i is increasing. As the coefficient with the lowest coefficient e_i is transmited, it will be a time n_0 where all coefficients are equal to c_0. Then we can apply the lemma 1. It will be a date n_1 where $Tps_{n_0+n_1} = Tps_{fair}$ and $e_i = c_0 + 1$.

We can easily prove by induction that $Tps_{n_0+k.n_1} = Tps_{fair}, \forall k \in \mathbb{N}^*$. Let be $p \in \mathbb{N}^*$ such that $p \le n_1$ we have:

$$Tps_{n_0+k.n_1+p} = \frac{T_{n_0+k.n_1} + T_p}{n_0 + k.n_1 + p}$$

where $T_{n_0+k.n_1}$ is the number of packets transmited until $n_0 + k.n_1$. T_p is the number of packet transmited in p time slot. As T_p is finite, we have

$$\lim_{k \to \infty} Tps_{n_0+k.n_1+p} = Tps_{fair}$$

This prove that

$$\lim_{n \to +\infty} Tps_n = Tps_{fair}$$

5.4   Convergence to $TpsW$

We have study the case where $TpsW \ge Tps_{max}$ and $TpsW < Tps_{min}$. Now we will study the case where $Tps_{fair} \le TpsW \le Tps_{max}$. We will establish that in this case the Tps from the network will tend to $TpsW$. This will give some control over the bandwith of our network.

Theorem 4. Let be $TpsW$ such that $Tps_{fair} \le TpsW \le Tps_{max}$, then we have:

$$\lim_{n \to \infty} Tps_n = TpsW$$

Proof. At the time 1 the coefficients e_i are initialised to the maximal clique number of f_i. Using corollary 1 we have $Tps_1 = Tps_{max}$.

As the coefficients e_i are increasing each time, f_i is in a transmission mode. There is a date n_0 where all coefficients will be equal. At this time, if we have no initialisation of the coefficients e_i, then $Tps_{n_1} = Tps_{fair}$ for some n_1 (see lemma 1).

This prove that it exists $n_2 \leq n_1$ such that $Tps_{n_2} \leq TpsW$ and then the coefficients e_i are initialised to the maximal clique number by the master.

There exists $k \in \mathbb{N}$ such that $Tps_{n_2+k} \geq TpsW$. If not, we have $\forall n \geq n_2, Tps_n \leq TpsW$. Therefore e_i is always initialised to the maximal clique number of f_i.

We have $lim_{n \to +\infty} Tps_{n_2+n} = Tps_{max}$ (see lemma 1). As $TpsW \leq Tps_{max}$, we have a contradiction.

We can reuse this arguments to prove that Tps_n is allways oscillating around $TpsW$.

Let's define $\overline{Tps}_n \geq TpsW$ to be the sequence of local maximum of Tps_n, and $\underline{Tps}_n \leq TpsW$ to be the sequence of local minimum of Tps_n. We order the term such that $\underline{Tps}_n$ is the first local minimum after $\overline{Tps}_n$.

By basic computation and by using the definition of Tps, we can prove that $lim_{n \to +\infty} \overline{Tps}_n = lim_{n \to +\infty} \underline{Tps}_n$. This implies that $lim_{n \to \infty} Tps_n = TpsW$.

6 Simulations

The simulations were done in C programming language. We use the Message Passing Interface (MPI) library to implement the internode communications.

We will apply our algorithm to the contention graph of fig.1. We should recall that this graph has $Tps_{fair} = 1.75$ and $Tps_{max} = 2$.

6.1 Let Be $TpsW = 1.8$

We have $Tps_{fair} \leq TpsW \leq Tps_{max}$ The graph at the left hand side of figure 2 represents Tps during the fist 100 time units, the one at the right hand side represents the packets transmit by each node during this time.

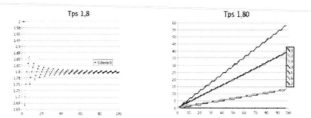

Fig. 2. Simulation with $TpsW = 1.8$

We see the oscillation of Tps around $TpsW$. In this case, there is no fairness of the transmission. We see that flow 4, the most contended flow, is never sending. The transmission flows can be classify in three sets of same transmission rate:

flow 7, flows 1-2 and flows 3-5-6. This is due to the equivalent position of the nodes in the contention graph.

6.2   Let Be $TpsW = 1.5$

In this case we have $TpsW < Tps_{min}$. The convergence will be to Tps_{fair}.

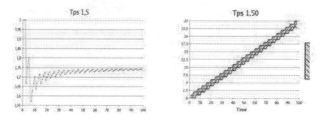

Fig. 3. Simulation with $TpsW = 1.5$

6.3   Let Be $TpsW = 2$

In this case we have $TpsW - Tps_{max}$.

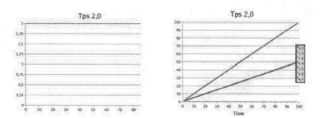

Fig. 4. Simulation with $TpsW = Tps_{max}$

We can notice that Tps is alway equal to Tps_{max} and that flows 7, 1 and 2 are in transmission mode. The flows 1 and 2 have a similar position in the contention graph.

7 Conclusion and Future Development

In this paper, we have presented an algorithm to manage the throughput and fairness parameters in a Ad-Hoc wireless network.

The theoretical study of the algorithm shows that the choice of $TpsW$ permit to get a throughput which can be Tps_{fair}, Tps_{max} or any value between them. The simulations show that the choice of $TpsW$ influences the fairness of the transmissions. The closest $TpsW$ will be from Tps_{fair} better will be the fairness of the network. We can although notice that the position of the flows in the contending graph is crucial and that the flows with equivalent positions will have the same transmission rate.

As the throughput is a parameter used to control our network, we need to centralise informations to compute this global information. This operation can be time consuming in large networks. To avoid this, it is possible to imaging the use of local masters to compute a local Tps. For each subgraph that will be controlled by a local master, the behaviour will be similar to the one describe in this paper.

References

1. J. G. Augustson and J. Radhakrishnan, *An analysis of some graph theoretical cluster techniques*, Journal of the Association for Computing Machinery, vol 17, no. 4, pp. 571-586, 1970.
2. S. Lu, V. Bharghavan and R. Srikant, *Fair scheduling in wireless packet networks*, IEEE Trans. Networking, August 1999.
3. H. Luo, S. Lu and V. Bharghavan, *A new model for packet scheduling in multi-hop wireless networks*, ACM MOBICOM'00, August 2000
4. H. Luo and S. Lu, *A topology-independent fair queueing model in ad hoc wireless networks*, IEEE ICNP'00, Nov. 2000.
5. P. Ramanathan and P. Agrawal, *Adapting packet fair queueing algorithms to wireless networks*, ACM MOBICOM'98, October 1998.
6. Tao Ma, Xinming Zhang, Guoliang Chen, *A Maximal Clique Based Packet Scheduling Algorithm in Mobile Ad Hoc Networks*, IEEE ICN'2004, February 29 - March 4 2004
7. L. Tassiulas and S. Sarkar, *Max-min fair scheduling in wireless networks*, IEEE INFOCOM'02, 2002
8. N. Vaidya, P. Bahl and S. Gupta, *Distributed fair scheduling in a wireless lan*, ACM MOBICOM'00 August 2000.
9. Xinran Wu, Clement Yuen, Yan Gao, Hong Wu, Baochun Li *Fair Scheduling with Bottleneck Consideration in wireless Ad-hoc Networks*, IEEE ICCCN'01, October 2001.
10. Yu Wang, J. J. Garcia-Luna-Aceves *Throughput and fairness in a hybrid channel access scheme for ad hoc networks*, WCNC 2003 - IEEE Wireless, Communication and networking conference, no. 1, March 2003 pp. 988-993
11. Syed Ali Jafar, Andrea Goldsmith *Adaptative multirate CDMA for uplink throughput maximization*, IEEE Transactions on wireless communications, no. 2, March 2003, pp. 218-228

12. Z. Y. Fang, B. Bensaou, *A novel topology-blind fair medium access control for wireless LAN and ad hoc networks*, ICC 2003 - IEEE International conference on communications, no. 1, May 2003 pp. 1129-1134
13. Li Wang, Yu-Kwong Kwok, Wing-Cheong Lau, Vincent K. N. Lau *On channel-adaptative fair multiple access control*, ICC 2003 - IEEE International conference on communications, no. 1, May 2003 pp. 276-280
14. John T. Wen, Murat Arcak *A unifying passivity framework for network flow control*, IEEE INFOCOM 2003 - The conference on computer communications, no. 1, March 2003 pp. 1156-1166
15. Yoshiaki Ofuji, Sadayuki Abeta, Mamoru Sawahashi *Fast packet scheduling algorithm based on instantaneous SIR with constraint condition assuring minimum throughput in forward link*, WCNC 2003 - IEEE Wireless communications and networking conference, no. 1, March 2003 pp. 860-865

A Correlated Load Aware Routing Protocol in Mobile Ad Hoc Networks

Jin-Woo Jung[1], DaeIn Choi[1], Keumyoun Kwon[1], Ilyoung Chong[2],
Kyungshik Lim[3], and Hyun-Kook Kahng[1]

[1] Department of Electronics Information Engineering, Korea University
#208 Suchang-dong Chochiwon Chungnam, Korea 339-700
jjw@korea.ac.kr
[2] Information and Communications Engineering dept. Hankuk University of FS
#207 Imun-dong, Dongdaemun-Gu, Seoul, Korea 130-790
[3] Computer Science Department Kyungpook National University
Taegu, Korea 702-701

Abstract. In this paper, we investigate the load distribution in ad hoc networks for reducing the possibility of power depletion and queuing delay at a node with heavy load environments. We examine the efficiency of dominant ad hoc routing protocols from a point of load balancing. Based on the results from various simulations, we design a new Correlated Load-Aware Routing (CLAR) protocol that considers the traffic load, through and around neighboring nodes, as the primary route selection metric. Simulation results show that CLAR outperforms other popular protocols and is better suited for the heady load networks with low mobility.

1 Introduction

In recent, the proliferation of portable devices with diverse wireless communication capabilities has made a mobility support on the Internet an important issue[1]. A mobile computing environment includes both infrastructure wireless networks and novel infrastructure-less mobile ad hoc networks (MANETs). A MANET is a self-organizing system of mobile nodes connected by multi-hop wireless links. Major challenges in such networks are to design an efficient routing protocol to establish and maintain multi-hop routes among the nodes, characterized by the frequent mobility, bandwidth limitation and power constraint. Several routing protocols have been proposed and can be broadly classified into proactive and reactive protocols [1]. While both proactive protocols and reactive protocols use the shortest route as the sole criteria to select a route, this shortest path may not always achieve the minimum end-to-end delay and fairly distributes the routing load among mobile nodes. Therefore, these unbalanced distributions of traffics may lead to network congestion and cause bottlenecks within the congested areas.

---

[1] This research was supported by University IT Research Center Project

M. Freire et al. (Eds.): ECUMN 2004, LNCS 3262, pp. 227–236, 2004.
© Springer-Verlag Berlin Heidelberg 2004

Several approaches which consider the traffic load of a path as the primary route selection criteria ([2],[3], and [4]) have recently been proposed. However, the metric used to choose the optimal route in these approaches is inaccurate due to the reflection of transient status and interference of radio channels.

In this paper, we examine the performance of dominant ad hoc routing protocols such as AODV (Ad Hoc On-demand Distance Vector Routing) and DSR (Dynamic Source Routing) from a view point of load balancing. Then, we propose a new load-balanced routing protocol, called as Correlated Load Aware Routing Protocol, to overcome a performance degradation of existing protocols. To achieve load balancing in ad hoc networks, the CLAR include new routing metrics and route discovery algorithm to determine the optimal route: path load, RREQ-delay and RREP-fast-forward.

2 Background and Related Works

2.1 AODV

AODV minimizes the number of broadcasts by creating routes on-demand. To find a path to any destination, the source node broadcasts a route request packet (RREQ). The neighbors in turn relay the packet to their neighbors until it reaches an intermediate node that has recent route information about the destination or until it reaches the destination. When a node receives the RREQ message it has two choices. If it knows a route to the destination or if it is the destination, they can send a Route Reply (RREP) message back to the source node. The RREQ allows AODV to adjust routes when nodes move around.

2.2 DSR

The key difference between DSR and other on-demand protocols is the routing information is contained in the packet header. Since all necessary routing information is contained in the packet header then the intermediate nodes do not need to maintain routing information. Unlike conventional routing protocols, there are no periodic routing advertisements. Therefore, DSR can reduce control overhead, particularly during periods when a little or no significant node movement takes place.

2.3 Related Works

Several approaches which consider the traffic load of a path as the primary route selection metrics ([2].[3], and [4]) have recently been proposed. However, the metric used to choose the optimal route in these approaches is inaccurate due to the reflection of transient status and the interference of radio channels. In [2] and [4], while the authors define the traffic load of a mobile node as the number of packets being queued in the interface, the metric is inaccurate due to the burstness of Internet traffic (especially, Transmission Control Protocol). A Load-Sensitive Routing (LSR) protocol proposed in [3] uses the nodal activity (the total number of active routes passing through the node and its neighbors) to determine the traffic load. Even though the load metric of LSR is more accurate than those of [2] and [4], it still does not consider the dynamic rate of traffic source and the characteristics of real network load (i.e. the traffic load of multiple connections for a source-destination pair may higher than that of one connection for a source-destination pair).

3 Correlated Load Aware Routing Protocol

3.1 Traffic Load Measurement

In a CLAR, traffic load at a node is considered as the primary route selection metric. The traffic load of a node depends on the traffic passing through this node as well as the traffic in the neighboring nodes [6]. The traffic load in a node is thus defined as the product of the *average queue size* at the node and the number of *sharing nodes*, which is that its average queue size is over one packet. The average queue size is calculated with an exponentially weighted moving average (EWMA) of the previous queue lengths. The EWMA is used in various research fields such as the calculation of average queue size in RED [6] and RTT estimation in TCP [7]. The advantage of an EWMA is to filter out transient congestion at the router. Because the short-term increases of the queue size that result from burst traffic or from transient congestion do not result in a significant increase of the average queue size. An exponentially weighted moving average is a low-pass filter and the equation of it is likely as below:

$$avg = (1 - w_q) \ avg + w_q q \tag{1}$$

The above formula is similar to the one used by TCP to estimate the round-trip delay. The q is current queue size and avg is the average queue estimate. The weight w_q (<1) determines the time constant of the low-pass filter. As mentioned in [6], if the w_q is too large, then the averaging procedure will not filter out transient congestion at the gateway. On the other hand, if the w_q is set to too low, the average responds too slowly to changes in the actual queue size. The calculation of the average queue size can be implemented efficiently when w_q is a (negative) power of two. The 0.2 value for w_q is recommended for representing more recent number of packets in a queue.

3.2 Traffic Load of Surrounding Nodes

For traffic load of surrounding nodes, if media access for each node is fair, the traffic load in a node is defined as the product of its average queue size and the number of sharing loads. For (a) and (b) in Figure 1, each node "A" has same number of neighboring nodes but each node "C" has different average number of packets in queue. If the traffic load in node "A" is the summation of the number of packets in the interface at the node "A" and its neighboring nodes [2], the total traffic load of node "A" in Figure 1 (a) is 9 and that of node "A" in Figure 1 (b) is 11 at the moment, respectively. However, the contention delay for node "A" in Figure 1 (a) and Figure 1 (b) is to be same if the number of packets in interface is constant. Thus, even if the total load of each node has different value, the contention delay at the node is same if the number of contention nodes in propagation range of the node is same and the contention nodes have always one more packets in a queue. Based on these observations, we define that traffic load of surrounding nodes as the number of sharing nodes.

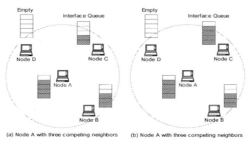

(a) Node A with three competing neighbors   (b) Node A with three competing neighbors

Fig. 1. Same contention delay with the different metric value

3.3 Obtaining Information About Sharing Node

In this section, we describe the operation for exchanging *sharing load* among neighboring nodes. The *sharing load* (SL) is notated as the average queue size exceeding the *threshold queue size* (TQS). That is, if any node has the average queue size over the *threshold queue size* (i.e., $AVG_q > TQS$), it is the *sharing node* (SN) that affects the traffic load of all nodes within its local transmission range. In CLAR, one packet as TQS value is used (i.e., in normal cases, when packets arrive and depart at the same rate, the queue changes from empty to one packet and the queue remains at one packet.)

If a mobile node and its neighboring nodes broadcast HELLO messages periodically within its transmission range, the SL information can be easily distributed into all nodes in any given range. When a node in ad hoc networks broadcasts its HELLO message to its neighbors, it inserts its SL information to HELLO message. Otherwise, the mobile node can obtain this information from RREQ message, which is received from its neighboring node. All nodes in ad hoc networks add their SL information to RREQ message whenever they generate (or forward) the RREQ message and rebroadcast their neighbors.

The acquired SL information is used to calculate total traffic load at a node. The estimation of total traffic load TL_k at node k is computed as follows:

$$TL_k = L_k + w_L * L_k * SN_k \qquad if \ SN_k > 0 \qquad (2)$$

$$TL_k = L_k \qquad\qquad\qquad\qquad if \ SN_k = 0 \qquad (3)$$

where L_k and SN_k to denote the average queue size of node k and the number of the sharing nodes of node k, respectively. The w_L is the load weight, which determines the dependency of sharing nodes (SNs), and is used to make the computation flexible. If the w_L is too large, then the TL will more depend on the number of SNs (i.e., if the media access is fair for each node without considering traffic pattern, high w_L value increases the performance of CLAR). In CLAR we use 0.5 as w_L value (i.e. because the load information of SN is periodically broadcasted to the neighboring nodes of SN and media access is unfair for each node according to traffic patterns, it is less accurate than traffic load at a node). In the definition of TL_k, it is clear that CLAR uses the traffic load at node k as the primary elementary for the path comparison and the traffic load of SNs as an secondary elementary. Following comparison illustrates the effect of SNs when CLAR chooses the path which gives high throughput. We consider few neighbors with $AVG_q > TQS$ and large number of neighbors with $AVG_q > TQS$. The TL_k value for the second case is higher than that for the first case. So, the former is preferred to the latter. The reason is that the probability of the node selected for transmitting a packet is higher than that in the latter since the former has very few active neighbors.

For a given path $P = \langle v_1, v_2, ..., v_n \rangle$, if we use $PL(P)$ that denotes the total path load of P and $TL(v_k)$ denotes the traffic load at a node $v_k (1 \leq k \leq n)$, then the $PL(P)$ is computed as follows:

$$PL(P) = \sum_{k=1}^{n} TL(v_k) \qquad (4)$$

3.4 Route Discovery and Maintenance

CLAR is on-demand routing protocol and consists of two phases: Route Discovery and Route Maintenance. Most of functionality for route discovery and maintenance in CLAR are inherited from Ad Hoc On-Demand Distance Vector (AODV) [8] since CLAR is an extension of AODV. Therefore, CLAR inherits all the advantages of AODV in addition to the capability of load balancing. In this section, we begin the explanation by describing the key differences between CLAR and the traditional AODV.

- The use of Hello messages in a CLAR is modified to allow mobile nodes to exchange their Sharing Load (SL) information. When a node sends a Hello message, it informs its neighbors whether it is the Sharing Node or not.
- CLAR adds a new total load field to the RREQ packet specified in AODV. It is initialized to zero by the source node before broadcasting a RREQ. Every intermediate node receiving the RREQ packet adds its current Total Load (TL) to the value of the total load field on the incoming packet. The result of the addition is assigned to the total load field before the node rebroadcasts the packet.
- CLAR does not allow the intermediate nodes to generate a RREP packet to the source node to avoid delivering unstable path load information using their cached path load. This prohibition guarantees the utilization of recent load information.
- CLAR allows intermediate nodes to support disjoint multipath. If the intermediate node receives duplicated RREQ packets, which includes lower path load than path load in previous received RREQs, it rebroadcasts the received RREQ instead of dropping that. While this mechanism may increase the routing overhead, it allows the destination to choose the optimal route.
- When an intermediate node receives new RREQ, it delays the received RREQ by the uniform load delay based on node' total load and rebroadcasts it through high priority queue.

When a source node desires to send packets to some destination node and does not have a valid route to that destination, it broadcasts a RREQ packet to its neighbors Once the destination node receives the RREQ, it first searches its forwarding route table for the originator. If the matching route is not found, it inserts the forwarding route entry to its routing table. Otherwise, it compares the path load in new RREQ with that in its route cache [see 3.4]. If the path load in the RREQ is less than that in route cache, it updates routing table and responds by unicasting a RREP packet back to the neighbors from which it received the RREQ. As the RREP is routed back along the reverse path, intermediate nodes along this path set up forwarding route entries in their route tables. When an originator receives the RREP, it can begin to transmit data packets to the destination through the received route.

If a source node moves, it is able to re-initiate the route discovery protocol to find a new route to the destination. If a node along the route moves, its upstream neighbor notices the movement and propagates a link failure notification message to each of its active upstream neighbors to inform them of the failure of that part of the route.

3.5 Route Selection Algorithm

As illustrated in previous sections, the destination node must select the best route among multi-paths since CLAR supports multi-paths between the source and the destination. CLAR uses following algorithm in selecting the least loaded route: When the RREQ reaches the destination node, it selects the path with the least sum among multi-paths as its best route. If there are one more routes, which have same traffic load, the destination selects the route with the shortest hop distance. When there are still multiple paths that have the least load and hop distance, the earliest path arrived at the destination is chosen. Figure 2 shows an example network to describe this algorithm. In Figure 2, route X has the sum of 21 (i.e., 3 + 3 + 6 + 9, hop count = 3), route Y has the sum of 22(i.e., 3 + 4 + 6 + 9, hop count = 3), route Z has the sum of 16(i.e., 3 + 4 + 9, hop count = 2). Therefore, route Z is selected and used as the optimal route for transmitting packets.

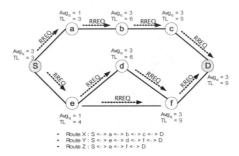

Fig. 2. Example Network

3.6 RREQ-Delay and RREP-Fast-Forward

Both RREQ-delay and RREP-Fast-Forward are based on the priority queuing at an output interface. When each intermediate node receives a RREQ packet, it first holds the packet during a period of time which is proportional to its total load. After waiting for some period, the intermediate node queues the received RREQ packet to the high priority queue. Then, this RREQ packet is immediately forwarded through packet scheduler. This simple delay mechanism is based on the fact that each node accepts only an earlier RREQ packet and discards other duplicate RREQs. This mechanism reduces the number of redundant path since RREQs with the non-optimal route more likely to be discarded than the RREQ message from nodes with the optimal route. On the other hand, the aims of RREP-fast-forwarding are to reduce the route discovery delay by delivering RREP generated from the destination to the source node as soon as possible.

4   Performance Evaluation

In this paper, all simulation models use the ns-2.27 as a base simulator, which is latest version in ns2, except the AODV model [9]. AODV-UU-0.8 developed at Uppsala university is used for simulating AODV protocol since it is designed and implemented based on RFC 3651 [8][9]. We add the CLAR module to ns2 by modifying the AODV-UU-0.8 to support the proposed mechanism.

4.1 Simulation Model

Simulations are run for 600 simulated seconds for various scenarios. We use the recommended values in [8] and [11] as values of simulation parameters for each protocol, respectively. We use 1500m x 300m with 50 nodes, which is the most popular network topology for simulating ad hoc routing protocols, as network topology. For generating random scenarios, random scenario generator developed by CMU is used and the movement of the nodes follows the "random way-point" model. In our simulations, the pause time, which affects the relative speeds of the mobile nodes, is varied: 0, 40, 80, 150, 300, 450, and 600 seconds. A zero pause time means that nodes are always moving while a 600 second pause time means that nodes have no motion for the entire simulation duration. Also, the source-destination pairs are randomly over the simulation network and the number of the source-destination pairs is varied to change the offered load in the network. And all traffic sources are CBR (constant bit rate): data rate is 4 packets per second and 512 bytes per packet.

The performance metrics which are used for evaluating the performance of each routing protocol are listed below:

- Packet delivery ratio: the ratio between the number of packets originated by the CBR sources and the number of packets successfully received by the CBR sink at the final destination.
- Average end-to-end delay: the average of the time difference between the time sent from the source and the time successfully received by the destination.
- Routing overhead: the ratio of the sum of all bytes of routing control packets sent during the simulation and the number of successfully received packets.

4.2 Simulation Results

Figure 3 shows the packet delivery ratio with different numbers of CBR sources and varying pause times. The packet delivery ratios for all routing protocols go down under the high mobility scenarios (that is, low pause times). This is because paths are more likely to break with high mobility. The CLAR shows the highest packet delivery ratio especially in more pause time and higher load scenarios since both AODV and DSR do not consider the load information in path selection and always choose a shortest path that may include congested nodes. In (a) and (b) in Figure 3, CLAR incurs the largest drop of more than 25% at high load scenarios since it can be attributed its aggressive route caching without any expiration mechanism and can be worsen due to the congestion. In Figure 4, as expected, the CLAR shows the lowest delay among three protocols as the rate of mobility decreases. This is because CLAR reduces the congested nodes by distributing traffic to several nodes. One interesting observation is that the delay of all protocols unexpectedly rises as the rate of mobility decreases. These results imply that mobility can somehow eliminate the congestion.

As illustrated in previous sections CLAR should generate larger routing overhead than both AODV and DSR since it just allows the destination to reply for the RREQs and more nodes are involved in the route discovery particularly in the route request

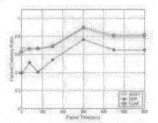

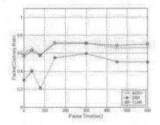

(a) Packet delivery Ratio: 20 sources   (b) Packet delivery Ratio: 25 sources

Fig. 3. Packet Delivery Ratio vs. pause time

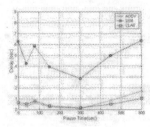

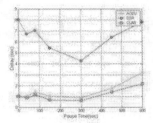

(a) End-to-End delay: 20 sources   (b) End-to-End delay: 25 sources

Fig. 4. End-to-End delay vs. pause time

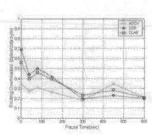

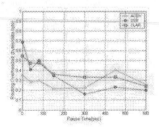

(e) routing overhead: 20-sources   (f) routing overhead: 25-sources

Fig. 5. Routing Overhead vs. pause time

propagation. However, note that load balancing to active destinations reduces the occurrence of congestion at some nodes. Therefore, the routing overhead of CLAR is acceptable since CLAR can maintain a higher packet delivery ratio than them of both AODV and DSR with high load scenarios.

5 Conclusions

In this paper, we proposed a Correlated Load Aware Routing (CLAR) protocol for distributing traffic load among nodes in ad hoc networks. The key concepts of CLAR are to provide more accurate metric for load distribution, the selection of light load path and the reduction of congested nodes in high load networks. We performed a simulation study on CLAR and compared it with AODV and DSR in terms of load balancing. The simulation results show that the CLAR protocol can improve packet delivery ratio, and reduce average end-to-end by keeping track of the average queue size at an interface and the traffic load of neighboring nodes when the network has some congested nodes. However, the CLAR is more useful to high load networks with low mobility such as conferences, lectures, and hotspot areas since CLAR needs more time to route discovery than other protocols (it just allows the destination to reply for RREQs).

References

[1] C. E Perkins, "Ad hoc networking", Addison Wesley, 2001
[2] S. J. Lee and M. Gerla, "Dynamic Load-Aware Routing in Ad hoc Networks", the Proceeding of ICC 2001, Helsinki, Finland, June 2001.
[3] H. Hassanein and A. Zhou, "Routing with Load Balancing in Wireless Ad Hoc Networks", in Proc. ACM MSWiM, Rome, Italy, July 2001.
[4] K. Wu and J. Harms, "Load-Sensitive Routing for Mobile Ad Hoc Networks", in Proc. IEEE ICCCN'01, Scottsdale, AZ, October 2001.
[5] J.H. Song, V.W.S Wong and V.C.M. Leung, "Load-Aware On-demand Routing (LAOR) Protocol for Mobile Ad-hoc networks", in Proc. IEEE VTC Spring'03, Jeju, Korea, Apr. 2003.
[6] Floyd, S., Jacobson, V., "Random early detection gateways for congestion avoidance", Networking, IEEE/ACM Transactions on Volume 1, Aug. 1993.
[7] M. Allman, V. Paxson, W. Stevens, "TCP Congestion Control", RFC2581 in IETF, April 1999.
[8] C. E. Perkins, E. M. Royer and S. R. Das, "Ad hoc On-Demand Distance Vector (AODV) Routing", July 2003, IETF RFC3651.
[9] The VINT Project, "The network simulator – ns-2", Available at http://www.isi.edu/nsnam/ns/
[10] H. Lundgren, "AODV-UU", http://user.it.uu.se/~henrikl/aodv/index.shtml, Uppsala University, Sweden.
[11] David B. Johnson and David A Maltz., Yih-Chun Hu "Dynamic source routing in ad hoc wireless networks", draft-ietf-manet-dsr-09.txt, Internet Draft in IETF, April 2003.

Trial-Number Reset Mechanism at MAC Layer for Mobile Ad Hoc Networks*

Dongkyun Kim and Hanseok Bae

Department of Computer Engineering,
Kyungpook National University, Daegu 702-701, Korea
TEL: +82-53-950-7571, FAX: +82-53-957-4846
dongkyun@knu.ac.kr, bae@monet.knu.ac.kr

Abstract. MANET(Mobile Ad Hoc Networks) routing protocols definitely need a route maintenance process due to frequent route breakage caused by node mobility, which takes advantage of the MAC layer's handshaking service (RTS-CTS-DATA-ACK exchanged packet transmission is considered in this paper). Reaching the limited number of trial of packet transmission, namely retry-limit, before a reception of a corresponding ACK packet indicating a successful data transmission, is used to notify the routing protocol of a link breakage, that is, a route disconnection. The packet, therefore, is discarded and the source is accordingly requested to perform a route recovery with producing many flooded route discovery packets. However, since a reception of a CTS packet during the MAC layer's handshaking means an availability of route, the current number of trial used to detect a link breakage should not increase, but it should be reset to 0, instead. Furthermore, the absence of an ACK packet during a given ACK timeout after a successful reception of CTS packet should not increase the current number of trial for the same purpose. We show the performance improvement with this trial-number reset mechanism by using GloMoSim simulator with AODV routing protocol.

1 Introduction

Recently, research interest in MANET (Mobile Ad Hoc Networks) has increased because of the proliferation of small, inexpensive, portable, mobile personal computing devices. MANET is a wireless network where all nomadic nodes are able to communicate each other within their wireless transmission ranges. Specially, since packet forwarding and routing is done via intermediate nodes, the MANET working group in IETF [1] has been trying to standardize its routing protocols and has promoted AODV (Ad Hoc On-Demand Distance Vector) [6], OLSR (Optimized Link State Routing) [2] and TBRPF (Topology Dissemination Based on Reverse-Path Forwarding) [3] to RFC documents. Ad Hoc routing protocols can be categorized into proactive and reactive routing protocols. In proactive

---

* This work was supported by grant No. R05-2004-000-10307-0 from Korea Science & Engineering Foundation.

M. Freire et al. (Eds.): ECUMN 2004, LNCS 3262, pp. 237–246, 2004.

protocols like OLSR and TBRPF, all nodes should maintain their routing tables for all possible destinations, without regard to the actual desire for the route between source and destination nodes[1]. However, in reactive routing protocols such as DSR (Dynamic Source Routing) [4], ABR (Associativity-Based Routing) [5] and AODV, only when a source node needs to send data packets to the destination node, it attempts to acquire the path in on-demand manner. Reactive approaches avoid the need of maintaining routing tables when there are no desires of routes between source-destination pairs. In addition, a hybrid routing protocol combining both proactive and reactive characteristics is the ZRP (Zone Routing Protocol) [7].

One of the most important processes for the MANET routing protocols is a route maintenance. In most of routing protocols, an intermediate node detecting a link breakage performs a partial path discovery from itself to the destination, or a source attempts to acquire a new path again after being notified of the route breakage from the intermediate node. To support an efficient route maintenance mechanism during the data transmission, the routing protocol takes advantage of the MAC layer's handshaking service where reaching the limited number of trial of packet transmission, namely retry-limit, before receiving a corresponding ACK packet for a successful data transmission, is used to notify the routing protocol of a link breakage, that is, a route dis-connection[2]. The packet, therefore, is discarded and the source is accordingly requested to perform a route recovery with producing many flooded route discovery packets, which also prevents the source from transmitting further data packet. However, since a reception of a CTS packet during the MAC layer's handshaking means an availability of route, the current number of trial used to detect a link breakage should not increase, but it should be reset to 0, instead. Furthermore, the absence of an ACK packet during a given ACK timeout after a successful reception of CTS packet should not increase the current number of trial for the same purpose. In other words, we increase the number of trial only when a CTS packet is not received for an RTS sent (CTS timeout).

This paper is organized as follows. We describe the AODV protocol, which IETF has standardized as one of on-demand routing protocol, and IEEE 802.11 MAC protocol [9] in Section 2 and Section 3, respectively. We present our idea based on trial-number reset mechanism in Section 4. We evaluate our proposed scheme in Section 5, which is followed by some concluding remark in Section 6.

2   Routing Protocol: AODV Protocol

In this section, we briefly describe AODV, one of on-demand routing protocols. When a source needs to send data packets to the destination, it broadcasts an

---

[1] In this paper, a source means a node sending packets to a destination in an end-to-end fashion and a sender is a node forwarding packets to a receiver over a link.

[2] The retry-limit is set to 7 if a packet transmission is activated through RTS/CTS handshaking. Otherwise, the value is set to 4. In this paper, we deal with the case using RTS/CTS handshaking.

RREQ (Route REQuest) message with a sequence number for acquiring the path toward the destination. When each node receives the RREQ message, it checks if the message with the same sequence number has reached the node before. If so, the RREQ message is discarded. Otherwise, it re-broadcasts the RREQ message and keeps track of the next-hop node towards the source.

During this flooding of the RREQ message, a destination or an intermediate node which knows the path toward the destination responds to the RREQ message by unicasting an RREP (Route REPly) message back to the source. This RREP message is transmitted to the source by using the information on the next-hop node toward the source which is kept in each intermediate node. During this unicasting of this RREP message, each intermediate node over the path becomes aware of and keeps the next-hop node toward the destination in the routing table. After acquiring a route, the source starts sending the data packet and each intermediate node can forward the packet by looking up its routing entries. At each wireless link over the path, the sender forwards the data packet to a next-hop node through its MAC layer service. When the MAC layer cannot succeed in transmitting the required MPDU (MAC Protocol Data Unit), the routing protocol is notified of the failure and sends an RERR (Route ERRor) message to the source, which requires the source to acquire a new path using the same procedure mentioned above.

3   MAC Protocol: IEEE 802.11 Mechanism

Besides the routing protocols at network layer, a medium access control protocol at link layer is needed to support the data transmission over the common radio channel which has the collision problem of access to the shared medium among the contending nodes. In general, the carrier sense multiple access (CSMA) protocols have been used in the packet radio network. However, since carrier sensing is location-dependent of nodes in MANET, the well-known hidden terminal problem can occur, resulting in collision on data transmission. For the purpose of resolving this problem, various approaches such as MACA (Multiple Access with Collision Avoidance) [8] have been developed by introducing the exchange of RTS (Request-To-Send) and CTS (Clear-To-Send) messages before data transmission. Furthermore, because of the absence of mechanism to enable a reliable data transmission in MACA, DFWMAC (Distributed Foundation Wireless MAC) protocol used in IEEE 802.11 [9] adds the transmission of ACK packet to this basic MACA protocol, that is, four-way exchange, RTS-CTS-DATA-ACK. Due to the usage of this RTS/CTS exchange, exposed terminal problem should still be addressed.

As shown in Figure 1, a node needs to sense whether the channel is idle or not for a DIFS (Distributed Inter-Frame Space) interval before attempting an RTS transmission and an SIFS (Short Inter-Frame Space) interval before sending an ACK packet and a CTS packet, respectively. After the idle time of DIFS, the sender transmits an RTS packet and waits for a corresponding CTS packet from a receiver, which requires the sender's neighboring nodes to defer their

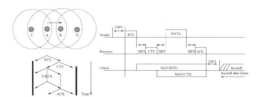

Fig. 1. The RTS-CTS-DATA-ACK handshake in IEEE 802.11 DFWMAC.

transmission until a DATA packet transmission is completed[3]. When a receiver receives the RTS packet successfully, it sends a CTS packet to the sender after an SIFS interval, which requires the receiver's neighboring nodes to defer their transmission through their NAVs. Receiving the CTS message allows the sender to transmit its DATA packet and awaits an ACK packet for the transmitted DATA packet. Receiving the ACK packet means the completion of a successful transmission.

4   Description of Our Protocol

In most of routing protocols, a MAC's service for detecting a link breakage is utilized. As mentioned in Section 3, the IEEE 802.11 protocol tries to transmit the same packet until retry-limit times in case that it does not transmit the packet successfully. Whenever the sender can not perform a successful packet transmission, it increases the current number of trial. When the trial number reaches the retry-limit, the sender discards the packet and assumes that a link failure occurred. Therefore, the sender makes an effort to get a partial path toward the destination, or it sends an RERR message toward the source, which requires the source to obtain a new path again by flooding an RREQ message in the network.(see Algorithm 1).

The counting activity of trial number is triggered when one of following events occurs.

1. When an RTS or DATA packet is corrupted: When an RTS or DATA packet sent by a sender is corrupted at the receiver due to the collision with other transmissions, the sender experiences a timeout event (i.e., CTS timeout or ACK timeout) before a corresponding CTS or ACK packet is received.
2. When a CTS or ACK packet is corrupted: Even though the receiver sent its CTS packet or ACK packet, the CTS or ACK packet can be corrupted at the sender due to the collision with other transmissions. In this case, the sender also experiences a timeout event (i.e., CTS timeout or ACK timeout) before a corresponding CTS or ACK packet is received.

---

[3] The duration for which neighboring nodes should be silent is set through NAV (Network Allocation Vector).

Algorithm 1 Existing Retry-Limit Mechanism

---

  if (ACK Received) **then**
    A transmission is completed; num_of_trial = 0;
  end if
  if (CTS TIMEOUT ‖ ACK TIMEOUT) **then**
    num_of_trial = num_of_trial + 1;
  end if
  if (num_of_trail = RETRY_LIMIT) **then**
    discard the packet; send an RERR packet to the source;
  end if

CTS timeout can be better used to detect a link breakage than ACK timeout because the transmission of DATA packet is initiated after recognizing an availability of route.

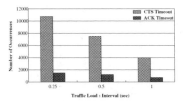

Fig. 2. CTS timeouts and ACK timeouts.

We performed a simple simulation with GloMoSim simulator using a random network configuration with node mobility [10]. We generated CBR source traffic with a fixed interval of packet arrival. We measured the average occurrences of CTS timeout and ACK timeouts according to traffic load represented by the inter-arrival times of packets from application layer. As shown in Figure 2, we could observe that many ACK timeout events can also occur in addition to CTS timeout. However, in the existing retry-limit approach, both timeout events are poorly utilized to increase the number of trial for detecting a link failure and performing a route re-discovery process. For the purpose of detecting a link failure, that is, route breakage, it is very rough to consider both of two cases without differentiation when computing the number of trial. If the retry-limit is reached without this differentiation, the source will try to acquire a new path although a DATA packet could not arrive at the receiver due to a collision instead of a broken link. Note that receiving a CTS packet actually means that the link between two nodes is available. Therefore, we need to reset the current number of trial to 0 when receiving a CTS packet successfully. Moreover, the sender should not use the ACK timeout to compute the number of trial for detecting a route breakage. In other words, only occurrences of CTS timeout event should be counted for the purpose. To summarize, our trial-number reset mechanism is shown in Algorithm 2.

Algorithm 2 Our Trial-number Reset Mechanism

---
  if (ACK Received) **then**
    A transmission is completed; num_of_trial = 0;
  end if
  if (CTS Received) **then**
    num_of_trial = 0; /* A link is still alive */
  end if
  if (CTS TIMEOUT) **then**
    num_of_trial = num_of_trial + 1; /* When a CTS packet is not received */
  end if
  if (num_of_trail = RETRY_LIMIT) **then**
    discard the packet; send an RERR packet to the source;
  end if

As shown in Figure 3(a) using the existing retry-limit approach, although the sender recognized that the link is available through a successful reception of CTS, the sender sends an RERR packet to the source for a new route discovery because the number of trial reached the retry-limit. However, our trial-number reset mechanism as shown in Figure 3(b) allows the sender to consider that the link is still alive and continue trying to send the packet using the link.

Apart from a view of route failure, although it seems that the existing retry-limit scheme treats a network congestion earlier than our trial-number reset mechanism and our proposed scheme permits the source to continue using the route even when the route is congested with other competing nodes, the number of trial due to continual CTS timeouts will be still able to reach the retry-limit if the network is actually congested. As a result, the source is allowed to acquire a new path avoiding the congested route.

5   Performance Evaluation

We compared our trial-number reset technique against the normal retry-limit mechanism by using GloMoSim simulator. For simulation, we assumed that all mobile nodes are equipped with 11 Mbps IEEE 802.11 network interface cards. As a MAC protocol, we used DFWMAC CSMA/CA protocol. In the simulations, AODV was utilized as our underlying routing protocol[4].

We adopted "random waypoint" model to simulate nodes movement, where the motion is characterized by two factors: the maximum speed and the pause time. Each node starts moving from its initial position to a random target position selected inside the simulation area. The node speed is uniformly distributed between 0 and the maximum speed[5]. When a node reaches the target position,

---
[4] AODV is chosen as an example and our approach can be applied to most routing protocols.

[5] For the node mobility, we simulated the proposed approach by varying this maximum speed.

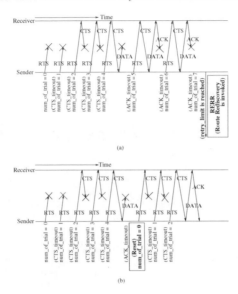

Fig. 3. (a) Existing Retry-Limit Mechanism and (b) Our Trial-number Reset Mechanism

it waits for the pause time, then selects another random target location and moves again. In addition, we avoided the disruption of data transmission caused by network partitions by spreading 100 nodes in a square area of 1000 x 1000 meters and enabling each node to reach other nodes in multi-hop fashion. For the applications, we used CBR traffic sources which was already implemented in GloMoSim simulator.

We measured the average number of the DATA packets received by each of the destination (i.e., throughput) and the average number of RERR packets received by each of the sources. A large number of RERRs will contaminate the network with the RREQ packets flooded for route re-discovery, consuming very scarce network resources. In addition, due to the stopping of transmission while making an effort to acquire a new route, a low throughput is obviously obtained.

First, we compared our scheme against the existing retry-limit approach according to traffic load. For this simulation, we set the pause time to 10 seconds and maximum speed to 5 m/s. Since the applications used CBR traffic sources, the traffic load can be represented by various inter-arrival times of packets from application layer. Five connections were selected in a random pair of source and destination nodes and we obtained the average results for all connections after

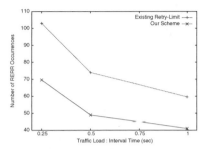

Fig. 4. The number of RERRs according to traffic load.

Table 1. Throughput comparison according to traffic load in terms of the number of received packets.

Interval time (sec)	Existing retry-limit scheme	Trial-number reset scheme
0.25	3898	4690
0.5	2305.4	2333
1	1132	1150

100 runs. As shown in Figure 4, the lower traffic load was, the lower total RERRs we got due to the reduced number of packet transmission. We also observed that the lower traffic load was, the lower throughput we achieved due to the same reason. Table 1 shows a clear difference between two schemes under a high traffic load. Compared to the existing retry-limit approach, our proposed mechanism shows the better performance in terms of two performance measures, namely throughput and the number of RERRs, because the existing retry-limit invokes the route rediscovery even though there still exists a link available between two nodes.

According to AODV mechanism [6], the routing layer which has failed to get a route a given threshold times notifies its upper-layer of the failure and the upper-layer finally disconnects the end-to-end flows. In this simulation, we did obtain an exact result that at the interval time of 0.25 second, our scheme showed a high achievement in terms of throughput because the existing retry-limit scheme allowed the sources to produce too many consecutive RERRs under a high traffic load and the scheme made the upper-layers finally disconnect the connections.

Second, we evaluated the performance according to node mobility. Also, in order to avoid the case where a high traffic load forces the sources to give up their connections due to the reaching of the retry-limit threshold through many consecutive RERRs as mentioned above, the inter-arrival time of packets from application layer was set to 1 second. We varied the maximum speed from 5 to

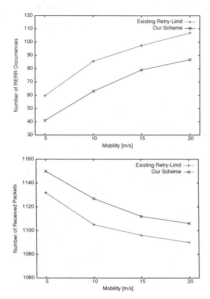

Fig. 5. The number of RERRs and Throughput according to node mobility.

20 m/s with the same pause time of 10 seconds. As expected, the more the node mobility increased, the lower throughput we got as well as the more RERRs the sources produced. In the existing retry-limit scheme, if the number of trial reaches the retry-limit even though there still exists a link, the source tries to acquire a new route, while stopping its transmission and contaminating the network with its new flooded RREQ packets. However, because of our process resetting the trial-number on a detection of an availability of link, our scheme produced better performance than the existing retry-limit scheme, irrespective of node mobility (see Figure 5).

6   Conclusion

In this paper, we improved the performance of MANET using the trial-number reset technique. This technique allowed a MAC protocol to differentiate between CTS timeout and ACK timeout for detecting a link failure, that is, route break-age. Irrespective of a type of timeout, the existing retry-limit approach discards a packet when the current number of its transmission reaches the retry-limit and a source, furthermore, is notified of a route breakage. In this paper, we increase the number of trial only when CTS timeout occurs because the reception of CTS packet means that there still exits an available link between two nodes. To reflect this effect, when the sender receives a CTS packet, it resets the current number of trial to 0. By simulation using GloMoSim with AODV routing protocol, we proved the performance improvement, compared with the existing retry-limit mechanism. The performance comparison when applied to other routing protocols using our trial-number reset technique is our exciting future work.

References

1. Internet Engineering Task Force, "Manet working group charter," http://www.ietf.org/html.charters/manet-charter.html.
2. T. Clausen, P. Jacquet, A. Laouiti, P. Minet, P. Muhlethaler, A. Qayyum, and L. Viennot, "Optimized Link State Routing Protocol(OLSR)", IETF RFC 3626.
3. R. G. Ogier, F. L. Templin, B. Bellur, and M. G. Lewis, "Topology Broadcast Based on Reverse-Path Forwarding (TBRPF)", IETF RFC 3684.
4. D. B. Johnson, D. A. Maltz, and Y. Hu, "The Dynamic Source Routing Protocol for Mobile Ad-hoc Networks (DSR)", IETF Internet Draft, draft-ietf-manet-dsr-08.txt, February 2003.
5. C. K. Toh, "Associativity based Routing for Ad Hoc Mobile Networks", Wireless Personal Communications, vol.4, no.2, 1997.
6. C. E. Perkins, E. M. Belding-Royer, and S. R. Das, "Ad-hoc On-Demand Distance Vector (AODV) Routing", IETF RFC 3561.
7. Z. J. Haas, M. R. Pearlman, and Prince Samar, "The Zone Routing Protocol (ZRP) for Ad Hoc Networks", IETF MANET Internet Draft, July 2002.
8. P. Karn, "MACA-a New Channel Access Method for Packet Radio", ARRL/CRRL Amateur Radio 9th Computer Networking Conference, pp. 134-140, ARRL, 1990.
9. IEEE Computer Society LAN MAN Standards Committee. Wireless LAN MAC and PHY Specification, IEEE Std 802.11-1997.
10. http://pcl.cs.ucla.edu/projects/glomosim/

Operator Services Deployment with SIP: Trends of Telephony over IP

Emmanuel Bertin, Emmanuel Bury, and Pascal Lesieur

France Telecom, Division R&D, 42 rue des Coutures, BP 6243, 14066 Caen Cedex 4, France
{emmanuel.bertin,emmanuel.bury,pascal.lesieur}@francetelecom.com
http://www.rd.francetelecom.com/

Abstract. Nowadays Telephony on IP is de facto a reality. SIP Protocol has reached a sufficient maturity for the development of advanced multimedia services. The goal of this paper is to study the potentialities of SIP architectures to develop and deploy advanced SIP services, through a service case study, the Customized Call Routing service, developed and trialed in R&D labs of France Telecom. This paper suggests a deployment model of SIP services driven by the requirements of the different actors and details a generic architectural framework enabling to position various functions needed for NGN multimedia services.

1 Introduction

SIP is often seen as THE protocol for the next-generation services, the standard that would pave the way to integrate voice and data services in the business field and in the mass market, for example with the integration of community tools and office tools. SIP offers openness to the Internet technologies, using standard service creation mechanisms, and thus a better integration of voice and Web services. SIP is specified by the IETF, as an adaptation of concepts and architectures which made the success of Internet (Web, mail, endpoints intelligence, distribution …) for the Telephony over IP.

Traditionally, operators have always cooperated within ITU-T for the definition of protocols suites used in their networks. Associated architectures were also specified within this scope. This is not really the case with SIP and manufacturers are today leading the SIP standardization effort. The emergence around SIP of a whole environment for real-time communication services over IP network potentially impacts the operator businesses and processes, from both protocol and architecture points of view.

The goal of this paper is to study, from an operator's point of view, fast the SIP potentialities for advanced services developments and then actor models enabled by SIP. We will first describe the experience of France Telecom in the SIP field with

M. Freire et al. (Eds.): ECUMN 2004, LNCS 3262, pp. 247–256, 2004.

development and trial of new services presenting an example of a value-added service: the CCR service, which was developed and patented in France Telecom R&D labs. It relies on industry products and standard SIP service creation mechanisms. Then we will detail a possible deployment model of SIP services, based on an intermediation model. Finally, we will conclude with the description of an architectural framework which can address the general lack of modeling for service architecture in next-generation networks (NGN).

2   Service Implementation Examples: The France Telecom's Experience

The R&D Division of France Telecom has led several projects on SIP to study it from several points of view:
- Study the relevance of SIP for advanced services,
- Benefit from flexibility in terms of routing,
- Study the service triggering mechanisms, especially towards third-party service providers,
- Position a whole actors' model improving the notion of intermediation,
- Elaborate a generic architectural framework.

2.1   Key for the Voice and Data Convergence

SIP offers the opportunity to develop really new services (not just the duplication of the current PSTN services), which provide a true added-value for the user.

As Web Services, SIP enables to build together, in a same application, services or components from various sources, like a construction set. Contrary to H323, the SIP community has made an important effort to standardize service creation mechanisms: CPL (Call Processing Language) and SIP CGI has been standardized by the IETF, the SIP Servlets by the Java community.

The CPL is a script language based on XML which is an easy but limited way to create services based on forwarding or filtering rules. Indeed it's impossible to create complex services only with CPL. The SIP CGI language is transposed from HTTP CGI: scripts are triggered on reception of messages; the content of the message is delivered to a software component (script developed in C, C++, perl, ...), which returns a set of actions to be done by the proxy. SIP servlets is a adaptation of web servlets (Java applications): they enhance the interface to support SIP functions. They offer a lot of functions and allow developing advanced services.
CPL is designed more to be used directly on the SIP proxy (and the operator may choose to give end-users access to the CPL rules). SIP CGI and SIP servlets are designed more to be used on an application server, even if lots of AS support all the 3 languages.

In the same time, manufacturers have developed VoIP terminals which more than the classical call handling, offer users new services. These terminals have advanced capacities, for example a Java virtual machine, and allow to put in the periphery of the network a big part of the service logic.

Therefore, R&D Labs of France Telecom has developed original services on terminals (having suitable capacities) and in the network. We will focus here on complex SIP services. By complex SIP services, we mean here services that go beyond routing configuration (e.g. PSTN least cost routing) or server consultation (e.g. media server, presence server); we mean typical operator services, relying on a network-based session control to ensure the global service coherence for the end-user.

2.2   A Service Example: The CCR Service (Customized Call Routing)

The aim of the patented CCR service is to improve the reachability of every subscriber. All subscribers must be equipped with a SIP phone (connected to a SIP architecture) and a desktop computer with Internet access, on which a specific client is installed.

First, the subscriber sets up the media he will authorize to be joined by his contacts (inside a buddy list and for each contact), depending of multiple criteria: time and date, presence information (using SIMPLE or Jabber clients), phone behaviour (busy, no-answer).... Then, when somebody calls the subscriber, the Caller will be proposed every media the subscriber has authorized for this Caller. The Callee must have subscribed to the CCR service.

The list of media is proposed to the Caller through a dedicated application on his desktop computer (if he's a CCR subscriber) or through a vocal server (if he hasn't subscribed), and includes typically choices like:
  – To be redirected to another phone (e.g. secretary or cell phone),
  – To leave a voice message,
  – To browse a web page defined by the Callee,
  – To send an email or an SMS to the Callee.

The originality of this service resides in the fact that the mean of contact is chosen by the Caller. For example, he will choose the cell phone for an urgent call and the email or the voicemail for an informative call. One service scenario is described in the figure 1.

The CCR service is a complex service, which integrates data applications (email, presence, web ...) and relies on a control of the user session. It combines various technologies (SIP Servlet, Web Services, EJB, presence, office applications ...), deployed over a SIP network. It has been developed using standard mechanisms and is implemented on a SIP distributed architecture, entirely basing on the protocol mechanisms.

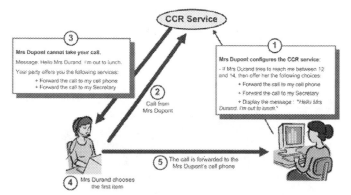

Fig. 1. CCR service scenario. Mrs. Durand tries to reach Mrs. Dupont that has put a rule in the CCR service.

According to the structure of a NGN network, the CCR implementation is divided in 4 domains (as shown in the figure 2):

- Client: endpoints and private network (e.g. Home GW),
- Periphery: access network and platforms,
- Core: backbones and core platforms,
- Applications: service providers' platforms.

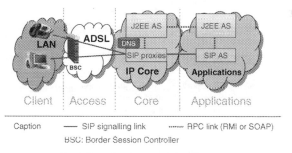

Fig. 2. Network architecture of the CCR service

The figure 3 shows the functional architecture of the service where all the components are divided into 5 levels. Only two modules are developed in SIP servlets and run on a SIP Application Server (AS). The other modules are developed as EJB and run on various servers: most of them are not specific to the CCR service but are reused from other services.

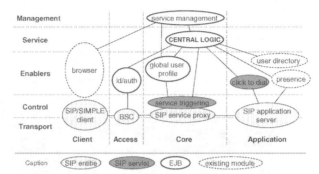

Fig. 3. Component architecture of the CCR service

An enabler is a re-usable building block which has 2 main roles: it offers an abstraction and a federation of network or media control and shares common functions. In the CCR service, 6 enablers have been developed or integrated: Global user profile, Notification (through a browser or a specific client), Identification/authentication, User directory, Click to dial and Presence. A service triggering component has been developed to act as an interface between the SIP proxy and the Central service logic. The service triggering component could not be designed as a generic building block because the trigger of a service mostly need some specific parameters (here parameters contained in the SIP signalling).

Relying on building blocks, the central logic plays only an orchestration role between enablers and other components: it routes queries to the appropriate module and treats their results.

The CCR implementation validated the relevance of the SIP protocol and of an IETF distributed call control architecture for the control plane. The distributed architecture allows flexible actors models to be designed, with different providers for SIP routing, SIP session control and basic or advanced services. The CCR service also validated an architectural framework which is particularly interesting for converged multimedia voice and data services.

3   Deployment Model of SIP Services

3.1   A Distributed Model …

One of the SIP foundations is the Internet concept of non hierarchical intelligence distribution. In some SIP guru visions, a pure Internet telephony, fully distributed, where each customer will choose at his own convenience between all SIP real-time communication applications, is presented like a bright future.

SIP allows increasing intelligence in periphery of the network: domestic, terminal facilities, service platforms... Indeed, as the call-handling intelligence is localized in the terminal (that registers itself to the network), every user of a SIP terminal which is connected to Internet could freely choose equipments to which he is going to connect.

3.2   ... With a Need of Federation

But this anarchical distribution doesn't probably fit all the market expectations: security, reliability, consistency of the customer relationship... apart from perhaps for highly techno-centric users. Indeed, the end-user wants to benefit from a much wider set of services with new tools (for instance the Instant Messaging service) and to access these services easily whatever access network and whatever terminal he might use, wherever he is located (either from home or on the move). But he also wants a good quality service without having to manage the complexity of both the service (simplicity, set of services with the same ergonomics) and the service interactions.
The operators' challenge, especially incumbent operators, is therefore to use SIP to associate multiple applications, while assuring a global consistency.

Lots of operators consider using the SIP protocol for the evolution of their network towards the NGN (Next Generation Network). By its openness and its flexibility, SIP allows a formal separation of the layers (service, call control, network) and offers normalized interfaces (related to the protocol and applications) between these layers.

3.3   A New Actors Model: The Intermediation Model

For the supply of telecom services, lots of actors models are possible but 2 of them come up against each other:

- − The intelligent network model where operators control all the value chain and don't offer any openness for 3rd party service providers.
- − The Internet model where service providers will provide not only services, but also their own control infrastructure, independently of any incumbent operator. Incumbent operators are limited to the supply of the access and of the link to the legacy telephony networks.

Between these extremes, France Telecom defines a model which takes into account all the actors' expectations: the intermediation model. In this model, the incumbent operator provides value-added services and control framework and is opened, not only to other services, but also to other control infrastructures.

3 main roles were identified (one actor can play several roles):

- SIP Network Connectivity Provider (NCP): The network connectivity provider furnishes end-users, service providers and enterprises with SIP connectivity. He also supplies global access connectivity: Internet, ADSL, mobile, PSTN…
- Global service provider (GSP): The GSP furnishes end-users with a whole SIP services environment: provisioning, execution and charging (at least for the applications). He may invoke a USP for the execution of specialized services or services that he doesn't directly offer but this invocation is transparent to the end-user and done using the SIP protocol.
- Unitary service provider (USP): The USP provides one or many specialized SIP services (like conferencing, unified messaging, presence, click to dial…). The service is provisioned by the GSP through web services mechanisms and the GSP is charged for this. The motivation to be a USP could stem from the business requirements for customized solutions or from a niche market for consumer communities, which both require a reactivity that a GSP cannot offer.

4   A Proposal of an Architectural Framework

The definition of a global architectural framework for multimedia real-time services delivery is not a topic widely addressed in standardization.

On one hand, standard activities around APIs like OSA/parlay aim at providing an interface between multiple actors for service building and triggering but do not provide a general view of all elements involved in service delivery. For instance, next-generation applications will be more and more distributed between terminals, home or corporate networks, access networks, service platforms… A superposition of APIs is not sufficient to model the service delivery, but a global architectural framework where all cooperating entities may be modelled is needed.

On the other hand, standard activities around networks at ETSI, ITU or 3GPP aim at building next-generation networks, but do not actually directly address the service architectures. Even if services are studied, it is in most cases as requirements for network entities standardization, and not as an independent field for architectural studies.

A standardization effort to define a global framework for service architecture was driven at ITU in 1998 with the Y-Series recommendations, especially in the Y.110 recommendation. This recommendation was quite accurate, but need to be updated to fit with the current technologies.

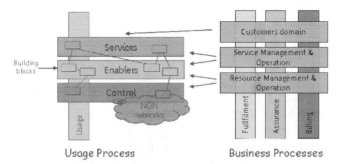

Fig. 4. Architectural framework

Starting from the Y.110 model, we propose to position all functional entities involved in multimedia voice and data communication services using the following architectural framework (figure 4).

The architectural framework consists of 3 main planes:

- Control plane contains basic entities for computing or networking, and especially transport entities for data transfer (e.g. routing, admission, firewalls, NAT, QoS enforcement) and control entities for the control of the network (e.g. admission control, QoS control, firewall control, call control, service triggering);
- Enablers plane includes all functional building blocks (also called enablers), which assure the independence between Control and Service planes and allow common building blocks to be shared between applications (e.g. presence, localization, calls states, terminal capabilities, content adaptation).
- Services plane contains all service logics; service being understood from the end-user point of view (i.e. directly perceptible by the end-user, unlike most of the middleware building blocks).

Finally, management entities include the management of enablers and control entities, and the global application management (business management, service provisioning, service accounting, global O&M), following e-TOM processes.

This framework could be profitable in the telecom services conception process in order to simplify global architectures, to determine best interfaces between functions and to re-use entities in many services. This framework could also help for a clearer positioning of various actors (Service Providers, Operators, Network Providers...) and for identifying interfaces to openness and security needs. For example, an application invocation could either be done in the control plane through the SIP protocol, or in the middleware plane through RMI.

Regarding the enablers plane, we propose to look at this plane not only as a software distribution mechanism, but also as a group of building blocks, like in the IN SIB (Intelligent Network Service Independent Building Block) notion. A building block is an independent component, involving equipment from the control plane and that is invoked from the services plan to obtain specific information or to do specific actions on control entities. Building blocks could be for instance localization, notification, presence, user profile, content adaptation, reachability, third-party call establishment... One building block may also invoke other building blocks (for instance reachability may use presence and localization information).

Moreover, building blocks may be combined to build a global service. This combination does not consist in reusing code but in the invocation of service building blocks from the global service (through typical remote invocation mechanisms like SOAP or RMI). Each building block may be used by many services (this re-usability need was already the source of the IN SIB notion).

Parlay/OSA specifications provide examples of building blocks' implementation. We believe that this concept could be generalized to all NGN services, whatever the underlying technologies (SIP, OSA, SOAP or EJB components). This would assure a true separation between the technologies used at the control, enabler and service planes. Architectures following this concept could flexibly evolve, for example when replacing a proprietary identification building block with a Liberty identification module.

5   Conclusion

Our studies allow us to prove the relevance and the flexibility of the SIP protocol for the service development and its potentialities for enabling multi-actors models. The framework proposal is fully suitable for service implementation, as done with the CCR service: it can be used on transverse projects, especially for coordination between various components.

The deployment of new generation network equipments will probably lead operators to reconsider the intelligence distribution between network and periphery. More than the only supply of network resources, the added-value of a telecom operator probably resides in an intermediation role between different actors and a federation role for the final customer with SLAs. So, operators have obviously to optimize the using of their resources (network localization, resources control, customer relationship ...) but it's their interest to position actors models where all actors of the telecommunication industry have the assurance to stare the benefits (operators, third-party service providers, manufacturers, content providers...).

References

1. IETF, J. Rosenberg, et al., "SIP: Session Initiation Protocol", RFC 3261, June 2002
2. ITU-T, Recommendation H.323, "Packet based Multimedia Communication Systems", version 4, November 2000
4. E. Bertin, E. Bury, P. Lesieur, "Next generation architectures: which roles for an incumbent operator?", Proc. Eurescom summit 2002, October 2002
5. E. Bertin, E. Bury, P. Lesieur, "Operator services deployment with SIP: Wireline feedback and 3GPP perspectives", ICIN 2003, Bordeaux, April 2003.
6. E. Bertin, E. Bury, P. Lesieur, "Intelligence distribution in next-generation networks, an architectural framework for multimedia services", ICC 2004, Paris, June 2004
7. G. Fromentoux et al., "IP networks towards NGN", Proc. IP networking&mediacom-2004 ITU-T Workshop, April 2001
8. J.Y. Cochennec, "Activities on Next-Generation Networks Under Global Information Infrastructure in ITU-T", IEEE Commun. Mag., Jul 2002
9. R. Copeland, "The Place of SIP in Network Intelligence", ICIN 2003, Mar-Avr 2003
10. ITU-T, Recommendation Y.110, "Global Information Infrastructure: principles and framework architecture", June 1998
11. A.R. Modarressi and S. Mohan, "Control and Management in Next-Generation Networks: Challenges and Opportunities", IEEE Commun. Mag., Oct 00
12. http://www.parlay.org

Network Resource Management: From H.323 to SIP

Marie-Mélisande Tromparent

Institute of Communication Networks, Munich University of Technology, Arcisstrasse 21,
80290 Munich, Germany
MM.Tromparent@tum.de
http://www.lkn.ei.tum.de/lkn/mitarbeiter/marie

Abstract. IP Telephony has become a hot topic in the last few years. There are
currently 2 major protocols for signaling IP telephony, namely SIP and H.323.
The first version of H.323 has been published in 1996, whereas SIP is much
younger. Since it is probable that both signalization protocols will coexist for a
certain time, there is a need to port existing H.323 systems into a SIP environment. We propose in this paper a solution for porting a Quality of Service Architecture, namely the Resource Management Architecture, from H.323 to SIP.
As an introduction, we provide a short overview of both protocols H.323 and
SIP focusing on the call setup procedure and the capability exchange process.
We then propose a solution for this concrete problem as well as technical possibility for implementing the porting of the architecture from H.323 to SIP.

1 Introduction

Since a few years, IP telephony has become a hot topic. Making a call on internet is
already possible, but today's networks do not provide the desired transmission quality
for real-time applications. Therefore, a lot of researchers have been working on Quality of Service (QoS) in IP networks over the last few years.

Currently, there are 2 major protocols for IP telephony: the H.323 protocol [1],
whose first version has been issued in 1996, and the SIP protocol [2] which is much
younger. The H.323 protocol was the first broadly accepted protocol enabling real-
time communications over IP networks. It encountered a big success, so that a lot of
companies and universities worked on developing H.323 compatible devices. However, a new protocol for multimedia communication has emerged in 1999, namely
SIP, which is becoming always more important. Therefore, there is a need nowadays
to port systems, which have been initially developed for the H.323 protocol, into a
SIP environment. In this paper, we present an example of adaptation of a QoS architecture developed in a H.323 context to the SIP protocol. For this purpose, we provide
a precise comparison between the call setup procedures in H.323 and SIP which is
rarely mentioned in papers comparing H.323 and SIP. Moreover, we propose in this
paper an alternative to the RFC 3312 "Integration of Resource management and SIP"
[3]. As an additional contribution, we mention some concrete realisation possibilities
using existing open source projects and libraries.

M. Freire et al. (Eds.): ECUMN 2004, LNCS 3262, pp. 257–266, 2004.

The rest of this paper is divided as follows: In section 2, we give a brief overview of the *Resource Management Architecture*, which should be ported from H.323 to SIP. We present the H.323 protocol briefly, and the integration of the *Resource Management Architecture* in the H.323 environment. Section 3 is the main part of this paper, it provides an overview of the SIP protocol as well as a comparison of the call setup procedures of H.323 and SIP. We then present our proposal for porting the *Resource Management Architecture* from H.323 to SIP and some hints for a technical realisation. Section 4 concludes this paper, listing some open issues.

2   Resource Management Architecture

2.1 Overview

The *Resource Management Architecture* (see Fig. 1 and references [4], [5]) is a QoS architecture on top of the IP-layer, which is meant to provide hard and soft QoS guarantees for real-time traffic in the context of enterprise networks. It is based on the principle of aggregating traffic into service classes on network and data link layer. Network resources (e.g. buffers in the nodes or bandwidth of the links) are explicitly assigned to each service class via configuration management and/or policy-based management. Specific service classes are exclusively dedicated to the transmission of real-time traffic.

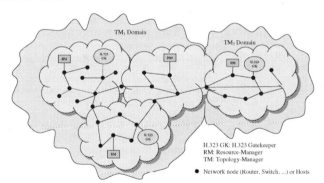

Fig. 1. Overview of the *Resource Management Architecture*

Resource-Managers (RMs) are special servers of the *Resource Management Architecture*, which are actually responsible for the management of the network resources. They operate on particular domains, named *RM Domains*. Each time, a terminal wants to establish a communication, the RM, whose domain contains this terminal, has to be contacted and it has to decide whether the call should be accepted or not

(Call Admission Control function). The RM has the complete knowledge of its domain (topology, service class configuration, load situation), so that it is able to make an appropriate decision according to the level of quality the initiating terminal wishes. If a connection is accepted, the resources required for this connection are virtually reserved by the *Resource-Manager*.

Given the topology and configuration of the network, a RM is able to maintain the map of its domain's load, updating it each time a connection is established or torn down. However, it must receive the topology and configuration information from another entity. These special entities of the *Resource Management Architecture* are referred to as *Topology-Managers* (TMs) [6]. As the RMs, the TMs are responsible for limited domains called *TM Domains*. For scalability and flexibility reasons, one TM Domain can contain several RM Domains. In order to get the knowledge of the network, the TM performs a topology discovery (layer 2 & 3 of the OSI model) within its domain, using standard protocols like SNMP ([7]) and ICMP ([8]).

2.2 Integration of the RM Architecture into a H.323 Environment

The *Resource Management Architecture* including the RM and the TM has been verified by a prototype implementation in a H.323 environment.

H.323 – Brief overview. H.323 is an umbrella recommendation (see [1]) from the International Telecommunications Union (ITU) that sets standards for multimedia communications over Local Area Networks (LANs) that do not provide a guaranteed Quality of Service (QoS). H.323 defines four major components for a network-based communication system: Terminals, Gateways, Multipoint Control Units and Gatekeepers. Terminals are the client endpoints on the LAN that provide real-time, two-way communications. Gateways provide many services, the most common being a translation function between H.323 conferencing endpoints and other terminal types. The Multipoint Control Unit (MCU) supports conferences between three or more endpoints. The gatekeeper (GK) is the most important component of an H.323 enabled network. It acts as the central point for all calls within its zone and provides call control services to registered endpoints. Gatekeepers have four important call control functions: address translation, call admission control, bandwidth control and zone management. The communication protocol used between terminals and Gatekeepers is called the RAS (Registration Admission Status) protocol or H.225.0 protocol. It is composed of a set of messages allowing following operations:

 – Gatekeeper discovery and terminal registration
 – Call Admission Control: Using the ARQ message (Admission Request), a terminal can ask its GK for the authorization to establish a call. The GK may answer with an ACF (Admission Confirm) or ARJ (Admission Reject) message.
 – Call Termination, which may be initiated either by a terminal involved in a call or by a GK (messages DRQ/DCF Disengage Request/Confirm)
 – Modification of the properties of an established call (messages BRQ/BCF/BRJ Bandwidth Request/Confirm/Reject)
 – Location service (messages LRQ/LCF/LRJ Location Request/Confirm/Reject)

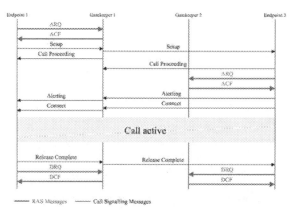

Fig. 2. Call setup procedure - Both endpoints registered - Both gatekeepers routing call signaling

Fig. 2 shows the message flow for a call establishment and termination.

In order for the *Resource Management Architecture* to be easily deployable, the RM has been implemented as a backend-service behind the H.323 gatekeepers (thus no changes in client applications are required). The H.323 gatekeeper has been lightly modified to allow the communication GK – RM. Instead of taking decisions itself, the gatekeeper asks the RM for authorization each time it receives an Admission Request (ARQ) message. In addition, the RM is informed by the gatekeeper each time the properties of an existing call are changing (modification of the properties of an established call or call termination).

The protocol between GK and *Resource-Manager* is very simple, it contains 3 messages and their corresponding responses:

- *ReservationRequest* sent by the gatekeeper when receiving a ARQ message. It is answered by a *ReservationConfirm* or *ReservationReject* depending on the RM's decision
- *ReservationUpdateRequest* sent by the Gatekeeper when one of the endpoints participating to an active call wants to change the resources used by the call (response: *ReservationUpdateConfirm* or *ReservationUpdateReject*)
- *ReservationRelease* sent by the Gatekeeper when a call is terminated (no response)

The parameters contained in each message reflect the information contained in the initiating RAS message. Each message contains a globally unique call identifier, a call reference value and a rough description of the traffic to be sent in the case of the *ReservationRequest* and *ReservationUpdateRequest* messages. Using this information, the *Resource-Manager* is able to perform virtual resource reservation and take a sensible decision concerning the admission of a new call.

3 Porting the RM Architecture from H.323 to SIP

SIP – Brief overview. Just as H.323, the Session Initiation Protocol (SIP) is a signaling protocol for initiating, managing and terminating voice and video sessions across packet networks. It is being developed by the SIP Working Group, within the Internet Engineering Task Force (IETF) and is described in [2]. SIP defines 2 major types of components: User Agent (UA) and network servers. An UA corresponds merely to the H.323 Terminal, and is composed of a client part (UAC, User Agent Client) which issues requests to the network servers, and a server part (UAS) which answers requests received from the network. 3 major network servers are defined in the SIP standard: Proxy Server, Redirect Server and Registrar. The most generic SIP operation involves a SIP UAC issuing a request, a SIP Proxy Server acting as end-user location discovery agent and a SIP UAS accepting the call. Fig. 3 shows the message flow for a call establishment and termination involving 2 Proxy Servers. With an INVITE message, a terminal notifies its Proxy Server, that it wants to establish a call. The INVITE message is forwarded hop-by-hop by some SIP Servers until it reaches the callee, who may accept the call using the 200 OK message. The data exchange may begin after the caller has sent an acknowledgement (ACK) message.

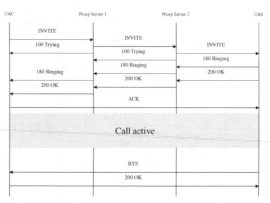

Fig. 3. SIP Session setup example

A Redirect Server may be involved as well. It is meant to map the SIP address of the called party into zero or more new addresses and to return them to the client. Unlike Proxy Servers, Redirect Servers do not pass the request on to other servers. A Registrar is a server that accepts REGISTER requests for the purpose of updating a location database with the contact information of the user specified in the request.

3.1 Comparison Between SIP and H.323 Regarding Call Setup Procedure

In order to perform resource reservation for a connection path, a precise description of the traffic to be exchanged is required. This description should be available at a time where it is still possible to reject the call if there are not enough resources free.

As the first steps of the call setup procedure for H.323 and SIP have already been described in the previous sections, we concentrate here on the capability exchange which allows 2 terminals wishing to establish a call to agree on the data format they will exchange during the communication (channel configuration, media coding schemes, etc.). The result of this negotiation phase is of primary importance for the *Resource-Manager*, which can only perform a precise resource reservation according to it.

In H.323 the capability exchange takes place within the call setup procedure after the preliminary exchange of H.225.0 messages. The protocol responsible for this exchange is the H.245 protocol from the standard protocol sets of H.323. Endpoint system capabilities are exchanged by transmission of the H.245 *terminalCapabilitySet* message. It provides for separate receive and transmit capabilities, as well as a method by which the terminal may describe its ability to operate in various combinations of mode simultaneously. After each terminal has announced its own capabilities, media streams may be exchanged using the corresponding codecs.

On the contrary, in SIP, endpoints normally negotiate their capability outside the setup procedure using the SIP message OPTIONS. In the OPTIONS message, each terminal lists its supported codecs so that the other party can propose appropriate coding schemes when sending the INVITE request. However, there is currently some work in the IETF proposing an alternative to this using an Offer/Answer Model with the Session Description Protocol [9]. In that case, the capability negotiation procedure may be included into the call setup procedure.

In our implementation, we extended the Gatekeeper so that it can communicate with the *Resource-Manager* each time it receives an ARQ or BRQ message. At this point of the call setup procedure, no precise information concerning the encoding of the streams is available, but the ARQ (resp. BRQ) contains a *bandwidth* parameter describing the total amount of resources needed for that call. We base on this parameter to reserve the resources using some mapping functions between the total amount of bandwidth requested and the corresponding codecs used. However the resulting resource reservation must be reviewed as soon as the capability exchange has taken place. If the RM notices that the effective resource occupation is too large, it may order the GK to clear the call.

The next section presents a solution for the SIP context.

3.2 Proposal

In order to port the *Resource Management Architecture* into a SIP environment, several aspects have to be considered: Firstly, it is necessary to define interfaces between the *Resource Management* and the SIP service architectures. This comprises the determination of the SIP components, which are going to interact with the *RM*, as well as the communication protocol to be used. Then we have to consider the modifications to be performed in the *RM* implementation in order to allow the communication SIP component – *RM*.

As mentioned previously, the interaction between the *Resource Management Architecture* and the H.323 service architecture is enabled thanks the extension of the Gatekeeper, which is provided with an interface to the *Resource-Manager*. This is motivated by the fact, that the Resource Management Architecture can be more easily deployed, when it does not produce any change on the client side. Since we want to have the same property for the SIP environment, the SIP component which is going to interact with the RM cannot be an User Agent. SIP defines 3 different functional servers: Proxy Server, Redirect Server and Registrar. Since Registrars are not concerned by the call setup procedure, they won't play any role in the integration RM Architecture / SIP. Proxy and Redirect servers can both be involved in call setup procedures. However, Redirect Servers are only asked for location information, when the initial INVITE request is issued, but they won't be contacted within the rest of the call setup procedure. Therefore the only candidate for interfacing the RM and the SIP architectures is the SIP Proxy Server. Therefore, we propose to extend the SIP Proxy Server with an interface to the Resource-Manager.

The *Resource-Manager* should be contacted each time a new call is set up or an existing call is modified, this means actually, each time an INVITE or BYE request is issued. However, many Proxy Servers may be involved in the call setup procedure, but not all of them should contact the *Resource-Manager*. Since the *RM* needs to know the IP address of the callee in order to perform the resource reservation for the call, it should not be contacted by a SIP Proxy Server which does not have this information. The address translation procedure in SIP is performed sequentially: Each Proxy Server of the path is somehow closer to the called party, but does not always know the exact IP address of the called endpoint. Therefore, in order to identify in a deterministic and systematic way, the SIP Proxy Server which should contact the RM, we decide to assign this task to the last SIP Proxy Server of the signalization path. Fig. 4 illustrates the resulting message flow considering as example: Terminal 1 firstly sends an INVITE message to its dedicated SIP Proxy Server (message 1). It is forwarded until SIP Proxy Server 3 (messages 2,3). SIP Proxy Server 3 could contact its RM at this point, but it is actually cleverer to wait until Terminal 2 accepts the call (messages 4,5). Consequently, RM3 won't start the resource reservation procedure if Terminal 2 refuses the call. Moreover, RM3 may perform directly the resource reservation for both directions (caller → callee and callee → caller) since the information of the traffic to be sent on both direction is now available. At this point RM3 starts the resource reservation procedure, which may require the coordination between several RMs in the case of a multidomain call. After that, RM3 sends its final decision to SIP Proxy Server 3 which in turns forwards the decision until Terminal 1.

Assuming that SIP Proxy Server 3 waits until the resource reservation has been performed induces an additional delay in the call setup procedure which is undesirable. We are currently working on optimizations allowing the parallelization of the SIP signalization and the resource reservation procedure (see future work in section 4).

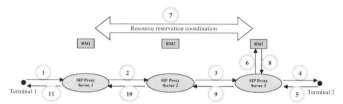

Fig. 4. Interaction RM Architecture / SIP

Let us consider the content of the messages depicted in Fig. 4: Messages 1, 2, 3, 4, 5, 9, 10, 11 are standard SIP messages (INVITE or 200 OK); messages 6 and 8 are described below; message 7 represents actually a set of messages exchanged between neighbor RMs in order to coordinate the resource reservation for the call on the complete data path. They are not in the scope of this paper since they are independent from the signalization protocol.

The messages exchanged between a Proxy Server and the *Resource-Manager* may be the same as for the H.323 case: *ResourceReservation* for reserving resources for a new call, *UpdateResourceReservation* for modifying the resource occupation of an existing call, and *ReleaseResources* for terminating a call and releasing the corresponding resources. However, the content of these message should reflect the content of the initiating SIP message, i.e. the parameters will differ from the content of the H.323 derived message. This is due to the fact that the *Resource-Manager* should be able to answer the Proxy Server with messages containing all information enabling the SIP Proxy Server to reconstruct the response to the initial SIP message without keeping any track of it.

Fig. 5 shows the modified message flow for a call setup and termination. SIP Proxy Server 2 is not represented since it only forwards messages.

Depending on the type of Proxy Server we are using, the *UpdateResourceReservation* message may or may not be used: In order to modify the properties of an existing call, SIP uses the INVITE message with the identifier of the existing call. If the Proxy Server receiving this INVITE message is a stateful Proxy Server (i.e. a Proxy Server keeping track of all active calls), it may be able to detect that this INVITE message refers to an existing call and thus send an *UpdateResourceReservation* to its *Resource-Manager*. However if the Proxy Server receiving the INVITE message is stateless, it doesn't keep any track of existing calls and will then send a *Reservation-Request* message to the RM as if it were a new call. In this case, the RM has to detect itself that it is actually a modification of an existing call. It can be easily realized using the call identifier provided in each INVITE request.

3.3 Results

We have implemented the *Resource Management Architecture* in a H.323 environment. For programming the *Resource-Manager*, we used the open source project

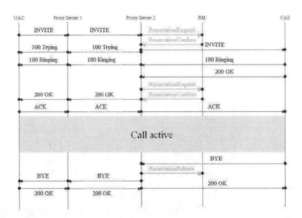

Fig. 5. Modified message flow accounting the SIP Proxy Server - RM communication

openH323 [10], which provides a lot of useful libraries for receiving and parsing the H.323 messages. For the Gatekeeper, we used the opengatekeeper from [11]. We built a testbed in order to test our implementations. In addition to one real IP telephony application, we developed a H.323 call generator based on the openH323 libraries. With this testbed we have demonstrated the feasibility of our approach and performed some performance tests.

We are currently working on the SIP approach, looking for open source version of SIP Proxy Server and client applications. We found several open source projects dealing with SIP, and installed some of them in order to compare their features. We finally decide to use for our testbed the Partysip Server [12], which is an implementation of a SIP Proxy Server together with the Ubiquity User Agent [13]. We now have to modify the implementation of the Partysip Proxy Server in order to enable the communication Proxy Server – *Resource-Manager*.

4 Conclusion and Future Work

Since both major protocols for multimedia communication H.323 and SIP will coexist for a certain time, there is obviously a need for porting existing H.323 systems into a SIP environment. We presented in this paper a concrete example, where this work should be done: The *Resource Management Architecture*. In this example, special server of the Architecture called *Resource-Manager* need to be integrated into the signalization flow of the SIP call setup procedure. We propose here to capture the INVITE and BYE request by lightly modifying SIP Proxy Server: Each Proxy Server receiving an INVITE request directed to a client with known IP address should forward it to the *Resource-Manager* so that it can perform the virtual resource reservation. The call is only confirmed if enough free resources are available in the network. Moreover we proposed some possible open source project to be used for that purpose. However we are still working on the implementation, so that one current issue is the verification and evaluation of our concept in a testbed.

Each *Resource-Manager* is responsible for a limited domain. If the network becomes bigger, it may be required to have several *Resource-Manager*. In this case, the resource reservation procedure takes longer, since *Resource-Managers* have to communicate with each other in order to coordinate the resource reservation over the different domains. One open issue consists in optimizing the call setup delay.

Acknowledgements. This work is supported by Siemens within a project called CoRiMM (Control of Resources in Multidomain Multiservice networks).

References

1. ITU-T Rec.: H.323, Packet-Based Multimedia Communications Systems, Geneva, Switzerland, July 2003; http://www.itu.int/itudoc/itu-t/rec/h (link to substandards)
2. J. Rosenberg, H. Schulzrinne, G. Camarillo, A. Johnston, J. Peterson, R. Sparks, M. Handley, E. Schooler: SIP: Session Initiation Protocol, IETF RFC 3261, June 2002
3. G. Camarillo, W. Marshall, J. Rosenberg: Integration of Resource Management and Session Initiation Protocol (SIP), IETF RFC 3312, Oct. 2002
4. C. Prehofer, H. Müller, J. Glasmann: Scalable Resource Management Architecture for VoIP, Proc. of PROMS 2000, Cracow, Oct. 2000.
5. J. Glasmann, H. Müller: Resource Management Architecture for RealtimeTraffic in Intranets, Networks 2002, Joint IEEE International Conferences ICN and ICWLHN, Atlanta, USA, August 2002.
6. J. Glasmann, M. Tromparent: Topology Discovery in the Context of Resource Management in IP-Networks, SoftCOM'02, Split, Croatia, Oct. 2002.
7. J. Case, M. Fedor, M. Scho_stall, J. Davin: A Simple Network Management Protocol (SNMP), RFC 1157, Mai 1990.
8. J. Postel: Internet Control Message Protocol (ICMP), RFC 792, September 1981.
9. J. Rosenberg, H. Schulzrinne: An Offer/Answer Model with the Session Description Protocol (SDP), IETF RFC 3264, June 2002
10. OpenH323: http://www.openh323.org
11. Opengatekeeper: http://opengatekeeper.sourceforge.net/
12. Partysip Server: http://www.partysip.org/
13. Ubiquity UA: http://www.ubiquity.net/products/SIP/

Distributed Multimedia Networking Services: Dynamic Configuration and Change Management

Martin Zimmermann

University of Applied Sciences Offenburg,
77723 Gengenbach, Germany
m.zimmermann@fh-offenburg.de

Abstract. In this paper, we propose a new streaming media service development environment comprising of a streaming media service model, a XML
based service specification language and several implementation and configuration management tools. Our approach is based on a high level streaming service
specification language, which allows specifying a service in terms of media objects, QoS, and distribution policies. Driven by such a streaming service specification and a streaming component library implemented with Java Media
Framework, the required distributed application infrastructure is generated
automatically by a service manager. To support flexible instantiation and termination of services as well as change management during runtime, e.g. migration or substitution of streaming components, we introduce instantiation and
termination rules, and reconfiguration rules.

1 Introduction

Multimedia networking services, also referred to as continuous streaming media services are getting more and more popular in business scenarios, e.g. video on demand,
IP telephony, distance learning, and business TV.

Long-running streaming services are a major challenge because they require specific mechanisms for reconfiguration due to operational and evolutionary changes.
Changes should be executable during runtime, they should cause minimal disturbance
to the running service and must be performed in such a way that leaves the service in
a consistent state.

Fig. 1 shows a realistic example of a streaming service delivered via a hierarchy of
media servers. The stream is composed of two live objects (audio and video of a
speaker) and an image file object. Temporal relationships define, when the different
media objects must be delivered, e.g. in parallel. QoS aspects include the delivery
strategy (push / pull), bitrates, codecs, and used protocols. Our configuration approach is based on a clear separation of a streaming service specification, and its
implementation by a distributed Java Media Framework (JMF) application. It enables
simplified specification of streaming services, automatic generation of the required
infrastructure, as well as rule based instantiation and reconfiguration of services.

M. Freire et al. (Eds.): ECUMN 2004, LNCS 3262, pp. 267–277, 2004.

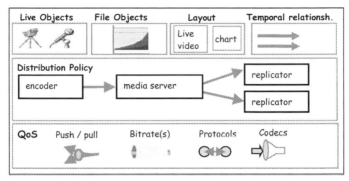

Fig. 1. Basic building blocks of a streaming service

The central purpose of this paper is to present a new framework supporting the flexible configuration, the implementation, and the change management of media streaming services. It provides an integrated service development environment comprising of a streaming service model, a service specification language and several implementation, and configuration management tools. The proposed framework is part of a multimedia project, financed by the state government of Baden-Württemberg in Germany

2 Related Work

The Session Description Protocol (SDP) [1] allows specifying sessions in terms of session identifiers, protocols, media objects. However, it does not support distribution policies and QoS specifications.

In [6] a component-based approach is described, which reflects the devices being used by the application. In the approach an explicit configuration is used, based on component types encoder, filter, network connector, and display.

In [5] stream interface descriptions based on compatibility rules are introduced to determine if two or more devices (their interfaces) are capable to interoperate (bind to each other). If interfaces are incompatible (unable to bind) media processing (filters) are inserted which transcodes between the incompatible interfaces. In [4] a concept for association management is introduced which enables passing of streamed movies with synchronization & control.

Some related work aims at integrating streaming services with middleware platforms based on remote method invocations such as CORBA. The CORBA Telecoms specification [13] defines stream management interfaces, but not the data transmission. Infopipes [9] provide a high-level interface tailored to information flows and flexibility in controlling concurrency and pipeline setup. However it does not provide a service specification technique.

In [10] a mobile streaming media CDN (Content Delivery Network) architecture is presented in which content segmentation, request routing, prefetch scheduling, and session handoff are controlled by SMIL (Synchronized Multimedia Integrated Language) modification. The approach concentrates on the segmentation aspect, which is important for mobile users. In [11] also an XML based specification technique for streaming services is introduced. However, it does not enable the specification of quality of service attributes and distribution policies.

3 Streaming Service Specification

A media service can be described by its media objects, distribution and replication policy, and the quality of service used for content delivery. Media streams can also be categorized according to how the media is delivered. In pull services data delivery is initiated and controlled from a client whereas in push services a server initiates data transfer and controls the flow. We use XML for the specification of streaming services: media objects, QoS, and distribution policies [12].

3.1 Media Objects

The specification of media objects is based on SMIL [2]. In addition to SMIL, the type of a media object (live / file), and in case of live streams, the nodes where the lives stream comes from can be specified. SMIL is a collection of XML elements and attributes that we can use to describe the temporal and spatial coordination of one or more media objects.

```
<StreamingService><MediaObjects>
    <par>
        <image id = "i1" type = "file"  src = "chart1.gif" />
        <video id = "v1" type = "live"  node = "c1.fh-offenburg.de" />
        <audio id = "a1" type = "live"  node = "c1.fh-offenburg.de" />
    </par>
</MediaObjects>
```

Fig. 2. Media objects specification

A SMIL presentation is a structured composition of autonomous media objects, e.g. <video> and <audio> objects. The <par> tag or parallel time container is used to specify that all children objects are all rendered in parallel. In our example, the two live streams (audio and video) are played in parallel (Fig. 2).

3.2 Quality of Service

Multimedia content can be delivered to the users as push and / or pull services (Fig. 3). In case of a push delivery, we can additionally define when the media objects must be delivered (date and time), together with the available bit rates. Moreover, the used protocols and codecs can be specified.

```
<StreamingService><QoS>
   <ServiceMode>
      <push codec = "MPEG" protocol = "RTP"  date = "20034-03-21" time = ... >
                     <bitrate>100</bitrate>
      </push>
   </ServiceMode>
   <MaxBandwidth>1000</maxBandwidth>
</QoS>
```

Fig. 3. QoS specification

Date and time attributes are also used in case of pull services. However, in case of a pull service these attributes define, when a media service is available, i.e. can be requested by a client.

Finally, one can specify the maximum bandwidth which can be used ("consumed") by the related service. The element <maxBandwidth> restricts the overall bandwidth which can be consumed by all the service users.

3.3 Distribution Policy

Fig. 4 illustrates the specification of a distribution policy, which is related to a media service specification. Each media service specification must be available on a *primary server*. As an optional part a set of replicators and spitters can be introduced. *Splitters* and *replicators* are two components of distribution, which enables us to deliver streams to multiple media servers.

```
<StreamingService><DistributionPolicy>
   <primaryServer node = "s1.fh-offenburg.de" creationOn = "static"/>
   <spitter node = "s2.domainA.de"  strategy = "pull" creationOn = "serviceRequest"/>
      <spitter node = "s3.domainB.de"/>
      <spitter node = "s4.domainC.de"/>
   <spitter node = "s5.domainD.de"/>
</DistributionPolicy>
```

Fig. 4. Distribution policy specification

At its most basic level, splitting is a technology that enables one component to deliver a live stream to another component. In such a server-to-server splitting, a primary server transmits a live stream to one or more additional servers. This allows distributing the same stream to multiple media servers across a network. In the conventional, or "push" form of server- to-server splitting, the transmitter initiates the connection to the receiver. In case of pull splitting, the initial media server (primary) does not deliver the stream to the receiver until the first media player makes a request. There is a slight delay as the receiver requests, receives, and delivers the stream. After that, the stream is live on the receiver, and subsequent player requests do not involve the session setup delay.

The attribute creationOn in Fig. 4 defines when a component (e.g. a splitter) should be created. In case of value "serviceRequest", the component is created when the service must be provided (e.g. in case of push services depending on the value of time and date). For on-demand streaming a content caching specification enables the definition of a cache size (not shown in the example). Each replicator maintains a portion of physical disk space for caching on-demand content originally deployed on the primary server. Such caching improves playback quality by propagating the content closer to the player. It also reduces delivery cost by caching content at the network's edge.

4   Implementation

In the following, we describe the management of services, components, and configurations as well as the mapping of service specifications onto JMF implementations in more detail.

4.1   Component Library

Available streaming components are stored in a separate component library. The properties of existing components (e.g. an encoder component) are described by a XML specification.

Component Property	Encoder
Name	encoder (avi2mpeg)
Source	(Reference to JMF code)
Processor Speed Req.	2 Ghz
Operating System	...
Input Formats	avi, ...
Output Formats	mpeg1, ...
Bitrates	12k,...,1.5M
Sure Stream	Yes
Service Modes	Push/Pull

Fig. 5. Component Library

Our current development environment uses the following categories: encoder, media server, replicator, and splitter [12]. Fig. 5 illustrates the properties of an encoder component.

4.2   Management of Services, Components, and Configurations

Configuration management for distributed streaming service covers all phases during the lifetime of a streaming service: service specification, service creation as well as change management.

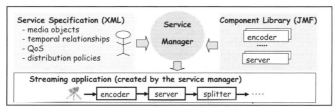

Fig. 6. Service Manager

Multimedia services as well as properties of streaming components are stored in a XML database. Services are managed by a service manager (see Fig. 6), which provides in the current Java implementation a simple JSP based interface to search, and request services. An additional management interface enables creation of new services, and deletion of existing services. Before a new service is stored, a consistency checker validates availability of nodes, and media objects according to the service specification. A specification is then analyzed by the service manager, which will select appropriate JMF components.

We distinguish two categories of configuration management activities. The first category contains management activities regarding creation of new streaming services and termination of running streaming services. These types of activities are specified using the concept of ***instantiation and termination rules*** (IT rules). IT rules are similar to the concept of constructors and destructors as used in programming languages like C++ or Java to express instantiation and termination activities associated with object creation and deletion. Similarly to IT rules constructors and destructors are implicitly invoked each time a class object is allocated or deleted.

The second category of management activities results from the fact that specific events from the application and network management, such as a crash of a computer node or the overload of a server component may lead to a reconfiguration activity. For this purpose, we provide a concept to specify a reconfiguration activity as a set of event driven ***reconfiguration rules***. Rather than describing the full flavour of both concepts, some important constructs are presented in the context of examples.

Service Creation

During runtime of a streaming service, the basic configuration management activities are the creation and termination of streaming components, the establishment of interconnections between components as well as the initialization of streaming components. Initialization actions include setting of parameter values, e.g. the used protocol, formats, bitrates, etc. Fig. 7 illustrates such an instantiation rule for creation of a schedule-based push service. In the first part of the rule (conditions) the set of push services is determined which must be started. The related actions are specified in the second part. Each identified service must first be configured, and then started.

```
CONDITIONS
NOT EMPTY serviceSet = { s: Service WITH {
            s hasStartTime NOW AND
            s hasServiceMode PUSH AND
            s hasCreationMode ON_DEMAND }}
ACTIONS
FOR ALL s: serviceSet {
            createConfiguration (s)
            startService (s) }
```

Fig. 7. IT rule example: creation of a schedule based push service

In the example, each push service is created automatically according to the specified date and time values. Creating a service means to create the required JMF components, e.g. encoder and media server components as well as to establish the communication relationships by construction of appropriate JMF media locators.

Each JMF component offers a RMI based management interface which allows defining its behavior by related method calls. E.g., the operation connect() is used to establish a communication relationships between two streaming components.

Change Management

A change management policy is a pattern of the form "condition → actions" (when? what? how?), where the condition specifies when something, namely the specified action, has to be done to preserve the policy. Each rule describes a condition and a sequence of related management actions to be executed when the condition is observed. A condition is a predicate expressed on the streaming service model, i.e. component properties, e.g. memory requirements (see Fig. 8), and information available as part of a service specification, e.g. location of media objects, values of service modes, etc...

Currently, we concentrate on change management related to distribution policy issues (an approach which focuses on QoS change management is described in [3]). Change management related to distribution aspects include:

- migration / substitution of media objects / streaming components
- creation / termination of streaming components
- modification of caching / splitting policies

The following rule (Fig. 8) describes the conditions and actions related to the substitution of a primary server by a new media server.

```
Rule: substitutePrimaryServer (NODE destination)
    CONDITIONS
    EXISTS ms: MEDIA_SERVER WITH {
            operatingSystem (destination) IN supportedOperatingSystems(ms)
            AND memory(destination) > memoryNeeds(ms)
            AND protocols(this) IN protocols(ms)


    ACTIONS
    moSet = {mo: MEDIA_OBJECT WITH
            mo hasType FILE
            AND mo IN MEDIA_OBJECTS(this)
    FOR ALL mo: moSet {
        copy mo TO destination }
```

Fig. 8. Reconfiguration rule example

The new media server must be created at a new node. The conditions test if there is an alternative media server component available. In the ACTIONS part all media object files are copied to the destination. The identifier "this" indicates a reference to the current service to which the rule is applied. Thus, the specification mediaObjects(this) in the action part determines all media objects of the current service.

However, the rule in Fig. 8 can only be applied to services which deliver media objects of type live. In case of "live services" another rule must be used, because live objects are normally not directly connected to a media server. They require a separate encoder component as an intermediate component between a live source object and a media server.

We distinguish between periodic events generated by a scheduler, and change events to reflect a state change of some application or network entity. The latter includes events from the network management (e.g. shutdown of a node) and/or application management (e.g. overload of a component). For example, depending on the current load of a server, a reconfiguration rule could describe that replication has to be initiated.

From the network management point of view, there are operations to indicate events relevant for the application management, e.g. an event prepareForShutdown indicating a shutdown of a computer node. In the opposite direction the network management component provides operations to receive events from the application management and operations to acquire resource properties of computer nodes. The latter operations are essential for determining alternative nodes for automatic placement at component creation time and migration of components during runtime. Application events are generated by notifications of service components. Performance monitoring events are of special interest, because they can result in the need to reconfigure the service. Therefore, metric objects are integrated into the service components. They are responsible for notifying a management component when relevant value changes take place.

A metric object consists of a metric element such as a counter or a gauge, and a description when to send an event notification. Hereby, the filter mechanism is incorporated into the application component.

```
Rule: shutDown (NODE n)


CONDITIONS                                      ACTIONS

NOT EMPTY serverSet = {                         //see actions Figure 8

  s : MediaServer WITH {                        readyForShutdown (n)

    s hasLocation n AND

    NOT EMPTY nodeSet = { m: NODES  WITH {

    memory(m) > memoryNeeds(s)

    AND  protocols(this)  IN  protocols(ms) AND ....
```

Fig. 9. Reconfiguration rule example

Figure 9 illustrates a reconfiguration rule which is triggered by an event from the network management. In the example, the first clause of the shutdown condition is a notification prepareForShutdown for a specific node which is indicated by the network management. If this clause becomes true, the next clause checks all other conditions required for migration for all servers running on that node. Then, there also must be an alternative node to migrate to which fulfils the resource requirements of the server.

Now, whenever a shutdown condition becomes true, the corresponding actions are performed. For each server of the set computed in the condition, an appropriate node is selected for migration. Finally, the shutdown request will be confirmed by a readyForShutdown notification sent back to the network management system.

Component Configuration Based on Java Media Framework (JMF)

Java Media Framework (JMF) [8] provides a unified architecture and messaging protocol for managing the acquisition, processing, and delivery of time-based media data. JMF uses a simple streaming processing model. A *data source* encapsulates the media stream much like a video tape and a *player* provides processing and control mechanisms similar to a VCR. Playing and capturing audio and video with JMF requires the appropriate input and output devices such as microphones, and cameras. A DataSource is identified by either a JMF MediaLocator or a URL (universal resource locator). A *processor* component provides control over what processing is performed on the input media stream. In addition to rendering media data to presentation devices, a Processor can output media data through a DataSource so that it can be presented by another Player or Processor, further manipulated by another Processor, or delivered to some other destination, such as a file.

The JMF API consists mainly of interfaces that define the behavior and interaction of objects used to capture, process, and present time-based media. Implementations of these interfaces operate within the structure of the framework. By using intermediary objects called *managers*, JMF makes it easy to integrate new implementations of key interfaces that can be used seamlessly with existing classes. The following implementation rules are used to map a service specification onto a JMF architecture (see Fig. 10):

- Media objects in XML are implemented by Media Locators in JMF
- Streaming components in XML are implemented by a JMF "pipeline": DataSource, Processor, etc.
- The functionality of a component (e.g. encoding) is implemented by an appropriate JMF processor object
- Communication relationships between server components are implemented by DataSource and DataSink objects

Further details are described in [12].

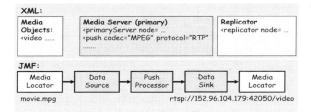

Fig. 10. Mapping of XML onto a JMF based architecture

5 Conclusion

A new approach for specification, implementation, and configuration management of streaming services has been introduced. The major contributions and extensions aim to provide a high level specification of streaming services in terms of media objects, quality of service, and distribution policies. Based on such a service specification, a set of tools support automatic creation of a distributed streaming service based on a component library containing JMF streaming components.

Separation of specification and implementation enables modularity, reusability and extensibility. Existing SMIL specification, which only defines the media objects together with temporal relationships, can be extended to get a complete service specification.

We report the current status of our prototype. The prototype has shown that the object model behind JMF is well suited for automatic generation of implementations. At implementation level, the different aspects are integrated in a general object-oriented architecture supporting modularity and reuse of software.

To support the implementation and management of distributed streaming applications, we developed a set of tools for the mapping of application specifications onto an object-oriented implementation model. This includes tools for consistency check of a service specification as well as selection of components and generation of distributed application configurations based on a JMF component library.

The clear separation of policies into instantiation rules, termination rules, and reconfiguration rules has several advantages. First, it supports modularity, reusability and extensibility of the service management rule data base. Additionally, our approach supports more flexibility in combining different rules taking into account the specific application requirements. Forming an integral part of a distributed application specification, generic policies can be defined for different streaming paradigms, e.g. for live services.

Future work will concentrate on the extension of the service model to support also reconfiguration driven by QoS changes. Another important aspect will be to allow the specification of service templates, e.g. a service template for video live streaming.

References

1. Schulzrinne, H., Rosenberg, J.: The Session Initiation Protocol: Internet-Centric Signaling, IEEE Communications Magazine, October 2000
2. SMIL: http://www.w3.org/TR/smil20/
3. Atallah, S.-B.: Dynamic Configuration of Multimedia Applications, INRIA, SARDES Project
4. Kahmann, V., Wolf, L.: A Proxy Architecture for Collaborative Media Streaming, Workshop on Multimedia Middleware, October, 2001, Ottawa, Canada, ACM Press
5. Rafaelsen, H. O., Eliassen, F.: Towards support for ad-hoc multimedia bindings, Workshop on Multimedia Middleware, October, 2001, Ottawa, Canada, ACM Press
6. Naguib, H., Coulouris, G.: Towards Automatically configurable multimedia applications Workshop on Multimedia Middleware, October, 2001, Ottawa, Canada, ACM Press, New York
7. Ayers, J., et al.: Synchronized Multimedia Integration Language (SMIL) 2.0, World Wide Web Consortium Recommendation, Aug. 2001, http://www.TR/2001/REC-smil20-20010807/.
8. http://java.sun.com/products/java-media/
9. Black, A. P. et al: Infopipes: An abstraction for multimedia streaming, Multimedia Systems 8:406 .419 (2002)
10. Yoshimura, T.: Mobile Streaming Media CDN Enabled by Dynamic SMIL, WWW2002, May 7-11, 2002, Honolulu, Hawaii, USA, ACM 1-58113-449-5/02/0005
11. ipdr.org: Service Specification – Streaming Media (SM), www.ipdr.org
12. Zimmermann, M., Althun, B.: Streaming Services: Specification and Implementation based on XML and JavaMediaFramework, in FIDJI'2003 Springer Proceedings, FIDJI 2003 Luxembourg, Luxembourg, 2003

The Dynamic Cache-Multicast Algorithm for Streaming Media

Zhiwen Xu, Xiaoxin Guo, Yunjie Pang, and Zhengxuan Wang

Faculty of Computer Science and Technology, Jilin University,
Changchun City, 130012, Jilin Province, China
xuzhiwen@public.cc.jl.cn

Abstract. The transmission of large capacity and high bit rate for streaming media becomes a challenging study problem for the Web application. The proxy cache for streaming media is an efficient method to solve this problem. In this paper, we proposed the algorithm of dynamic cache-multicast based on partial cache method such as the prefix cache, segmentation-based cache et al. The idea of the algorithm is using a dynamic cache to keep partial media object requested by users in an interval time. The advance is that reference frequency of partial media in interval time is higher than that of full media requested by users. The dynamic cache-multicast algorithm enhances the efficiency of proxy cache for streaming media, mitigates network burden from content server, and saves the traffic resource for network backbone. Event-driven simulations are introduced to evaluate this algorithm is very efficient.

1 Introduction

At present, web proxy cache has broad application, since the proxy cache technology for caching text and image objects does not apply for streaming media. The reasons are follows: First of all, video files require large memory volume. A single file may demands 10 M to 10 G memory volume, which is determined by the quality and length of the video. Most content cached should be stored in the hard disc, and hard disc for proxy cache and memory cache must be organized with great care. Secondly, the fact that real-time media transmission requires obviously large disc volume, bandwidth of network and support within a long period of time, which demands that effective cache algorithm should be used in proxy cache so as to avoid using too much disc volume for caching new content. The characteristic of the streaming media determines that streaming media is its caching objects instead of web objects. The research of streaming media caching technology is a challenging subject.

2 Related Work

2.1 Prefix Cache

As to the study of proxy cache of streaming media, we are still at the stage of theoretical and experimental research, needing further development. In order

M. Freire et al. (Eds.): ECUMN 2004, LNCS 3262, pp. 278–286, 2004.

to solve the problem of startup delay and realize smooth data transmission, Z.Miao proposed the method of prefix cache/suffix cache [1]. When transmitting the media, they divide the media into two parts. The smaller preceding part is called prefix cache, while the latter part is called suffix cache. Generally prefix Cache is stored in proxy cache. When the users make an application, what we store in the prefix cache will be played first. Meanwhile, the part stored in suffix cache is transmitted from the content server to the proxy cache. When media stored in prefix cache is over, the part stored in the suffix cache starts to play. In this way, the problem of startup delay is solved effectively. The literatures [9.10.11] give a introduction to the research in such areas as the management and organization of proxy cache based on prefix cache, the connecting schemes of the server, the schedule of batch and patch in the proxy cache.

2.2   Segmented Cache

The initial portion of a media stream in the proxy cache is more important than the latter portion. The result that the initial portion is of great importance and the media objects should be cached partially applies to most media objects, which leads Wu Kun-Lung [2] and his team to develop a segmented approach to cache media objects. Blocks of a media object received by the proxy cache are grouped into variable-sized, distance-sensitive segmentations. In fact, the segmented size increases exponentially from the beginning segmentation. For simplicity, the video i is 2^{i-1} blocks and contains media blocks $2^{i-1}, 2^{i-1}+1, \cdots, 2^i - 1$ (i∈(1,2,···,M)). The motivation for such exponentially-sized segmentation is that we can quickly discard large blocks of media objects cached. Caching managed in this way can make a quick adjustment on partial objects cached. For example, the proxy cache can release 1/2 of media object cached. The number of segmentations for each cached object is dynamically determined by the cache admission and replacement policies. Different segmentations are given different caching values.

2.3   Segmented Caching Admission and Replacement Strategies

We should take two set of important caching management strategy into consideration in proxy cache. One is the caching admission strategy and the other is caching replacement strategy. These two are closely related. The caching admission control is to judge whether we should give permission to cache a certain media according to its popularity. We apply a two-tiered admission control approach, which is based on the segmented number, to cache the segmentation with a segmented number smaller than a threshold. The initial segmentation of the media object, K_{min}, is cached as a basic unit. The value of K_{min} should be determined by the network conditions and should ensure that the non-cached media can be transmitted in time from the content server to the proxy cache so as to meet the requirements of the users who apply for continuous media play. What's more, it should be determined by the network delay between the proxy cache and the content server and the load conditions of the content server as well.

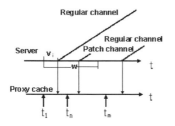
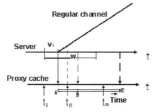

Fig. 1. Process of the server and proxy   **Fig. 2.** Process of cache-multicast

3   The Dynamic Cache-Multicast Algorithm of Proxy Cache

In the proxy cache of streaming media based on segmentation, the size of the media cached varies with the changes of the requested frequency of the certain media. In general, only a part of the media object is stored in the proxy cache, whereas the other part is not stored. When a user requests this media object, the part not stored should be released from the content server to proxy cache and then can be transmitted to the client. When the media has request by many clients at the same time, it should be transmitted from the content server to the proxy cache for several times. We put forward a dynamic efficient organizing algorithm of cache-multicast, utilizing the characteristic of proxy cache for segmented streaming media and dynamic caching technology. Within a certain duration of time, if two or more users request the same video material, we may cache a segmentation of this video dynamically to meet those users' requirements. It is unnecessary for the proxy cache to apply for the video material frequently. Once is enough to satisfy all the users. The length of this duration of time should be shorter than that of the video material and the permitted over caching threshold.

Fig.1 shows the process how the server and proxy handle the clients' requests. At time t_1, t_n, t_m, different clients' request to the same media. If $t_n - t_1 <$ the width of largest patch window W, represented by W request at time t_n will be handled by patch channel. If $t_m - t_1 > W$, the request at time t_m will be handled by the regular channel. In both cases, clients at time t_n, t_m both take up the network and server resource. Fig.2 is the process in which the proxy cache makes use of dynamic cache-multicast algorithm. ABC denotes the dynamic caching windows moving with time. The length of AB equals to $t_n - t_1$. In order to handle the request at time t_n, we cache this segmentation so as to cache the resource that patch channel takes up in the process of web transmission and the server. The length of AC equals to $t_m - t_1$. In order to handle the request at time t_m, we cache this dynamic duration for streaming media so as to save the resource that usual channel takes up in the process of web transmission and the server.

How should we apply the algorithm of the dynamic cache-multicast so as to ensure he highest byte hit ratio the length of video i with popularity λ is L_i. And than cached length in the proxy cache is V_i. Clients make request respectively at time $t_0, t_1, \cdots, t_j$. Setting $t_0 = 0$, then when j=1, 2, $\cdots$ and $(t_j - t_{j-1}) \leq$ the threshold of dynamic cache-multicast, we make used of cache-multicast for all the requests from the clients so as to ensure that the proxy cache has the benefits for every user's video request. That is to say we make use of cache-multicast for clients' requests. The method of cache-multicast will help to save network resource and the load of server. J represents the times of clients' applications managed by cache-multicast. Proxy cache saves server's management and network transmitted resource for j times. $\Delta = t_j - t_0$ is the length of time between two requests, which equals to the time's length of cache-multicast. When $\Delta \leq V_i$,the proxy cache uses cache-multicast at time $t_0 + V_i$. The time length of the cache-multicast is Δ. When $\Delta > V_i$, the proxy cache saves, at time $t_0 + V_i$, dynamically the resource of network to transmit media. The time length of the cache-multicast is Δ, and the length of time of prefetch caching in the proxy cache is $\Delta - V_i$. Suppose this is the Poisson process. The users' average request time is $1 + V_i\lambda_i$.The j represents the average request time in the dynamic cache-multicast, and j $= \Delta\lambda_i$. L_i represents the length of the media. C_s and C_p are respectively the transmitted value coefficients, from the server to the proxy cache and from the proxy cache to the clients. The media segmentation, $[V_i, L_i]$, should be taken out from the original server. $C_i(V_i)$ is the average value of delivering video i. Then

$$C_i(V_i) = (C_s \frac{L_i - V_i}{1 + \lambda_i V_i} \frac{1 + V_i\lambda_i + \Delta\lambda_i}{1 + V_i\lambda_i} + C_p L_i)\lambda_i B_i \qquad (1)$$

The former item is the delivering value from server to proxy cache. The latter item is the delivering value from proxy cache to the clients. Our main aim is to reduce the resources of the backbone network, that is to say the smaller the value of the first item, the better. We considered if two or more applicants apply for the same video within time V_i, we may save the resource of the backbone network for j times on average, and improve the byte hit ratio of proxy cache, by making those latter applicants not take up network resources of the backbone.

In the segmented strategy, according to the application frequency for the certain media, we cache video based on segmentation with different length. This strategy considers the significance of the initial part of a requested media, and ensures higher efficiency of the proxy cache. however, it doesn't consider adequately the effect imposed on it by the media for requested clients in duration. We design the algorithm of dynamic cache-multicast with the main intention to consider adequately the case when many clients apply for the same media within the duration of time and on the basis of segmented-cache, dynamically cache the segmentation of the media on demand by multi-user so as to ensure that it is necessary for only the very first applicant to take out the part of media not cached in the proxy cache from the server and cache this segmentation in the proxy cache. This length of time is regarded as the length of the media cached. We adopt FIFO replacement policy to make this part of cached media meet the need of multi-user within the duration of time. The efficiency of proxy cache will be influenced by the determination of this time length. If we determine this time length by calculating users' behavior, then the proxy cache allocates the caching length according to this time length. The method mentioned above will bring some troubles to the management of proxy cache, and therefore, permission control and replacement strategy will become more complex. Moreover, calculating the most efficient time interval according to the users' behavior, will on the one hand bring burden to the proxy cache, and on the other hand save the resources of the network and server.

4   Performance Evaluation

4.1   Methodology

We utilize an event-driven simulator to stimulate the proxy cache service and furthermore to evaluate the algorithm of the dynamic cache-multicast based on variable-size. The algorithm of LRU and FIFO stack were used to keep track of media objects in the cache. LRU stack was to track the initial segmentation and the segmented-cache, whereas the FIFO was to track the cache-multicast. C_{init} represents the initial segmentation, and C_{multi} is used to represent dynamic cache-multicast segmentations. We calculate byte hit ratio and request time with startup delay. The byte hit ratio measures the total bytes of the objects cached over the total bytes of objects requested. When a request arrives but the initial K_{min} segmentation is not cached in the proxy cache, then there will be a startup delay. Let's suppose that the media objects are videos and the size of these videos are uniformly distributed between 0.5 B and 1.5 B blocks, where B represents video size. The default value of B is 2,000. The playing time for a block is assumed to be 1.8 seconds. In other words, the playing time for a video is between 30 minutes and 90 minutes. The size of cache is expressed on the basis of the quantitative description of media blocks. The default cache size is 400,000 blocks. The inter-arrival time distributes with the exponent λ. The default value of λ is 60.0 seconds. The requested video titles are selected from a total of the distinct video titles. The popularity of each video title M accords to the Zipf-like distribution. The Zipf-like distribution brings two parameters, x and M. the former has something to do with the degree of skew. The distribution is given by $p_i = C/i^{1-x}$ for each $i \in (1, \cdots, M)$, where $C = 1/\sum_{i-1}^{M} 1/i^{1-x}$ is a normalized constant. Suppose x = 0 corresponds to a pure Zipf distribution, which is highly skew. On the other hand, suppose x = 1 corresponds to a uniform distribution with no skew. The default value for x is 0.2 and that for M is 2,000. The popularity of each video title changes with time. It is very likely that a group of users may visit different video titles at different periods of time and the users' interest may be different. In our simulations, the distribution of the popularity changes every request R. The correlation between two Zipf-like distributions is modeled by using a single parameter k which can be any integer value between 1 and M. First, the most popular video in the first Zipf-like distribution finds its counterpart, the r_1-th most popular video in Zipf-like distribution 1, where r_1 is chosen randomly between 1 and k. Then, the most popular video in the second Zipf-like distribution finds its counterpart, the r_2-th most popular video. r_2 is chosen randomly between 1 and min (M, k+10), except r_1. The rest may be deduced by analog. When k represents the maximum position in popularity, a video title may shift from one distribution to the next. k = 1 expresses perfect conformity, and k = M expresses the random case or unconformity.

We compared the cache-multicast algorithm with the full video approach, the variable-sized segmented approach, and the prefix schemes in terms of the impact they imposed on byte hit ratio and startup delay from the following aspects: the cache size, the skew of the video popularity, users' viewing behavior and other related system parameters.

4.2   Impact of Cache Size

We study the impacts imposed by the cache size on the byte hit ratio and startup delay. For a fairly wide range of cache size, the dynamic cache-multicast method has the highest byte hit ratio and the lowest fraction of requests with startup delay, whose byte hit ratio is higher than the variable-sized segmented approach and the prefix schemes with the same startup delay. Fig.3 shows the impact cache size imposes on the byte hit ratio. Fig.4 presents the impact imposed by cache size on the fraction of requests with startup delay. The full video approach and the prefix have comparable byte hit ratio, with the full video approach having a slight advantage over the prefix scheme. For a smaller cache size, the advantage of byte hit ratio managed by the variable-sized segmented approach is quite evident. The dynamic cache method proves to have the highest byte hit ratio. Even though the full video and the prefix approaches perform almost equally in byte hit ratio, they differ dramatically in the fraction of requests with startup delay. The full video approach has a significantly higher fraction of requests with startup delay (Fig.4). For example, for a cache size of 400,000 blocks, 0.60 of the requests cannot start immediately using the full video approach. However, only 0.156 of applicants encounter startup delay using dynamic cache, variable-size segmentation and prefix approaches. Within the whole range of cache size, the effect of the dynamic cache-multicast approach, variable-size segmented method and the prefix strategy are basically the same.They all effectively solve the problem of startup delay.

4.3   Impact of Video Popularity

Let us examine the impact that the video popularity imposes on the byte hit ratio and startup delay. The dynamic cache-multicast method has the highest byte hit ratio when the video popularity makes changes of wide scope. The dynamic cache-multicast approach, the variable-sized segmentation and the prefix schemes all have the same fewest request time with startup delay, which is superior to the whole video. Fig.5 shows the impact of skew in video popularity on byte hit ratio, while Fig.6 shows its impact on the startup delay. In addition to the parameter of Zipf, x, we also studied the changes of the popularity distribution and the impact of the maximum video shifting position k. The request R of

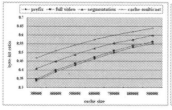 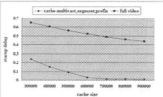

Fig. 3. Impact of byte-hit ratio   **Fig. 4.** Impact of startup delay

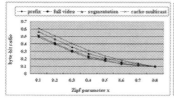 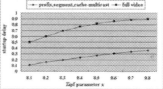

Fig. 5. Impact of byte-hit radio   **Fig. 6.** Impact of startup delay

the video shift was set to be 200. Fig.7 shows the impact of the maximum shifting position of a video. When the maximum shifting distance increases, the byte hit ratio of the dynamic cache-multicast, the variable-sized segmentation and the prefix approaches will fall, but only very slightly. The dynamic cache-multicast method is always better than the variable-sized segmentation and the prefix approach, which is closely related with the popularity distributions of the video titles and with the range of K, which is from 5 to 40. We also change the value of R, but the result is quite similar. Its byte hit ratio has only very slight impact on R.

4.4   Impact of Other System Parameters

Fig.8 shows the impact of video length imposes on the byte hit ratio. In general, as the size of the media file increases, the byte hit ratio will fall, this is true for all the four approaches. When the size of a media file is very large, the dynamic cache-multicast algorithm can ensure higher byte hit ratio than the segmentation and other two approaches. As to a video with the length of 3000 blocks, the byte hit ratios of dynamic cache strategy and variable-sized segmented strategy are respectively 0.32 and 0.28. If the length falls to 1000 blocks, the byte hit ratios may reach 0.61 and 0.59 respectively. No matter which approach we use, dynamic cache strategy or variable-size segmented approach, caching large media will cause the byte hit ratio in proxy cache to fall. However, dynamic cache-multicast strategy is better than variable-sized segmented strategy.

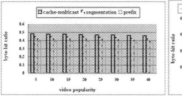

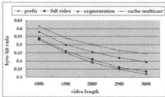

Fig. 7. Impact of popularity   **Fig. 8.** Impact of video length

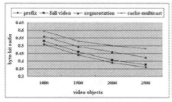

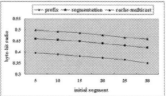

Fig. 9. Impact of video objects   **Fig. 10.** Impact of initial segmentation

Besides the size of a media file, the efficiency of proxy cache can also be affected by the distinct media objects. On the Web there exist many distinct media objects. As the demand on such objects increases, the proxy cache becomes less effective. Fig.9 shows the cases of applicants for distinct media objects from the angle of quantity. Once again, the dynamic cache-multicast strategy and the variable-sized segmented approach have much more advantage over the other two approaches, even when the conditions for caching are less favorable for both of them. Comparatively speaking, the advantage of the cache-multicast strategy is more outstanding. Fig.10 examines the percentage of caching dedication for storing the initial segmentation. Because the cache for the suffixes is reduced, the byte hit ratio falls with the increase in using initial segmentation. This slight decrease in byte hit ratio can be offset by increasing benefits substantially by the means of reducing start delay. For example, let us compare these two cases,0.05 and 0.15. The byte hit ratio is barely decreased, but the fraction of delayed startup drops substantially. However, no more benefits can be derived once the percentage of the initial segmentation cached increases beyond 0.20.

5   Conclusion

In this paper, we put forward the algorithm for dynamic cache-multicast, based on the variable-sized segmented approach. The segmented approach groups media blocks into variable-sized segmentations. This method differs from the way we handle a web object, which is usually handled as a whole. The algorithm of dynamic cache-multicast considers adequately the users' request behavior. While maintaining the advantage of the variable-sized segmentation, it provides the multi-user within a period of time with the same media they request by using dynamic cache. The algorithm of cache-multicast greatly saves traffics resource on the backbone of network, and enhance the byte hit ratio and efficiency of the proxy cache. Event-driven simulations evaluation was introduced to compare the dynamic cache-multicast with the variable-sized segmented approach, the full video approach and the prefix caching approach from the following angles: the cache size, the skew of video of popularity, the length and the quantity of a certain video. Therefore, the algorithm for dynamic cache-multicast effectively saves the network resource and enhances the byte hit ratio of proxy cache.

References

1. Z.Miao, A.Ortega: Proxy caching for efficient video services over the Internet.In Proc. Of Int, Web Caching Workshop, Apr. 1999.
2. K.L.Wu, P.S.Yu: Segment-Based Proxy Caching of Multimedia Streams. In: Proc. of IEEE INFOCOM, May, 2001
3. K.A.Hua and S.Sheu: "Skyscraper broadcasting: A new broadcasting scheme for metropolititan video-on-demaond system.": In Proc. Of ACM SIGCOMM 97 conference,pages 89-100,Sept. 1997.
4. K. Hua, Y. Cai and S. Sheu: "Patching: A multicast technique for true video-on-demand services.": in Proc. ACM Multimedia, September 1998.
5. D. Eager, M. Vernon, and J. Zahorjan:Minimizing bandwidth requirements for on-demand data delivery: in Proc. 5th Inter.Workshop on Multimedia Information Systems, October 1999.
6. S.Sen, J.Reforrd, and D.Towsley: Proxy prefix caching for multimedia streaming: In Proc. Of IEEE INFOCOM, Mar.1999.
7. B.Wang, S. Sen, M. Adler, and D. Towsley: Proxy-based distribution of streaming video over unicast/multicast connections: Tech. Rep. 01-05, Department of Computer Science, University of Massachusetts, Amherst,2001.
8. S. Ramesh, I. Rhee, and K. Guo: Multicast with Cache (MCache): Anadaptive Zero-Delay Video-on-Demand Service. in: Proc. IEEE INFOCOM,April 2001.
9. J. Almeida, D. Eager and M. Vernon: A hybrid caching strategy for streaming media files: in Proc. SPIE/ACM Conference on Multimedia Computing and Networking, January 2001.
10. O. Verscheure, C. Venkatramani, P. Frossard and L. Amini: Joint server scheduling and proxy caching for video delivery: in Proc. 6th International Workshop on Web Caching and Content Distribution, June 2001.
11. S. Sen, L. Gao and D. Towsley: Frame-based periodic broadcast andfundamental resource tradeoffs: in Proc. IEEE International Performance Computing and Communications Conference, April 2001.
12. G.K.Zipf: Hurnan Behaviour and the Principles of Least Effort: Addison-Wesley. Cambridge,MA,1949.
13. D.L.Eager, M.C.Ferris and M.K.Vernon: Optimized regional caching for on-demand data delivery: In Proc. Of multimedia Computing and Networking. Jan,1999.

IP Traffic Classification via Blind Source Separation Based on Jacobi Algorithm*

Walid Saddi, Nadia Ben Azzouna, and Fabrice Guillemin

France Telecom, Division R&D, 2 Avenue Pierre Marzin, F-22300 Lannion, France
Fabrice.Guillemin@francetelecom.com

Abstract. By distinguishing long and short TCP flows, we address in this paper the problem of efficiently computing the characteristics of long flows. Instead of using time consuming off-line flow classification procedures, we investigate how flow characteristics could directly be inferred from traffic measurements by means of digital signal processing techniques. The proposed approach consists of classifying on the fly packets according to their size in order to construct two signals, one associated with short flows and the other with long flows. Since these two signals have intertwined spectral characteristics, we use a blind source separation technique in order to reconstruct the original spectral densities of short and long flow sources. The method is applied to a real traffic trace captured on a link of the France Telecom IP backbone network and proves efficient to recover the characteristics of long and short flows.

1 Introduction

A common approach to modeling Internet traffic relies on the concept of flow (see for instance [1,2,3]). Short and long flows are frequently referred to in the technical literature as mice and elephants, respectively even if a clear definition does not exist (see [4] for a detailed discussion on the difficulty of defining mice and elephants). From a practical point of view, the major issue when using the concept of flow is that a complete traffic trace has to be captured and then off-line analyzed in order to properly classify the different flows. To know whether a flow is indeed a mouse or an elephant, the complete history of the flow has to be known and flows have to be sorted according to some criteria. Because of the very high transmission rate of network links, off-line analysis requires huge storage capacities and sorting flows is highly CPU consuming. Moreover, an exhaustive description of flows in not always needed, since only some metrics of interest have to be estimated. For instance, in the case of elephants, their bit rate and their duration give some insights into the way end users utilize the network.

Extracting metrics directly from traffic traces can be done, in principle, by using digital signal processing techniques. The major problem, however, is that the spectral characteristics of the different types of flows may be intertwined.

---

* This work has been partially supported by the RNRT project Metropolis

M. Freire et al. (Eds.): ECUMN 2004, LNCS 3262, pp. 287–296, 2004.

Hence, computing metrics for a given flow type first requires the isolation of the spectral characteristics of the component under consideration.

In this paper, we propose a method of processing packets instead of flows in order to construct two signals, one for mice and the other for elephants by using a very simple definition for these two types of flows. Packets are classified on the basis of their size, using the fact that large packets have a good chance of belonging to elephants, while small packets are fairly distributed among mice and elephants. This classification yields two processes, which can reasonably be assumed Gaussian. We then employ a blind source separation method for recovering elephant and mice traffic. This method is based on minimizing the mutual information between the two Gaussian processes under consideration, provided that elephant and mouse processes are to a large extent independent. This method has been designed by Pham [5,6] and we illustrate its use on ADSL traffic observed on an IP backbone link of the France Telecom network.

The organization of this paper in as follows: In Section 2, we present the experimental setting and recall some modeling results established in an earlier study [7]. The blind source separation method is presented in Section 3 and is applied in Section 4 to real traffic traces, where some elements for inferring traffic parameters are presented, especially for elephants. Some concluding remarks are presented in Section 5.

2   Experimental Setting and Preliminary Results

2.1   Experimental Setting

Throughout this paper, we consider measurements from a 1 Gbit/s link of the France Telecom IP backbone network. We observe TCP traffic from the backbone network in direction to several ADSL areas. Traffic is mainly due to ADSL customers and is thus quite different from LAN or Tiers One traffic usually analyzed in the technical literature [1,8,9,10]. The total load of the link (including TCP and UDP traffic) is about 43.5%.

Traffic is observed by means of a measurement device, which performs a copy of the headers of both TCP segments and IP datagrams. We are thus able to identify those packets with the same 5-tuple composed of the source IP address, the destination IP address, the source port, the destination port and the protocol type (namely TCP). Packets with the same 5-tuple are said to belong to the same flow (a.k.a. micro-flow in the technical literature [2]).

Measurements were performed in October 2003 during the time period between 9:00 pm and 11:00 pm, which usually corresponds to the highest daily activity period by ADSL customers. In the following, we evaluate the "instantaneous" bit rate by computing the number of bits arriving in time intervals of length $\Delta = 100$ ms. Let X_n denote the bit rate evaluated over the nth time interval. In the following, we are interested in the properties of the process $\{X_n\}$, which can also be seen as a time series, assumed to be stationary in the wide sense, characterized by its mean and spectral density function $\psi(x)$ defined by

$$\operatorname{cov}(X_n, X_{n+k}) = \int_{-\pi}^{\pi} e^{ikx} \psi(x) dx,$$

where $\operatorname{cov}(X_n, X_{n+k})$ is the covariance of the random variables X_n and X_{n+k}. (This covariance depends only upon k because of wide sense stationarity.)

2.2   Modeling ADSL Traffic

A mathematical model for the global bit rate process has been established in [11,7] by using a flow-based approach and in particular, the mouse/elephant dichotomy with the following definitions.

Definition 1 (Mouse). A mouse is a flow comprising less than 20 packets and is terminated when no packets of the flow has been observed for a time period of 5 seconds.

The 5 second timer is introduced in order to concentrate on the transmission phase of the mouse. This prevents mice from being unduly stretched because some segments (typically SYN or FIN segments) arrive too far from other segments.

Definition 2 (Elephant). An elephant is a flow with more than 20 packets.

The above definitions may at first glance appear very crude, but they are actually sufficient to describe the global bit rate. This is in particular due to the fact that the majority of traffic is due to peer-to-peer (p2p) applications, generating 80% of global traffic: elephants related to large data transfer tend to be very long, making a clear difference between long and short flows and p2p mice due to signalling are very numerous and of small size.

Mice, which represent 95% of the flows, contribute only 6% to the global load. To describe the bit rate of mice, we are led to aggregate them according to some criteria. In particular, mice related to p2p protocols are not aggregated in the same way as other mice associated with classical applications (web, ftp, nntp, etc.). By aggregating mice, it is possible to introduce new entities (referred to as macro-mice), which have the key property of arriving according to a Poisson process. Moreover, their duration D can be well described by a two-parameter Weibullian distribution, i.e.,

$$\mathbb{P}(D > x) = \exp\left(-(x/\eta)^\beta\right),\tag{1}$$

where η and β are the scale and skew parameters, respectively. Finally, their fluid bit rate Y, defined as the total amount of bits of the macro-mouse divided by its duration, lightly depends on the duration, so that the conditional expectation $\mathbb{E}[Y^2 \mid D]$ can be approximated by a constant κ.

Concerning elephants, we first note that some elephants are mainly composed of ACK segments. This corresponds to the fact that the terminals of some ADSL customers play the role of servers for p2p applications, revealing a symmetric usage of the Internet under the impetus given by the massive deployment of p2p services. The corresponding elephants are referred to as ACK elephants and give rise to a bit rate, which can be assimilated to a white noise. Other elephants are referred to as data elephants.

These latter elephants are not nicely transmitting but composed of bursts separated by time periods of weak activity. To model the bit rate, we decompose elephants into mini-elephants and elephant mice. These two new objects enjoy the same properties as macro-mice introduced above. In particular, they arrive according to Poisson processes, their durations can be well approximated by two-parameter Weibullian distributions and the conditional moment of the squared fluid bit rate can be taken equal to a constant (denoted by κ_e for mini-elephants).

In the following, we assume that the processes describing the bit rates of the different components are Gaussian (see [7] for details). Moreover, the spectral density of the global bit rate is dominated by that of elephants in low frequencies and that of non p2p mice in high frequencies.

2.3   Blind Source Separation Method

Blind source separation consists of recovering components from a set of observed mixtures. In our case, components are the sources generating mice and elephants. The goal of source separation is to obtain two distinct signals, one for mice and the other for elephants, so that the characteristics of the two signals can be analyzed separately.

Assuming that elephants and mice are linearly superposed, blind source separation can be formalized as follows: The observed process is modeled as

$$\mathbf{X}(t) = \mathbf{A}\mathbf{S}(t),$$

where $\mathbf{X}(t)$ is the vector of observation ($\mathbf{X}(t) = (X_1(t), ..., X_K(t))$), $\mathbf{S}(t)$ is the vector of original sources ($\mathbf{S}(t) = (S_1(t), ..., S_K(t))$), and $\mathbf{A}$ is a $K \times K$ non singular matrix. In the case considered in this paper, $K = 2$.

In a first step, what we have to do is to construct the two observed processes $X_1(t)$ and $X_2(t)$. For this purpose, we adopt the following strategy based on the packet size: We fix a threshold T (expressed in bytes) and two probabilities p_1 and p_2; if a packet has a size smaller than T, then the packet belongs to an elephant with probability p_1 and to a mouse with probability $1 - p_1$; similarly, if the size of the packet is greater than the threshold T, the packet belongs to an elephant with probability p_2 and to a mouse with probability $1 - p_2$. With this stochastic packet classification, packets of mice could be considered as packets of elephants and vice versa, giving rise to a classification error.

Instead of dealing with continuous time signals, we fix in practice an integration interval of length Δ (in the following $\Delta = 100$ milliseconds) and the quantities $X_1(n)$ and $X_2(n)$ represent the total amount of bytes of packets classified as belonging to mice and elephants, respectively. Moreover, in the experiments reported below the probabilities have been taken equal to $p_1 = 0.5$ and $p_2 = 0.9$.

The signals under consideration are altered by noise caused by different phenomena (e.g., discrete packet arrivals, ACK elephants, etc.) and the magnitude of noise is very large. In the following, we shall eliminate noise by filtering the observed signals by means of a wavelet filter (see [12] for details in the case of a $1/f$ signal). Moreover, we shall assume that processes are centered. The two resulting signals are denoted by $\{X_e(n)\}$ and $\{X_s(n)\}$, corresponding to elephants and mice, respectively. With a small abuse of notation, the filtered mouse and elephant sources are still denoted by $\{S_1(n)\}$ and $\{S_2(n)\}$.

With the above assumptions, we have the system

$$\begin{pmatrix} X_s \\ X_e \end{pmatrix} = \begin{pmatrix} a_{11} & a_{12} \\ a_{21} & a_{22} \end{pmatrix} \begin{pmatrix} S_1 \\ S_2 \end{pmatrix}, \tag{2}$$

where a_{ij} are unknown mixing coefficients, which quantify the classification error and which entail that the signals $\{X_e(n)\}$ and $\{X_s(n)\}$ are not exactly equal to the sources $\{S_1(n)\}$ and $\{S_2(n)\}$.

3   Blind Source Separation Based on Gaussian Mutual Information

In this section, we review the method proposed by Pham [5] and based on the minimization of Gaussian mutual information. This method seems to be well adapted to the problem considered in this paper since we deal with Gaussian processes, associated with mice and elephants.

The mutual information between Gaussian processes is defined by [5,6,13]

$$I_g[Y_1, ..., Y_K] = \frac{1}{4\pi} \int_{-\pi}^{\pi} \log \det \left[\mathrm{diag}[\mathbf{f_Y}(\lambda)] \right] - \log \det[\mathbf{f_Y}(\lambda)] d\lambda, \tag{3}$$

where for some matrix A, $\mathrm{diag}(A)$ denotes the diagonal matrix with the same diagonal elements as the matrix A and $\mathbf{f_Y}(\lambda)$ is the spectral density matrix of the process $(Y_1, \ldots, Y_K)$, the Y_i being real Gaussian processes.

The mixing model is given by equation (2). The goal of the source separation method is actually to find a matrix B minimizing $I_g[Y_1, Y_2]$, where

$$Y = \begin{pmatrix} Y_1 \\ Y_2 \end{pmatrix} = B \begin{pmatrix} X_s \\ X_e \end{pmatrix}$$

Y_1 and Y_2 represent the reconstructed sources (denoted by S_1 and S_2 in the following). It is worth noting that the separation is achievable if and only if the sources have spectral densities, which are not proportional.

Ideally, the matrix B minimizing the mutual information, should be equal to the right inverse of A so that $BA = \mathbb{I}$, where $\mathbb{I}$ is the identity matrix. Denoting by B^t the transpose of the matrix B, the mutual information (3) is then equal to

$$I_g[Y_1, ..., Y_K] = \frac{1}{4\pi} \int_{-\pi}^{\pi} \log \det \left[\mathrm{diag}[B\mathbf{f_X}(\lambda)B^t] \right] - \log \det[B\mathbf{f_X}(\lambda)B^t] d\lambda.$$

Minimizing the mutual information can be viewed as reducing the deviation from diagonality of the matrices $B\mathbf{f_X}(\lambda)B^t$ for $\lambda \in [-\pi, \pi]$. In practice, the matrices $B\mathbf{f_X}(\lambda)B^t$ are evaluated at some point λ_l for $l = 1, \ldots, L$, which are distributed in the interval $[0, 2\pi]$. The mutual information then reads

$$\frac{1}{2L} \sum_{l=1}^{L} \log \det \left[\text{diag}[B\mathbf{f_X}(\lambda_l)B^t\right] - \log \det[B\mathbf{f_X}(\lambda_l)B^t] \tag{4}$$

The algorithm proposed by Pham [6] consists of jointly diagonalizing the matrices $B\mathbf{f_X}(\lambda_l)B^t$ for $l = 1, \ldots, L$ and relies on the Jacobi diagonalization method: Let

$$\mathbf{B} = \begin{pmatrix} b_{11} & b_{12} \\ b_{21} & b_{22} \end{pmatrix} - \begin{pmatrix} B_1 \\ B_2 \end{pmatrix}$$

and define recursively

$$\begin{pmatrix} B_1 \\ B_2 \end{pmatrix}_{k+1} = \begin{pmatrix} B_1 \\ B_2 \end{pmatrix}_k - \alpha_k \text{ where } \begin{cases} \alpha_k = \frac{2}{1+\sqrt{1-4h_{12}h_{21}}} \begin{pmatrix} 0 & h_{12} \\ h_{21} & 0 \end{pmatrix} \begin{pmatrix} B_1 \\ B_2 \end{pmatrix}_k, \\[2mm] \begin{pmatrix} h_{12} \\ h_{21} \end{pmatrix} = \begin{pmatrix} w_{12} & 1 \\ 1 & w_{21} \end{pmatrix}^{-1} \begin{pmatrix} g_{12} \\ g_{21} \end{pmatrix}, \end{cases}$$

with

$$g_{ij} = \frac{1}{L} \sum_{l=1}^{L} \frac{[B_k \mathbf{f_X}(\lambda_l)B_k^t]_{ij}}{[B_k \mathbf{f_X}(\lambda_l)B_k^t]_{ii}} \quad \text{and} \quad w_{ij} = \frac{1}{L} \sum_{l=1}^{L} \frac{[B_k \mathbf{f_X}(\lambda_l)B_k^t]_{jj}}{[B_k \mathbf{f_X}(\lambda_l)B_k^t]_{ii}}.$$

The above operations are iterated until the quantity $c = \sqrt{g_{12}^2 + g_{21}^2}$ is less than a given threshold $\varepsilon \ll 1$.

The algorithm described above actually achieves source separation only up to a scaling matrix and a permutation. In other words, the output matrix B is not exactly the right inverse of the matrix A but is such that

$$BA = P\mathbb{D}, \tag{5}$$

where P is a permutation matrix and $\mathbb{D}$ is a diagonal matrix.

In addition, the matrix $\mathbf{f_X}(\lambda)$ appearing in the above algorithm is the theoretical spectral density matrix. In reality, this matrix has to be evaluated by using an empirical set of samples with limited size. In the following, the spectral density matrix is evaluated by using periodograms or the discrete Fourier transform of autocorrelation functions [14,15]. Finally, to obtain a smooth version of a spectral density function S_X associated with a process X, we can take the convolution of the function S_X with a Parzen kernel K_M, i.e., a 2π-periodic function depending upon a parameter M and converging to the Dirac comb when $M \to \infty$. The Parzen kernel used in the experimental results reported below is the kernel K_M given in [16], namely

$$K_M = \begin{cases} 1 - 6(\frac{u}{M})^2(1 - |\frac{u}{M}|) & \text{if } 0 \leq |u| < \frac{M}{2}, \\ 2(1 - |\frac{u}{M}|)^3 & \text{if } \frac{M}{2} \leq |u| < M, \\ 0 & \text{otherwise.} \end{cases}$$

4   Application of the Source Separation Algorithm and Inference of Traffic Parameters

4.1   Blind Source Separation of Mice and Elephants

We apply the blind source separation algorithm presented in the previous section to the processes $\{X_e(t)\}$ and $\{X_s(t)\}$ associated with elephants and mice, respectively (see Section 2.3). In the experimental data reported below, we have used the following parameter values: The threshold T for discriminating packets has been set equal to be 200 bytes and mixing probabilities $p_1 = 0.5$ and $p_2 = 0.9$. The autocorrelation and intecorrelation functions are evaluated with a sample size $N = 45056$ over 4096 points (4096 values of lag k). Since we have real processes, the autocorrelation and intercorrelation functions are even. Therefore, we can extend the evaluation to 8192 points. Thus, we have applied the discrete Fourier transform and obtained 8192 samples for the power spectral density. The smoothed version of the spectral density is obtained with $M = 16$ for the Parzen kernel.

Starting with the identity matrix as initial value for the matrix B, the Jacobi algorithm is run in order to diagonalize jointly $L = 8192$ matrices. The algorithm converges in a few iterations.

The normalized spectral densities of the observed processes are displayed in Figure 1(a). We note that they are intertwined and very close one to each other. The former property is also verified by the actual spectral densities of mice and elephants as illustrated in Figure 1(b). The spectral density of mice is of the same order of magnitude as that of elephants for high frequencies, and the latter is dominating in low frequencies.

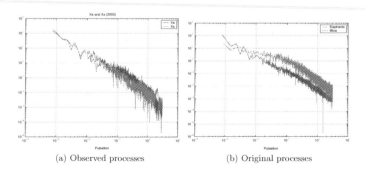

(a) Observed processes   (b) Original processes

Fig. 1. Spectral densities of observed processes and real sources.

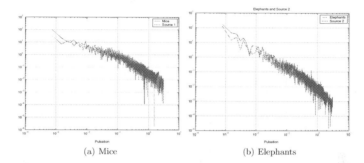

(a) Mice   (b) Elephants

Fig. 2. Comparison between the spectral densities of the reconstructed and real sources.

When comparing in terms of spectral densities the observed processes against the real processes, we note that the observed elephant process is close to the real one. This is due to the fact the probability of classifying a packet of large size as belonging to an elephant is quite large, equal to $p_2 = 0.9$. Hence, the correlation structure of elephants due to large packets arriving close one to each other is preserved in spite of the random drawing of packets. This property is not enjoyed by mice. These flows are with small durations, and their structure is then altered by the random packet drawing.

The spectral densities of the reconstructed signals are displayed and compared with those of the real sources in Figures 2(a) and 2(b). It clearly appears that the blind source separation algorithm is very efficient to recover from the observed signals the normalized spectral densities of the real sources. This can be used to infer the traffic characteristics of the different components, especially those of elephants, which represent almost the totality of traffic. This point is addressed in the next section.

4.2   Inference of Traffic Parameters

The normalized spectral densities can be used to estimate the mean duration of mice and elephants. Indeed, for a given type of flow, the normalized spectral density has the property of being close to $1/(\pi \mathbb{E}[S]x^2)$ for a large range of frequencies x (see [7] for details). It can be checked that the spectral densities of the reconstructed sources indeed decay as c/x^2 for high frequencies. By estimating the coefficient c for each curve, we can deduce the mean duration of mice and mini-elephants. Note that this yields a criterion for determining which output signal of the source separation algorithm corresponds to mice or elephants, since the mean duration of mini-elephants is of course greater than that of mice. This allows us to fix the permutation matrix appearing in equation (5).

Table 1. Estimated values of the traffic parameters compared against real values.

parameter	estimated value	real value
λ_e	35.0	40.01
κ_e	9.2e8	1e9
$\mathbb{E}[D_e]$	164.6	192.95
$\mathbb{E}[D_s]$	3.05	3.249

To simplify the notation, assume that $S_1 = S_s$ and $S_2 = S_e$. Moreover, let α_1 and α_2 denote the diagonal elements of the diagonal matrix $\mathbb{D}$ appearing in equation (5). Then, we have

$$\alpha_1 S_1 = b_{11} X_s + b_{12} X_e, \tag{6}$$
$$\alpha_2 S_2 = b_{21} X_s + b_{22} X_e. \tag{7}$$

Recall that all the random variables in the above system are centered.

If we increase the integration interval Δ up to 4 seconds, then mice appear more or less as noise for elephants, which represent the dominant part of traffic. Using the general relations established in [7]

$$\text{var}[X] \approx \lambda_e \kappa_e \mathbb{E}[D_e] \quad \text{and} \quad \mathbb{E}[X] \approx \lambda_e \sqrt{\kappa_e} \mathbb{E}[D_e],$$

where D_e is the duration of elephants, we can deduce the quantity κ_e (the squared of the fluid bit rate of mini-elephants) and then the mini-elephant arrival rate λ_e. By using equation (6), we can compute α_2.

Concerning mice, by using the process X evaluated over time interval of 100 ms, we have $\text{var}[X] = \lambda_s \kappa_s \mathbb{E}[D_s] + \lambda_e \kappa_e \mathbb{E}[D_e]$, which allows us to deduce the quantity $\lambda_s \kappa_s$. Equation (7) can be used to compute the diagonal element α_1.

The numerical values obtained by using the above heuristic parameter inference method are given in Table 1 and compared against real parameters. Estimated values for elephants are obtained by approximating the behavior of the spectral density of the reconstructed process in low frequencies. The mean duration of macro-mice is computed by using the reconstructed mouse source. The estimated values are of the same order of magnitude as the real ones and can be used to qualitatively estimate the behavior of mini-elephants.

5   Conclusion

We have investigated in this paper a method for rapidly inferring the spectral characteristics of short and long flows giving rise to global traffic on a link connecting several ADSL areas to an IP backbone network. Analyzed traffic has the particularity that it can be described by simple $M/G/\infty$ models with Weibullian service times under heavy traffic conditions so that all processes appearing in the analysis can be assumed to be Gaussian. The proposed method relies on the classification of packets according to their size and on minimizing mutual information between Gaussian processes by using the algorithm designed by Pham [5,6].

This method turns out to be very efficient to recover from observed data the spectral characteristics of long and short flows. The approach followed in this paper opens the door to an on-line estimation of traffic parameters instead of storing huge amounts of data and performing tedious off-line analysis. Designing on-line parameter estimation methods will be addressed in further studies.

References

1. Barakat, C., Thiran, P., Iannaccone, G., Diot, C., Owezarski, P.: A flow-based model for Internet backbone traffic. In: Proc. ACM SIGCOMM Internet Measurement Workshop, Marseille (2002)
2. Fredj, S.B., Bonald, T., Proutiere, A., Regnie, G., Roberts, J.: Statistical bandwidth sharing: A study of congestion at flow level. In: Proc. ACM Sigcomm. (2001)
3. Olivier, P., Benameur, N.: Flow level IP traffic characterization. In: Proc. ITC'17, Salvador de Bahia, Brasil (2001)
4. Papagiannaki, K., Taft, N., Bhattachayya, S., Thiran, P., Salamatian, K., Diot, C.: On the feasibility of identifying elephants in Internet backbone traffic. Technical Report TR01-ATL-110918, Sprint Labs, Sprint ATL (2001)
5. Pham, D.: Blind separation of instantaneous mixture of sources via the Gaussian mutual information criterion. Technical report, Laboratory of Modeling and Computation, IMAG, C.N.R.S, France (2001)
6. Pham, D.: Mutual information approach to blind separation for stationnary sources. In: Proc. ICA'99 Conference, Ausois (1999) 215–220
7. Ben-Azzouna, N., Clérot, F., Fricker, C., Guillemin, F.: A flow-based approach to modeling ADSL traffic on an IP backbone link. Annals of Telecommunications (2004)
8. Claffy, K., Miller, G., Thompson, K.: The nature of the beast: Recent traffic measurement from an Internet backbone. In: Proc. of Inet. (1998)
9. Leland, W., Taqqu, M., Willinger, W., Wilson, D.: On the self-similar nature of Ethernet traffic. IEEE/ACM Trans. Net. (1994) 1–15
10. Z.L.Zhang, Ribeiro, V., Moon, S., Diot, C.: Small time scaling behavior of Internet backbone traffic: An empirical study. In: Proc. Infocom 2003. (2003)
11. Ben-Azzouna, N., Guillemin, F.: Analysis of ADSL traffic on an IP backbone link. In: Proc. Globecom 2003, San Francisco (CA) (2003)
12. Wornell, G.: Signal processing with fractals. A wavelet-based appoach. Prentice Hall Signal Processing Series (1995)
13. Cover, T., Thomas, J.: Elements of Information Theory. Wiley-Interscience Publication (1991)
14. Castanié, F.F., ed.: Analyse spectrale. Hermès Science Publications (2003)
15. Priestley, M.: Spectral Analysis and Time Series. Academic Press, London (1981)
16. Kunt, M.: Traitement Numérique des Signaux. Presses Polytechniques Romandes, Lausanne (1984)

Multi-time-Scale Traffic Modeling Using Markovian and L-Systems Models

Paulo Salvador[1], António Nogueira[1], Rui Valadas[1], and António Pacheco[2]

[1] University of Aveiro / Institute of Telecommunications Aveiro
Campus de Santiago, 3810-193 Aveiro, Portugal
{salvador, nogueira}@av.it.pt, rv@det.ua.pt
[2] Instituto Superior Técnico - UTL / Department of Mathematics and CEMAT
Av. Rovisco Pais, 1049-001 Lisboa, Portugal
apacheco@math.ist.utl.pt

Abstract. Traffic engineering of IP networks requires the characterization and modeling of network traffic on multiple time scales due to the existence of several statistical properties that are invariant across a range of time scales, such as self-similarity, LRD and multifractality. These properties have a significant impact on network performance and, therefore, traffic models must be able to incorporate them in their mathematical structure and parameter inference procedures.

In this work, we address the modeling of network traffic using a multi-time-scale framework. We evaluate the performance of two classes of traffic models (Markovian and Lindenmayer-Systems based traffic models) that incorporate the notion of time scale using different approaches: directly in the model structure, in the case of the Lindenmayer-Systems based models, or indirectly through a fitting of the second-order statistics, in the case of the Markovian models. In addition, we address the importance of modeling packet size for IP traffic, an issue that is frequently misregarded. Thus, in each class we evaluate models that are intended to describe only the packet arrival process and models that are intended to describe both the packet arrival and packet size processes: specifically, we consider a Markov modulated Poisson process and a batch Markovian arrival process as examples of Markovian models and a set of four Lindenmayer-Systems based models as examples of non Markovian models that are able to perform a multi-time-scale modeling of network traffic. All models are evaluated by comparing the density function, the autocovariance function, the loss ratio and the average waiting time in queue corresponding to measured traces and to traces synthesized from the fitted models. We resort to the well known Bellcore *pOct* traffic trace and to a trace measured at the University of Aveiro.

The results obtained show that (i) both the packet arrival and packet size processes need to be modeled for an accurate characterization of IP traffic and (ii) despite the differences in the ways Markovian and L-System models incorporate multiple time scales in their mathematical framework, both can achieve very good performance.

Keywords: Traffic modeling, Markovian arrival processes, L-Systems.

M. Freire et al. (Eds.): ECUMN 2004, LNCS 3262, pp. 297–306, 2004.
© Springer-Verlag Berlin Heidelberg 2004

1   Introduction

In the Internet, the complexity associated to mechanisms for traffic generation and control, as well as the diversity of applications and services, have introduced several peculiar behaviors in traffic, such as self-similarity, long-range dependence and multifractality, which have a significant impact on network performance. These behaviors have in common a property of statistical invariance across a range of time scales. Thus, a suitable traffic model must be able to capture statistical behavior on multiple time scales.

In order to completely characterize a traffic model, we have to specify the model structure and its parameter inference procedure. Multi-time-scale characteristics can be incorporated in the parameter fitting procedure or can be embedded in the model structure. Moreover, accurate modeling of IP traffic requires the characterization of both the packet arrival and packet size processes. Assuming that packet size is fixed and equal to the average packet size of the measured data trace may lead to large errors when packets have variable size.

In this paper, we address the modeling of network traffic under a multi-time-scale framework. Our main goal is to evaluate and compare the performance of two classes of traffic models (based on Markovian and Lindenmayer-Systems models) that incorporate the notion of time scale using different approaches: directly in the model structure, in the case of Lindenmayer-Systems based models or indirectly via the fitting of the second-order statistics, in the case of Markovian models. In addition, we also address and evaluate the importance of modeling packet size in IP traffic, an issue that is frequently misregarded. Thus, we evaluate in each one of the selected classes, models that are intended to describe only the packet arrival process and models that are intended to describe both the packet arrival and packet size processes.

The Markovian models considered in this study are a discrete time Markov modulated Poisson process (dMMPP), intended to describe only the packet arrival process, and a discrete time batch Markovian arrival process (dBMAP), that is intended to describe both the packet arrival and packet size processes. The use of Markovian models benefits from the existence of several mathematical tools for assessing queuing behavior, such as average waiting time and packet loss ratio. Besides, using appropriate parameters inference procedures it is possible to incorporate in their structure the multi-time-scale characteristics of the traffic while still maintaining their analytical tractability.

The second class of traffic models considered in this paper is based on stochastic Lindenmayer-Systems (hereafter referred to as L-Systems). L-Systems are string rewriting techniques which were introduced by biologist A. Lindenmayer in 1968 as a method to model plant growth [1]. They are characterized by an alphabet, an axiom and a set of production rules, and can be used to generate fractals [2]. The alphabet is a set of symbols; the production rules define transformations of symbols into strings of symbols; starting from an initial string (the axiom), an L-System constructs iteratively sequences of symbols through replacement of each symbol by the corresponding string according to the production rules. If the production rules are random, the L-System is called a stochastic L-System. Stochastic L-Systems are a method to construct recursively random sequences with multi-time-scale behavior. Four different variants of L-Systems based traffic models will be proposed, one intended to describe only the packet arrival process and three destined to describe both the packet arrival and packet size processes.

The performance of the different traffic models is evaluated by comparing the first and second order statistics and the queuing behavior (in terms of packet loss and average waiting time) of (i) original measured data traces and (ii) traces generated via discrete event simulation of the traffic models whose parameters are inferred from the measured data.

The results obtained show that both the packet arrival and packet size processes need to be modeled in order to obtain an accurate and complete fitting of the measured data. Moreover, despite the differences in the way Markovian and L-System models incorporate multiple time scales, both can achieve very good performance. This can be attributed, on one hand, to the detail that is used in the fitting procedures of the Markovian models and to the technique used in the matching of the autocovariance function and, on the other hand, to the fact that the construction process of the L-System models is inherently performed on a time-scale basis.

The paper is organized as follows. Section 2 presents the traffic models and inference procedures considered in this paper; in Section 3 we discuss the results of applying the proposed models and fitting procedures to measured and synthesized traffic traces. Finally, Section 4 presents the main conclusions.

2   Traffic Models

2.1   Discrete-Time Markovian Models

The first Markovian model considered is a discrete-time Markov Modulated Poisson Process (dMMPP) with a parameter fitting procedure that leads to accurate estimates of queuing behavior for network traffic exhibiting LRD behavior. It addresses only the characterization of the packet arrival process. The model and the respective parameter inference procedure were proposed in [3]. The inference procedure matches both the autocovariance and marginal distribution of the counting process. A major feature of the model is that the number of states is not fixed a priori, but can be adapted to the particular trace being modeled. In this way, the procedure allows establishing a compromise between the accuracy of the fitting and the number of parameters, while maintaining a low computational complexity. The MMPP is constructed as a superposition of L 2-MMPPs and one M-MMPP. The 2-MMPPs are designed to match the autocovariance and the M-MMPP to match the marginal distribution. Each 2-MMPP models a specific time scale of the data. Here, the specific behavior of each time scale is incorporated through the fitting of the second-order statistics of the data. The procedure starts by approximating the autocovariance by a weighted sum of exponential functions that models the autocovariance of the 2-MMPPs. The autocovariance tail can be adjusted to capture the long-range dependence characteristics of the traffic, up to the time-scales of interest to the system under study. The next step is the inference of the M-dMMPP probability function and the stationary probabilities of the L 2-dMMPPs from the empirical probability function of the original data trace, within the constraints imposed by the autocovariance matching. These parameters can be obtained through

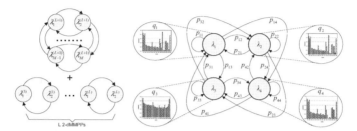

Fig. 1. Superposition of one M-dMMPP and L 2-dMMPP processes.

Fig. 2. Example of a 4-dBMAP.

a constrained minimization process. At this point, all parameters of the 2-dMMPPs have been determined. The M-dMMPP probability function is then approximated by a weighted sum of Poisson probability functions, and the number of states (M) is determined as the number of Poisson functions necessary to perform the approximation. The mean and weight of the Poisson probability functions determine, respectively, the states arrival rates and the transition probabilities of the M-dMMPP. The final MMPP with $M2^L$ states is obtained by superposing the L 2-MMPPs and the M-MMPP (figure 1).

The second Markovian model considered is a discrete-time batch Markovian arrival process (dBMAP), that is able to characterize both the packet arrival and packet size processes of IP traffic. This model was proposed in [4]. In the dBMAP model, packet arrivals occur according to a dMMPP and each arrival is characterized by a packet size with a general distribution that may depend on the phase of the dMMPP. Figure 2 illustrates an example of a dBMAP with four states. The fitting procedure is designed in order to provide a close match of both the autocovariance and the marginal distribution of the packet arrival process, using the dMMPP previously described; then, a packet size distribution is individually inferred for each state of the dMMPP in a such a way that gives the best fit to the size distribution of all packets that arrived during the time slots associated to that state. As in the case of the dMMPP, the number of states of the fitted dBMAP is not fixed a priori; it is determined as part of the inference procedure itself. The packet size characterization is independently performed for each state of the inferred $M2^L$-dMMPP. This task involves two steps: (i) association of each time slot to one of the $M2^L$-dMMPP states and (ii) inference of a packet size distribution for each state of the $M2^L$-dMMPP. In the first step, we scan all time slots of the empirical data. A time slot is randomly assigned to a state, according to a probability vector that is calculated from the number of packets that arrived in that particular time slot and from the inferred dMMPP parameters. The inference of the packet size distribution in each state resorts to histograms. The construction of each histogram is only based on the packets that arrived during the time slots previously associated with the corresponding state of the dMMPP.

2.2   L-System Models

The application of stochastic L-Systems in the traffic modeling context was first introduced by the authors in [5]. The model proposed in that work only addressed the characterization of the packet arrival process. In this model, which will be called here Single L-System, iterations are interpreted as time scales, the symbols are arrival rates that are associated with time intervals and the production rules generate two arrival rates from a single one. Starting at the coarsest time scale with a single interval and a single arrival rate, the construction process iteratively generates two arrival rates from a single one, according to the stochastic production rules and, at the same time, halves the width of the time intervals associated with the new arrival rates. We allow the grouping of time scales in time scale ranges and the definition of different sets of production rules for each time scale range. The traffic process construction is illustrated in Figure 3, for the simple case of considering only four scales ($j = 0, 1, 2, 3$) grouped in two scale ranges ($r = 1, 2$). The inference procedure can then be divided in three major steps: (i) determination of the L-System alphabet and axiom, (ii) identification of the time scale ranges and (iii) inference of the L-System production rules. The alphabet of the L-System will consist in L equidistant arrival rate values, ranging from the minimum to the maximum values present in data. The axiom is inferred as the average arrival rate of the empirical data, rounded to the closest alphabet element. The identification of time scale ranges is based on wavelet scaling analysis. We use the method described in [6], which resorts to the (second-order) logscale diagram. A (second-order) logscale diagram is a plot of the energies against scales, together with confidence intervals about the energies, where these energies are a function of the wavelet discrete transform coefficients at a specific scale. The time scale ranges correspond to the set of time scales for which, within the limits of the confidence intervals, the energy values fall on a straight line, i.e., the scaling behavior is linear in a time scale range. The last step is the inference of the L-System production rules. First, data is rounded to the closest alphabet element in order to define the data sequence for each time scale. Finally, the production rules probabilities can be directly inferred from data inspection, specifically from the number of times that a particular pair of arrival rates was originated from a particular arrival rate located at the upper timescale.

Three subsequent extensions addressed the characterization of both the packet arrival and packet size processes. The first extension corresponds to a model with two independent L-Systems, one for the packet arrival and the other for the packet size process, called Double L-System model [7]. Due to the independence of the two L-Systems, this model is not able to capture the correlations between packet arrivals and sizes, although it captures multifractal behavior on both the packet arrival and packet sizes processes. The second extension is based on a single bi-dimensional L-System and is called Joint L-System [8]. In this model, the alphabet elements are pairs of arrival rates and mean packet sizes and the production rules generate two pairs, each one consisting of an arrival rate and a mean packet size, from a single pair. Opposite to the previous model, this one is able to capture correlations between arrivals and sizes. The traffic process construction is illustrated in Figure 4, once again for the simple case of considering only four scales ($j = 0, 1, 2, 3$) grouped in two scale ranges ($r_b = 1, 2$). One potential disadvantage of this model is that it may require a large number of parameters. The third extension, which was proposed in [9], was devised in order to allow a lower number of

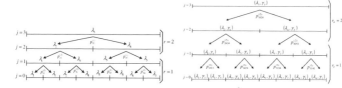

Fig. 3. Construction of a Single L-System based traffic model.

Fig. 4. Construction of a Joint L-System based traffic model.

parameters and also to provide a more detailed modeling of the packet size. In this case, only the packet arrival process is modeled through an L-System and the characterization of the packet size is performed by associating, at the finner time scale, a probability mass function (PMF) of packet sizes to each packet arrival rate. In this way, the model is able to capture correlations between packet arrivals and packet sizes, and multifractal behavior on packet arrivals (but not on packet sizes). Note that in this extension the packet sizes are characterized individually, whereas in the previous ones only the mean packet sizes were modeled. This model is called L-System with PMFs.

3   Numerical Results

We have applied our fitting procedures to two traces of IP traffic: (i) the well known *pOct* Bellcore trace and (ii) one trace measured at the University of Aveiro (UA). The UA trace is representative of Internet access traffic produced within a University campus environment. University of Aveiro is connected to the Internet through a 10 Mb/s ATM link and the measurements were carried out in a 100 Mb/s Ethernet link connecting the border router to the firewall, which only transports Internet access traffic. The main characteristics of the used traces are summarized in Table 1.

The parameter estimation took, in all cases, less than 2 minutes, using a MATLAB implementation running in a 1.2 GHz AMD Athlon PC, with 1.5 Gbytes of RAM. This shows that all fitting procedures are computationally very efficient.

The suitability of the traffic models and the accuracy of the fitting procedures was assessed using several criteria. For the original data traces and for traces synthesized according to the inferred models, we compare both the probability mass and autocovariance functions of the packet arrival process. For the same two traces,

Table 1. Main characteristics of measured traces.

Trace name	Capture period	Trace size (pkts)	Mean rate (byte/s)	Mean pkt size (bytes)
pOct	Bellcore trace	1 million	362750	568
UA	12.41pm to 14.27pm, July 6^{th}2001	7 millions	654780	600

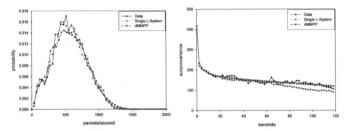

Fig. 5. Density functions approximation, *pOct*   **Fig. 6.** Autocovariance approximation, *pOct*

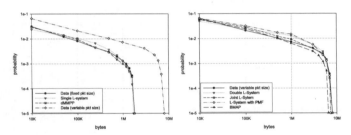

Fig. 7. Packet loss ratio, trace *pOct* $\rho = 0.7$ **Fig. 8.** Packet loss ratio, trace *pOct* $\rho = 0.7$
(without packet size modeling)   (with packet size modeling)

we compare queuing behavior as assessed by packet loss ratio and average waiting time estimated through trace-driven simulation. In the case of the Single L-System and dMMPP models, we have considered a fixed packet length equal to the mean packet size of the original data.

Figure 5 shows that, for the case of the *pOct* trace, all models were able to match relatively well the PMF of the original data. This agreement is not so good in the case of the autocovariance function (Figure 6). The dMMPP was able to closely match the autocovariance function and the Single L-System has also performed quite well. Similar results were obtained for the other trace.

To assess queuing behavior, the buffer size was varied from 10 Kbytes to 70 Mbytes. The service rate was 518 Kbytes/s for the *pOct* trace and for an utilization ratio (ρ) of 0.7, 403 Kbytes/s in order to have an utilization ratio of 0.9 for the same trace, and 666 Kbytes/s for the UA trace (corresponding to an utilization ratio of 0.98). Figures 7, 9, 11, 13, 15 and 17 show that the Single L-System and the dMMPP model are able to reproduce the queuing behavior of the empirical traffic when the packet size is considered as a fixed value, but both models failed to reproduce the queuing behavior of the traffic when the variable packet size of the IP traffic is considered. In Figures 8, 10, 12, 14, 16 and 18 it is

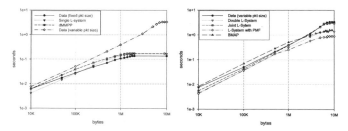

Fig. 9. Average waiting time, trace $pOct$ $\rho =$ 0.7 (without packet size modeling)

Fig. 10. Average waiting time, trace $pOct$ $\rho =$ 0.7 (with packet size modeling)

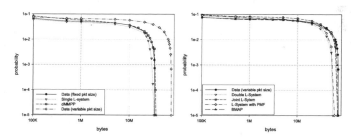

Fig. 11. Packet loss ratio, trace $pOct$ $\rho = 0.9$ (without packet size modeling)

Fig. 12. Packet loss ratio, trace $pOct$ $\rho = 0.9$ (with packet size modeling)

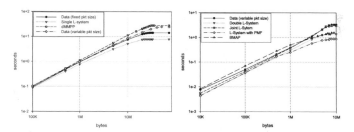

Fig. 13. Average waiting time, trace $pOct$ $\rho =$ 0.9 (without packet size modeling)

Fig. 14. Average waiting time, trace $pOct$ $\rho =$ 0.9 (with packet size modeling)

possible to observe the queuing results corresponding to the traffic models that include the ability to reproduce the arrival and packet size processes. The fitting of the queuing behavior was very good for all considered models.

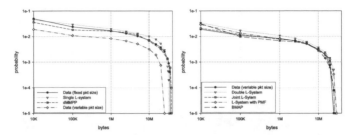

Fig. 15. Packet loss ratio, trace UA $\rho = 0.98$ (without packet size modeling)

Fig. 16. Packet loss ratio, trace UA $\rho = 0.98$ (with packet size modeling)

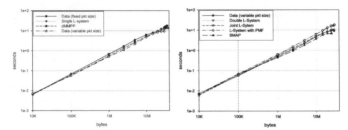

Fig. 17. Average waiting time, trace UA $\rho = 0.98$ (without packet size modeling)

Fig. 18. Average waiting time, trace UA $\rho = 0.98$ (with packet size modeling)

From these results, we first conclude that it is important to characterize both the packet arrival and packet size processes. In general, the three L-System based traffic models and the BMAP model that incorporate these characteristics achieve good fitting performance. The best one, which was able to track almost perfectly the queuing behavior of all traces, is the Joint L-System. This can be attributed to its ability of capturing correlations in and between the packet arrival and packet size processes, as well as multifractal behavior on the byte arrival process. The Double L-System is somewhat impaired by its inability to capture the correlation between arrivals and sizes and the L-System with PMFs by its inability to capture correlations and multifractal behavior on the packet size process. However, the L-System with PMFs has a lower number of parameters and therefore can be considered a good alternative.

4   Conclusion

Time scales are an important ingredient for the modeling of today's Internet traffic. In this paper, we addressed the modeling of network traffic using a multi-time-scale framework by evaluating and comparing two classes of traffic models: Markovian models and models based on Lindenmayer-Systems. We also addressed the importance of modeling the packet size for IP traffic, an issue that is frequently misregarded. Our results indicate that both the packet arrival and packet size processes need to be modeled for an accurate characterization of IP traffic and that, despite the differences in the way Markovian and L-System models incorporate multiple time scales in their mathematical frameworks, both can achieve very good performance.

Acknowledgements. This research was supported in part by Fundação para a Ciência e a Tecnologia, the project POSI/42069/CPS/2001, and the grant BD/19781/99.

References

1. Lindenmayer, A.: Mathematical models for cellular interactions in development II. Simple and branching filaments with two-sided inputs. Journal of Theoretical Biology **18** (1968) 300–315
2. Peitgen, H., Jurgens, H., Saupe, D.: Chaos and Fractals: New Frontiers of Science. Springer-Verlag (1992)
3. Salvador, P., Pacheco, A., Valadas, R.: Multiscale fitting procedure using Markov modulated Poisson processes. Telecommunications Systems **23** (2003) 123–148
4. Salvador, P., Pacheco, A., Valadas, R.: Modeling IP traffic: Joint characterization of packet arrivals and packet sizes using BMAPs. Computer Networks Journal **44** (2004) 335–352
5. Salvador, P., Nogueira, A., Valadas, R.: Modeling multifractal traffic with stochastic L-Systems. In: Proceedings of GLOBECOM 2002. (2002)
6. Abry, P., Flandrin, P., Taqqu, M., Veitch, D.: Wavelets for the analysis, estimation and synthesis of scaling data. in Self-Similar Network Traffic Analysis and Performance Evaluation, K. Park and W. Willinger Eds (1999)
7. Salvador, P., Nogueira, A., Valadas, R.: Modeling multifractal IP traffic: Characterization of packet arrivals and packet sizes using stochastic L-Systems. In: 10th International Conference on Telecommunication Systems, Modeling and Analysis. (2002) 577–587
8. Salvador, P., Nogueira, A., Valadas, R.: Joint characterization of the packet arrival and packet size processes of multifractal traffic based on stochastic L-Systems. In: 18th International Teletraffic Congress, ITC 18. (2003)
9. Salvador, P., Nogueira, A., Valadas, R.: Framework based on stochastic L-Systems for modeling IP traffic with multifractal behavior. In: SPIE Conference on Performance and Control of Next Generation Communication Networks, ITCom 2003. (2003)

Performance Analysis of Congestion Control Mechanism Using Queue Thresholds Under Bursty Traffic

Lin Guan, Irfan Awan, and Mike E. Woodward

Department of Computing, School of Informatics
University of Bradford
Richmond Road, Bradford, West Yorkshire, BD7 1DP, UK
{L.Guan, I.Awan, M.E.Woodward}@Bradford.ac.uk

Abstract. Performance modeling and analysis of the Internet Traffic Congestion control mechanism has become one of the most critical issues. Several approaches have been proposed in the literature by using queue thresholds. This motivates the analysis of a discrete-time finite capacity queue with thresholds to control the congestion caused by the bursty traffic. The maximum entropy (ME) methodology has been used to characterize the closed form expressions for the state and blocking probabilities. A GGeo/GGeo/1/{N_1, N_2} censored queue with external compound Bernoulli traffic process and generalised geometric transmission times under a first come first serve (FCFS) rule and arrival first (AF) buffer management policy for single class jobs has been used for the solution process. To satisfy the low delay along with high throughput, a threshold, N_1, has been incorporated to slow the arrival process from mean arrival rate λ_1 to λ_2 once the queue length has been reached up to the threshold value N_1 (N_2 is the total capacity of the queue). The source operates normally, otherwise. This is like an implicit feedback from the queue to the arrival process. The system can be potentially used as a model for congestion control based on Random Early Detection (RED) mechanism. Typical numerical results have been presented to show the credibility of ME solution and its validation against simulation.

1 Introduction

With the enormous growth in the Internet traffic, the control of congestion has become one of the most critical issues in present networks to accommodate the increasingly diverse range of services and types of traffic [1]. Congestion control to enable different types of Internet traffic to satisfy specified Quality of Service (QoS) constraints is becoming significantly important. Many systems in the network environments require the queue to be monitored for impending congestion before it happens [2].

The traditional technique for managing router queue lengths is to set a maximum length for each queue, usually equal to the buffer capacity, and then accept packets until the queue becomes full. The subsequent arrivals will be blocked until some space becomes available in the queue as a result of departures. This technique is known as "tail drop", since the packet that arrived most recently (i.e., the one on the tail of the

M. Freire et al. (Eds.): ECUMN 2004, LNCS 3262, pp. 307–316, 2004.

queue) is dropped when the queue is full. This method has been used for several years in the Internet, but it has two important drawbacks: 'Lock-Out' and 'Full Queues' [3]. In order to solve the problems, some active queue management (AQM) mechanisms have been proposed and implemented to manage the queue lengths, reduce end-to-end latency, reduce packet dropping, and avoid lock-out phenomena so that the control of congestion can be achieved by the use of appropriate buffer management schemes. These mechanisms include random early detection (RED) [4], random early marking (REM) [5, 6], a virtual queue based scheme where the virtual queue is adaptive [7, 8, 9] and a proportional integral controller mechanism [10], among others. Of these schemes to implement AQM, RED is the default mechanism for managing queue lengths to meet these goals in a FIFO queue, which is recommended by the Internet Society in RFC 2309 [3].

In contrast to tail drop, RED [4] drops arriving packets probabilistically depending on setting thresholds in the queue. When the average queue length is less than a minimum threshold, all incoming packets are allowed to the queue. If the mean queue length is greater than a maximum threshold, every arriving packet is dropped. Between the minimum and maximum thresholds, incoming packets are dropped with a probability that increases linearly as a function of the mean queue length, reaching a maximum dropping probability at the maximum threshold. Since RED was proposed by Floyd and Jacobson [4] in 1993, most researchers have used simulation tools as the choice of modelling to examine the performance of various aspects of the RED algorithm. Only a few publications, such as [11-14], have attempted to theoretically evaluate the performance of RED. To our knowledge, there is no clear description of the parameter settings and exact information being measured, So it is very important and necessary to use an analytical approach to address the more fundamental aspects of the RED algorithm.

This paper presents a ME-based approximate analytical solution to model the GGeo/GGeo/1/{N_1, N_2} censored queue for implementing AQM. The external bursty traffic has been modelled using compound Bernoulli process while the generalised geometric (GGeo) process represents the transmission process. A threshold N_1 has been incorporated in the discrete-time finite buffer with full capacity N_2 to control the external arrivals. RED depends on setting thresholds in the queue and our research uses the principle and looks at this in a simplified way where we have one threshold N_1 at the moment which define changes in the arrival rate so that this system can be potentially used as a model for RED.

The paper is organised as follows: The maximum entropy (ME) methodology and GGeo-type distribution are described in Section 2. ME solution for a stable GGeo/GGeo/1/{N_1, N_2} censored queue with threshold N_1 and full buffer capacity N_2 is characterised in Section 3. Numerical validation results against simulation, involving generalised geometric (GGeo) interarrival and service time distributions, are included in Section 4. Section 5 follows conclusions and future work.

2 Preliminaries

2.1 The Maximum Entropy Methodology

The maximum entropy (ME) method is an approximate technique for finding probability distributions using information theory. It is alternative way to solve quite complex queues instead of traditional balance equations. The principle of maximum entropy (PME) [15, 16] provides a self-consistent method of inference for charactering an unknown but true probability distribution, subject to the mean value constraints supplied by the given information. The ME solution can be expressed in terms of a normalising constant and a product of Lagrangian coefficients corresponding to the constraints. In an Information theoretic context [15], the ME solution corresponds to the maximum disorder of system states and, thus, is considered to be the least biased distribution estimate of all solutions that satisfy the system's constraints. In sampling terms, it has been shown [16] that, given the imposed constraints, the ME solution can be experimentally realised in overwhelmingly more ways than any other distribution. Shore and Johnson [17] showed that the PME is a uniquely correct method of inductive inference when information is given in the forms expected values. Tribus [18] used the principle to derive a number of known probability distributions by using a variety of mean value constraints in terms of moments. More details on PME and its applications can be found in [19].

2.2 The GGeo Distribution

The GGeo-type distribution is an interarrival-time or interdepature-time distribution of the form [20] (c.f. Fig.1)

$$f_n = \Pr(W = n) = \begin{cases} 1 - \tau, & n = 0 \\ \tau\sigma(1-\sigma)^{n-1}, & n \geq 1 \end{cases} \tag{1}$$

$$\tau = 2/(C^2 + 1 + v), 0 \leq \tau \leq 1 \tag{2}$$

$$\sigma = \tau v, 0 \leq \sigma \leq 1 \tag{3}$$

where W is a discrete-time random variable (rv) representing the interarrival-time or interdepature-time of individual job in a steady single sever queue, while $1/v$ and C^2 are the corresponding mean and squared coefficient of variation (SCV) of rv W. And v is mean event rate (arrivals or departures per slot). The GGeo distribution is stochastically a true probability distribution when $C^2 \geq |1 - v|$.

The batch arrivals or departures are allowed in the GGeo-type distribution when rv W achieve zero value, it is also implied that a batch interevent pattern according to a compound Bernoulli process (CBP) with rate σ, while the number of events (e.g. arrivals or departures) in a slot (i.e. batch size) is geometrically distributed with parameter τ. And the GGeo pattern is generated by a sequence of batch Bernoulli independent and identically distributed non-negative integer values rv's $\{W_k\}$, where W_k, k=1, 2,

..., is the number of the events occurring at the kth slot, with a fixed probability distribution given by [21]

$$g_l = \begin{cases} P(W_k = 0) = 1 - \sigma, & l = 0 \\ P(W_k = l) = \sigma\tau(1-\tau)^{l-1}, & l \geq 1 \end{cases} \tag{4}$$

The GGeo distribution is versatile, possessing pseudo-memoryless property which makes solutions of many GGeo-type queueing systems and networks analytically tractable. And it is an extremal member of a family of two parallel-phase Geo distributions with the same mean and SCV, where one of the phase rates bursts mathematically to $+\infty$ with probability $(1-\tau)$. More details can be found in [20, 21].

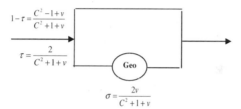

Fig. 1. The Generalised Geometric (GGeo) distribution with parameters σ and τ $(0 < \sigma, \tau \leq 1)$

The choice of the GGeo distribution is further motivated by the fact that measurements of actual traffic or service time maybe generally limited and so only few parameters can be computed reliably. Typically, only the mean and variance maybe relied upon. In this case, the choice of distribution which implies least biased (i.e. introduction of arbitrary and, therefore, false assumptions) within a discrete-time domain is that of GGeo-type distribution. [22]

3   Analysis of a GGeo/GGeo/1/{N1, N2} Censored Queue

This section presents the ME analysis of a stable single server GGeo/GGeo/1/{N$_1$, N$_2$} censored discrete-time finite capacity queue with:
N$_1$ threshold value, N$_2$ full buffer capacity, FCFS scheduling discipline, arrival first (AF) buffer management policy and censored arrival process for single class jobs.

Notation
Let
        S be the state of the queue
        Q be the set of all feasible states of S
        Λ be the arrival rate and $\Lambda = \lambda_1$ if mean queue length (MQL) reaches the threshold value N$_1$ or $\Lambda = \lambda_2$ after MQL exceed the threshold N$_1$
        μ be the service rate

π be the blocking probability that an arrival finds the queue full

P(S) be the stationary state probability

Analyse this queue, we consider two kinds of jobs (i=2) in the queue; jobs arrived with rate λ_1 (i=1) and these arrivals after the threshold with rate λ_2 (i=2). For each state S, S ϵ Q, the following auxiliary functions are defined:

$n_i(S)$ = the number of jobs present in state S,

$$s_i(S) = \begin{cases} 1, & if \quad n_i(S) > 0 \\ 0, & otherwise \end{cases}$$

$$f_i(S) = \begin{cases} 1, & if \quad n_i(S) = N_i \\ 0, & otherwise \end{cases}$$

Suppose that the following mean value constraints about the state probability P(S) are known to exist:

(i) Normalisation,

$$\sum_{S \epsilon Q} P(S) = 1 \tag{5}$$

(ii) Utilization,

$$\sum_{S \epsilon Q} s_i(S)P(S) = U_i, 0 < U_i < 1, \quad i = 1,2 \tag{6}$$

(iii) Mean queue length,

$$\sum_{S \epsilon Q} n_i(S)P(S) = L_i, U_i < L_i < N_i, \quad i = 1,2 \tag{7}$$

(iv) Full buffer state probability,

$$\sum_{S \epsilon Q} f_i(S)P(S) = \phi_i, 0 < \phi_i < 1, \quad i = 1,2 \tag{8}$$

where ϕ satisfies the flow balance equations, namely

$$\lambda_i(1 - \pi_i) = \mu_i U_i, \quad i = 1,2 \tag{9}$$

The choice of mean values (5)-(8) is based on the type of constraints used for the ME analysis of stable single class FCFS G/G/1/N queue [23]. If additional constraints are used, it is no longer feasible to capture a closed-form ME solution at the building block level, with clearly, adverse implications on the efficiency of an iterative queue-by-queue decomposition algorithm for general QNMs. Conversely, if one or more constraints form the set (5)-(8) are missing, it is expected that the accuracy of the ME solution will be generally reduced.

3.1 ME Solution

The form of the state probability distribution, P(S), S ϵ Q, can be characterised by maximising the entropy functional

$$H(P) = -\sum_S P(S) \log P(S)$$

subject to constraints (5)-(8). By employing Lagrange's method of undetermined multipliers the following solution is obtained

$$P(S) = \frac{1}{Z} \prod_{i=1}^{R} x_i^{n_i(S)} g_i^{s_i(S)} y_i^{f_i(S)}, \quad \forall S \in Q$$

where Z is the normalising constant given by

$$Z = \sum_{S \in Q} \left(\prod_{i=1}^{R} x_i^{n_i(S)} g_i^{s_i(S)} y_i^{f_i(S)} \right)$$

and x_i, g_i, y_i are the Lagrangian coefficients corresponding to constraints (6)-(8), respectively.

3.2 State Probabilities

Defining the sets
$$S_0 = \{ S / S \in Q : s(S) = 0 \}$$
$$Q_k = \{ S \in Q : n(S) = k \ \& \ k \geq 1 \}$$
It is implied , after some manipulation, that the aggregate ME state probability distribution is given by

$$P(S_0) = \frac{1}{Z} \tag{10}$$

$$P(k) = \frac{1}{Z} \sum_{i=1}^{R} x_i g_i y_i^{\delta_i(k)} C(k-1), \quad 1 \leq k \leq N_2 \tag{11}$$

$$Z = 1 + \sum_{k=1}^{N_2} \sum_{i=1}^{R} x_i g_i y_i^{\delta_i(k)} C(k-1) \tag{12}$$

where $i = 1,2$ $R = 2$; $\delta_i(k) = 1$,   if $k \geq N_i$,   or   0,   otherwise;   and $C(v) = x_2^{\delta(v+1-N_1)} (x_1 + x_2)^{\delta(v)}$, at which $\delta(v+1-N_1) - 0$, if $v+1-N_1 \leq 0$, or $v+1-N_1$, otherwise; and $\delta(v) = v$, if $v < N_1$, or $N_1 - 1$, otherwise.

3.3 Blocking Probability

The flow balance condition (9), which is used in deriving the Lagrangian coefficient y_i, is characterised by the blocking probabilities π_i of a censored GGeo/GGeo/1/$\{N_1, N_2\}$ queue under AF buffer management policy. These blocking probabilities can be approximated based on a censored GGeo/GGeo/1/N queue [24] as follows:

$$\pi_i = \sum_{k=0}^{N_2} \delta_i(k)(1 - \tau_{ai})^{\xi(k)} P(k) \tag{13}$$

where $\delta_i(k) = \begin{cases} \dfrac{\tau_{si}}{\tau_{si}(1-\tau_{ai}) + \tau_{ai}}, & i = 1 \ \& \ k = 0 \\ 1, & \text{otherwise} \end{cases}$, $\xi(k) = \begin{cases} 0, & i = 2 \ \& \ k \leq N_1 \\ [N_1 - k]^+, & \text{otherwise} \end{cases}$

and $[N_1-k]^+= N_1-k$, if $k<N_1$, or 0, otherwise. Where $\tau_a = 2/(1+C_a^2 + \Lambda)$, and $\tau_s = 2/(1+C_s^2 + \mu)$, where C_a^2 and C_s^2 are the squared coefficients of variation for the interarrival and interdeparture times, respectively.

3.4 The Lagrangian Coefficients $\{x_{i,} \quad g_{i,} \quad y_i\}$

The Lagrangian coefficients x and g can be approximated analytically by making asymptotic connections to an infinite capacity queues. Assuming x, g and y are invariant to the buffer capacity size N, it can be established that

$$x_i = \frac{<n_i>-\rho_i}{<n>} \tag{15}$$

$$g_i = \frac{(1-X)\rho_i}{(1-\rho)x_i} \tag{16}$$

where $X = \sum_{i=1}^{R} x_i$, $<n>=\sum_{i=1}^{R}<n_i>$ and $<n_i>$ is the asymptotic marginal mean queue length of a multiple class queue. Note the statistics $<n_i>$, $i=1,2$ can be determined by (c.f. Kouvatsos et al. [25]) under FCFS rule:

$$<n_i>=\frac{\rho_i}{2}(C_{ai}^2 +1)+\frac{1}{2(1-\rho)}\sum_{j=1}^{R}\frac{\Lambda_i}{\Lambda_j}\rho_j^2(C_{aj}^2 +C_{sj}^2) \tag{17}$$

where $\rho=\sum_{i=1}^{R}\rho_i$ and $\rho_i =\frac{\Lambda_i}{\mu_i}$

By substituting the value of state probabilities, P(k), k=0,1,..., N_2, and the blocking probabilities, π_i, into the flow balance condition (9), the Lagrangian coefficient y_i, lead to the following expression:

$$y_i =\frac{\rho_i}{\sum_{v=N_i-1}^{N_i-1}x_ig_iC(v)}\left[1+\sum_{j=1}^{R}x_jg_j \sum_{v=\phi(i)}^{\delta_j(v)}y_j^{\delta_j(v)}C(v)-\frac{1}{\rho_i}x_ig_i\sum_{v=0}^{N-2}C(v)-\xi(i)-\sum_{j=1}^{R}x_jg_j \sum_{v=\theta(i)}^{N-1}(1-\tau_{ai})^{N-v}C(v-1)\right] \quad i=1,2 \quad R=2$$

$$\tag{18}$$

4 Numerical Results

This section presents the credibility of the proposed ME-based approximation solution against simulation based on QNAP-2 [26] via typical numerical experiments involving GGeo-type queue (c.f. Fig. 2-6). It is clearly shown that the comparison results between ME-based approximation and simulation matched very well as we expected.

Figure 2 and 3 show the comparison results of MQL and throughput against a range of threshold values respectively. Similarly, Figure 4-6 demonstrate the comparison results of delay, server utilization and blocking probability for different threshold values

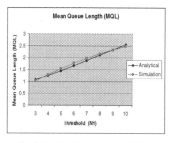

Fig. 2. Results Comparison for MQL

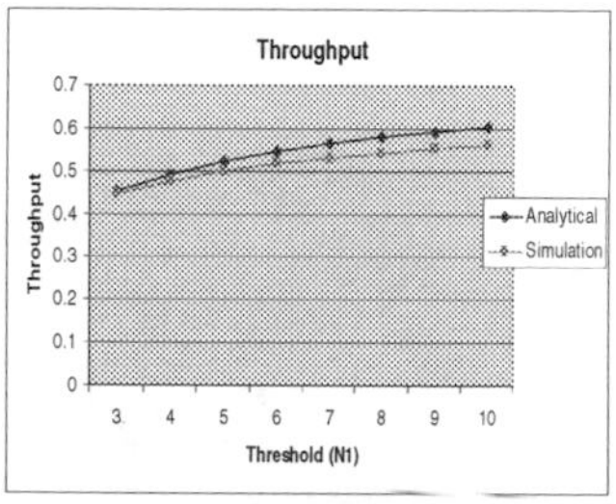

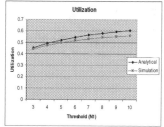

Fig. 3. Results Comparison for Throughput

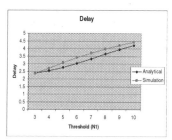

Fig. 4. Results Comparison for Delay

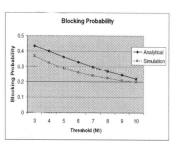

Fig. 5. Results Comparison for Utilization

Fig. 6. Results Comparison for Blocking Probability

against simulation respectively. It can be seen that MQL, throughput, delay and server utilization increase as we increase the threshold value, whilst, blocking probability decreases. In all experiments, we won't consider the situation when threshold value are very small, e.g. N_1=1 and 2 since they have no exact meaning in real-time system.

5   Conclusions and Future Work

An analytical solution, based on the principle of maximum entropy methodoloy, has been presented to model the congestion control mechanism for implementing the AQM scheme. In this context, a stable single server GGeo/GGeo/1/$\{N_1, N_2\}$ censored discrete-time finite capacity queue with N_1 threshold value, N_2 full buffer capacity, FCFS scheduling discipline and arrival first (AF) buffer management policy is analysed. Closed form analytical expressions for the state probabilities have been presented. This work has focused on single class of jobs for the external bursty arrival process. The traffic sources slow down the arrival process as soon as the number of packets in the queue reaches the threshold and jobs are blocked once the queue becomes full. Different QoS requirements under various load conditions can be satisfied by adjusting the threshold value. Typical numerical results were included to demonstrate the credibility of ME solution against simulation results. The system can be potentially used as a model for congestion control based on RED mechanism. Future work includes the extension of present work to develop analytical model incorporates multiple thresholds for multiple classes of jobs so that the technique of variable thresholds and blocking can be applied as an effective congestion control mechanism.

References

1. Y. Atsumi, E. Kondoh, O. Altintas, T. Yoshida, "A New Congestion Control Algorithm for Sack-Tcp with RED Routers" Ultra-high speed Network & Computer Technology Labs. (UNCL), Japan.
2. Jian-Min Li, Indra Widjaja, Marcel F. Neuts: "Congestion Detection in ATM Network". Performance Evaluation 34, 147-168, 1998.
3. B. Braden et al, "Recommendations on Queue Management and Congestion Avoidance in the Internet" IETF RFC 2309 April 1998.
4. S. Floyd and V. Jacobson, "Random Early Detection Gateways for Congestion Avoidance",   IEEE/ACM Transaction on Networking, Vol 1, No 4, pp397-413, August 1993.
5. D. Lapsley and S. Low, "Ransom Early Marking for Internet Congestion Control", Proceedng of GlobeCom'99, pp1747-1752, Dec 1999.
6. S. Athuraliya, D. Lapsley and S. Low, "An Enhanced Random Early Marking Algorithm for Internet Flow Control ", Infocom, 2000.
7. R. Gibbens and F. Kelly, "Distributed Connection Acceptance Control for a Connectionless Network", Proceeding of the 16th Intl. Teletraffic Congress, Edinburgh, Scotland, June,1999.
8. S. Kunniyur and R. Srikant,"End-to-End Congestion Control: Utility Function, Random Losses and ECN Marks", Proceedings of Infocom 2000, Tel Aviv, Israel, March, 2000.
9. S. Kunniyur and R. Srikant, "A Time-Scale Decomposition Approach to Adaptive ECN Marking", Proceeding of Infocom 2001, Alaska, Anchorage, April, 2001.
10. C. Hoolot, V.Misra, D.Towlsey and W.Gong, "On Designing Improved Controllers for AQM Routers Supporting TCP Flows", UMass CMPSCI Technical Report 00-42, 2000.
11. H. Alazemi, A. Mokhtar, and M. Azizoglu, "Stochastic modelling of random early detection gaeway in tcp networks" in IEEE Global Telecommunications Conference, 2000. GLOBECOM '00, VOL. 3, PP. 1747-1751.

12. M. May, T. Bonald, and J. Bolot, "Analytical evaluation of RED Prformance," in Proceedings of INFOCOM ' 2000, 2000.
13. A. Mokhtar and M. Azizoglu, "A random early discard frame work for congestion control in ATM networks," in IEEE ATM Workshop '99 Proceedings. Tokyo. Japan, 1999, pp. 45-50.
14. H. Alazemi, A. Mokhtar, and M. Azizoglu, "Stochastic approach for modelling random early detection gateways in TCP/IP networks" IEEE International Conference on Communications, v 8, 2001, p2385-2390.
15. E.T.Jaynes, Information Theory and Statistical Mechanics, Phys. Rev 106, (1957), pp. 620-630.
16. E.T.Jaynes, Information Theory and Statistical Mechanics, II, Phys. Rev 108, (1957), pp. 71-190.
17. Shore, J.E., and R.W. Johnson, "Axiomatic Derivation of the Principle of Maximum Entropy and the Principle of Minimum-cross Entropy", IEEE Transactions of information Theory, VOL. IT-26, pp.26-37, 1980.
18. Tribus, M, "Rational Descriptions, Decisions, and Designs", Pergamon Press, New York, 1969.
19. D.D. Kouvatsos, Entropy Maximisation and Queueing Network Models, Annals pf Operation Research 48, (1994), pp. 63-126.
20. D.D. Kouvatsos, and N.M. Tabet-Aouel, "GGeo-type Approximations for General Discrete-time Queueing Systems", Proc. of IFIP Workshop on Modelling and Performance Evaluation of ATM Technology, H.G. Perros et al (editors), La Martinique, January 1993.
21. D.D. Kouvatsos, N.M. Tabet-Aouel and S.G. Denazis "ME-based Approximations for General Discrete-time Queueing Models" Journal of Performance Evaluation 21 (1994) 81-109.
22. D.D. Kouvatsos, Irfan U. Awan, R.J. Fretwell and G. Dimakopoulos " A Cost-effective Approximation for SRD Traffic in Arbitrary Multi-buffered Networks" Journal of Computer Networks, 34 (2000) pp.97-113.
23. D.D. Kouvatsos, Maximum Entropy and the G/G/1/N Queue, Acta Informatica, Vol. 23, (1986), pp. 545 565.
24. D.D. Kouvatsos, N.M. Tabet-Aouel and S.G. Denazis "Approximation Analysis of Discrete-time Networks with or without Blocking" High Speed Networks and their Performance (C-21) H.G. Perros and Y. Viniotis (Editors), 1994.
25. D.D. Kouvatsos, P.H. Georgatsos and N.M. Tabet-Aouel, "A Universal Maximum Entropy Algorithm for General Multiple Class Open Networks with Mixed Service Disciplines" in Modelling Techniques and Tools for Computer Performance Evaluation, eds. R. Puigjaner and D. Potier, Plenum, 1989, pp. 397-419.
26. M. Veran and D. Potier; QNAP-2: A Portable Environment for Queueing Network Modelling Techniques and Tools for Performance Analysis, D. Potier, Ed., (North Holland, 1985).

Modeling Traffic Flow Using Conversation Exchange Dynamics for Identifying Network Attacks

Sudhamsu Mylavarapu[1], John C. McEachen[2], John M. Zachary[3],
Stefan L. Walch[2], and John S. Marinovich[2]

[1]Department of Computer Science and Engineering
University of South Carolina, Columbia, SC USA
mylavara@engr.sc.edu
[2]Department of Electrical and Computer Engineering
Naval Postgraduate School, Monterey, CA USA
{mceachen, slwalch, jsmarino}@nps.navy.mil
[3]Innovative Emergency Management, Inc.
Baton Rouge, LA, USA
john.zachary@ieminc.com

Abstract. We present a novel approach to identifying anomalous network events Specifically, a method for characterizing and displaying the flow of conversations across a distributed system with a high number of interacting entities is discussed and analyzed. Results from from attacks contained in the DARPA Lincoln Lab IDS test data and from operational network traffic are presented. These results suggest that our approach presents a unique perspective on anomalies in computer network traffic.

1 Introduction

Understanding network behavior for the purposes of diagnosis and intrusion detection is currently a major effort in the quest to build secure, robust and dependable computing systems. Specifically, intrusion detection systems (IDS) are detection security mechanisms that monitor a computer system or network, attempt to detect malicious activity, and raise an alarm to system or security administrators. IDSs can be classified as either anomaly-based detection or signature-based detection [1]. The former approach detects anomalous behavior, which may be a superset of undesirable behavior, and generally suffers from high false alarm rates. The latter signature-based approach may reduce false alarm rates but generally depends on a well-defined security policy to base detection on. Furthermore, signature-based intrusion detection systems are unable to detect events for which a signature is not defined in their signature database.

We present a novel approach to modeling distributed system with a high number of interacting entities. This problem is notoriously complex, and our model seeks to provide some level of data reduction so as to distinguish what is an anomaly from what is typical network activity. Consequently, this technique has the potential to find applicability to a wide variety of systems (beyond computer network systems).

Efforts related to our approach can be found in [2], [3] and [4]. These approaches all consider the problem from the abstraction of determining statistical properties of

M. Freire et al. (Eds.): ECUMN 2004, LNCS 3262, pp. 317–326, 2004.

the network. In this paper, however, we intend to focus on the analysis of the underlying state descriptors with the hope of extending these to global properties in the future.

2 Describing Network Conversation Flow

As was stated before, the goal of our approach is to reduce standard network data into a useful, reproducible, and meaningful form, ultimately to allow accurate detection of network anomalies. The notion of state will be more carefully defined below, but quickly summarizes into activity levels of various conversation groups within the network. One end product is a real time graphical description of the configuration of the network.

We model the computer network as a closed dynamical system. An implicit assumption of such a system is that information flow is conserved between nodes in a network. The network is constructed as a state space of information sources and sinks. As information quanta move throughout a network, the state space is updated accordingly.

States are represented as a vector $\vec{v}$ of sources and sinks. The analogy used is that of *buckets* and *balls*. Information moves between nodes represented as buckets as indivisible balls. As the network moves information around, this is represented as balls being passed between buckets. For example, a series of n packets transmitted from a node N_a to another node N_b would be modeled as n balls moved from bucket X and placed in bucket Y. The association of node N_a with either bucket X or Y depends on the nature of the conversation. The bucket can be defined using any combination of conversation characteristics including the affiliation of who is talking (individual hosts or networks), the language they are speaking (TCP, UDP, or ICMP), or the job they are performing (client or server).

In its simplest form, each node in a network is associated with one or more buckets and the total number of packets exchanged between nodes is modeled as moving balls from bucket to bucket. The collection of all buckets together with the allowable distribution range of balls forms a *bucket state space*.

The bucket state space includes an initial distribution of balls among the buckets (corresponding to a computer network with an expected distribution of information). This initial distribution forms an *initial condition* for the bucket state space. Likewise, the allowable bucket state spaces are assumed to be represented by a set of *boundary conditions*. From the bucket state space, the initial conditions, and the boundary conditions, the set of thermal properties representing entropy, energy, and temperature may be ultimately derived.

A state transition causes a shift in the distribution of information between the buckets. In other words, this model translates network behavior into bucket state transitions by selecting a ball from the bucket matching the source characteristics of the packet and moving that ball into the bucket matching the destination characteristics of the packet, thereby redistributing the information and transitioning the state.

Figures 1 and 2 show the importance of the state walk, and how the model provides more information than a simpler model. Both graphs represent a state space made up of two conversation groups, bucket A and bucket B, and they start with 5 balls each. As stated before, a network exchange can cause a state transition in three ways: by moving a ball from bucket A to bucket B, by moving a ball from bucket B to bucket A, or by either moving a ball from bucket A to bucket A or bucket B to bucket B, essentially resulting in no change for that period.

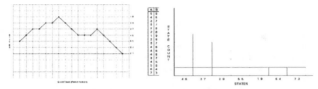

Fig. 1. A sample state walk for a two bucket model. (Left) A plot of the state walk over time. (Middle) A plot of the bucket sizes over the course of the state walk. (Right) A ranked histogram of the bucket states over the entire time period of this state walk.

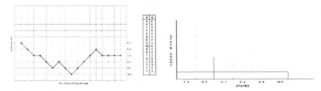

Fig. 2. An alternative state walk for a same two bucket model as shown in fig 1. (Left) A plot of the state walk over time. (Middle) A plot of the bucket sizes over the course of the state walk. (Right) A ranked histogram of the bucket states over the entire time period of this state walk. Note how the histogram varies for this state walk even though both examples end in the same state.

The paths shown in the figure 1 and 2 share a few notable similarities. The paths begin and end at the same state and undergo the same number and types of state changes, including five changes up, seven down, and four no changes. However, the transitions that make up those changes produce remarkably different paths, shown on the left hand side of the figures, and those paths produce remarkably different state counts, shown on the right hand side of the figures. Thus it is shown that this model provides a number of unique discriminators including number and type of state changes, starting and ending point and the general path in-between, and the state counts.

By examining the manifold, or canyon, developed by collecting various state histograms over time, anomalies can be easily spotted as perturbations in the normal flow of the canyon. The cause of such dramatic changes in practice ranges from a single transition to many thousands of packets.

An anomaly can cause one of two effects in the Thermal Canyon graph. Either new states are visited, or previously visited states are seen more often. The first effect will cause a spike oriented along the z-axis and the latter along the y-axis. An anomaly that is orthogonal to the normal traffic flow will tend to cause a spike oriented along the z-axis, due to the new states visited. An anomaly that is parallel to the normal traffic flow will tend to cause a spike oriented along the y-axis, due to the revisiting of previously visited states. The magnitude of the potential spike is what determines the ability of the operator to detect the anomaly. The orientation of the anomaly with respect to the normal traffic flow will determine the magnitude of the perturbation. A single packet anomaly that is orthogonal to the normal traffic flow will cause a large perturbation in the graph, where a packet that is parallel to the normal traffic flow will cause a relatively small perturbation. The less orthogonal the anomaly is to the normal traffic flow, the larger the number of anomalous packets required to cause a noticeable perturbation in the graph.

For example, figure 3(a) represents all of the possible bucket states that are contained in the bucket state space for a system consisting of three buckets (a, b, and c) each containing four balls. Each of the blue nodes represents a different bucket state. The number of balls in a given bucket is given by the lines parallel to the side opposite the vertex of interest. Each of the vertices corresponds to the case where all of the balls are in the associated bucket. The purple node represents the initial ball distribution, or initial bucket state of {4, 4, 4}. In figure 3(b), the number of balls in bucket 'c' is constant at four. The thick green line represents the nine possible bucket states based on a conversation between the remaining two buckets.

An example of the results of a single packet anomaly, that is orthogonal to the normal traffic flow, can be seen in figure 4. In this case the packet caused a ball to move from bucket 'c' into the conversation between buckets 'a' and 'b'. The result is a new line of possible bucket states. This new line contains ten possible bucket states. Given that the data from any given SL is averaged over a WLs of time there are now nineteen possible bucket states, which is more than double the original number of nine. This results in a run out (in the z-axis direction) of the Thermal Canyon graph. This type of anomaly is very easy to detect even though it was caused by a single packet.

Given that the Thermal Towers graph displays the average ball count per bucket per SL time period, it is less sensitive to anomalies that represent only a small percentage of the traffic. The Thermal Towers graph shows significant changes in the traffic flow. Therefore, if an anomaly is to be noticed in the Thermal Towers, it must comprise an appreciable percentage of the total traffic in the affected buckets. For example, a few packets that are parallel to the normal traffic flow of 500 pps will not be seen, but if the packets are orthogonal to the normal traffic flow, regardless of traffic rate, they will be seen.

The more orthogonal anomalous traffic is to the normal traffic flow, the greater effect the anomaly will have on the system graphs. Since it is not possible to know all the expected anomalous traffic in advance, the key is to create a bucket space that provides tight classification of critical traffic. For example, traffic should be parsed by functional group, like web servers, as opposed to grouping servers and clients together. There is a limit to the number of buckets a configuration can have, ideally, multiple instances of The system should be run concurrently to allow for smaller bucket spaces. This is also beneficial in reducing the complexity of interpreting the graphs which increases with the number of buckets. The next section presents some of the analysis on actual network traffic.

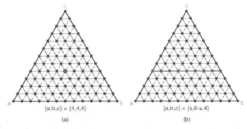

(a)   (b)

Fig. 3. Graphics that depict the total bucket state space for a system containing three buckets each with four balls. Each node corresponds to a different bucket state. – (a) The purple node corresponds to the bucket state of {4,4,4}. (b) The green line represents the range of possible bucket states for a conversation between buckets 'a' and 'b'.

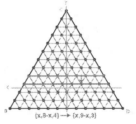

Fig. 4. A graphic depicting the results of an anomalous packet that is orthogonal to the normal traffic flow. The anomalous packet causes a ball to move from bucket 'c' into the conversation between buckets 'a' and 'b'. The results is the state walk moves from the line at $c = 4$ to the line at $c = 3$.

3   Experimentation and Analysis

This section is divided into two parts: controlled experiments conducted laboratory test equipment and results from real traffic on operational networks.

A   Laboratory Experiments

The purpose of these experiments was to show how the bucket state histogram varies as the bucket space description deviates from the actual network configuration. Two categories of experiments are discussed. The first shows the effect of an increasing number of rogue web servers on the bucket space histogram. The second shows the effect of a single out-of-profile packet on the bucket space histogram.

The simulation network consisted of a trusted subnet of 10 web servers connected to an untrusted internet of 1000 clients. In order to simulate typical network traffic, Spirent TeraMetrics traffic generators running TeraCaw software were used. This system is capable of simulating millions of client/server sessions using various application level protocols, beyond the initial three-way handshake. The system was configured such that any where between 400 and 800 users (using the 100 client machines) would randomly access the 10 web servers. Each web access consisted of establishing a TCP connection with the server, an HTTP GET message and then a 64-byte HTTP Response message. The connection was then terminated with a RESET.

The bucket space definition included a list of "authorized" web servers, identified an address space associated with trusted users, and considered ports below 1024 to be service ports. Consequently, for figures 5 through 8 the bucket space was partitioned as follows:

1. A trusted IP address, a web server, and a service port
2. A trusted IP address, a web server, and not a service port
3. A trusted IP address, not a web server, and a service port
4. A trusted IP address, not a web server, and not a service port
5. Not a trusted IP address and a service port
6. Not a trusted IP address and not a service port

The number of possible bucket states, N, can be determined as:

$$N = \binom{n+k-1}{n-1} \tag{1}$$

where n is the number of buckets and k is the number of balls in the system. Thus for the experiments above, with six buckets and initially four balls each ($k = 24$), there are theoretically $N = 118,755$ possible bucket states.

In the control run the bucket space description matches the actual network topography. Figure 5(a) illustrates the average bucket sizes and bucket space histograms over a period of two minutes (120 seconds). The typical load on the network was approximately 1000 packets per second. The bucket histogram is shown at two scales in this example.

Since the bucket space definition is aligned with the actual traffic patterns, the number of non-zero bucket states is small. Hence the histogram tail is very short.

Perhaps more importantly, this display also illustrates the smoothing effect defined by the central limit theorem on traffic that has been shown to be highly self-similar in nature on a per-client basis ([5], [6]). In other words, even though all clients arguably have the same heavy-tailed exchange characteristics, the actual distribution of states as shown by the bucket state histograms is highly normal and smooth, particularly when the bucket definitions are aligned with the configuration (i.e. all web servers in the web server list).

Next an anomalous packet was injected into the network for each during the above scenario. The anomalous packet was an UDP packet from one of the web servers to one of the clients. The packet originated from an ephemeral port (1025) with a destination service port on the client (53). Figure 5(b) shows the effect on the bucket state histogram for this packet a scales comparable to figure 5(a).

Fig. 5. The bucket state histogram over time. Decreasing frequency of bucket states comes out of the graphic. (Left) Traffic is confined to external clients visiting internal web servers so the number of bucket states is small. (Right) The bucket state histogram with a single UDP injected into the over 200,000 web traffic packets. Compare to (a).

The difference caused by the anomaly of the UDP packet should be readily apparent when comparing figures 5(a) and (b). Keep in mind that during this two minute period, over 200,000 packets were exchanged. The reason for the significant protrusion is because the UDP packet forces a ball to be transferred to a state that would not otherwise be visited, reducing the counts of the "normal" buckets and altering the frequency of the histograms (hence the notch in the graphics.)

The second experiment illustrates the effect when one web server removed from the configuration file. This web server is now in effect an unauthorized or rogue web server. The top graphic in figure 6 shows the effect of pulling the single server out of the web server group.

One immediately notes the significant variation in the bucket state histogram when compared to figure 5. This is due to the self-similar nature of the requests made to the single rogue web server. This hypothesis becomes more evident in lower graphics of figure 6 where two, four and six web servers are removed from the "web server" list. As we might expect, the increasing number of rogue servers invokes the central limit theorem, producing an increasingly smooth display. We do see and increasingly larger number of active bucket states, however, due to the larger variation in traffic characteristics.

Fig. 6. Bucket state histograms after the removal of one (top, left), two (top, right), four (bottom, left) and six (bottom, right) web servers from the web server list. Note the difference between these and the middle of figure 7. As we remove more web servers from the "authorized" list we see increasing smoothness in the display but also and increase in frequency of bucket states.

Fig. 7. Bucket state histograms for the traffic in figure 6 with a single UDP injected amongst the 200,000 packets. Histograms shown after the removal of one (top, left), two (top, right), four (bottom, left) and six (bottom, right) web servers from the web server list.

Next a single UDP packet was injected into all of the scenarios of figure 6. The re sulting displays are found in figure 7. Note how the UDP packet gets completely lost in the variable traffic of a single server removed, but is still quire prominent in all the others.

B DARPA Lincoln Lab IDS Test Data

For analysis and configuration of the system we utilized the *Tcpdump* files that were collected by Lincoln Labs using their 1999 Simulation Network. Per references [7], the simulation network was created to conduct evaluations of intrusion detection systems by measuring detections and false alarm rates.

Figure 8 shows the response to attack #41213446, an ICMP flood or "Smurf" attack. This attack is very common and generally easy to detect. Figure 9 displays the response to a Mailbomb attack, #42155148. This is a denial of service attack directed against the sendmail program. This is accomplished by sending a unique set of strings to the sendmail server. A similar type of attack to the Mailbomb is the Apache2 attack, #51140100. The response of the system to this attack is shown in Figure 10. Finally, the response of the system to a sweep of IP addresses is shown in figure 11. The shape of this response is particularly worth noting.

C Operational Network Observations

Figure 12 depicts a flood of ICMP packets originating inside a monitored operational network after normal working hours. This flood increased the state count; thereby producing a large spike on the GUI. This flood consisted of 6,032 ICMP echo re-

Fig. 8. Response of the system to an ICMP flood, #41213446.

Fig. 9. Response of the system to a Mailbomb attack, #42155148.

Fig. 10. Response of the system to an Apache2 attack, #51140100.

Fig. 11. Response of the system to an IP sweep, #52211313.

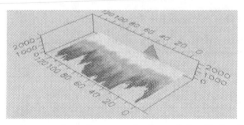

Fig. 12. An anomalous event observed on an operational network. Specifically a flood of ICMP packets was released from an internal client over a four second period.

quests/replies within a four second time frame. ICMP echo requests and replies are not necessarily anomalous, however the owner of the system was logged off and at home requiring notification of the local CERT for further investigation.

4 Conclusion

We have presented a novel approach to characterizing the conversation flow of a computer network. This approach associates traffic of certain characteristics with a category or bucket and observes the transfer of information from one category to another. Revealing behavior is identified by viewing the histogram of the bucket states, or category arrangements, over time. The approach has been evaluated on both simulated laboratory traffic and operational network traffic.

References

1. S. Axelsson, "Intrusion detection systems: A survey and taxonomy", Chalmers University Technical Report 99-15, March 2000.
2. Burgess, M., "Thermal, nonequilibrium phase space for networked computers," *The American Physical Society*, Volume G2 Number 2, pgs. 1738-1742, August 2000.
3. Evans, S. C. and Barnett, B., "Network security through conservation of complexity," *Proceedings of IEEE Military Comms. Conf. (MILCOM 2002)*, pgs. 1133 – 1138, Los Angeles, October 2002.
4. Donald, S. D., McMillen, R. V., Ford, D. A., and McEachen, J. C., "Therminator 2: A real-time system for patternless intrusion detection", *Proc. of the IEEE Military Comms. Conf. (MILCOM 2002)*, pgs. 1498 – 1502, Los Angeles, October 2002.
5. Crovella, M. and Bestavros, A. "Self-Similarity in world-wide web traffic: Evidence and possible causes," *Proc. ACM Sigmetrics Conf. on Meas. And Mod. Of Comp. Sys.*, May 1996.
6. Arlitt, M. and Jin, T. "A workload characterization study of the 1998 world cup web site," *IEEE Network*, May/June 2000.
7. Massachusetts Institute Of Technology, Lincoln Laboratory, DARPA Intrusion Detection Evaluation. [http://www.ll.mit.edu/IST/ideval/index.html]. 14 September 2003.

Models of Cooperation in Peer-to-Peer Networks – A Survey

Fares Benayoune and Luigi Lancieri

France Telecom R&D
42 Rue des coutures 14000 Caen
{fares.benayoune, luigi.lancieri}@francetelecom.com

Abstract. Peer to peer systems (P2P) have more and more success from individual users point of view as well as industrial uses. P2P bring original capacities by giving to the end-users a more important role than a simple "customer". Although many works addressed the various techniques of exchange in these networks, a few explored it under the angle of co-operation with end-users. This seems necessary since it is obvious that networks data flows take their roots in human exchanges and cooperation. Thus, the aim of this article is to analyze the modes of co-operation in P2P systems, under the light of more general model of co-operation in networks taking in account the influence of the human factor. We also try to clarify the purpose of each application by proposing several mode of classification.

1 Introduction

The evolution of powerful PC with great storage capacities and fast access to Internet and the increase of available multimedia contents (video, audio, etc.) allowed developing direct forms of information exchange between users without using central servers. The example of direct FTP and P2P exchanges are included in this category. Although reflecting a recent change of uses, these practices are not new with reference to the beginning of Internet. Contrary to traditional mechanisms of communication such as the "client/server" model, these systems give more interest to individuals who freely share their resources.

Apart from technical constraints, the human factor is a fundamental element in order to understand networks mode of operation. It is clear, for example, that the availability of an object will be higher if it is replicated on several network nodes. Actually, the rate of contents replication is directly linked to its level of availability and will be dependent on several forms of co-operation between individuals trough, for example, implicit or explicit form of cooperation. The explicit co-operation consists in formulating contents accessibility as a deliberate choice from the person who replicates contents and the one who decides to download it (both actors of a more or less direct interaction). The implicit co-operation implies that the factors of the accessibility are less controllable. Indeed the availability does not imply the accessibility that takes into account the facility of data access. Content will be faster to download if it is popular but this popularity is not directly controlled by the

M. Freire et al. (Eds.): ECUMN 2004, LNCS 3262, pp. 327–336, 2004.
© Springer-Verlag Berlin Heidelberg 2004

individual who takes the initiative to put a content on the network nor even by who decides to download it. The popularity of contents is the consequence of complex phenomena of collective actions not easy to evaluate.

The purpose of this article is to analyze modes of cooperation in P2P networks in the light of more general mode of co-operation in the networks especially in systems based on contents replication and on the human factor. That will enable us to describe various academic works and commercial products. First of all we give an overview of the mode of cooperation in general networks technology and more specifically in web oriented replication systems (e.g CDN, caches networks, etc). Then we survey the main known P2P technologies under several angles and classification. The first is a functional segmentation describing the use of the technology (distributed computing, data sharing, etc). The second view categorizes mainly technologies with regards to the size of the community (few users, middle range community, intranet, Internet). We also survey P2P application under the angle of the centralization level, the structure of the information system and the replication strategies.

2   Cooperation in Networks and Replication Systems

The co-operation is a well-known phenomenon in living world and more specifically in the human activities. Most of the time, it is a factor of progress that encouraged researchers to investigate co-operation based technologies. Internet is an interesting example of co-operation between individuals through technological system (information sharing, exchange ideas in news groups, etc). We gave in the introduction section the example of implicit and explicit action of individuals in order to show the human factor impact but the network as an intermediary between humans action also bring its part in the global resulting complexity. In this section, we will analyze the co-operation in networks in general then we focus on contents replication as in P2P networks. In each time, we try to emphasize the impact of the human factor.

2.1   Cooperation in Networks

The communication networks are conceived on the principle of exchange between individuals. They are in some extent like a black box that relay human interactions. Inside this box, several components inherit from primary individual actions to engage other combined forms of cooperation that contribute to the good operation of the communication. Even if the human activity is less discernible in this secondary in box interaction it is not less present combined with the technical constraints of the network. We will see that according to different parameters, the mode of co-operation can be more or less implicit.

In Internet for example, a whole of nodes (routers) are connected the one to the others through transmission links. Packets from the same source towards the same destination can take various paths. Indeed, a router acts locally and can update his routing table according to the new knowledge transmitted by his neighbors or from the histories of the routing requests. Although the users require these destinations, what implies that they play a part in the update of these routing tables, no user has knowledge of the used path. This highlights a rather implicit human factor impact on the behavior of the network. The same logic can be applied with other mechanisms in operation in the Internet like DNS for example. From the other hand, we find systems where the co-operation between users or the human factor impact on the network is rather explicit. It is the case of the collaborative editions in which, a set of users takes part in the creation of a common object. For example, CVS [31] (Concurrent Versions System) is one of the most used tools for software collaborative development. It is a solution for sharing and managing resources based on a star architecture i.e. the users are organized around a server which uses a tree structure called the repository in which are stored the files, as well as the history of their modifications. Here the participants cooperate explicitly because the contribution of each user can be easily evaluated. It is also the case of the networks collaborative drafting, or interpersonal e-mail exchange. Other technologies involves a more equilibrate mixture of implicit and explicit impact of the human factor. This is the case of forums (e.g. Usenet [32], Weblog, etc.) that involve communication in larges groups of users. In the Usenet system, the users cooperate first explicitly by posting articles and answering questions. The implicit form of co-operation intervenes by the combined action of the group that will support certain topics rather than others. The lifespan of the posts will be more or less long according to this shared interest. For example, even if it is very old, a post can "goes up" when one of the members answers what could encourages the other members to answer, etc. The popularity of a topic thus is strongly correlated to the lifespan of the messages.

We see that the level of complexity of these interactions depended on several factors like the level of observation (application level of the end user or low level data flow) or the quantity of user implied in the interaction (interpersonal, group, etc.). In some technique as in P2P the mode of replication of the objects is a key factor that adds complexity in users' interactions.

2.2   Cooperation in Replication Systems

The Web replication systems are software, hardware or architectures making it possible to copy objects, references to objects (e.g. URL hyperlinks), or both in localizations that facilitates its use.

One of the main advantages of objects replication systems is to optimize the speed of access to information by bringing it closer to the users. This principle is already used for a long time in Web mirrors or proxy-caches for examples. These two systems work differently with regards to human factor impacts. In the Mirrors, the administrator must explicitly decide to replicate or eliminate contents. In the proxy-cache the replication is implicit. Indeed, a stable memorization of contents and consequently the cache effectiveness involve that users often request objects. In fact, no user has a vision on the re-use frequency; each one takes part in a blind or implicit way in the phenomenon of replication that results from co-operation.

In the case of references replication systems, the advantage is not any more the speed of access but the data selectivity. The reference replication takes its root in the hyperlinks principle that is the heart of the Web (i.e. several links target the same object). Search engines are other examples of systems based on references replication.

Some specialized engines like Citeseer [36] make a classification of results according to references popularity. As we will see, in some P2P architecture, the availability of an object can be know or not depending on the mode of reference replication.

Quite naturally, the mixed approach inherits advantages of the first two systems. For example, the Google search engine cache (replicate) both the objects and its references. In an Intranet context of use, Google allows to obtain a good information selectivity as well as a fast access. The CDN (Content Delivery Network) are also examples of combination of references and objects replication . Based on the location of the user or the object, the references can be replicated and modified in order to redirect the user request to a more accessible copy of the object. Moreover, some cache techniques can be used in order to keep contents most frequently required. In addition to the profit in term of bandwidth and latency time, CDN networks offer a solution to balance the load of a data server. We will see that P2P can also be based on such approach.

3   Models of Cooperation in P2P Networks

The term Peer-to-Peer (P2P) refers to any exchange system characterized by direct interaction and data exchange between its entities (called nodes or peers). Peer-to-Peer Working Group [17], a consortium dealing with the development of this technology, defines the P2P as "the sharing of computer resources by direct exchange". These systems were popularized thanks to the applications of file sharing but they also give other functionalities. The P2P working group identifies five operational models: firstly the "*atomistic model*" that consists of a direct access between nodes without any mediation (e.g. GNUtella). The second model is "*user centered*" with a directory used to identify the users (e.g. Groove [40]). The third model "*data centered*", rather take into account the indexing of the contents (e.g. Napster). The fourth called "*Web Ml2*" is a mixture of the previous. Finally the model of "*distributed processing*" with computing sharing applications (e.g. Seti@Home [12]). According to the criterion used (functionality, architecture, etc.) we can consider several other segmentations. In order to have a more precise view of the diversity of P2P concept we present now 5 kinds of more typical categories.

3.1   Functional Segmentation

From a functional point of view, P2P applications can be subdivided in three basic sub categories: management and contents-sharing applications, distributed processing and finally, collaboration and communication However, there are also platforms, such as JXTA [29], and Globus, that aims at facilitating the development of these applications by offering a set of common basic services such as the authentication or research and routing services. The table 1 summarizes this classification. The file-sharing applications are extremely popular on the Internet and have a large user base. Recent statistics show that the activities of these applications consume more than 60% of ISP's (Internet Service Provider) traffic. Since there is about 70 different P2P applications and in order to have a reasonable reference section size, it is not possible to give all the references of the presented applications but a basic request in a regular search engine will provide the necessary details.

Table 1. Functional segmentation of P2P application

Management and contents sharing	Distributed processing	Collaboration and Communication
Napster, Audiogalaxy , GNUtella, FastTrack (KaZaA, Grokster, Morpheus), Blubster, DirectConnect, BitTorrent, Freenet, Aimster, iMesh, eMule, eDonkey2000, OpenNap (WinMX), LimeWire, Shareaza, XoLoX, Chord, Tapestry, Pastry, Tornado, CAN ...	Seti@Home, Genome@Home, Folding@Home, Evolutionary@Home, XPulsar@home, Life Mapper, ChessBrain , FightAIDS@Home Avaki, Jivalti, Axceleon, Entropia, GridSystems...	Groove, NextPage, Kanari, Magi, Jabber, AIMster, MSN, AOL Chat, NetMeeting ...

3.2   Level of Users' Cooperation

We can classify P2P applications according to the potential number of participants in four sub classes. We have first the basic communication where two users exchange contents directly from PC to PC, to several million of users. This quantitative factor is important since, more the number of contributors is high more the implicit part in the cooperation is also high. Interpersonal exchange involves explicit cooperation since each user is conscious of other contributions. Applications based on a low level of participation such as Groove make it possible users to create their own communities (near 10 members). These small communities can be a set of friend, family members or colleagues and are characterized by contents with strong common interest such as a family album photo. Applications with a broader level of co-operation are dedicated to a more specific use for example in companies (CDN based on P2P [28]). The members of such network are more numerous (e.g. hundreds of employees). Other more known applications are intended for public exchange between million of participants (Kazaa, e-mule, etc.). Thus, the most interesting objects will be duplicated in a not very controllable way. As we saw, this level of popularity is directly related to their level of availability.

3.3   Degree of Decentralization

Today, Internet is largely based on the Client/Server model but the use of central servers inevitably waste resources, creates bottlenecks and makes the system sensible to failure. The P2P systems, by their decentralization, remedy these limits by increasing the scalability and fault tolerance. We can categorize these systems in three categories through their degree of decentralization.

1-Purely decentralized systems: These systems, also called *pure peer to peer*, were popularized with GNUtella and Freenet (also, Plaxton, Blubster, etc.). The nodes within such system communicate together without any intermediate central point. In fact, each node termed *Servent* (**serv**er and cli**ent** at the same time) performs research

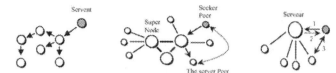

Fig. 1. Purely decentralized   **Fig. 2.** Partially decentralized systems   **Fig. 3.** Hybrid Systems
systems

and routing functions. This model is more robust than centralized one because the failure of any particular node does not impact the system resulting in high availability of the network (fault tolerance) and reduced costs. Unfortunately, this model presents two mains drawbacks. First, the localization of an object is not guaranteed because of the directory decentralization. Second, the mechanism of research by flooding (set of broadcast) wastes high amounts of network bandwidth. Indeed, the increase of the number of the peers generates an exponential increase of requests what pollute the network, slow down downloads and poses scalability problems.

2-Partially decentralized systems: A new wave of peer-to-peer systems is born from purely decentralized system combined with a centralized one (e.g. FastTrack : KaZaA, Morpheus, Grokster, iMech, etc.). This model is based on a Super-Node (or Super Peer) that acts as a centralized server for a set of nodes. The nodes send their queries to the Super-Node that provide the content or relay requests to other nodes. Such a system solves the problem of networks extensibility of purely decentralized systems by keeping the efficiency of the centralized ones. However, in case of failures of the primary Super-node, others are defined to automatically replace it. On the other hand, the mode of selection of the Super-Nodes is still complex. They can be elected according to their own capacities, in particular in term of bandwidth and persistence in the system (time of connection) but each user can decide to be or not a Super-Node. Thus, the fact of being Super-Node can cause security problems.

3-Hybrid decentralized systems: In this model, a directory centralizes references whereas shared-contents are decentralized (e.g. Napster, BitTorrent, etc.). Before using the network, a peer must connect to a central server, which manages a central directory of shared resources and users. Searches requests are sent to the server but files are downloaded directly from a node. This mode is a very efficient way to locate resources and to have a complete view of the network. Even if shared-files are more easily managed, there is a single access point that poses various problems (failure, overload, copyright, security, etc). Moreover, the central server limits the extensibility of the network.

3.4   Structure of the Information System

P2P Systems use various techniques of references publication, contents search and routing requests. Thus, another segmentation can be done according to the mode of information (or meta information) management in order to discover distributed data or evaluate its level of accessibility.

1-Unstructured systems: In some P2P systems like GNUtella or KaZaA, (also Morpheus, DirectConnect, BitTorrent, etc), nodes have no information on the location of the required files. In fact, the placement of these files is completely independent of the network topology. Thus, the research will be random and the used discovery methods are known as blind. Nodes relay requests from node to node towards all the neighbors in order to seek the maximum of objects on the maximum of nodes. Thus, the network will be flooded by duplicated message and the partial answers will be limited to a zone of locality defined by the TTL (the node's horizon).

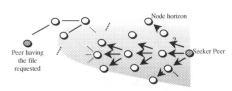

2-Structured systems: These hash-based systems (e.g Chord [24], CAN [25], Tapestry [14] or Tornado) are supposed to correct the lack of scalability of the prior systems. The objective is to add more dynamism to P2P network by proposing at the same time an algorithm of localization and routing in an entirely distributed environment. In these networks the files (or their references) are placed on quite precise places. Node and files have identifiers independent of their localization and contents semantics. Each node of the network has a routing table containing two parts: set of neighbors and set of pointers towards the nodes. In the case of Plaxton [23], the first system built in order to meet these aims, the table of the neighbors contains s=log (N)/b levels where N the number of peers and B the base of the identifiers. The major weakness of these systems is that it is difficult to maintain the structure in a context of a changing population where users often join and leave the system.

3-Loosely structured systems: Networks like Freenet [38] can be classified between the two preceding systems. Indices are provided on the localization of the files to select the next peer to query. Thus the searches will be guided but without guaranties of success. Furthermore, Freenet is also fault tolerant, thanks to the replication of files, every time a file is requested, "closer" to nodes that frequently make requests. The two last segmentations can be summarized on the following diagram:

3.5   Replication Strategies

A last segmentation can relate to the explicit strategies of replication [22] of objects or references. We can distinguish three categories: The first called *"owner replication"*, used by systems which has no proactive replication policy such as GNUtella, Napster, KaZaA, Morpheus, etc. Only the source and the applicant node keep a copy of references or objects locally. In the second category called *"path replication"*, a strategy of proactive replication is used in order to facilitate the access to more required objects. The principle is to store requested object from each node

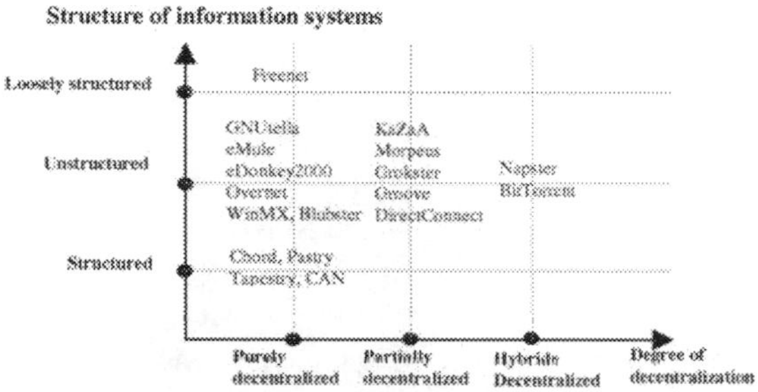

Structure of information systems

Fig. 4. Summary of the two last segmentations

along the route between the applicant node and the server one. Finally, a strategy known as *"random replication"* is a combination of the two priors. In this technique, replicas are stored in a random way on the path between the applicant and the owner of the object. It is the case of Freenet for example. Therefore, in the two last methods, more contents are asked, more they will be replicated closer to the node which claims it. This last mode is strongly influenced by the method of replication used and illustrate implicit co-operation. In fact, in the case of "path replication", most required objects will be duplicated everywhere through the network thus it will be more accessible.

4   Discussion

The future of a technology depends on its use and acceptance by the various actors. P2P Systems offer a new capacity to the users who are not any more simple passive consumers but also a more active element within complex system integrating the network. This strong implication of users will encourage without any doubt and support the development of P2P technology in the future. Our state of the art showed that P2P networks have many advantages but also limits. Most of current implementations do not offer a good selectivity (difficult to find what one seeks). In addition, even if they make it possible to highlight the availability of contents, which is not always the case, they do not guarantee a good accessibility (i.e. the access remains very slow). Another important aspect to take into account relates to the limits of ethical and legal nature concerning contents spread. The future systems must bring solutions to these limitations and answer new needs and requirements.

It seems to us that one of the solutions to this problem is in taking into account more largely the impact of the human factor. Other technologies as collaborative filtering (e.g. used in amazon.com to suggest books to customers) showed that it is possible to re-use phenomena of collective intelligences to optimize services with a very low algorithmic complexity. In particular thanks to users models of behavior, it is possible to anticipate downloading near the end users so as to increase accessibility. Several works were completed from this point of view (see [01] for a state of the art).

References

1. L. Lancieri, Reusing Implicit Cooperation: A Novel Approach to Knowledge Management, In tripleC International Journal 2004.
2. A.L. Soller, Supporting Social Interaction in an Intelligent Collaborative Learning System. Int. Journal of Artificial Intelligence in Education, 12(1), 40-62, 2001.
3. S. Saroiu, K. Gummadi, S. Gribble, A measurement study of peer-to-peer file sharing systems. In Proceedings of Multimedia Conferencing and Networking, San Jose, 2002.
4. K. Aberer et al,"An Overview on p2p Information System", http://lsirpeople.epfl.ch/hauswirth/papers/WDAS2002.pdf
5. C. Shirky. What is P2P And What Isn't, Nov 2000.; http://www.openp2p.com/ pub/a/p2p/2000/11/24/shirky1-whatisp2p.html
6. I. Clarke, T.W. Hong, S.G. Miller, O. Sandberg, and B. Wiley. Protecting Free Expression Online with Freenet, IEEE Internet Computing 6(1), 40-49, 2002.
7. M. Ripeanu, I. Foster and A. Iamnitchi, Mapping the Gnutella Network: Properties of Large-Scale Peer-to-Peer Systems and Implications for System Design, in IEEE Internet Computing Journal special issue on peer-to-peer networking, vol. 6(1) 2002.
8. M. Kelaskar, V. Matossian, P. Mehra, D. Paul and M. Prashar, A Study of Discovery Mechanisms for Peer-to-Peer Applications, Proceedings of the 2nd IEEE/ACM International Symposium on Cluster Computing and the Grid (CCGrid'02). 2002.
9. M.K. Ramanathan, V. Kalogeraki and J. Pruyne, Finding Good Peers in Peer-to-Peer Networks, Proc. of the International Parallel and Distributed Processing Symposium (IPDPS'02), IEEE 2002. PP 24-31
10. B. Yang and H. Garcia-Molina, Improving Search in Peer-to-Peer Networks, Proceedings of the 22nd International Conference on Distributed Computing Systems (ICDCS'02),
11. H. Stockinger, A. Samar, S. Mufzaffar, and F. Donno. Grid Data Mirroring Package (GDMP). Journal of Scientific Programming, 2002.
12. E. Korpela, D. Werthimer, D. Anderson, J. Cobb, and M. Lebofsky. SETI@home: Massively Distributed Computing for SETI, Scientific Programming.
13. The Peer-To-Peer Working Group, http://www.peer-to-peerwg.org.
14. B.Y. Zhao, L. Huang, J. Stribling, S.C. Rhea, A.D. Joseph, and J. Kubiatowicz, Tapestry: A Resilient Global-scale Overlay for Service Deployment, IEEE Journal on Selected Areas in Communications, Vol. 22, No. 1, Pgs. 41-53, Jan 2004.
15. J..Gao and P. Steenkiste. Design and Evaluation of a Distributed Scalable Content Discovery System. IEEE Journal on Selected Areas in Communications (JSAC), 22(1):54-66, Special Issue on Recent Advances in Service Overlay Networks, Jan 2004.
16. K. Calvert, J. Griffioen, B. Mullins, A. Sehgal, and S. Wen. Concast: Design and implementation of an active network service. IEEE Journal on Selected Areas in Communications, March 2001.
17. M. Castro, P. Druschel, A.M. Kermarrec, A. Rowstron, Scribe: A large-scale and decentralized application-level multicast infrastructure, IEEE Journal on Selected Areas in communications (JSAC), 2002. http://cse.ogi.edu/~krasic/cse585/castro02scribe.pdf.
18. D. Goldschlag, M. Reed, and P. Syverson. Anonymous Connections and Onion Routing. In IEEE Journal on Selected Areas in Communication - Special Issue on Copyright and Privacy Protection, 1998.
19. S. Shenker. Fundamental design issues for the future internet. IEEE Journal of Selected Areas in Communication, 13(7) 1176-1188, September 1995.
20. D. Wessels and K. Claffy. ICP and the Squid Web Cache. IEEE Journal on Selected Areas in Communications, Vol. 16, No. 3, April 1998.
21. T. Imielinski and C. Navas. Geographic addressing, routing, and resource discovery with the global positioning system. Communications of the ACM Journal (CACM), 1997.
22. Lv. Qin, C. Pei, C. Edith, L. Kai, and S. Scottr. Search and Replication in Unstructured Peer- to-Peer Networks. In Proceedings of ACM ICS, 2002.

23. C. Plaxton, R. Rajaraman, and A. Richa, Accessing nearby copies of replicated objects in a distributed environment, in Proc. of the ACM Symposium on Parallel Algorithms and Architectures, 1997.
24. I. Stoica, R. Morris, D. Karger, M.F. Kaashoek, and H. Balakrishnan, Chord: A scalable peer-to-peer lookup service for internet applications. In Proc. of SIGCOMM, August 2001.
25. S. Ratnasamy, P. Francis, M. Handley, R. Karp and S. Shenker, A Scalable Content-Addressable Network, Proceedings of ACM SIGCOMM 2001.
26. A.Rowstron and P.Druschel, Pastry: Scalable, decentralized object location and routing for large-scale peer-to-peer systems, 18th IFIP/ACM International Conference on Distributed Systems Platforms, 2001.
27. S. Iyer, A. Rowstron and P. Druschel, Squirrel: a decentralized p2p web cache, In Proceedings of Principles of Distributed Computing (PODC), 2002.
28. E. Turrini and F. Panzieri, Using P2P Techniques for Content Distribution Internetworking: A Research Proposal, in Proceedings of P2P'02, Sept. 2002, pp. 171-172.
29. L. Gong. "JXTA: A Network Programming Environment", IEEE Internet Computing, 88-95, Juin 2001. http://www.jxta.org/project/www/docs/JXTAnetworkProgEnv.pdf.
30. Jini, How Jini Technology Works, http://www.sun.com/2000-0829/jini/works.html.
31. B. Berliner. CVS II: Parallelizing Software Development. In Proceedings of the USENIX Winter 1990 Technical Conference, p. 341–352, Berkeley, CA, 1990. USENIX Association.
32. R. Salz, InterNetNews: Usenet transport for Internet sites, Open Software Foundation, Usenix Summer Conference, June 1992.
33. C. Gkantsidis, M. Mihail, et A. Saberi, Random Walks in Peer-to-Peer Networks, http://www.cc.gatech.edu/ fac/Milena.Mihail/rwp2p04.pdf
34. Q.Lv, S.Ratnasamy, and S.Shenker. Can heterogeneity make gnutella scalable? In Proceedings of the 1st International Workshop on Peer-to-Peer Systems (IPTPS '02), http://www.cs.rice.edu/ Conferences/IPTPS02/165.pdf
35. O'Reilly's website, http://www.openp2p.com/p2p/.
36. Citeseer web site http://citeseer.ist.psu.edu/.
37. FastTrack, The FastTrack Protocol, http://www.fasttrack.nu/.
38. Freenet web site, http://freenet.sourceforge.net/freenet.pdf.
39. DirectConnect web site, http://www.neo-modus.com/.
40. Groove web site, www.groove.net.

Localization of Mobile Agents
in Mobile Ad Hoc Networks

Youcef Zafoune and Aïcha Mokhtari

Computer Science Department
USTHB, Bp 32 El Alia, BEZ, Algiers, Algeria
{yzafoune, mokhtari}@wissal.dz

Abstract. The mobile agent technology is promising in many fields of application, especially in the field of nomadic computer science. Mobility and the new mode of communication used in ad hoc mobile networks cause problems proper to a mobile environment. In this paper, we study the localization of mobile agents in ad hoc networks; an object which has a double mobility is to be located, first, in relation to the mobility of its physical support in a dynamic environment, the station, and that of the code through the network stations. Two localization protocols of mobile codes in ad hoc mobile networks are proposed. The first is of the dispersed type and the second of the centralized type, based on a multi-agent mobile server, called the hurricane's eye, constituting an alternative of the absence of a fixed infrastructure.

1 Introduction

With the development of network, it is common that the code and the data are not in the same place. In traditional architectures (for example client/server) the data are moved to the code. However, for many reasons (the volume of the data, connection, etc.) the reverse is sometimes preferred [18]: it is the notion of mobile code.

The mobile code technology, issued of artificial intelligence domain under the mobile agent designation, is actually growing fast. This promising in many fields of application, especially in the field of nomadic computer science and in the distributed applications in large scale network (such as that of information retrieval and electronic commerce in the web). Problems of performances, heterogeneity of the execution supports, and the critical manner, the safety and the security (confidentiality and integrity) are then identified (held).

The objects localization mechanism (agent or mobile station) in mobile environment, used generally in the first step of information routing, allow to ensure the communication between this objects in spite of the change of their positions. In this paper, we are interested by the problem of the mobile agent localization in ad hoc networks. It is the matter to localize an object of twice mobility, first of its physic support mobility in a dynamic environment, the station, and the code's one through the network stations.

M. Freire et al. (Eds.): ECUMN 2004, LNCS 3262, pp. 337–348, 2004.

In the first section of this paper, we present different definitions and characteristics of the mobility code. In the second section on the mobile network, we recall briefly the two sorts of network on mobile environment. Then, the problematic of mobile agent localization is presented according to three different contexts. Finally, two proposed mobile agents localization protocols in the ad hoc mobile network are described in the last section.

1.1   The Mobile Agent Notion

A simplistic definition of agent concept is the following: « an agent is a logician or physic entity to which is attributed certain mission which is able to accomplish with autonomous manner and in cooperation with other agents » [6].

A mobile agent is defined as an agent conform to previous definition and able to move from one site to the other in the network to accomplish its task.
The mechanism of mobile agent allows three not negligible gains [12]:

• The client machine is not continually waiting for intermediate results and can be disconnected;

• One part of the application executes itself in the network and can benefit of important resources which constitute them;

• The activity which is executing in the network can move near distributed data avoiding the transfer of intermediate data.

Thus, this new communication paradigm is interesting in the following situations:

• The client machine has not any resources to do an action in the network (processor, memory, battery, connection).

• The action is long and does not require the user interaction, allowing thus disconnected and asynchronous mode.

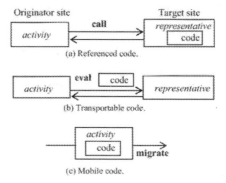

Fig. 1. The execution models distributed.

Classification. A remote action results of the execution of one part of the code on target site. We identify three execution models distributed according to the question: *where does the action code come from and which realizes it?* [12]

Referenced Code. The activity sends a message and activates an action code which is already on the target site and which is executed by a remote representative (call of remote procedure) (Fig. 1.a);

Transportable Code. The activity sends a message and activates an action code which is given by the activity and which is executed by a remote representative (Fig.1.b);

Mobile Code. The activity moves on the target site and performs itself the action (Fig.1.c).

We differentiate also in this execution model the migration from the mobility process itself.

Migration. The migration activity in distributed systems conceived for local network, is defined as the move of the running entity on the remote machine.
It gives a mechanism introduced to solve three sorts of problems [4]:
   • The charge repartition: it is the matter to optimize the global computer resource by varying the change of available processor, we move a running activity on a very busy processor to a lesser one [8].
   • The fault tolerance: it is the matter to manage the fault by moving the activity to a redundant processor to ensure a constant service availability [13].
   • The information sharing: The information sharing between distributed activities is set by moving the activity to the machine where the data are in memory rather to manage the data duplication in many memories [3].
In this case of mobility, it is the system which decides the activity migration.

Mobility. The difference between a mobile code with a classical activity migration is the activity itself (i.e. the code or the agent) which decides when, why, and where to move. One migrates to a remote site, a mobile agent becomes an autonomous entity which can according to the code running, move on one site, accumulate result, communicate with other agents, indicate events, etc, without maintaining a permanent connection with the initial site.

Structure. An instance of mobile agent is composed of three parts: data, code and execution context [16].
   • The data are the parameter values defined by the agent model. Among the data of a mobile agent, we find his name, his itinerary, the task program to execute and a folder destined to receive its execution result.
   • The code sets the access primitives to the parameters values.
   • The state execution context reflects mobile agent current execution (registers values, execution pile).

Properties. Two kinds of mobility mechanisms are proposed in mobile agent environments: the weak mobility and the strong mobility.

• For the strong mobility, the agent execution context is moved at the same time than its code and its data to the destination place; the agent execution rerun on the remote site at the instruction following immediately the migration primitive in the code.

• On the other hand, the weak mobility of an agent consists in moving only its code and its data; a new context is eventually constructed with the initial data.

Communications. We identify three kinds of communication in a mobile agent system [12]:
  • The communication of a mobile agent with a static agent,
  • The communication between mobile agents,
  • The communication between an agent and the client environment.

The static agent represents a service on the server machine. The communication between mobile agents poses the problem of mobile entities appointment and requires localization and notification mechanisms for realizing the connection of the agents. The communication with the proprietor, identified in the agent attributes, must be asynchronous.

1.2 Mobiles Execution Environments

An execution environment of mobile agents is constituted of a set of static programs, called places or agencies, which are executed on the sites of systems capable of receiving agents. The places are programs which offer to the agents the basic infrastructure for their execution and allow them to move.

Execution environments of mobile agents offer many basic services allowing the execution of a mobile agent on a site. These services are [16]:
  -Creation and migration of a mobile agent,
  -Reception of the agent and activation of the code,
  -Communication between the mobile agents,
  -Access to local resources by the mobiles agents,
  -Laying out of mobile agents in a process of execution.

Different execution environments or mobile agent platforms have been published. The *Telescript* system of *General Magic* is originally of the first mobile agent systems, born in the first half of the 90'. It has since given rise to numerous descendants, extensions of Java language, the *Aglets* of IBM, *D'agents* or *Agents-Tcl* of Dartmouth university, *Voyager* of ObjectSpace society of Dallas, etc.

1.3 Motivations and Applications

The first motivation of the utilization of mobile agents is generally the minimization of remote communications. It is generally less expensive to move the code rather than the data, which can be much more voluminous. It is the case of information retrieval applications.

Another motivation is the case of nomadic computer science. The mobile agents can be used for programming applications destined to nomadic computer science environment, especially it overcomes the problems caused by the stations disconnection. Indeed, a mobile agent created for the initiative of mobile station can comply in the system even if the mobile creative station work in disconnect mode after the migration of the agent towards other sites [16].

Another interest of mobile codes is their ability to accomplish remote operations, in an autonomous manner, without depending on the network state. Indeed, the mobile agents can allow remote interactions more robustness on non safe networks (for example, failure of communication channel or of servers). It allows to envisage new mechanisms of faults tolerance, to assure an accrue viability of executions.

2 Mobile Networks

Contrary to static environment, mobile environment allow to computing units a free mobility and do not pose any restriction on the users localization.

A mobile network is a system composed of mobile sites which allows its users to have access to the information independently of their geographic positions. The mobile or wireless networks, can be classified in two categories: the networks with fixed infrastructure which use generally the cellular communication model, and the networks without infrastructure or ad hoc networks [2].

2.1 Networks with Fixed Infrastructure

They are composed of two distinct sets of entities: the fixed sites of network wired and mobile sites. Some fixed sites, called basic stations are equipped of wireless communication interface for the direct communication with the mobile sites localized in a geographic limited zone, called cellular.

2.2 Ad Hoc Mobile Networks

An ad hoc network can be defined as a collection of mobile entities interconnected by a wireless technology, forming a temporary network without the help of any administration or of any fixed support. This class tries to extend the mobility notions to all the composites of the environment.

The absence of the fixed infrastructure, oblige the mobile units to behave as routers which participate to the discovery and the maintenance of the paths for the network hosts. These mobile hosts themselves which form, in an ad hoc manner, a way of global architecture which can be used as an infrastructure of the network.

Shortly, we can say that the Ad hoc mobile networks are characterized by:
-a dynamic topology,
-a limited bandwidth,
-the constraints of power,
-a limited physical security and
-the absence of infrastructure.

3   Localization Problem of Mobile Objects

The localization mechanism is used generally in the first phase of routing, more exactly in the roads discover procedure, which allow any object in a network to discover dynamically a path toward any network object. Three different contexts are presented above, in which is posed the localization problem of mobile objects in networks:

3.1   Mobile Code Localization in Fixed Network

The mobile object describes here an agent or mobile code which moves in a network with fixed stations. Two mechanisms of localization in this case are principally used [1]:

-The first mechanism creates dynamically a chain of forwarders in order to localize a mobile agent. Every time an agent leaves a station, it leaves a special reference, called forwarder, which point to the next destination, (i.e. the mobile agent on a reception site). This creates dynamically a chain of forwarders which allow to localize the mobile agent; when a forwarder receives a message, it sends it to the next destination (it is possible that is the mobile agent).

The second mechanism set up a centralized server to accomplish this task. The server keeps a trace of the mobile agent's position in a data base. Every time an agent migrates, it informs the server of its new position. When the source site wants to reach its mobile agent, it sends a localization request to the server.

3.2   Mobile Station Localization in an Ad Hoc Mobile Network

The mobile object is a station which moves in a mobile environment and without a fixed infrastructure (ad hoc network). The localization mechanism in this case, is used generally in the first phase of information routing, allowing to ensure communication between the stations despite the permanent change of their positions.

Despite the absence of fixed infrastructure and the mobility of stations, each node can be used to participate to the search of the stations position [2]. This phase presents mostly a cycle of request diffusion and response wait:

The source diffuses a localization request in all the network by accomplishing inundation. Each station not destined (not concerned) receiving the request, propagate it toward its neighbors in order that it reaches the destination station. Once the destination reached, it can send a response using the path traced by the request during its propagation in the network. Each request vehicles a single identifier which allows the avoidance multiple diffusions of the same request by the same site and guarantees that the single copy propagation of the request cannot occur through loops of the nodes.

The inundation (the pure diffusion) done in small size networks is not expensive. By contrast, in a voluminous network, the lack of information concerning the destination positions can involve an enormous diffusion in the network (the traffic caused by the diffusion, in this case, is added to the traffic already existing in the network which can degrade considerably the transmission performances of the system characterized principally by a weak bandwidth) [2].

In order to reduce the traffic caused by the diffusion, some routing protocols use a localization technique based on the spatial locality, which consists to limit the search of mobile station in a small region of the network (near the last position of the node), with a big probability to find the destination node. In the opposite case the request is diffused to neighboring regions.

3.3   Mobile Code Localization in Ad Hoc Mobile Network

It is the matter to localize an object of double mobility, first its physical support mobility in a dynamic environment, the station, and the code's one through the network stations.

In order to solve this problem, which to our knowledge, hasn't been dealt with so far, this double mobility makes us think to combine one of the mobile code localization techniques in fixed network with the mobile stations localization technique in ad hoc network based on diffusion. That consists for example, to adapt one of the two mobile code localization mechanisms presented previously and which are widely used, to a mobile environment.

We remark that the working of the first distributed protocol based on the chain of forwarders in the mobile network, pose the problem of the breaking of the chain, caused by the mobility of the marked stations. Whereas the second centralized protocol based on a server, appears to be not adapted to an ad hoc mobile network, due to the absence of a fixed infrastructure and the permanent stations mobility in the network.

4   Proposed Solutions

4.1   Distributed Protocol in a Mobile Environment

The mobile code localization in a distributed protocol must go through the reestablishment of the chain of forwarders when it is broken. We propose two solutions to this problem of the breaking of the chain:

Proactive Protocol. The first solution (proactive type) is based on the maintaining of the chain of forwarders in an established state, to do that, it must every time be established immediately after its breaking:

1-The neighbors marked stations exchange periodically and continuously check messages, in order to assure the maintaining of the chain in an established state.

2-When a contact is lost between two neighbors marked stations, from i to j, the station i emits a diffusion for the localization of j in order to re-establish the contact with it. The chain is thus re-established by adding new stations between i and j (Fig.2).

3-The marked stations of a higher order which intercept the diffusion request, answer the station i in order to optimize the length of the chain, and this in the case where there is such a station which is nearer of i compared with j. The chain is re-established and updated (Fig. 3 or 4).

4-The mobile code localization will be established then in the same manner of a fixed network case, since the chain is supposed to be established all the time.

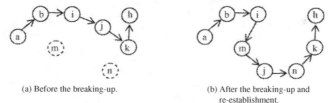

    (a) Before the breaking-up.   (b) After the breaking-up and
                                                        re-establishment.

Fig. 2. Chain re-established without optimization.

The figure 2.a shows an example of marked stations chain in an ad hoc network, from a to h order. With reference to a relative move if the station j, the chain is broken at the level of the station i which emits the localization process of j. At the same time, this latter emits the same process for the localization of k since it has lost the contact with its next marked station.

The re-establishment of the chain is shown in the figure 2.b which is done by inserting of new stations in the chain (m and n in the example).

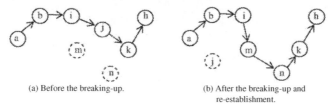

    (a) Before the breaking-up.   (b) After the breaking-up and
                                                        re-establishment.

Fig. 3. Chain re-established without the station causing its breaking.

The figure 3.b shows a re-establishment of the chain of the figure 3.a broken at the level of the station i. The marked station k, of a higher order than j, having established contact with i without passing by j, allows the re-establishment of the chain without this latter but by inserting new stations.

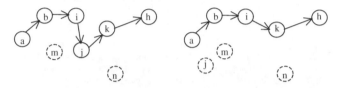

(a) Before the breaking-up.   (b) After the breaking-up and
                                                    re-establishment.

Fig. 4. Chain re-established with optimization.

The re-establishment of the chain of the figure 4, always broken at the level of j, is achieved in optimal manner since the station k of a higher order than j is within reach of i.

Reactive Protocol. The second solution (reactive type) is based on the re-establishment of the chain, when it is broken, just when the agent is localized:

1-To localize a mobile agent, the mother station of the agent contacts the first marked mobile station in the chain, which in its turn contacts the next marked station, and so on until the mobile agent is localized.

2-When a marked mobile agent j is not accessible by a station i (break of the chain), a limited diffusion is emitted for the search of marked stations of a higher order in the chain and which is located within reach of station i.

3-If there are answers to its request, the station i chooses the station of a higher order, update the chain, and to which it hands then the mobile agent localization request (Fig. 4).

4-Otherwise, the step 2 and 3 of earlier protocol (proactive) are executed (Fig. 2 or 3).

Discussion of solutions. The proactive solution allows the optimum reduction of the localization time of the mobile code, as the chain of forwarders is always in an established state, but it is expensive as it consumes bandwidth due to check messages exchanged periodically. On the other side, the reactive solution does not use many bandwidth, but the response time can be important enough due to the localization step of marked stations, which precede the mobile code position search.

4.2 Centralized Protocol in a Mobile Environment

To palliate the fixed infrastructure absence, we propose a multi-agents mobile server, called *the hurricane's eye*, in charge of the management of the database containing information about the localization of mobile agents in ad hoc networks.

Hurricane's Eye Protocol

1-The localization database (of positions) of mobile agents is distributed on several sub databases through a sub set of mobile stations.

2-A mobile agent is placed in each station of the sub set which has as a task the management of one of the sub database.

3-The sub set of the mobile agents constitutes the centralized server of the database. Initially, the server or *the hurricane's eye* is formed by only one station, positioned in the centre of the network and has to be designed as being the leader of the group.

Due to the fact that the radium of the propagation of the transmission of the stations in an ad hoc network is limited, and in order to keep connected the agents making up the server:

4-When a station is beyond reach of communication of the zone regrouping the sub set of mobile stations, another station in the same zone is elected for its replacement. The election principle is based on a criterion given, such as that of the relative degree of mobility or of connectivity of the stations.

5-The agent placed on the replaced station migrates towards the station elected, where it continues the management of the database.

6-The mobile code localization will be take place then by sending a request to one of the mobile server agents (found in the nearest station) which by consulting the database, gives an answer by sending a message conveying the looked for position.

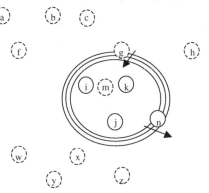

Fig. 5. Mobile multi-agents server in an ad hoc network.

In this figure, the eye of the hurricane includes four (04) mobile stations, i, j, k and n, thus making up a dynamic and centralized and relatively mobile server. Station n which is about to be out of reach of the other stations of the group, will be replaced by station m which is already within reach of the group. The agent associated with station n moves towards m.

Discussion. We think that the performance of the centralized protocol depends on the level of interconnection of at least a sub-group of stations in the network, that is to say, a group of stations where the connectivity has a long life span. The weaker this level of interconnection is, the smaller the number of stations becomes in the eye of the hurricane, moving towards one station only, which should then shelter several agents of the server which could cause its saturation. An alternative to this would be to allow the diameter of the hurricane to be variable, increasing thus the number of stations for the server.

5 Conclusion and Prospects

The mobility and new mode of communication used in the mobile networks cause problems proper to a mobile environment. The problem we have dealt with is the localization of a mobile agent in an environment characterized mostly by a lack of a fixed infrastructure. As an alternative to this, we have proposed a centralized protocol called the eye of the hurricane, based upon a mobile multi-agent server. We have equally adapted the distributed protocol, based upon a chain of forwarders to the ad hoc network. Two solutions have been proposed for the treatment of the breaking down of the chain, the first of the proactive type and the second reactive. A comparative study between the two distributed and centralized protocols, based upon a simulation, being in process of realization, it is not yet possible to proceed to evaluation of the performances.

Having dealt with the problems of localization of mobile agents, we are considering studying next the problem of routing in ad hoc networks.

References

1. Alouf, S., Huet, F. and Nain, P. Forwarders vs. Centralized server : An evaluation of two approaches for locating mobile agents. Rapport de recherche N°4440, Avril 2002 INRIA.
2. Badache, N. et Lemlouma, T. Le routage dans les réseaux mobiles Ad hoc. Available at opera.inrialpes.fr/people/Tayeb.Lemlouma/Papers/ AdHoc_Presentation.pdf
3. Balter, R. et al. Architecture and Implementation of Guide, an Object Oriented Distributed Systems. Computing Systems, 1991.
4. Balter, R., Banâtre, J.P. and Krakowiak, S. Construction des systèmes d'exploitation répartis. Collection didactique INRIA.

5. Benmammar, B. et Krief, Francine. La technologie agent et les réseaux sans fil. Le 17ᵉᵐᵉ congrès DNAC, De Nouvelles Architectures pour les Communications. Paris, Octobre 2003.
6. Briot, J.P. et Demazeau, Y. Introduction aux agents «Principes et architecture des systèmes multi-agents», collection IC2, Hermès, 2001.
7. Castañeda, R., Das, S.R. and Marina, M.K. Query Localization Techniques for On-demand Routing Protocols in Ad Hoc Networks. Available at citeseer.nj.nec.com/neda99query.html
8. Douglis, F. and Ousterhout, J. Process migration in the Sprite operating system. Proceedings of the 7ᵗʰ International Conference on Distributed Computing Systems, 1987.
9. Duda, A. et Perret, S. Une architecture d'agents mobiles pour des réseaux de stations nomades. Available at duda.imag.fr/ftp/cfip97.pdf
10. Guy, B. Apport des agents mobiles à l'exécution répartie. 4ᵉᵐᵉ Ecole d'Informatique des Systèmes Parallèles et Répartis (ISYPAR'00), Toulouse, France, 1-3 Fev.2000.
11. McGuire, M., Plataniotis, K.N. and Venetsanopoulos, A.N. Estimating position of mobile terminals from path loss measurements with survey data. Wireless communications and mobile computing. 3:51-62, August, 2003.
12. Perret, S. Agents mobiles pour l'accès nomade à l'information répartie dans les réseaux de grande envergure. Thèse de doctorat en sciences, Université de Grenoble-I, Novembre 19, 1997.
13. Popek, G.J. The Locus Distributed System Architecture. M.I.T. Press, 1985. Ramamurthy, N. Role of mobile agents in mobile computing environment. Available at http://crystal.uta.edu/~kumar/cse6392/termpapers/Naveen_paper.pdf
14. Ramanathan, S. and Steenstrup, M. A Survey of Routing Techniques for Mobile Communications Networks. Available at citeseer.nj.nec.com/ramanathan96survey. Html
16. Sahai, A. Conception et réalisation d'un gestionnaire mobile de réseaux fondé sur la technologie d'agent mobile. Thèse de doctorat, Université de Rennes 1, Janvier 25, 1999.
17. White, T., Bieszczad, A. and Pagurek, B. Distributed Fault Location in Networks Using Mobile Agents. Available at citeseer.nj.nec.com/2653.html
18. Conception et vérification de code mobile pour des besoins télécom. Available at http://lifc.univ-fcomte.fr/~bergerot/These.html

Multimodal Access Enabler Based on Adaptive Keyword Spotting

Sorin-Marian Georgescu

Ericsson Research Canada, System Management, 8500 Decarie Blvd.,
H4P2N2 Montreal, Canada
Sorin.Georgescu@ericsson.com

Abstract. The Multimodal Access Enabler is a Service Layer HTTP/Speech Proxy, which extracts keywords from HTML/XHTML pages given a set of predefined rules. The keywords are highlighted, in this way providing an indication to the user on which words to use in speech commands, when selecting a specific hyperlink. Synchronisation of the HTTP User Agent and the Speech Agent, is achieved using a "Push" module located in the HTTP/Speech Proxy. This module triggers page reload command execution in MS User Agent, once the page requested by voice command has been fetched. For unrecognised voice commands, the Multimodal Access Enabler uses a TTS module to synthesise speech dialogs/prompts, which either ask the user to select a command from a given set, and/or to repeat his command.

1 Introduction

Recent architectural developments in the Service Layer of 3G Wireless Networks, have laid down the structural foundation for implementing services which combine the Voice and Data Paradigms (MMS, weShareImage etc.). Depending on the technology used to transport voice, these services can be categorised into services using only Mobile Station Packet User Agent (VoIP/Streaming for voice and TCP/UDP transport for data), and services using concurrent voice and data sessions. In the later case, both the Mobile Station (MS) and the network have to support concurrency of voice and data sessions. So far, the Wireless Networks which support this feature are GPRS class A network and UMTS network. In this paper, we investigate an architecture based on concurrent voice and data sessions, enhanced with an adaptive mechanism capable to extract keywords from browsed content. Such architecture is well suited to applications featuring a Client/Server model, similar to the Web programming model we now see in the Internet. When instantiating a Client/Server service, the user accesses through the client module running in the terminal, the server part hosted by a Portal, or an Application Server. Similarly to the Internet, the Client/Server communication protocol is HTTP, while content is usually encoded in HTML/XHTML.

Apart from protocol related issues (WTCP vs TCP), another important requirement which differentiates Mobile Internet from the fixed Internet, is the interaction paradigm between the user and the User Agent running in the MS. Due to wireless terminals' small size, Mobile Internet users should not be restricted to use

M. Freire et al. (Eds.): ECUMN 2004, LNCS 3262, pp. 349–357, 2004.
© Springer-Verlag Berlin Heidelberg 2004

only MS keypad. Instead, combinations of keypad input and/or voice commands should be supported. This interaction paradigm is known as Multimodal Interaction or Multimodal Access.

Multimodal Access is a user friendly method for invoking network services. In essence, it refers to a combination of various types of input methods (keypad input, touch-screen input, voice commands etc.) synchronised internally in the MS, or externally in a network node. Synchronisation of inputs is probably the most challenging task in Multimodal Access, as it requires either new standardised interfaces in the terminals, or standardised multimodal network nodes/protocols.

1.1 Multimodal Architectures

Several architectures have been proposed so far to implement multimodality in Telecom networks. Among proposed architectures, VoiceXML and SALT are probably the fundamental architectures which established the reference basis for almost all subsequent proposals.

In the VoiceXML architecture [2], an application is defined as a set of input/output voice dialogs. The output voice dialogs are audio files/streams and text-to-speech prompts, while input dialogs are touch-tone keys (DTMF) and automatic speech recognition. A typical VoiceXML system consists of an Application Server hosting VoiceXML content, the VoiceXML Gateway (Interpreter) and the Mobile Station. The VoiceXML Gateway can further be divided into Speech Browser (VoiceXML Client) and Speech/Telephony Platform. The interaction with the application is done through a voice menu/form, to which the user provides his selection/input by voice. Speech recognition, text-to-speech conversion and DTMF functionalities are implemented by the Speech/Telephony Platform, which converts to/from speech the VoiceXML dialogs. The Speech Browser, uses the voice dialogs interpreted on-the fly, to control the sequence of events. It is helpful to stress out that VoiceXML systems make use only of MS voice part, when interacting with the user. Although this does not necessarily mean that MS User Agent (Web/WAP Browser), can not be used concurrently with the voice session. Most VoiceXML application developers provide both HTML/XHTML and VoiceXML versions of their application. However, as VoiceXML does not provide any synchronisation mechanism to Web browsers, status of commands executed in a voice session, can not be passed to MS Browser. Therefore the user can not mix input methods during same session. It is just like having two separate terminals and running one session using a voice terminal, and the next session using the data terminal. This important limitation of the VoiceXML architecture, was solved by the SALT architecture. Another weakness of VoiceXML architecture, is that the user experience when accessing applications through voice commands, is very much different from what is seen when browsing.

SALT architecture overcomes many of the limitations of VoiceXML architecture. The philosophy behind SALT [3] is to use a small set of XML elements (listen, prompt, dtmf, smex) which after being inserted into original HTML/XHTML page, provide a speech interface to it. To allow the interpretation of new XML tags, SALT architecture defines a SALT Voice Browser located in the Telephony Server (voice access through normal phones) and a SALT Multimodal Browser for true Multimodal clients. It is not specified in the architecture where should the merge of HTML/XHTML content and SALT tags be done. Possible scenarios to do the merge are either to post-process accessed pages directly in the Application server, or use a Proxy in the data path. The speech processing involved when interpreting SALT tags, namely the Speech Recognition and the Text To Speech processing, is performed in the architecture either locally in the MS, or remotely in the network. SALT does not define the network interfaces for the remote speech processing use case.

As mentioned earlier, SALT architecture solves some of the issues with VoiceXML architecture. Thus, SALT architecture provides true multimodal experience as speech browsing tags are inserted into original XML page. Also, because SALT extends and enhances markup languages, rather than altering the behavior, SALT documents can be viewed by visual-only browsers, by simply omitting SALT tags. Above statement is not typically applicable to old terminals which usually crush, when new language tags are encountered. Among the drawbacks of SALT architecture, one should mention that very often the vocabulary recognised by local Speech Recogniser (MS internal), and even by the network Speech Recogniser, is different from the one used when inserting SALT tags. Actually in case the content is post-processed by the Application Server, two different business entities deal with producing content, and viewing it. Another issue is the possible misalignment of vender specific MS vocabulary, in case local Speech Recognition is used. This could lead to acccess failure to page links associated to unrecognised words. Lastly, SALT architecture does not provide any paradigm/rules to follow up when inserting the SALT tags into XML page. Practice shows that due to complexity of content, simple queries often lead to numerous unrecognised commands, whereas the use of a hierarchical rule based paradigm, can significantly reduce the number of prompts asked by the system.

2   Multimodal Access Enabler Architecture

The Multimodal Access Enabler is a Service Layer entity which acts as a HTTP/Speech Proxy . As HTTP Proxy, the Enabler extracts browsing keywords from received HTML/XHTML pages, based on predefined rules. Then, the keywords are emphasised in the original page, in order to provide an indication to the user on words to utilise in speech commands, when selecting a specific hyperlink.

Due to the adaptive spotting method used to extract keywords, the Multimodal Access Enabler does not require the use of a large vocabulary Automatic Speech Recogniser (ASR). The Enabler may very well function with a middle size vocabulary Speech Recogniser. The ASR is the node which converts input speech to text, text afterwards compared to highlighted keywords. Usually, middle size vocabulary ASRs are capable of recognising continuous, speaker independent speech. Therefore, no preliminary training/adaptation is required to set-up the system. Customisation of recognised vocabulary should however be considered, to ensure that most frequently used keywords are in the set. The rules used to extract speech keywords can be grouped into following categories:

1. Syntactic rules: rules like "use the subject and the predicator from the paragraph associated to targeted hyperlink". Several syntactic rules, prioritised on availability of keywords in the vocabulary, may be used.

2. Simple rules: rules like "select a unique keyword from hyperlink's body or from the paragraph associated". Speech commands associated to simple rules may look like "Go to X" or "Go to the paragraph containing X".
3. Numeric rules: this refers to numbering hyperlinks in the page/paragraph. Can be used as well for menu item selection.

In Figure 1, we present the network nodes of proposed architecture, as well as the protocols/interfaces used for intercommunication.

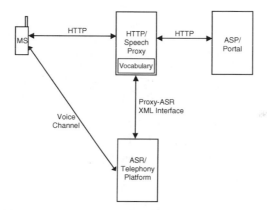

Fig. 1. Multimodal Access Enabler Architecture. HTTP protocol is used over all data interfaces, while voice may be carried either using VoIP, or normal circuit switched channels.

The Speech Proxy uses speech dialogs/prompts whenever received commands are ambiguous. Standard text messages containing words from synthesised vocabulary, are forwarded to the Text To Speech (TTS) block in the Telephony Platform. The TTS block converts these messages to speech dialogs, which are then sent to user's MS through the voice channel. The speech dialogs may look like: "Did you select the paragraph containing keyword X?". No particular requirements are put on the technology used to transport voice. For backward compatibility reasons, in this paper we suggest the use of circuit switched voice channels. However, as 3G/4G terminals can use IP bearer to transport voice, it is just a matter of upgrading the signalling and trunking sub-system of the Telephony Platform, with required protocols. Referring to 3G networks, the Telephony Platform needs to support SIP protocol for signalling, and RTP/RTCP protocols for voice transport.

2.1 Network Interaction

The interaction at network level between Multimodal Access Enabler nodes, is described in below sequence diagram.

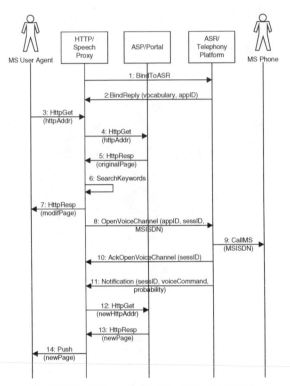

Fig. 2. Multimodal Access Enabler Network Sequence Diagram.

Two actors have been defined to model the MS, e.g. MS User Agent (WAP/Web browser) and MS Phone (voice sub-system). There is no synchronisation between these actors, the HTTP/Speech Proxy being the entity to implement this functionality. Here is how nodes interact:

1. HTTP/Speech Proxy connects to the ASR/Telephony Platform, specifying preferred application/vocabulary.
2. ASR/Telephony Platform returns in the reply, the vocabulary specified (recognised words in text format) and the features supported by Telephony Platform (call-back activated, number of voice ports etc.).
3. Subscriber opens a normal browsing session.
4. The Proxy authorises subscriber access to Multimodal Access service. The HTTP request is then forwarded to an ASP or a Portal.

5. ASP/Portal replies to HTTP GET request, by returning a HTML/XHTML page.
6. HTTP/Speech Proxy analyses the page, finds keywords applying defined rules and highlights the keywords (ex. through underscoring).
7. HTTP/Speech Proxy forwards modified page to MS Browser.
8. HTTP/Speech Proxy asks the ASR/Telephony Platform to open a voice session.
9. The Telephony Platform calls the MS, using specified MSISDN. A voice channel concurrent with the data session channel is opened between ASR and MS.
10. After opening the voice channel (call answered), the ASR sends an acknowledgement message to the Proxy.
11. Every time the ASR receives and interprets a speech command, it forwards recognised words to the Proxy. The Speech Proxy tries to match these words to keywords selected in step 6.
12. In case of match, the link obtained in step 11 is used as URL in the HTTP GET request sent to ASP.
13. After receiving the reply from ASP, the page is processed as in step 6.
14. The Proxy "pushes" the page to MS User Agent using the mechanism described in chapter 2.2.

For simplification reasons, we have presented the use case in which recognised words match the keywords associated to just one hyperlink. However, it is very possible that none of recognised words belong to highlighted set, or they correspond to several links. The Speech Proxy could then use word recognition probabilities for building up a set of choices, from which user should select. When recognition probabilities are lower than a predefined threshold, the Speech Proxy may either ask the user to repeat his selection, or may modify the page by replacing all keywords with numbers. As mentioned in chapter 2, speech synthesis is performed by the TTS module. The Speech Proxy has only to send the prompt message in clear text to the Telephony Platform, via the XML interface. No "intelligence" resides in the Telephony Platform. In fact, all recognition and speech prompt selection rules are implemented in the Speech Proxy. We should mention in conclusion of this chapter, that main principle for designing speech prompts, is to minimise the number of prompts per selection dialog. It is reasonable to expect that selection should end after around 1-2 speech prompts. Otherwise, the interaction becomes cumbersome, determining the user to continue his access only through the MS User Agent (Web/WAP Browser).

2.2 "Push" Module

Implementation of multimodality as access method to Web content, typically involves the existence of a synchronisation mechanism between User Agents running in different network nodes or application contexts. This synchronisation mechanism could either be centralised as the Interaction Manager from the W3C Multimodal Interaction Framework [1], or distributed in several nodes. In this paper we propose a distributed synchronisation mechanism, based on a simulated network semaphore.

This synchronisation module, also named "Push" module, is implemented in both the MS User Agent and Speech Proxy nodes. It is mainly based on a semaphore object (ClientSemaphore) inserted into original HTML/XHTML page, by the HTTP/Speech Proxy. A similar semaphore, called ProxySemaphore, initially having the same value as the ClientSemaphore, is stored in the proxy. The ProxySemaphore is set ON, whenever a new page needs to be "pushed" towards MS. Synchronization of MS User Agent and Speech Proxy, is achieved through periodic updates of ClientSemaphore with the value of ProxySemaphore. Very limited bandwidth is required for this periodic update due to just one small object being reloaded, instead of the entire page. On the client side, the ClientSemaphore is periodically checked if set ON, by a small script downloaded together with original HTML/XHTML page. When ClientSemaphore switches to ON, this means that new content is available in the Proxy. Therefore, the client needs to execute a HTTP GET, to update the page. This GET Request actually represents the means to simulate Proxy's "Push" of voice browsed content. On the proxy side, once the proxy update script switches ProxySemaphore ON, semaphore's value remains unchanged until ClientSemaphore has been updated. Then, the script may reset its value waiting for another voice command to be parsed in and executed.

From a language implementation perspective, the Semaphore object is not supported by any XML language. This is why we will refer in the followings to a specific implementation, more precisely to WML 2.0 [4]. WMLScript Standard Libraries [5] are used as well in proposed implementation.

The ClientSemaphore is modeled by means of a WML script variable. This script, downloaded from the proxy, triggers the HTTP GET method which fetches voice browsed page/card. The proxy stores two instantiations of the script. One with the semaphore set ON, and another with the semaphore set OFF. Only the instantiation which contains the actual value of ProxySemaphore will be stored in the URL directory, from where the MS downloads the updated client script. Below, we present a possible implementation of the script:

```
extern function updateSemaphore()
{
  var semaphore = "semaphoreValue";
  if (semaphore = "ON")
  {
    var url =
    "http://speechProxy.ericsson.se/wml/getPage.wml";
    WMLBrowser.go(url);
  }
}
```

Periodic invocation of updateSemaphore client script, is implemented using a timer element inserted by the proxy into original WML page/card. After timer expiry, the updated binary script updateSemaphore will be fetched from the proxy, and executed. Should the semaphore be ON, a HTTP GET is issued to fetch voice browsed page/card. The proxy will re-map the URL from client's request, to the one resulted from voice command interpretation, and issue a HTTP GET to the ASP. Voice browsed content can thus be downloaded to MS User Agent, without any intervention from the user. Here is how updateSemaphore script can be called from a WML card:

```
<card>
  <onevent type="timer">
    <go href="http://browsingProxy.ericsson.se/scripts
        /semaphore.wmls#updateSemaphore()"/>
  </onevent>
</card>
```

2.3 Architectural Review

Relatively to VoiceXML and SALT architectures, the Multimodal Access Enabler is first of all, a lot easier to implement. This is because no new XML tags are defined and therefore, any WAP/XHTML terminal which supports a scripting language (WMLScript, JavaScript etc.), will work straight away. So, there is no impact on existing terminals and as expected, no special requirements on coming ones (3G or 4G). In terms of multimodality, our architecture is much closer to SALT than to VoiceXML in the sense that, it implements a synchronisation mechanism which provides true multimodal experience. And because keywords are extracted on-the fly from accessed content, using simple/complex rules and speech prompts, browsing experience remains the same, irrespective of selected input method (voice commands or keypad input). In SALT architecture, by using a few simple elements inserted into the XML document, we can define speech prompts and sentences which should be used during voice interaction. This implies that the user has to ask specific questions and can not use his own expressions. The Multimodal Access Enabler does not have this drawback due to the use of keywords directly selected from browsed page.

Another improvement featured by the Multimodal Access Enabler versus SALT, is it does not require new standardised interfaces between terminals and other network nodes. Both architectures use a network speech recogniser, although the information passed to the recogniser is different. The Multimodal Access Enabler sends to ASR, the speech commands in clear voice, using an open voice channel (VoIP or E1/T1 trunking), whereas SALT suggests sending pre-processed speech data.

Due to the keyword spotting paradigm used by our Multimodal Access Enabler in the recognition phase, we could conclude during tests, that middle size vocabulary ASR is good enough. This type of ASR can recognise continuous, speaker independent speech. Hence, no special vocabulary adaptation is required when new users are subscribed to this service. We could not see in SALT architecture same design pattern, e.g. the use of isolated speech keywords. Therefore, our best guess was they consider using large size vocabularies. This type of ASR is usually speaker dependent and implementation costs are much higher than middle size vocabulary ASR.

Lastly, in case SALT implementation is based as described in the standard, on a local ASR (MS internal), there is no specified interface to allow terminal vocabulary download to the network node handling SALT tag insertion. This could result in the use of speech commands which do not match terminal's vocabulary. Referring to the Multimodal Access Enabler, tests showed that a customised recognition vocabulary, selected based on the keyword occurrence rate given a set of frequently accessed pages/services, leads to a fairly good overall hit rate.

3 Discussion

Multimodal Access Enabler is a new Service Layer node which allows subscribers to access Web content and Web applications, by using voice commands and/or MS keypad. Proposed architecture is based on the keyword spotting recognition paradigm, known to have very good performance in the category of continuous, speaker independent speech recognition applications. Through tests, we were able to identify a set of around 3000 words, which optimise the overall hit rate, e.g. 90% of links in browsed documents could be associated to vocabulary keywords. Remaining 10% of links, were associated to numbers.

The architecture, provides also a solution to the issue of backward/forward terminal compatibility. Thus, there are no special requirements on the technology used to transport the stream of speech commands to ASR node (E1/T1 trunking, VoIP or any other technology for Voice over Data Networks). So, 2.5G, 3G or coming 4G terminals work seamlessly. Furthermore, as there is no new element definition in the markup language used for voice browsing, all old HTML browsers, as well as the new XML browsers, work. The only module that needs to be upgraded in case the markup language for visual content changes, is the HTTP Proxy. But this is same issue with any HTTP Proxy Server on the market.

Among known paradigms to access information over Private/Public Data Networks, browsing is probably the most widely used method. This is why our paper has mainly focused on Voice Browsing application. Nevertheless, our future research is now targeting other Service Layer applications, such as MMS and 3GPP IP Multimedia Subsystem (IMS). As ideas for future work, we foresee the possibility to integrate proposed "Push" mechanism into MMS architecture. Main gain from this research would be the significant reduction of MMS network complexity, due to removal of Push Gateway and SMS infrastructure from the architecture. In the IMS area, we are planning to investigate the possible ways of integrating a speech engine, with IMS SIP Application Server. A new family of IMS services based multimodal interaction could thus be defined, making the systems we interact with everyday, a little bit closer to our natural way of communication.

References

1. Bodell, M., Johnston, M., Kumar, S., Potter, S., Waters, K.: W3C Multimodal Interaction Framework. World Wide Web Consortium (2003)
2. McGlashan, S., Burnett, D.C., Carter, J., Danielsen, P., Ferrans, J., Hunt, A., Lucas, B., Porter, B., Rehor, K., Tryphonas, S. (eds.): Voice Extensible Markup Language (VoiceXML) Version 2.0. World Wide Web Consortium (2004)
3. SALT Forum Companies (eds.): SALT Speech Application Language Tags (SALT) 1.0 Specification. SALT Forum (2002)
4. WAP Forum Companies (eds.): Wireless Markup Language Version 2.0. WAP Forum (2001)
5. WAP Forum Companies (eds.): WMLScript Standard Libraries Specification. WAP Forum (2001)

Effects of the Vertical Handover Policy on the Performance of Internet Applications

Andrea Calvagna and Giuseppe Di Modica

Dipartimento di Ingegneria Informatica e delle Telecomunicazioni
Università di Catania
Viale A. Doria, 6 95125 Catania – Italy
Andrea.Calvagna@unict.it,
Giuseppe.DiModica@diit.unict.it

Abstract. One of the major challenges posed by 4th generation (4G) network concept is that of integrating heterogeneous wireless networks technologies to build a globally interoperable system. In this context, issues like mobility management and session management have been widely addressed by researchers. In this paper we attempt to design effective vertical handover schemes applied to mobile users roaming across heterogeneous wireless access domains. The focus is put on how the performance of the Internet applications are affected by vertical handovers. We modeled a scenario of two interworking (WLAN and GPRS) access systems and run, by simulation, an analyses of how the connection discontinuity provoked by frequent vertical handovers between the two heterogeneous networks impacts on the performance of the application's sessions.

1 Introduction

Both researchers and vendors are currently expressing a growing interest in the 4th Generation (4G) networks concept, which should support wide-band data and telecom services for mobile users roaming across multiple, wireless and wired, integrated access networks. Basically, the purpose is that of combining all the existing heterogeneous wireless networks into a single, interoperable system, being IP protocol the "glue" between the set of underlying radio access and physical layers. Nevertheless, several technological and administrative issues arise, that researcher and developers have to face: access network integration, mobility support, authentication, authorization and accounting (AAA) "one-stop" support, QoS support, etc. Schemes for integrating heterogeneous wireless and wired networks must be developed and improved. Within this specific topic, several standardization activities have been already started, like the 3GPP project [1] and ETSI [2] efforts. Mobility management in heterogeneous wireless cum wired networks currently has MobileIP (MIP)[3] as to be the best candidate protocol to support mobile users' roaming across different administrative domains. In particular, its enhanced version HMIPv6 [4] solves some drawbacks of MobileIP and provides better scalability. Providing end-to-end QoS for user applications in heterogeneous wireless environments [5] is a challenging task, since applications should someway be aware of the underlying

M. Freire et al. (Eds.): ECUMN 2004, LNCS 3262, pp. 358–366, 2004.
© Springer-Verlag Berlin Heidelberg 2004

access technologies, which may happen to be quite different from network to network.

Thus, in this work, the focus is on the valuating strategies for vertical handovers scheduling, and in particular we propose an analysis of the impact that vertical handovers have on the performances of applications widely accessed in the Internet..

The paper is organized as follows. In Section 2 the related works and the motivation of our work are discussed. Section 3 depicts the scenario where simulations have been run. We report simulation results in Section 4 and conclude our work in Section 5.

2   Related Works and Motivation

Several schemes have been proposed to handle vertical handovers between heterogeneous wireless networks. When designing a vertical handover scheme, two aspects are to be taken into account. On one hand, the handover should be as smooth as possible: when being handed over, the MH should only experience a change in the perceived bandwidth and end-to-end packet delay. On the other hand, an decision algorithm is needed to schedule the time for the handovers, according to some QoS criteria.

In the BARWAN project [8] the performances of handover are improved by employing multicast and buffering techniques, but the requirements imposed by this scheme currently seems to prevent any feasible deployment. The solution proposed in HOPOVER [9] relies on a resources early-reservation scheme and on buffering techniques as well, claiming to solve the problems incurred by the BARWAN's solution. Buffering techniques are widely employed by all schemes that aim at guaranteeing smooth and seamless handovers. In [10] authors have investigated the upward vertical handover in wireless overlay networks, and have determined buffer requirements for lossless handovers. In [11] a seamless vertical handover procedure is presented, together with an effective algorithm for handover decisions based on the requirements imposed by the running application sessions whose continuity is to be granted. In [12] authors propose another optimization scheme for mobile users performing vertical handoffs. The profitability of a vertical handover is stated according to performance parameters such as mean throughput and handoff delay.

As far as vertical handover decision algorithms are concerned, all the schemes proposed in literature, as well as the one we use, borrow from techniques widely employed in the wireless cellular networks. Of particular interest are the ones based on the averaging window and hysteresis margin [13], those using pattern recognitions [14][15], and the fuzzy-based [16][17] [18]. The common target of these techniques is to minimize the so called "unnecessary handovers" while maintaining the QoS constraint. In wireless cellular network, minimizing the number of handovers is important as each handover increases the signal and processing load, and causes traffic management problems, thus affecting the QoS. In general, for such networks the QoS constraint related to the mobility management is the handoff failure probability, also referred to the forced termination probability of handover calls.

This work proposes an analysis of the impact that such fluctuations have on the performances of main internet applications schemes have been proposed to speed up the performances when vertical handovers are triggered [19][20][21][22][23]. We do not introduce new techniques for a smooth vertical handover, neither propose a new handover decision algorithm. Instead, we report an analysis of how the tuning of the hysteresis margin of a very straightforward handover decision's algorithm influences the performances of applications' sessions.

3   Scenario Description

We have developed a model implementing a subset of the requirements for the interworking between WLAN and GPRS. A MIP-like distributed mobility protocol has been designed and integrated in the model to support the roaming of mobile hosts (MH) in the WLAN and in the GPRS domain.

The protocol does not deal with authentication or security issues. Rather, it focuses on the support of the continuity of MH's ongoing transport sessions while it changes wireless access due to its frequent movements. Smart strategies have been devised to handle both horizontal and vertical handovers, aiming at guaranteeing the MH's ongoing connections against sudden disruptions due to the MH's crossing the boundaries of adjacent radio covered regions. No matter whether the MH traverses the boundary of regions served by homogeneous or heterogeneous wireless access points, the protocol is in charge of scheduling the time for triggering the handover, based on the assessment of the MH's future movements and on the evaluation of the wireless channels' conditions.

We have modeled the interworking between two different wireless access technologies: WiFi and GPRS. The MH is then equipped with two network cards. The WiFi card is configured to work on promiscuous mode, meaning that at the same time signals coming from more different WiFi APs can be sensed. The GPRS card is always on, and a permanent connection is established and kept alive. This is to avoid re-running of authentication procedures each time the MH access the GPRS network

As long as the MH moves within a WiFi domain, a Cellular IP (CIP) [24] derived protocol is used to take care of micro movements management. In particular, an hysteresis-based strategy has been conceived for horizontal handovers triggering. A decision algorithm, running on the MH device, builds up statistics form the beacon packets sensed from the WiFi network card. Whenever an handover is to be triggered, the strongest access point is chosen among the monitored ones. The performances of the horizontal handoff strategy are out of the scope of this work, so we will not go into further details.

3.1   Modeling of Vertical Handovers

According to [8], we refer to a WiFi-to-GPRS handover as an upward vertical handover; such an handover is in general triggered when the MH moves out of reach from a narrow-coverage (but broadband) network while already inside an overlaying wide-coverage (and narrowband) network. The common upward vertical handover strategy is very straightforward. Beacon packets from every available Access Points (APs) are constantly monitored, with particular attention to the ones collected by the AP that is currently serving the MH. Whenever the signal strength level of these beacons falls below a given threshold, meaning that the current radio signal is going to be lost, a new AP is looked for, among the monitored ones, in order to hand over to it all MH's ongoing sessions (in this case, an horizontal handover would be triggered). If none is available, or their signal strength is too weak, an upward vertical handover is triggered to the GPRS network. The MH's data connections might be heavly affected by this kind vertical handover, since (by definition) it is just the consequence of a lack of connectivity (radio "silence") experienced by the MH, whose handling over to the wide-coverage network is not immediate but requires also another, so called, handover latency.

4 System Evaluation

4.1 Web Traffic Models

In [26] a classification of the traffic running over the Internet is given. *Elastic traffic* includes TCB-based applications like Telnet, FTP, P2P file sharing, Email and Web browsing. Even though reliability is a crucial QoS parameter for these application, throughput may be considered a performance metric as well. *Inelastic traffic* is generated by real time services (voice and video) and, in general, by all the data services to which both timing and throughput are relevant parameters in order to meet the QoS requirements.

As far as the elastic traffic is concerned, a further classification can be introduced, based upon the level of user interactivity imposed by the semantics of each application. The transfer of a long-sized file, as well as the downloading of an email with a big attachment, generates *long-lived* TCP transmissions with no interaction from the user. Conversely, applications like Web browsing and Telnet envisage a tight user interaction but generate a lot of *short-lived* TCP transmissions.

The inelastic traffic category enumerates applications like VoIP, MPEG and H.263 video sources. Given the strict timing requirements, they reside on top of the UDP transport layer and usually generate transmissions at constant or variable bit rate.

In the proposed analysis we make an effort to cover all the categories of Internet traffic, by employing several traffic source models.

4.2 Performance and Metrics

Let S_m be the minimum signal strength level in order for a packet to be correctly sensed by the MH when it roams in a WiFi domain. For each category of simulation, several simulations will be run, respectively setting *Th* to different values. In particular, one simulation will be run by setting the threshold to such a value that the MH will never abandon the GPRS connection (let us call this conservative settings "connection-safe"). Another simulation is run by instructing the handover algorithm to search for just WiFi access points, even if that means that the MH will experience "black-outs" in its connections (let us call this settings "bandwidth greedy"). Other simulations are run by setting the *Th* to values $1.1*S_m$, $1.4*S_m$, $5*S_m$ and $10*S_m$ respectively. A complete set of simulations is thus run, ranging from the most connection-conservative configuration to the most bandwidth-greedy. Results will show how varying the *Th* affects in different ways the performance of the running applications. We can monitor the performance of the TCP-based applications by observing how fast the number sequencing of the TCP segments increases and/or how big is the amount of transferred packets. At the end of a simulation, the greater the sequence number, the better the performance of the TCP are. As far as the UDP-based applications are concerned, the relevant metrics that we will monitor are the total number of lost packets and the packet delay as a function of the time. In order for UPD to perform well, the packet loss rate has to be as low as possible, and also the packet delay has to be kept constant as much as possible.

4.3 TCP-Based Simulations

4.3.1   FTP Session

The first battery of simulations if focused on the study of the dynamics of the behavior of the TCP protocol, when a file transfer is activated, during vertical handovers, both upward (from WiFi to GPRS) and downward (GPRS to WiFi). An FTP session is set up between the MH and the CH. It is started at time t=1.0 s and stops, without any user intervention, at the end of the simulation. The FTP application generates a typical non-interactive long-lived TCP transmission.

In Fig. 1 the sequence number progression of the TCP segments received by the CH is plotted for different configurations. The simulation that performs better is the one with $Th=1.1*S_m$. We notice that its performance does not much differ from the simulation obtained with the "bandwidth-greedy" (BG) configuration. Even if the latter experiences long "black-out" periods, the TCP protocol is able to greatly evolve whenever the WiFi access is available (thanks to the high bandwidth and low RTT value that characterize such access). Conversely, the simulation with "connection-safe" (CS) settings does not experience connection black-outs, but can only benefit from a connection with limited bandwidth and a high RTT value (the GPRS one). We can conclude that for long-lived TCP transmissions (like, i.e., a file transfer), "bandwidth-greedy" handover strategies seem to perform better. For this kind of transmissions it is more important an even intermittent connection with high bandwidth and low RTT, rather than a permanent but poor connection.

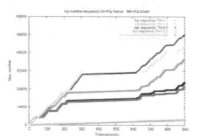

Fig. 1. The TCP number sequencing

4.3.2   Http Session

In this set of simulations an Http session is activated between the MH (the http client) and the CH (the http server). The client's page request generation process follows the Pareto distribution model. This model has been used in order to simulate the typical behavior of the user that browses the Web: a new page request is issued after the relevant information contained in the just downloaded page have been read. The average time spent by the user to read a page (i.e., the average *Toff* in the Pareto distribution) is set to 15 seconds, while the average web page size (i.e., the one that determines the *Ton* in the Pareto distribution) is set to 8 Kb. The traffic model that is being simulated is short-lived, and a loose user interaction is observed. In Fig. 2 the total amount of the downloaded traffic for each simulation is shown. Once again, the best performance is obtained by setting $Th=1.1*S_m$. This time, the simulation with BG settings (yellow curve) has the worst performance, while the one with CS settings (cyan curve) is not greatly penalized. The Http traffic model, in fact, gives rise to short-lived TCP connections. The application does not greatly benefit from the higher bandwidth available in the WiFi domain, given that the most of time is spent by the user to read the downloaded page, whilst the connection is exploited for a very little fraction of time. That is why the simulation in which the MH has a permanent connection to the GPRS network performs better than the simulation in which the MH connects only to the WiFi network. In fact, in the latter, from time t=270 to time t=620 the MH's connection gets stuck because of the WiFi black-outs (i.e., the MH is out of the range of any WiFi access point), while the former can benefit from the GPRS connection. The conclusion that we draw is that the connection parameters (i.e., the bandwidth and the RTT) have a low influence on the overall performance when short-lived TCP transmission are considered.

4.3.3   Telnet Session

We describe the results that we obtained by simulating a telnet session. This kind of application gives rise to a traffic model that resembles the one generated by a Http session. The user, in fact, interacts with his terminal, but this kind of interaction is tighter than the one observed for the Web browsing. Furthermore, the size of the packets exchanged between the telnet client program and the server one are much smaller than those exchanged in http sessions. We refer to this kind of traffic as a very short lived one with high user interaction. In Fig. 3 the measured performance for each simulation is reported.

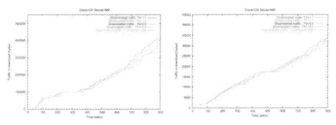

Fig. 2. Downloaded traffic during the http session   **Fig. 3.** Downloaded traffic during the telnet session

All the simulations, except the one with BG settings, seems to show equal performance. Once again, the strategy of searching at any cost for the WiFi access does not pay.

4.4 UDP-Based Simulations

Simulations have been run to evaluate the impact of vertical handoffs on the performances of applications built on top of the UDP protocol. In particular, RTP sessions have been set up between the MH and the CH.

An CBR application in the CH sends packet at the rate of 35 Kb/sec. Furthermore, let us assume that at most an end-to-end packet delay of 600 ms is tolerable for the application. Note that both the rate and tolerable delay can be sustained by the GPRS link. In Fig. 4 the total number of lost packets for each simulation are shown.

The graph refers to the number of packets that have been lost during handovers, both horizontal and vertical. The more the threshold increases, the less packets are lost. Of course, the simulation with CS settings gives the best result (no packet lost), while the one with BG settings experiences a great packet loss.

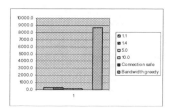

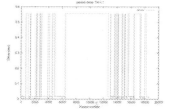

Fig. 4. Total number of packets lost during handovers

Fig. 5. Packet delay fluctuations for Th=1.1

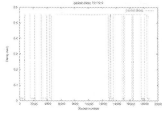

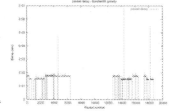

Fig. 6. Packet delay fluctuations for Th=10.0

Fig. 7. Packet delay fluctuations for BG

In Fig. 5, 6 and 7 the measured per packet delay is plotted for the simulations with $Th=1.1* S_m$, $Th=10.0* S_m$ and the one with bandwidth-greedy settings have been reported. During the 900 seconds of simulations, the MH switches several times from the WiFi network to the GPRS. The ongoing CBR session undergoes frequent packet delay fluctuations (from 0.018 seconds in the WiFi link to 0.55 in the GPRS link) as far as the experienced throughput and the RTT are concerned. In particular, for this specific simulation, only the end to end RTT is affected by the frequent vertical handovers, given that the transmission rate of the CBR session does not exceed the GPRS link capacity. The graphs show that the rate of the packet delay fluctuations lowers as soon as the threshold increases. As far as the BG simulation is concerned, no fluctuations can be found, since the packet delay is almost constantly set to 0.018 seconds; but of course, as showed in Fig. 4, a lot of packets are lost during the black-out periods. The results of the simulation with CS settings (whose graph is not reported) showed a constant packet delay almost equal to half the RTT of the GPRS link.

5 Conclusions and Future Works

In this paper we contributed to study of the impact of vertical handovers on the performances of well known internet applications, through modeling and simulation in the context of a typical scenario of user mobility. In our opinion, a new user-centric approach to the problem is required. The user himself should be given the means to define his own preferences; the handover algorithms would accordingly adapt their own strategy, thus acting on behalf of the user. This is only the first step to enable the customization of handover strategies, that is the target of our future work.

References

1. 3GPP, "Feasibility study on 3GPP System to WLAN Interworking" Tech. Rep. 3GPP TR 22.934 v1.2.0, May 2002
2. ETSI, "Requirements and Architectures for Interworking between HIPERLAN/3 and 3rd Generation Cellular Systems", Tech. Rep. ETSI TR 101 957, August 2001
3. C. Perkins, "IP Mobility Support for IPv4," IETF RFC 3220, Jan. 2002
4. HMPIv6, IETF Draft, http://www.ietf.org/internet-drafts/draft-ietf-mobileip-hmipv6-07.txt
5. Gábor Fodor, Anders Eriksson, and Aimo Tuoriniemi, Ericsson Research, "Providing Quality of Service in Always Best Connected Networks", IEEE Communications Magazines, Vol. 41, no. 7, July 2003
6. J. Mitola, "The Software Radio Architecture," *IEEE Communication Mag*azine, vol. 33, no. 5, May 1995, pp. 26–38
7. Eva Gustaffson and Annika Jonsson, "Aways Best Connected", IEEE Wireless Communications, February 2003
8. M. Stemm, R. H. Katz, "Vertical handoffs in wireless overlay networks," *ACM Mobile Networking and Applications (MONET)*, vol. 3, no. 4, pp. 335-350, 1998
9. Fan Du, Lionel M. Ni and Abdol- Hossein Esfahanian, "HOPOVER: A New Handoff Protocol for Overlay Networks", IEEE ICC2002, New York, New York, May 2002, pp. 3234–3239

10. Muhammed Salamah, Fatma Tansu, Nabil Khalil, "Buffering Requirements for Lossless Vertical Handoffs in Wireless Overlay Networks", IEEE VTC, Jeju, Korea, April 2003
11. Hyo Soon Park, Sung Hoon Yoon, Tae Hyoun Kim, Jung Shin Park, Mi Sun Do, Jai-Yong Lee, "Vertical Handoff Procedure and Algorithm between IEEE802.11 WLAN and CDMA Cellular Network" CDMA International Conference, 2002, pp. 103-112
12. M.Ylianttila, M. Pande, J. Mäkelä, P. Mähönen, "Optimization Scheme for Mobile Users Performing Vertical Handoffs between IEEE 802.11 and GPRS/EDGE networks", Proceedings of IEEE Global Telecommunications Conference 2001, San Antonio, Texas, USA, Vol. 6, pp. 3439 -3443
13. Senarath, N.G. and Everitt, D. - "Controlling handoff performance using signal strength prediction schemes & hysteresis algorithms for different shadowing environments", in Proc IEEE 46th Vehicular Technology Conference, Atlanta, pp 1510-1514, April 1996
14. K D. Wong and D. Cox. A handoff algorithm using pattern recognition. In IEEE International Conference on Universal Personal Communications (ICUPC), Florence, Italy, pp. 759-763, October 1998
15. K.D. Wong and D. Cox. Two-State Pattern Recognition Handoffs for Corner Turning Situations. IEEE Transactions on Vehicular Technology, 50(2), pp. 354-363, March 2001
16. George Edwards, Ravi Sankar, Microcellular handoff using fuzzy techniques, Wireless Networks, v.4 n.5, p.401-409, Aug. 1998
17. T . Onel, E. Cayirci and C. Ersoy, "Application of Fuzzy Inference Systems to the Handoff Decision Algorithms in Virtual Cell Layout Based Tactical Communications Systems", IEEE MILCOM 2002, October 2002
18. N. D. Tripathi, J. H. Reed, and H. F. VanLandingham, "An Adaptive Direction Biased Fuzzy Handoff Algorithm with Unified Handoff Selection Criterion," VTC '98 -- Vehicular Technology Conference, Ottawa, Ontario, Canada, May 18-21, 1998, vol. 1, pp. 127-131
19. S.E. Kim and J. A. Copeland, "TCP for Seamless Vertical Handoff in Hybrid Mobile Data Networks", GLOBECOM 2003
20. H. Balakrishnan, V. N. Padmanabhan, S. Seshan, R. H. Katz, "A comparison of mechanisms for improving TCP performance over wireless links," IEEE/ACM Trans. Networking, vol. 5, no 6, pp.756-769, Dec. 1997
21. S. Mascolo, C. Casetti, M. Gerla, M.Y. Sanadidi, R. Wang, "TCP Westwood: bandwidth estimation for enhanced transport over wireless links," in Proc. ACM MOBICOM., pp. 287-297, 2001
22. T. Goff, J. Moronski, D. S. Phatak, V. Gupta, "Freeze-TCP : A true end-to-end TCP enhancement mechanism for mobile environments," in Proc. IEEE INFOCOM., pp.1537-1545, 2000
23. A. Bakre, B. Badrinath, "I-TCP : Indirect TCP for mobile hosts," in Proc. IEEE ICDCS, pp. 136-143, 1995
24. A.T. Campbell A.G. Valko and J. Gomez, "Cellular ip", Internet Draft, draft-valko-cellularip-00.txt (1998)
25. UCB/LBNL/VINT. Network Simulator - ns (version 2). www.isi.edu/nsnam/ns/, software tool
26. Z. Sun, L. Liang, C. Koong, H. Cruickshank, A Sánchez and C. Miguel. "Internet QoS Measurement and Traffic Modelling". 2nd International Conference on Conformance Testing and Interoperability (ATS-CONF 2003). 20-21 January 2003

User-Empowered Programmable Network Support for Collaborative Environment

Eva Hladká[1], Petr Holub[1,2], and Jiří Denemark[1]

[1] Faculty of Informatics
{eva,xdenemar}@fi.muni.cz
[2] Institute of Computer Science
hopet@ics.muni.cz
Masaryk University, Botanická 68a, 602 00 Brno, Czech Republic

Abstract. We introduce a user-empowered UDP packet reflector to create virtual multicasting environments as an overlay on top of current unicast networks. The end-users' ability to fully control this environment by a specific communication protocol is the main advantage of our approach. Serializing the parallel communication schema for group communication allows us to introduce special features that are possible in unicast communication only. Similar to working with programmable routers, users can submit their own modules, which can be linked into the reflector and perform user-specific operations (filtering, transcoding etc.). The reflector is the basic element of the overlay network support for the user-empowered group communication in collaborative environments.

1 Introduction and Theoretical Background

In the current world, people are looking for systems supporting easy-to-use and inexpensive activities like video-seminars, tele- and video-consulting, and virtual meetings, which are specific forms of a virtual collaborative environment [1]. This paper focuses on both building a theoretical framework and creating a practical implementation of a network support system for communication among smaller groups of participants (up to 20 sites, usually bellow 10) that can be fully controlled by the participants themselves. The system is intended to be simple to use and yet flexible, capable of reacting to pre-defined as well as dynamic events such as changes in number and location of participants, network conditions (bandwidth, delay, security), etc.

Two basic principles are being adopted in a complementary way when transferring data over the networks: *connection oriented* and *packet oriented* approaches. Both approaches reached widespread use in different environments and nowadays we see a lot of effort dedicated to their convergence. This is also dictated by new applications and their requirements of scalability on one hand and transport quality control on the other hand. The packet based networks with rather "dumb" active elements targeted for only one function—data routing— won the field of high-speed networks, while sacrificing most of the control features

M. Freire et al. (Eds.): ECUMN 2004, LNCS 3262, pp. 367–376, 2004.

needed for advanced applications. A quality of service is offered on a statistical basis only (e. g. using DiffServ approach) and the users usually have no way of influencing or at least monitoring the transport of "their" data over the network. As reaction to these problems, we are developing a novel approach based on following cornerstones:

- Active elements within the network, programmable directly or at least indirectly by the users and their applications. These serve as the underlying technology for implementing the higher layers [2].
- Overlay networks as a framework for introducing specific services within the packet oriented network. The overlay networks allow minimizing the necessary overhead for advanced services without limiting their complexity [3].
- User empowered approach as a way to put the control plane into the hands of end users. The users can set up and tear down services, control and monitor their behavior while the services are well isolated so as to avoid any unwanted influence on other users.

The reflector is built as a special active node within a network, with full control by the user who uses it for group communication. The active router was modified to serve as the user controllable (user empowered) multimedia data reflector. The active node is implemented as a specific service within an ordinary computer. Fulfilling the requirement of full user control means overworking the active router and moving its functionality into the user space without any changes on the kernel level. This special implementation of an active router in user space was created and became the basic element for the overlay network for group communication.

2   Reflector

The reflector is a network element that replicates and optionally processes incoming data usually in the form of UDP datagrams and distributes this data to its clients in sequential manner using unicast communication only. If the data is sent to all listening clients, the number of data copies is equal to the number of clients. Our reflector is designed as a user-controlled modular programmable router, which can optionally be linked to special processing modules in run-time. The reflector runs entirely in user-space of the underlying operating system and thus it works without the need for administrative privileges on the host computer. The reflector architecture comprising the administrative part and data routing and the processing part is shown in Fig. 1.

The data processing and replication works as follows: the entry points of the reflector are network listener modules which are bound to one UDP port each. The received packet is placed into the shared memory and the listener adds a reference to a "to-be-processed" queue. A packet classifier reads the packets from this queue, checks with a routing AAA module whether the packet is allowed or not and determines a path through processor modules for each packet. After the processing, the data is distributed to clients by a packet scheduler/sender module

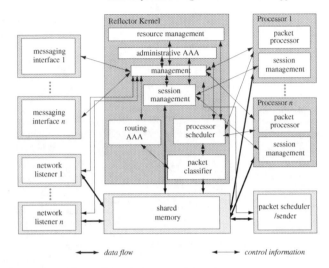

Fig. 1. Architecture of the reflector.

according to a distribution list obtained from a session management module. The number of copies of the data inside the reflector is minimized in order to boost performance: for simple scenarios the reflector works in the zero-copy mode.

The session management module is responsible for maintaining the distribution lists for each group, for adding new clients (usually after the client starts to send data), and removing inactive (dormant) clients. Simple client authorization is based on IP address restrictions. The access control list contains an "accept" or "deny" rule for each IP address or subnet record. The decision is made by the routing AAA module, rejected packets are dropped and an appropriate event is generated and can be logged if requested.

The administrative part of the reflector can be accessed via secure messaging channels such as HTTP with SSL/TLS encrypted transport or SOAP with GSI support[1]. The user can authenticate using various authentication procedures, e. g. combination of login and password, Kerberos ticket, or X.509 certificate. Authorization uses access control lists (ACLs) and is performed on per-command basis. Authentication, authorization, and accounting for the administrative part of the reflector is provided by an administrative AAA module. The actual reflector control is provided by a management module, which accepts commands in a

---

[1] Basic transport used for secure web services in Grid environments.
http://www.doesciencegrid.org/Grid/projects/soap/

specific messaging language, the Reflector Administration Protocol (RAP) [4]. All the events that occur in the reflector (users joining or leaving the reflector, exceptions etc.) can be logged for further inspection.

The data received by the reflector are replicated and sent back to all the connected clients and thus the limiting outbound traffic on the reflector grows with $n(n - 1)$ where n is the number of active (sending) clients. The scalability issue arises obviously which can be mitigated by creating networks of reflectors with tunnels connecting them (see Sec. 3.1). The network can be built in either a static way (pre-configured) or dynamic way (e. g. using distance-vector routing algorithms or some more efficient routing algorithms from peer-to-peer networks like Pastry [5]). Reflector networks can also be used for building overlay networks that are more resilient to network outages than the underlying network [3].

2.1 Advanced Reflector Scenarios

Because of the data replication for each individual client, it is possible to implement per-user processing which is impossible to do with multicast. The modularity of the reflector allows users to add and configure specific functionality in run-time. Examples of per-user processing are shown below:

- Multimedia data transcoding. Data processing modules can convert data between different formats (e. g. re-compress data from the DV format to the H.261 format). The reflector can be thus used as a gateway allowing clients with limited support of compression formats or insufficient network or processing capacity to join videoconference without forcing the rest of the communicating group to use low-quality or low-bandwidth multimedia formats.
- Video image composition. Composing several video images into a single image can be useful for a collaborative environment with a large number of participants in which there is not sufficient processing or display capacity to provide full video windows from all the clients simultaneously.
- Synchronization. When using parallel media streams encapsulated in RTP protocol, it is possible to synchronize such streams [6]. RTP packets contain relative time-stamps that can be converted to absolute local time on the sending machine by utilizing both relative and absolute time-stamps sent in complementary RTCP packets. When the synchronized streams originate on different computers, it is necessary to synchronize time on these computers, e. g. using NTP protocol.

By connecting reflectors with different functionality, it is possible to create an overlay network allowing users to connect to reflectors according to their needs.

Reflectors can be used for building strongly secured communication and collaboration environments. In the secured scenario each client must maintain a secured reliable connection to the reflector (usually an SSL encrypted TCP connection) that is used to exchange encryption keys between the client and the reflector. UDP datagrams are then sent encrypted from the client to the reflector, processed, and distributed to other clients encrypted again. Such reflectors however, requires modified MBone Tools to work with [7]. The reflector can also be used in an adverse networking environment restricted by firewalls and NAT deployment since it is possible to tunnel UDP data between reflectors through a TCP connection using some well-known ports that are enabled on the firewall [8].

3   Performance Evaluation of a Prototype Implementation

The reflector described above has been implemented and its performance has been evaluated in order to verify its usability. The testbed environment comprised three powerful machines used as a traffic generator (gerard), a reflector (test4), and a receiver (brand). The machines were connected via the HP ProCurve 6108 gigabit Ethernet switch. More detailed information on configuration of these machines is shown in Tab. 1.

Table 1. Overview of configurations of the testbed machines.

	test4	brand	gerard
brand	–	DELL PowerEdge	DELL PowerEdge
model		1600 SC	1600 SC
processor	2× Intel Xeon	2× Intel Xeon	2× Intel Xeon
	2.80 GHz	2.80 GHz	2.80 GHz
memory	1024 MB	1024 MB	1024 MB
NIC	Intel 82545EM	Broadcom BCM5701	Intel PRO/1000
	64 bit/66 MHz	64 bit/100 MHz	32 bit/66 MHz
operating system	Linux	FreeBSD	FreeBSD
	2.4.23	5.2-RELEASE	5.2-RELEASE

To evaluate the performance, clients sending 30 Mbps stream each were used thus emulating multimedia clients utilizing DV [9] video format sent in RTP packets over the IP network [10]. During the experiment the number of active (both sending and receiving) clients was increased and there was a single passive (listening only) client used as a measuring probe. The results summarized in Fig. 2 show that the system is usable for communication of up to five active clients working with very high quality video. For clients with lower bandwidth utilization, the number of clients that can get connected grows $n \propto 1/b$ where n is the maximum number of connected clients and b is the bandwidth used. It is also obvious from the results that the reflector is capable of fully saturating a gigabit Ethernet network link with limits imposed by the hardware and operating system used. The problem of scalability can be further tackled by building networks of reflectors (Sec. 3.1).

We have also evaluated the maximum forwarding throughput of the reflector which proves to be more demanding compared to common replication. This corresponds to the fact that more data is transmitted over the PCI bus compared to the replication mode. The results summarized in Fig. 3 show that it is possible to forward approximately 450 Mbps without significant packet loss.

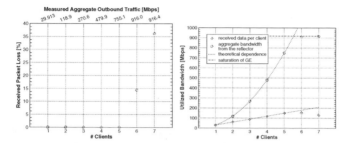

Fig. 2. Reflector prototype performance evaluation for 30 Mbps clients.

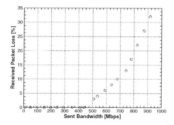

Fig. 3. Raw forwarding performance of the reflector.

3.1   Scalability Implications

As already mentioned in the Sec. 2 and 3, the scalability of the reflector-based communication environment can be further improved by creating networks of the reflectors connected via tunnels. The simplest model, which can also be used as the worst case estimate, is a complete graph in which each reflector communicates directly with all the remaining reflectors as shown in Fig. 4a. We call such model full mesh tunneling.

Let's assume a mesh of the reflectors in which each reflector has either n_r or $n_r - 1$ clients resulting in the most balanced population of reflectors with clients. It is possible to show that the number of inbound streams on each reflector is

$$\text{in} = n, \tag{1}$$

where n is the total number of clients. The number of outbound streams for reflector with n_r clients is

$$\text{out}_{n_r} = n_r(m + n - 2), \tag{2}$$

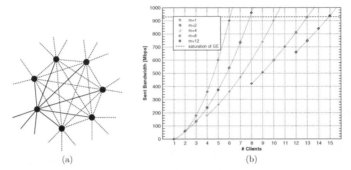

(a)   (b)

Fig. 4. (a) Full mesh reflector tunneling model. (b) Dependence of the number of 30 Mbps clients on the number of reflectors in the mesh.

where m is the total number of reflectors in the mesh. The ratio of outbound traffic for the reflectors with n_r and $n_r - 1$ clients is

$$\frac{\text{out}_{n_r-1}}{\text{out}_{n_r}} = \frac{n_r - 1}{n_r}. \tag{3}$$

Taking into account that $n_r = \lceil \frac{n}{m} \rceil$ and the number of streams on a single stand-alone reflector is $\text{out}_s = n(n-1)$, it is possible to show that the ratio between the limiting outbound traffic for the reflector participating in the mesh of reflectors out_{n_r} and a stand-alone reflector out_s is

$$\frac{\text{out}_{n_r}}{\text{out}_s} \approx \frac{m + n - 2}{m(n-1)}. \tag{4}$$

As illustrated in Fig. 4b for meshes with varying numbers of the reflectors, it is possible to increase the number of clients sending 30 Mbps to 15 when a mesh of 12 reflectors with gigabit network link is used.

4   End-User Applications

The reflector, capable of processing and distribution general purpose UDP packets, is able to support variety of end-user collaborative applications. Specific end-user requirements can be served by different versions, distinguished by specific processing capabilities (modules). Routinely, we use reflectors in connection with the `vic`, `rat`, and `wbd` tools from the MBone Tools package [11].

Different user groups used this environment during the last 3 years mostly for videoconferencing purposes, in very heterogeneous network conditions. While some groups enjoy rather homogeneous network environment, where all clients are connected through 100 Mbps network and the backbone runs on 1 Gbps and above speed, the reflector-based videoconferencing system is also regularly used in an environment where some clients are connected via cable TV. Even the network latency can be accommodated—we support a user group where most members are located in the Czech Republic (with clients both on high speed academic network and on a cable TV one) and one client connects via cable TV from Seattle (Washington, USA). Such an environment is rather hostile to the native MBone while reflector based data distribution works without problems. Another experimentally confirmed advantage is simultaneous support of different versions of MBone tools—again very problematic with native multicast.

As current high-quality videoconferencing tools tend to utilize high quality and high-bandwidth multimedia streams, we have also successfully tested the reflector with the DV over IP transmission tools from the DVTS project. We re-implemented the `xdwshow` tool (to overcome some problems encountered in its official implementation) and we have versions for PAL and NTSC formats under the Linux and FreeBSD operating systems. Our implementation uses robust thread architecture, where individual threads are used to input data from network, render them and display the resulting picture. This new implementation also support communication with reflector [12]. We plan to use this high quality video environment for teaching purposes, e.g. real time transmission from a neurological operation. Also, such high quality video streams can be used for the 3D video, using also the synchronization feature of the reflector.

5   Conclusion

The reflector is an active programmable network node providing all the necessary support for group communication in unicast networks. Our solution can simulate the multicast connectivity transparently, so the multicast clients can be used with ease, while keeping the advantages of unicast point-to-point communication lines. The reflectors function as multicast rendezvous points, allowing clients to connect and leave without any undesired influence on the rest of the group. The startup and shutdown of reflectors is a part of active network programming and as such it is fully user controlled. Users are also free to connect reflectors together in an ad hoc way and to specify behavior of each individual reflector, including possible security requirements and QoS parameters. A more general contribution is the method introducing new services into a network. The user empowered overlay network can be built a local scope (where needed), using only the features actually needed and within a limited time frame only. The network in this case is not overloaded by new protocols etc. and remains simple, robust, and fast.

While individual reflectors do not scale well and are able to support groups of tens of clients at most, their mesh is scalable enough to support a sufficient number of clients. Interesting direction of reflector development is the implementation of self-organizing and automatic discovery capabilities stemming from ideas of peer-to-peer networking. We will consider either the pure, hybrid or super-peer modes to define the best model for the reflectors self organization. The reflectors will be able to create overlay networks that can sustain even partial network disintegration without completely breaking overlay network.

As the reflectors can create the overlay networks, the reflector based solution does not depend on any specific functionality of the underlying network. All the advanced features are provided by higher user-controlled layers. Any group needing to collaborate can start its own reflector(s) and a unicast connectivity is the only required network capability from any client to the reflector. While the data routing and the data replication are automatically provided, user-specific services can be added as extensions (modules) to the reflector.

The future work will include control through grid service interfaces, as specified by an Open Grid Services Architecture framework [13]. This will enable easy integration with both next-generation collaborative Grid environments (e. g. AccessGrid version 2.x [14]) and with an optical network control plane, purposely built using the web/grid service approach. Thus the reflector will be able to cooperate easily with either the underlying network or other collaborative environment frameworks. The next direction in future work is concerned in development of user administration of reflector's data processing capabilities. Possible example of this administration is moderating data streams and creating communication environment for sub-group discussion inside the videoconferencing groups and fully moderated discussion like teaching in virtual classroom.

Acknowledgements. This research has been kindly supported by a research project "High Speed Research Network and its New Applications" (MŠM000000001) and "Optical Network of National Research and Its New Applications" (MŠM 6383917201). The authors would like to thank to Tomáš Rebok for helping with the implementation of performance evaluation tools.

References

1. Chin Jr., G., Myers, J., Hoyt, D.: Social networks in the virtual science laboratory. Communications of the ACM **45** (2002)
2. Psounis, K.: Active networks: Applications, security, safety and architectures. IEEE Communication Surveys (1999)
3. Andersen, D., Balakrishnan, H., Kaashoek, F., Morris, R.: Resilient overlay networks. In: 18th ACM Symp. on Operating Systems Princpiles (SOSP), Banff, Canada (2001)
4. Denemark, J., Holub, P., Hladká, E.: RAP – Reflector Administration Protocol. Technical Report 9/2003, CESNET (2003)
5. Rowstron, A., Druschel, P.: Pastry: Scalable, distributed object location and routing for large-scale peer-to-peer systems. In: IFIP/ACM International Conference on Distributed Systems Platforms (Middleware), Heidelberg, Germany (2001) 329–350
6. Rebok, T., Holub, P.: Synchronizing RTP Packet Reflector. Technical Report 7/2003, CESNET (2003)
7. Bouček, T.: Kryptografické zabezpečení videokonferencí. Master's thesis, Military Academy Brno (2002) Czech only.

8. Salvet, Z.: Enhanced UDP packet reflector for unfriendly environments. Technical Report 16/2001, CESNET (2001)
9. Internation Electrotechnical Commission: IEC 61834: Recording – Helical-scan digital video cassette recording system using 6,35 mm magnetic tape for consumer use (525-60, 625-50, 1125-60 and 1250-50 systems). (1998, 1999, 2001) Parts 1–10, http://www.iec.ch.
10. Ogawa, A., Kobayashi, K., Sugiura, K., Nakamura, O., Murai, J.: Design and implementation of DV based video over RTP. In: Packet Video Workshop 200. (2000) http://www.sfc.wide.ad.jp/DVTS/pv2000/index.html.
11. Hladká, E., Holub, P., Denemark, J.: Teleconferencing support for small groups. In: Terena Networking Conference '02, TERENA (2002) http://www.terena.nl/tnc2002/proceedings.html.
12. Hladká, E., Holub, P., Liška, M.: Modular communication reflector with dv transmission. In: VRS'04, PASNET (2004) Czech only.
13. Foster, I., Kesselman, C., Nick, J., Tuecke, S.: The Physiology of the Grid: An Open Grid Services Architecture for Distributed Systems Integration. Open Grid Service Infrastructure WG, Global Grid Forum (2002)
14. Childers, L., Disz, T., Olson, R., Papka, M.E., Stevens, R., Udeshi, T.: Access grid: Immersive group-to-group collaborative visualization. In: Proceedings of Immersive Projection Technology, Ames, Iowa (2000)

A Hybrid Overlay Topology for Wide Area Multicast Sessions

Rédouane Benaini, Karim Sbata, and Pierre Vincent

Institut National des Télécommunications, 91011 Evry Cedex, France
UMR 5157 Samovar. Département LOgiciels-Réseaux
{redouane.benaini,karim.sbata,pierre.vincent}@int-evry.fr

Abstract. MPNT (Multicast Proxies NeTwork) is an overlay architecture that was first conceived to provide multicast access to unicast-only users, like TutTelNet distant students. MPNT has been then improved by implementing multimedia access (mainly RTP, RTCP and SIP). MPNT original topology is spanning tree based. It is the most efficient topology in terms of switching delay (no routing) but the global delay variance can be important. To remedy this, we implemented an alternate hypercube-based topology. Indeed, hypercubes have several interesting properties - mainly compactness - that ensure a better scalability. Nevertheless, this logical hypercube optimization is valid provided that interconnecting links are comparable. To adapt the architecture to the network heterogeneity, we implemented thus a hybrid topology exploiting both spanning tree and hypercube advantages.

1   Introduction

The MPNT (Multicast Proxies NeTwork) platform [1] is an overlay architecture that was first conceived to provide multicast access to unicast-only users, like TutTelNet [2] distant students. Indeed, TutTelNet e-learning program aimed to be as accessible as possible. It needed a multicast support for ISP end-users, which are mostly unicast-only, and for LAN clients as well. For this purpose, we developed MPNT, an applicative broadcasting network that provides TCP, UDP and IP multicast. Afterwards, we improved our architecture by implementing multimedia access (mainly RTP, RTCP and SIP). P-STN access and multimedia applications have been then made possible. We used it since then for media-conferences, especially heterogeneous audio-conferences (P- STN + IP). MPNT original topology was spanning tree based. It is the most efficient topology in terms of switching delay (no routing) but the global delay variance can be important. To remedy this, we implemented an alternate hypercube-based topology [3]. Indeed, hypercubes have several interesting properties - mainly compactness - that ensure a better scalability. Nevertheless, this logical hypercube optimisation is valid provided that interconnecting links are comparable. To adapt the architecture to the network heterogeneity, we implemented thus a hybrid topology exploiting both spanning tree and hypercube advantages. The object of this paper is to expose this new topology. After a brief presentation

M. Freire et al. (Eds.): ECUMN 2004, LNCS 3262, pp. 377–385, 2004.

of the related works, we will give an overview of MPNT, through the brief description of its functionalities and topologies. We will deal afterwards with the presentation of the new hybrid topology and its benefits.

2 Related Works

Because of IP multicast limitations (hardware update, unreliability, etc.), a new approach for multicast communications appeared in the middle of the nineties, called applicative or overlay multicast technology. The purpose of this section is to present some of the existing overlay architectures. One of the first was RMTP [5]. It has been developed by Bell Labs around 1996. Its main innovation was the introduction of reliability. It is used for example for data transmission through RMFTP, a multicast reliable version of FTP. Another interesting solution is HyperCast [6], developed by the University of Virginia, which uses also a hypercube topology, but only for control transmissions. ScatterCast[7] and its reliable version, RMX[8], are also interesting, as they use a relay approach, in order to allow standard clients to join their sessions. Actually, there is a plethora of other interesting solutions. Among these, we can mention Narada[9], Almi[10], TBCP[11], Yoid[12] or SCRIBE[13].

3 MPNT Overview

This section aims to present briefly our architecture, through its functionnalities and initial topologies. Nevertheless, we suggest to report to [1] and [3] for more detailed information, especially for the algorithms.

3.1 How Does It Work?

MPNT architectures are composed of a set of relaying nodes (called MPNT proxies), interconnected as an overlay network (see 3.2 and 3.3). Each node provides access for a subset of the protocol stack given by figure 1. Providing an access means launching a listening server, with specific transport and application layer. Examples of accesses can be [RTP—UDP], [RTP—IP] or [SIP—TCP], where IP stands for IP multicast. Actually, according to the implemented stack, a node can provide up to 12 different types of access (4 application protocols and 3 transport protocols). Of course, some accesses are more natural than others (e.g. [RTP—UDP] vs. [RTP—TCP]) but all can be useful in some particular context. Most SIP phones, for example, use TCP as a spare solution when UDP flows are blocked. The last application protocol in the stack (called "none" in fig. 1) is actually an empty protocol. Launching a "none" server is equivalent to launch a raw transport server (no treatment is done on the received data). It is used by end-to-end applications that need only a generic carrier for their data. From end-users point of view, the MPNT network is seen through the application layer only. There are only four possible user groups (SIP, RTP, RTCP, none) instead of twelve. For example, SIP over TCP clients will communicate with SIP over UDP clients.

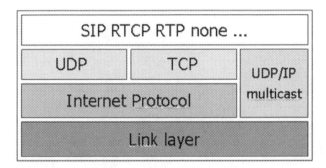

Fig. 1. MPNT protocol stack

3.2 Spanning Tree Topology

Interconnecting MPNT nodes has been introduced to ensure load scalability and provide more flexibility for conceiving architectures. To avoid routing cost, we first adopted a single-path interconnection model, based on a spanning tree topology. For each message, there is only one way to go from a node i to a node j. But avoiding routing does not mean no management at all. We need some mechanisms like interconnection tables update, loop avoidance and dynamic recovery.

Reservation algorithm. Actually, this algorithm is used for both spanning tree and hypercube topologies, and for both connection and disconnection processes. It is based on blocking the network(s) before any new (dis)connection process, unblocking it after its termination. This way, simultaneous connections are avoided. Let us assume that a node A from a network netA wants to create an interconnection with a network netB via a node B. First, A will perform an algorithm to check if netA is ready for a new interconnection. If the reservation process initiated by A is successful, A contacts B and asks it if the interconnection is possible (via ICX_REQUEST). Else, an error message is transmitted to the administrator. When B receives the interconnection request (ICX_REQUEST), it initiates the reservation algorithm in netB like A did in netA. If it is successful, a positive response is sent to A. Else, the response is negative and an error message is generated in A.

Building mechanism. For all topologies, each interconnection consists in creating two channels: one for management, and another for data. Both channels use TCP, for a better reliability. The first step (after reservation) of the mechanism concerns loop avoidance. When a node receives a request for performing loop avoidance, it has just to check if the request emitter exists or not in its interconnection table. If it exists already - it means that the emitter is already connected via another path - an error message is sent and the control channel is closed. Otherwise, the receiver authorizes the emitter to continue the connecting process (creation of the data channel and update of the tables). Concerning the tables, each node has got two: one containing connected nodes and their access path, and another one for available servers on the MPNT network. Both are updated in the previous algorithm. Another functionality of the control channel is to detect interconnection failures and remedy them. The mechanism used is detailed in the next section.

Dynamic recovery algorithm. The main drawback of spanning tree topologies is weak fault tolerance. When a node fails, the network is divided in several independent overlay subnets. To avoid this eventuality, we implemented a dynamic recovery algorithm that rebuilds the architecture after a failure. First, the failed node's neighbours broadcast the failure information (IP address of the failed node). When a node receives this message, it updates its interconnection table in order to remove the broken down node and its dependencies (i.e. the nodes that reach the current node through the failed one). The second step is the creation of a table containing all the peer nodes i.e. all the former neighbours of the failed node. The third step is the selection of the peer node that will recover the failure. The rule used to select the performing node is based on nodes load: the less interconnected node is chosen. In case of equality between nodes' loads, the lowest IP address is chosen. The final step is the connection of the selected node to the rest of the peer nodes. The meaning of connection here includes the connection itself (in fact, the double connection for control and data) and the interconnection tables updating process.

3.3   Hypercube Topology

The spanning tree topology has been chosen for its simplicity and low processing cost. But for a large number of nodes, it cannot ensure a low delay variance, necessary for real-time and synchronised applications. This is the reason why we implemented an alternative hypercube based topology. Indeed, whereas the maximum delay is proportional to the number of nodes (N) for spanning trees, it is limited to $E[\log 2(N)]+1$ for hypercubes, where $E[x]$ is the highest integer inferior to x.

Building mechanism. In our topology, each node of the hypercube corresponds to an MPNT proxy. We introduced converting tables that associate nodes IP addresses with Grey-encoded labels. To optimise the architecture and keep real distances between neighbours as short as possible, we classify IP addresses by subnets before labelling them. Thus, nodes belonging to the same network are necessarily neighbours in our topology. When a node is launched, the table is initialised with a single entry (the node itself). When a node A wants to connect to a node B, A sends its table to B. This latter checks first if there is no intersection between its table and the received one, and if no, concatenates the two tables and sends back the new table to A. Both A and B will broadcast the new table. When a node receives a new table, it rebuilds the hypercube and checks if it needs to change its interconnections. For avoiding simultaneous connections or disconnections at this stage, we impose that a node can connect and disconnect only to nodes with lower labels. This mechanism ensures a complete hypercube state awareness in each node. But problems can occur for simultaneous interconnection requests. In this case, some tables might be incomplete. We fixed this problem by introducing a reservation algorithm.

Broadcast algorithm. Whereas broadcast algorithm is trivial in the spanning tree topology, it is quite complex in this one. The algorithm we implemented is based on H. P. Katseff work[4] and is detailed in [3]. It uses a label-based transmission, in which every bit (called bi) indicates whether or not the message should be transmitted over the bound link (called Li). Note that this algorithm is used for both data and control transmission, and works for complete or incomplete hypercubes.

Dynamic recovery. The dynamic recovery process for this topology is a bit different from the previous one. Indeed, new links cannot be indifferent: they need to respect a hypercube topology. Thus, the recovery in this topology consists on rebuilding the hypercube without the failed node. Note that the hypercube is not rebuilt if the failed node has a label greater than all its neighbours labels. Actually, this occurs if and only if the failed node is the last node of the hypercube. In this case, we don't need to rebuild the hypercube but just to remove the failed node from the table.

3.4 Limitations

Both topologies have advantages and drawbacks. Concerning the spanning tree topology, it is the most efficient topology in terms of switching (no routing) but the global delay variance can be important. To remedy this, we implemented the alternate hypercube topology. Indeed, hypercubes have several interesting properties - mainly compactness - that ensure a better scalability. Nevertheless, this logical hypercube optimization is valid provided that interconnecting links are comparable. To adapt the architecture to the network heterogeneity, we defined thus a hybrid topology exploiting both spanning tree and hypercube advantages.

4 Hybrid Topology

4.1 Definition of the Model

The hybrid topology is a topology that combines spanning tree with hypercubes. Indeed, the resulting overlay networks are organized as spanning trees of

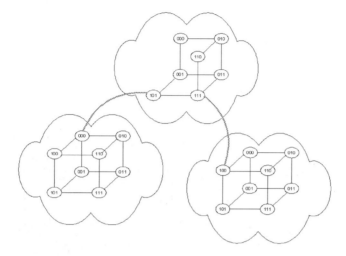

Fig. 2. An example of hybrid topology

hypercubes (cf. fig. 2). Hypercubes are defined as OAS (Overlay Autonomous Systems) and have just a partial knowledge (restricted to their nodes) of the overlay members. The notion of OAS is analog to that of AS (Autonomous System) of the Internet routing, the hypercube being an internal topology and the spanning tree an external one. Indeed, a spanning tree is used to interconnect all the hypercubes. Its membership knowledge is global. But unlike the topology described in section 3.2, this spanning tree does not connect proxies but hypercubes (OAS). Some changes were then necessary, in particular for the dynamic recovery algorithm (see section 4.3).

4.2   How Could It Be a Solution?

The hypercube topology was introduced to improve the scalability of the architecture. In fact, it does improve it provided that some hypothesis are respected. Indeed, building an efficient hypercube overlay network implies that links between nodes are comparable in terms of delay and bandwith. This is often the case while deploying the architecture over corporate or universitary networks. But over the Internet, this hypothesis is hardly respected. In some cases, the hypercube topology can even become more penalizing than the spanning tree (e.g. multiple overlay branches over a weak physical link). By introducing the hybrid topology, we tried to remedy this problem by dividing the underlying network in several OAS, inside of which the provided QoS should be homogeneous. Deploying hypercubes inside these OAS is then legitimate. To manage the heterogeneous links, a spanning tree will be used to connect all the OAS. This configuration should be efficient, as it takes advantage of both spanning tree and hypercube benefits. Indeed, the flexibilty and the simplicity of the spanning tree topology allow the administrator to choose the better links (or at least the less penalizing) to interconnect OAS and the hypercube topology optimizes the transmission inside them.

4.3   Implementation

The implementation of this new topology required several changes concerning the main processes (construction, data delivery and fault recovery). The only unchanged process concerns the reservation algorithm.

Hypercube identifier. To manage the duality of the topology, we introduced a new table (HCIDTable) in each node, containing all the identifiers (HCID) of the hypercubes to which it is connected. The HCID of a hypercube is generated by concating all its nodes' IP addresses (in a four-bytes format). Note that this identifier changes for each new hypercube connection or disconnection. In this case, a message containing both old and new identifiers is broadcasted to the whole architecture. Each proxy receiving this message will update its HCIDTable.

Building mechanism. This process is similar to those of the existing topologies. The only changes concern the management of the hypercube and spanning tree duality. When a node wants to connect to another through the spanning tree topology, it performs a process analog to that described in section 3.2.2, using HCIDs instead of IP addresses. In particular, HCIDTables are used for the loop avoidance algorithm. When a node wants to connect to another through the hypercube topology, it has to perform the algorithm described in section 3.3.1, the HCID loop avoidance process described above and the HCID updating process. This consists on generating a new HCID, updating its own table and broadcasting the new identifier. Note that two nodes of the architecture must not be directly connected through both hypercube and spanning tree topologies. This is why we perform the HCID loop avoidance for any connection, whatever its type.

Broadcast algorithm. In this new topology, there is two levels of broadcast: global and local (OAS-restricted). From a practical point of vue, global messages are transmitted over both topologies whereas local ones are limited to the hypercube topology. Local broadcast is useful for control and data. Indeed, there are some hypercube-specific control messages that does not need to - or even must not - be transmitted to other OAS. Concerning data delivery, local broadcast can be used for confidential or local interest information. Practically, rules for choosing local or global control broadcast are based on the type of message. They are specified in the software and cannot be changed. On the other hand, rules for data broadcast are specified by the administrator while creating a session (i.e. launching a new listening process). By default, new sessions are global.

Dynamic recovery. When a proxy fails, all its peer nodes perform a recovery algorithm. This algorithm depends on the topology used. If the failed proxy and the peer node were part of the same hypercube, the algorithm described in section 3.3.3 is performed, followed by the HCID updating process. If the peer proxy was connected to the failed one by a spanning tree link, a new algorithm is performed: the node extracts all the members of the failed node's OAS from its HCID and connects to another member of the OAS (arbitrarily the first extracted), performing also the HCID updating process. Note that interferences between hypercube and spanning tree recovery processes can occur. Mainly, an obsolete HCID can be transmitted over the spanning tree if the hypercube process was not ended. Nevertheless, this is not really detrimental: when the hypercube process will end, the correct HCID will be transmitted, updating the obsolete one.

4.4 Evaluation of the Model and Future Work

The hybrid version of MPNT is in its alpha phase. The deployment of several nodes over several networks (mainly the network of Institut National des Télécommunications, our institution, and the network of ENIC - Télécom Lille 1) is currently tested. This deployment is also being used to measure parameters like end-to-end delay and overall bandwith in real configurations, using different multicast applications (audio-conference and e-learning) as well as TrafGen [14], a traffic generator we developed.

In addition to real measurements, simulations are planned to compare the performance of the different topologies of MPNT while changing the scale of the architecture. According to the results obtained, we will define precise policies for choosing the most appropriate topology for each deployment.

5   Conclusion

The MPNT architecture implements two different topologies so far: a spanning tree topology, manually managed, and a hypercube-based one, completely automatic. To benefit from both topologies, we proposed to use them jointly, the hypercube for internal interconnections (between two nodes) and the spanning tree for external interconnections (between two OAS). This new topology - called hybrid - has been implemented and is currently in a testing phase. Measurements over small to average deployments and large-scale simulations are planned to compare it with other topologies and evaluate its scalability.

References

[1] K. Sbata, P. Vincent, R. Benaini, N. Naja, "MPNT: An Architecture for Mul-
     timedia Applications", IWCSE Workshop, part of 2nd IEEE ISSPIT (December
     2002 - Marrakesh/MAR), available at http://www-lor.int-evry.fr/~ sbata.
[2] JF. Colin, K. Sbata, P. Vincent, "TutTelCast: A Multipoint Based Architecture",
     5th IASTED International Conference CATE (May 2002 - Cancùn/MEX), avail-
     able at http://www-lor.int-evry.fr/~ sbata.
[3] K. Sbata, P. Vincent, R. Benaini, "Implementation of a Hypercube Based Topol-
     ogy in the MPNT Architecture", Proceeding of the 7th World Multiconference
     on Systemics, Cybernetics and Informatics SCI'03, Orlando (USA), July 2003,
     available at http://www-lor.int-evry.fr/~ sbata.
[4] H.P. Katseff, "Incomplete Hypercubes", IEEE Transactions on Computers, May
     1988, 604-608.
[5] S. Paul, K. K. Sabnani, J. C. Lin, S. Bhattacharyya, "Reliable Multicast Transport
     Protocol (RMTP)", IEEE Journal on Selected Areas in Communications, Vol. 15
     No. 3, April 1997, Pages 407-421.
[6] J. Liebeherr, Tyler K. Beam, "HyperCast: A Protocol for Maintaining Multicast
     Group Members in a Logical Hypercube Topology", Proc. First International
     Workshop on Networked Group Communication (NGC '99), July 1999.
[7] Y. Chawathe, "Scattercast: An Architecture for Internet Broadcast Distribution
     as an Infrastructure Service.", PhD thesis, University of California, Berkeley. De-
     cember 2000.
[8] Y. Chawathe, S. McCanne, E. Brewer, "RMX: Reliable Multicast in Heteroge-
     neous Environments", Proceedings of IEEE INFOCOM 2000, Tel Aviv, Israel,
     March 2000.
[9] Y-H. Chu, S. G. Rao, S. Seshan and H. Zhang, "A Case for End System Multicast",
     IEEE Journal on Selected Areas in Communication (JSAC), Special Issue on
     Networking Support for Multicast, October 2002.
[10] D. Pendarakis, S. Shi, D. Verma, M. Waldvogel, "ALMI: An Application Level
     Multicast Infrastructure", 3rd USENIX Symposium on Internet Technologies, San
     Francisco, March 2001
[11] L. Mathy, R. Canonico, D. Hutchison, "An Overlay Tree Building Control Proto-
     col", Proc. of 3rd Intl. COST264 Workshop on Networked Group Communication
     (NGC 2001), London, UK, November 2001.
[12] http://www.icir.org/yoid/
[13] M. Castro, P. Druschel, A-M. Kermarrec, A. Rowstron. "Scribe: A large-scale
     and decentralized application-level multicast infrastructure", IEEE Journal on
     Selected Areas in Communications, 2002.
[14] R. Benaini, K. Sbata, P. Vincent, "TrafGen : A Network Performance Analyzer",
     Proceedings of the 13th International Conference on Modelling Techniques and
     Tools for Computer Performance Evaluation, Chicago (USA), August 2003.

Fault/Attack Tolerant Recovery Mechanism Under SRLG Constraint in the Next Generation Optical VPN

Jin-Ho Hwang[1], Ju-Dong Shin[1], Mi-Ra Yun[1], Jeong-Nyeo Kim[2], Sang-Su Lee[2], and Sung-Un Kim[3],[*]

[1] Pukyong National University, 599-1 Daeyeon 3-Dong Nam-Gu,
Busan, 608-737, Korea
{jhhwang, jdshin, eggshape}@mail1.pknu.ac.kr
[2] Electronics and Telecommunications Research Institute,
161 Gajeong-Dong, Yuseong-Gu, Daejeon, 305-350, Korea
{jnkim, sangsu}@etri.re.kr
[3] Pukyong National University, 599-1 Daeyeon
3-Dong Nam-Gu, Busan, 608-737, Korea
kimsu@pknu.ac.kr

Abstract. A "Virtual Private Network (VPN) over Internet" has the benefits of being cost-effective and flexible. However, given the increasing demands for high bandwidth Internet and for reliable services in a "VPN over Internet," an IP/GMPLS over DWDM backbone network is regarded as a very favorable approach for the future "Optical VPN (OVPN)" due to the benefits of transparency and high data rate. Nevertheless, OVPN still has survivability issues such that a temporary fault can lose a large amount of data in seconds, moreover unauthorized physical attack can also be made on purpose to eavesdrop the network through physical components. Therefore fault/attack tolerant recovery mechanism that considers physical components is needed because optical network has vulnerabilities involved in the intrinsic characteristics, and these characteristics possibly menace reliable services in OVPN. Thus in this paper, with considering fault/attack in the next generation OVPN, we propose a recovery mechanism under shared risk link group (SRLG) constraint for network survivability by means of the classification of optical components and shared risk levels.

1 Introduction

As the Internet and optical network technology advances, the IP over DWDM has been envisioned as the most promising solution for the next generation optical Internet (NGOI). Especially, core transport networks in NGOI are currently in a transition period evolving from SONET/SDH-based time division multiplexed (TDM) networks utilizing a single wavelength to dense-wavelength

---

[*] Corresponding Author.

M. Freire et al. (Eds.): ECUMN 2004, LNCS 3262, pp. 386–396, 2004.
© Springer-Verlag Berlin Heidelberg 2004

division multiplexed (DWDM) networks with the multiple wavelengths strictly for fiber capacity expansion. On purpose to control for both optical and electronic networks, generalized multi-protocol label switching (GMPLS) has shown up and is currently under standardization at the Internet engineering task force (IETF)[1][2]. Therefore IP/GMPLS over DWDM network is emerging as a dominant technology for use in the next generation backbone network.

VPN is a private network that uses a public network (usually the Internet) to connect remote sites or users together. The primary advantages of "VPN over Internet" are cost-effectiveness and flexibility while coping with the exponential growth of Internet. However, the current disadvantages are the lack of sufficient quality of service (QoS) and provision of adequate transmission capacity for high bandwidth services. For resolving these problems, OVPN over the next generation optical Internet (NGOI)[3] has been suggested for supporting a variety of guaranteed high bandwidth-needed services in OVPN, but it still needs to provide optical QoS provisioning as described in[4], and the network survivability.

For the network survivability in OVPN, sequential procedures such as fault/attack detection, localization and recovery, are the most important issues because a short service disruption in DWDM networks carrying extremely high data rates causes loss of vast traffic volumes. In addition, the existing schemes for fault/attack management may no longer have access to the overhead bits that are otherwise used in legacy networks to transport supervisory information. Eventually, unlike in the case of the existing network, the new survivability mechanism considering intrinsic OVPN features is necessary to provide network survivability[5][6][7][8][9]. GMPLS can manage fault/attack in OVPN. It provides detection, localization, notification and recovery mechanism. When fault/attack or signal degradation is detected, the localization procedure gets started immediately by link management protocol (LMP) that runs between adjacent nodes. Thereafter, the notification procedure is started by resource reservation protocol (RSVP-TE), and determined recovery scheme transfers the traffic to a backup path.

Recently, a key feature of GMPLS is the backup path establishment that keeps physical-diversity (which is also called by physical-disjoint). It is also a dominant issue in OVPN backbone network. In OVPN, each link set up in one lightpath may cross one or more optical components, where the fault/attack of optical components may result in the potential failure of the link. A component here essentially presents any part or site involved in the integrity of the links and associated with a shared risk group (SRG) defined as resource groups having shared risk in common [10][11][12].

In this paper, we ramify SRG with considering the coverage of fault/attack and optical components in OVPN, and propose fault/attack tolerant recovery mechanism to guarantee the physical-diversity under SRLG constraint. Thereafter, we simulate and analyze the performance of the proposed recovery mechanism in the aspect of survivability[13][14].

The rest of this paper is organized as follows: section 2 describes OVPN structure and network survivability issues. In section 3, fault/attack classification for SRG is presented in the viewpoint of survivability, and SRG is defined in accordance with the coverage of fault/attack. In section 4, we illustrate the proposed recovery mechanism under SRLG constraint, and evaluate for the performance through simulation in section 5. Finally, some concluding remarks are made in Section 6.

2 OVPN Structure and Network Survivability Issues in OVPN

2.1 OVPN Structure

The suggested OVPN structure in figure 1 consists of the customer sites in the electric domain and the DWDM network in the optical domain. The external customer sites based on IP network aggregate (or de-segregate) IP packets at customer edge (CE) nodes and the internal OVPN backbone network composed of the provider edge (PE) nodes and the provider (P) core nodes forwards data traffic between the customer sites without electronic-optic-electronic (E-O-E) conversions[4].

An established lightpath between the CE1 and the CE2 may cross a number of intermediate P nodes interconnected by fiber segments, amplifiers and optional taps. The optical components that constitute a P node, in general, include an optical switch, a demultiplexer comprising of signal splitters and optical filters, and a multiplexer made up of signal combiners. A P node may also contain a transmitter array (Tx) and a receiver array (Rx) enabling local add/drop of the wavelengths.

In this structure, we can describe three management sections taking into consideration resource types (optical components) and the coverage of fault/attack effects.

- *Optical Channel Section(OCh)* : Channel management section for one lightpath established between CE nodes.
- *Optical Multiplexing Section(OMS)*: Link management section for one link between adjacent nodes. This includes Optical Amplifier Section (OAS) and Fiber Intrusion Section (FIS).
- *P(orPE) Node Section* : Node management section including demux, optical switch and mux that are divided and managed by sub management sections, i.e. Demultiplexing Section (DS), Switching Section (SS) and Multiplexing Section (MS).

2.2 Network Survivability Issues in OVPN

The ramification of network survivability in OVPN backbone network is depicted in figure 2. Fault survivability contains fault management for a sudden fault of

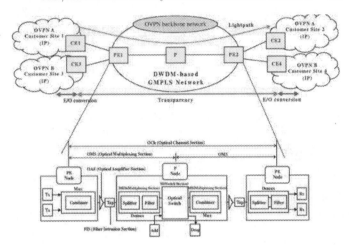

Fig. 1. OVPN Structure

optical components and signal degradation management. Also, attack survivability is divided into physical attack management and logical attack management depending on attack methods. Especially, physical attack in optical domain needs to be managed in optical layer because it causes signal degradation by maliciously using intrinsic characteristics of optical components[6][7][8][9]. However, logical attack is defined as an unauthorized person's network access on purpose to modify or to eavesdrop information, and has to be dealt by quantum-cryptography, but it is beyond the scope of this paper.

In order to manage fault/attack, the sequential mechanism is needed as follows: detect fault/attack as soon as possible (detection), separate fault/attack from normal traffic (localization), and notify fault/attack to network elements which are responsible for network management (notification) and recover traffic to avoid fault/attack (protection/restoration).

3   Fault/Attack Classification for SRG

3.1   Fault/Attack Classification

OVPN backbone network has many fault possibilities due to vulnerable characteristics of optical components used in DWDM network, so short and sporadic failures of network elements may cause a large amount of data loss. In fault survivability, the physical fault (or hard fault) on optical components has to be considered firstly. It causes failure in all optical channels that are going through

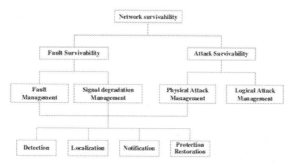

Fig. 2. Network Survivability in OVPN

a link or in a specified optical channel. The coverage of fault is specified depending on the optical components. Resource types and the coverage of fault are summarized in figure 3.

On the other hand, optical components such as optical fiber or erbium-doped fiber amplifier (EDFA) can be used by first attack point to cause signal degradation or to eavesdrop information. For example, gain competition attack causes signal degradation in optical channels that are going through a link by using intrinsic feature of EDFA as mentioned in[8]. With reference to the OVPN structure shown in figure 1, we categorize attack issues at two functional levels, and they are summarized in figure 3 [6][7].

• *Direct attack* : there are certain physical link elements with their own peculiar characteristics that are more likely to be exploited by an intruder as direct attack ports.

• *Indirect attack*: there are certain optical components (P or PE node) that are unlikely to be attacked directly either because a direct attack is too complicated to generate the desired effect or because the ports are not easily accessible to the potential intruders.

As the aspect of the above, in OVPN backbone network, a single fault or attack has various coverage of effect (OCh, OMS, node) depending on resource types or fault/attack types. Thus recovery mechanism needs to be done by making common risk group to avoid common fault/attack. In the following subsection, we define SRG according to the analysis of fault/attack coverage.

3.2   SRG Definition

A SRG is defined as a group of links or nodes that share a common risk component, whose fault/ attack can potentially cause the failure of all the links or nodes in the group. When the SRG is applied to the link resource, it is referred

Cate-gory	Resource type	Fault possibility	Attack possibility		SRG	
Path (OCh)	Transmitter	Laser or laser driver electronic problem	Signal Degradation with high power laser	Direct Attack	Channel	
		Pump laser temperature due to high current				
	Receiver	Out of range power or unacceptable input optical power	Unauthorized access to information			
Link (OMS)	Fiber (FIS)	Fiber damaging or cutting	Fiber cut or optical power reduction		Fiber	S R L G
			Tapping only or jamming only			
			Tapping & Jamming			
	Amplifier (OAS)	Amplifier optical path failure (due to fiber cutting)	Gain Competition due to local attack		Conduit	
		Passive component failure with in the amplifier	Gain Competition due to remote attack			
		Pump laser or Pump laser driver electronic problem	Crosstalk due to high power signal			
	Conduit	Conduit damaging or cutting	Conduit cut or optical power reduction			
Node (OXC)	Demux (DS)	Electronic driver failure at Demux or Optical fiber failure	Intentional crosstalk using high power signal	Indirect Attack		S R N G
		Out of range power or unacceptable input optical power				
	Switch (SS)	Electronic driver failure at switch, Miswrouting	Intentional crosstalk using high power signal			
		Input power is over/under threshold or out of range	Unauthorized access to information using crosstalk			
	Mux (MS)	Electronic driver failure at Mux or Optical fiber failure	Intentional crosstalk propagation from preceding devices			
		Out of range power or unacceptable input optical power				

Fig. 3. Fault/Attack classification for SRG

to SRLG. For example, all fiber links that go through a common conduit under the ground belong to the same SRLG, because the conduit is a shared risk component whose failure, such as a cut, may cause all fibers in the conduit to be broken. This SRLG is introduced in the GMPLS and can be identified by a SRLG identifier, which is typically a 32-bit integer. On the other side, the SRG is applied to the node, and it is referred to SRNG[15][16]. SRNG has to be controlled by a network manager, because it may affect the whole network survivability.

In this paper, in accordance with resource types and coverage of fault/attack effects, we suggest that the SRLG has three levels as follows:

• *SRLG in channel level* : sub-channels that are aggregated in one established channel (lightpath) have the same risk level. This SRLG information can be applied to routing constraint via multiple domains.
• *SRLG in fiber level* : a fiber that connects two nodes is composed of more than one optical channel, and these optical channels have the same risk level with failures in fiber level (such as FIS, OAS).
• *SRLG in conduit level* : a fiber group that connects different nodes can have physical structure bundled by a conduit. Thus fibers in conduit have the same risk level with failure in conduit level.

Figure 4 illustrates a simple example of the proposed SRLG concept. The upper plane is logical topology controlled by GMPLS and the lower plane is

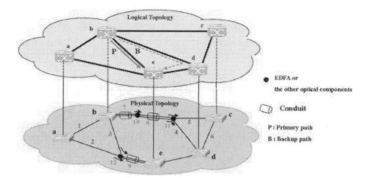

Fig. 4. SRLG example

the physical topology in which optical components (i.e., fiber, conduit, EDFA, etc.) are deployed. To provide physical-diversity of a primary path and a backup path, SRLG should be considered in the physical topology. Let us assume the physical topology consists of 5 nodes, 6 links, 3 conduits and 3 EDFAs. These uniquely have SRLG identifiers in fiber level and in conduit level. When there is a connection request between node b and node e, b-e (P in figure 4) can be a primary path by shortest path first (SPF) algorithm, and there could be two candidates for a backup path, b-a-e and b-d-e. If we only look at the logical topology, both the two backup paths can be allowed. However, If the backup path resolves b-a-e, the primary path and the backup path that go through the same conduit can fail at the same time by one single fault or physical attack in conduit 9, so the determined backup path is b-d-e (B in figure 4). Consequently, in order to make a network survivable against failure generated by fault or physical attack, the SRLG constraint should be imposed on the selection of backup path.

According to the description above, SRLG is the most important criteria concerning the constrained-based path computation of optical path. By applying the SRLG criteria to the constraint-based path computation, one can select a path taking into account diversity of physical resources and logical structure.

4   Recovery Mechanism Under SRLG Constraint

As discussed earlier, in order to achieve fault/attack tolerant recovery mechanism, the backup path has to keep physical-diversity with a primary path, because the failure of optical components as presented in figure 3 results in the potential failure of the link. Thus the recovery mechanism considering SRLG is essentially needed for network survivability in OVPN. On the other hand, a node failure is actually just a special case of SRNG where links are placed in groups based on whether or not they share a common node. A network manager should control it to avoid the service disruption of the whole network survivability.

In designing recovery mechanism under SRLG constraint, we raised three goals as follows: i) approximately 100% recovery capability and high-speed recovery. ii) almost the same blocking probability compared SPF algorithm with SRLG constraint with SPF algorithm without SRLG constraint. iii) less computational complexity for a backup path.

In order to achieve our goals, the recovery mechanism needs 1:1 path protection, SPF algorithm, SRLG constraint and First-Fit algorithm as wavelength assignment scheme.

The notations used in this paper are as follows:

- $G(N, L)$: The given network, where N is the set of nodes and L is the set of links.
- M: The set of source-destination connection request pairs.
- C_{ab}^{l} : The link cost between link pair (a, b) where (a, b)∈(s, d)/M.
- $srlg_{ab}^{l}$: The set of SRLG IDs in a link pair (a, b) where (a, b)∈(s, d)/M.
- $srlg_{sd}^{p/ab}$: The set of SRLG IDs used in a primary path (s, d) where (s, d)∈M.
- $R(l)$: The number of currently available wavelengths on a link l where l∈L.
- N_c : The total number of connection requests.
- N_f : The total number of failed connections.
- N_s : The total number of successfully established connections.
- N_d : The total number of service disruptions, when a primary path is failed, and there is no backup resource available, then the service on this primary path will be disrupted. If the failed path is a backup path in the running state, the service on this path will also be disrupted.

The procedure, how to find a primary path and a backup path, is described as follows:

STEP 1: Correlate the network resources and compute C_{ab}^{l} for all (a, b) included in G(N,L).

STEP 2: Wait for a request between a (s, d) pair as the current demand where (s, d)∈M.

  (a) If it is a connection request, go to STEP 3.

  (b) If it is a connection release request, go to SETP 7.

STEP 3: Route the request for a primary path between node s and node d selected by SPF algorithm.

SETP 4: SRLG identifiers are correlated between end nodes. Update $srlg_{sd}^{p/ab}$.

SETP 5: Update link cost : $C_{ab}^{l}:[(srlg_{ab}^{l} \in srlg_{sd}^{p/ab}) \cup (R(l)=0)]$.

STEP 6: Reserve the resource for the request as a backup path between node s and node d selected by SPF algorithm, go to STEP 8.

SETP 7: Release the primary path and the backup path pair (s, d).

STEP 8: Update the network resource states, go to STEP 1.

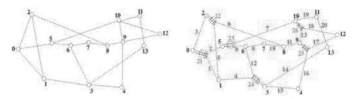

Fig. 5. NSFnet logical topology vs. physical topology considering arbitrary SRLG identifiers

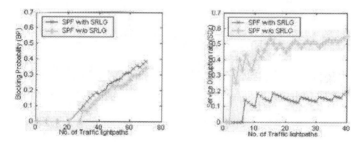

Fig. 6. Blocking probability vs. service disruption ratio as a function of number of traffic lightpaths

The STEP 5 is the most important step because it updates the cost of all links depending on the condition, $C_{ab}^l : [(srlg_{ab}^l \in srlg_{sd}^{p/ab}) \cup (R(l)=0)]$ which means whether or not the backup path has the same risk in common with the primary path and there are available wavelengths in the link. The recovery mechanism provides very fast and simple to satisfy the goals and is evaluated by blocking probability and service disruption ratio (SD_r) as the performance evaluation metrics, and these are defined as $BP = N_f/N_c$ and $SD_r = N_d/N_s$, respectively. The metrics show performance evaluation results in the viewpoint of survivability in the next section.

5   Numerical Results

In this section, simulation is carried out to evaluate the performance of the recovery mechanism described in section 4. To prove the efficiency, we analyze the results of blocking probability and service disruption ratio with or without SRLG constraint.

The topology used in simulation is NSFnet and we allocate arbitrary conduit level in the physical topology, which is composed of fiber groups. The numbers on a fiber or on a conduit are unique IDs for fiber level SRLG and conduit level SRLG.

The assumptions for the simulation are as follows: i) the physical topology consists of 14 nodes, 20 links and 6 conduits as shown in Figure 5. ii) links are bi-directional, each link has two fibers to different directions and the number of wavelengths per a fiber is 8. iii) all nodes have wavelength converters. iv) the topology is static and is not reconfigured during the simulation. v) connection requests arrive in sequence.

The numerical results of blocking probability and service disruption ratio are plotted as a function of the number of traffic lightpaths in figure 6.

In the case of the blocking probability, SPF with SRLG is slightly higher than SPF w/o SRLG. However, the service disruption ratio guarantees that the recovery mechanism recovers the failed traffic approximately 85% with failed backup paths for a single conduit level failure. Thus, the results of the recovery mechanism confirm better performance in the viewpoint of survivability. However, there is a tradeoff between blocking probability and service disruption ratio as considering SRLG constraint in physical topology.

Acknowledgment. This work was supported by grant No.(R01-2003-000-10526-0) from Korea Science and Engineering Foundation.

References

1. E. Mannie et al.: Generalized Multi-Protocol Label Switching (GMPLS) Architecture, draft-ietf-ccamp-gmpls-archtecture-07.txt, Internet Draft, Work in progress, May 2003.
2. A. Banerjee, et al.: Generalized multiprotocol label switching: an overview of signaling enhancements and recovery techniques, IEEE Commun. Mag., vol.39, no.7, pp.144-151, Jan. 2001.
3. Hamid Ould-Brahim et al.: Service Requirements for Optical Virtual Private Networks, draft- ouldbrahim-ppvpn-ovpn-requirements-01.txt, Internet Draft, Work in progress, July 2003.
4. Mi-Ra Yoon et al.: Optical LSP Establishment and a QoS Maintenance Scheme Based on Differentiated Optical QoS Classes in OVPNs, Photonic Network Commun., vol.7, no.2, pp.161-178. March 2004.
5. Jing Zhang et al.: A Review of Fault Management in WDM Mesh Networks: Basic Concepts and Research Challenges, IEEE Network, vol.18, no.2, pp.41-48, March/April 2004.
6. Sung-un Kim and David H. Su: Modeling Attack Problems and Protection Schemes in All-Optical Transport Networks, Optical Network Magazine, vol.3, no.4, pp.61-72, July/Aug. 2002.
7. Sung-un Kim and David H. Su: A Framework for Managing Faults and Attacks in All-Optical Transport Networks, DISCEX 2001, June 2001.
8. Muriel Medard et al.: Security Issues in All-Optical Networks, IEEE Networks, vol.11, no.3, pp.42-48, May/Jun 1997.

9. M. Medard, R. Chinn, Saengudomlert: Attack Detection in All-Optical Networks, Technical Digest of Optical Fiber Conference (OFC), 1998.
10. Panagiotis Sebos et al.: Auto-discovery of Shared Risk Link Groups, Optical Fiber Communication Conference, 2001.
11. D. Papadimitriou et al.: Inference of Shared Risk Link Groups, draft-many-inference-srlg-02.txt, Internet Draft, Nov. 2001.
12. Eiji Oki et al.: A disjoint path selection scheme with Shared Risk Link Groups in GMPLS Networks, IEEE Communications letters, vol.6, no. 9, pp.406-408, Sep. 2002.
13. Yun Wang et al.: Dynamic Survivability in WDM Mesh Networks under Dynamic Traffic, Photonic Network Commun., vol.6, no.1, pp.5-24, July 2003.
14. Guido Maier, Achille Pattavina, et al.: Optical Network Survivability: Protection Techniques in the WDM Layer, Photonic Network Communications, vol.4, no.3/4, pp. 251-269, July/Dec. 2002.
15. Sebos, P. et al.: Effectiveness of shared risk link group auto-discovery in optical networks, Optical Fiber Communication Conference and Exhibit, pp.493-495, 2002.
16. Haibo Wen et al.: Dynamic RWA Algorithms under Shared-Risk-Link-Group constraints, IEEE 2002 International Conference on, vol. 1, pp.871-875, July 2002.

A Flow-Through Workflow Control Scheme for BGP/MPLS VPN Service Provision*

Daniel Won-Kyu Hong[1] and Choong Seon Hong[2]

[1] Operations Support System Lab., KT
62-1 Hwaam-Dong Yuseong-Gu, Daejeon 305-718 KOREA
wkhong@kt.co.kr
[2] School of Electronics and Information, Kyung Hee Univerity
1 Seocheon Giheung Yongin, Gyeonggi 449-701 KOREA
cshong@khu.ac.kr

Abstract. In a competitive telecommunications marketplace, operators must provide customers with rapid access to both new and traditional service offerings. Market share is developed through greater customer satisfaction and by reducing dissatisfaction. To maximize profitability, an operator must reduce the costs of delivering new services while building revenues from existing services. This paper proposes a workflow-based service delivery architecture for telecommunications services. We describe an business process management in telecommunications and propose the workflow patterns that are commonly applicable for the workflow management of the proposed telecommunications architecture. In addition, this paper proposes an XML-based workflow control scheme incorporating the workflow patterns for providing telecommunications services. Finally, we describe a scenario for the delivery of the BGP/MPLS VPN service adopting the proposed workflow-based service delivery architecture.

1 Introduction

In the 1990s, workflow was often used as part of business process reengineering exercises to automate reengineered business processes. The emphasis was on technology, i.e. applications and systems, with less thought given towards human interactions within the process, and as a result, workflow developed a poor reputation. However, because of its ability to model and monitor business processes in real time and to easily change those processes in response to volatile market trends and technology, interest is again growing in business process management [1,2,4,5].

This has been a particularly difficult time for telecommunications businesses in most countries, with the saturated wired services, the depressed markets and the recession. On the other hand, by opening the door to competition, most telecommunications carriers need to act rapidly to differentiate themselves from

---

* This research was supported by ITRC Project of MIC.

M. Freire et al. (Eds.): ECUMN 2004, LNCS 3262, pp. 397–406, 2004.

others. Modern carriers face increasingly growing market challenges and must be able to fulfill even the most sophisticated customer demands. As a result, they must continuously invest in IT and telecommunications infrastructure and solutions, which help carriers to survive and win in the global market. However, networks and applications are becoming more and more complicated, so that keeping them effectively and efficiently operating requires additional effort.

Therefore, most telecommunications carriers should move quickly to set key technology solutions in place to streamline their service delivery operations. One of the breakthroughs is the automation of carriers' service delivery process, in whole or in part, during which documents, information or tasks are passed from one participant to another for action, according to a set of procedural rules.

2   The Business Process Management in Telecommunications

The integration of BSS and OSS is ultimately about business process support. There are many functional management components in the telecommunications architecture. These components are integrated with each other with the aim of supporting some overall set of defined business processes that together comprise the business model of the service provider [3]. An example of business process management in telecommunications is the Enterprise Application Integration (EAI) solution, which is a kind of middleware that integrates the various telecommunications components without any dependency on their technology and infrastructure to achieve the business goal.

Sales & customer care, billing, inventory management, activation and workflow management components are key ingredients in achieving the business goal of telecommunications service delivery. The sales & customer care component enables and supports all other components of the integrated architecture [3].

The sales & customer care component can be considered as a form of marketing - a happy customer will inform five people: an unhappy customer may inform ten. Unsatisfied customers will then go elsewhere and consequently, you lose a customer to your competitors and in turn, they may drag others with them. Create a customer relations policy instead of just making it up as you go along. Your policy is a guarantee to your customers that you are dedicated to achieving customer satisfaction.

The inventory management component is seen today as a service provider's resource platform rather than just a static list of network assets. Modern realtime inventory integrated with other OSS/BSS can deliver significant improvements in operational effectiveness and efficiency [3]. With in-depth understanding of network infrastructure resources and services, operators can reduce equipment costs and improve operating margins. The inventory management component can be composed of the following functions: network modeling & planning, inventory reconciliation, data migration service, inventory management application maintenance, and business process consulting & package implementation.

The business process management in telecommunications provides a full range of information management services, freeing communications providers to focus on revenue generation and customer care. The billing component can produce the bills in their final format, print them and mail them to our customers. Or we can provide them in electronic format for final processing by another billing fulfillment company.

The activation component provides various kinds of APIs that allow the development and integration of telecom management components, taking into account the device-specific and service-specific characteristics. From a single request, the activation component is able to provide multiple services across different network technologies. This greatly reduces the cost of service delivery while increasing the accuracy of deployed services. Rapid, fault-free deployment of new services is essential in maintaining your competitive advantage. The flexible service configuration of the activation component greatly reduces the time to market new services and simplifies the modifications to existing services. On the other hand, a fully automated service provision and activation platform is essential for increased customer satisfaction and generating maximum revenue from your network. An activation platform must be able to handle high volumes of transactions and provide flow-through activation for multiple services.

Enterprise Application Integration (EAI) has been a sorely misunderstood and misrepresented term. In the past, EAI applications have concentrated on middleware solutions aimed at connecting disparate applications together. Now, businesses are realizing that technical solutions alone cannot help us tame the legacy dragon. EAI has been and continues to be a technology driven by a real business need - to make effective use of existing and future data and application assets. It has been estimated that organizations spend 20-40% of their technology efforts on these types of integration tasks. By reducing this amount by just half, organizations can save millions.

Faced with limited resources, you are interested in rapidly planning for what must be done in order to have your EAI initiatives succeed. Data engineering principles enable us to formulate an approach to these types of challenges using structured techniques - where the form of the problem can be used to guide the form of the solution. The resulting framework-based solution provides a system of ideas for guiding the analyses: a means of organizing the project data and metadata: a data integration priorities decision-making framework: and a means of assessing progress toward project goals. Every XML-based initiative that begins with metadata recovery also begins with a clear understanding of the information content of the subject data. Thus, every time you wrap some of your data in XML, it provides an opportunity to contribute to overall EAI efforts.

Figure 1 shows a service delivery process with mutual interaction among the integrated telecom components using the EAI embedding workflow engine.

The process of Provisioning and Service Activation involves the acceptance of service orders from billing and customer care systems or separate order management systems. These service orders are translated into provisioning requests

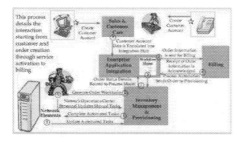

Fig. 1. Order to cash process

that need to be performed on one or more network elements. Typically, an operator will choose the best of the breed applications and network elements to suit its operations, and it is therefore not unusual that the service requests from the billing and customer care system are in a completely different format from that which is understood by the different network elements. In addition, there are situations where features supported in the billing and customer care system are not supported by the network element(s) or vice versa. Therefore, a service provisioning system needs to be flexible enough to handle these different scenarios.

3   The Workflow-Based Service Delivery Scheme

Workflow management consists of the automation of business procedures or "workflows" during which documents, information or tasks are passed from one participant to another in a way that is governed by rules or procedures [6,7]. Workflow software products, like other software technologies, have evolved from diverse origins. While some offerings have been developed as pure workflow software, many have evolved from image management systems, document management systems, relational or object database systems, and electronic mail systems.

Vendors who have developed pure workflow offerings have invented terms and interfaces, while vendors who have evolved products from other technologies have often adapted terminology and interfaces. Each approach offers a variety of strengths from which a user can choose. Adding a standard-based approach allows a user to combine these strengths in one infrastructure.

3.1   Workflow Patterns

There can be five basic workflow patterns [4,6] for the realization of workflow-based telecommunications service delivery.

Sequence is the most basic workflow pattern that is required when there is a dependency between two adjacent tasks. So, one task cannot be started before the adjacent task is finished. Sequence workflow pattern can be represented as , where '—' represents the sequence. Split is required when two or more tasks need to be executed in parallel. Split workflow pattern can be represented as , where ' ' represents the concurrent processing. Merge is required when a task can be started only when two or more parallel tasks are completed. Merge workflow pattern can be represented as , where ' 'represents the merge point. Cycle is required when a task has to wait until a number of tasks are finished. Cycle workflow pattern can be represented as , where ' ' represents the recurrent point. Branch is required when the next task is determined according to the status or condition of a task. Brach workflow pattern can be represented as , where 'if' represents the condition. We define a workflow pattern schema that can fully represent the above workflow patterns.

3.2   Workflow Control Scheme

In order to provide telecommunications services, we need a well-defined workflow control scheme besides the previously defined workflow pattern. The workflow control scheme is basically based on the workflow patterns. The service provision workflow can be described with the various combinations of workflow patterns.

Our workflow control scheme consists of two major modules of Workflow Control Module (WCM) and Work Process (WP), as shown in Figure 4. WP represents the application process and takes on the role of specific activities such as order receipt, order validation, device activation, order completion, etc. On the other hand, WCM takes on the roles of service order distribution and workflow administration and monitoring.

In WCM, the workflow is designed by the service designer via GUI interface using the Workflow Administration and Monitoring Tool (WAMT). The designed workflow is maintained in the Workflow Information Base (WIB) in eXtensiable Markup Language (XML). In addition, the service designer can dynamically

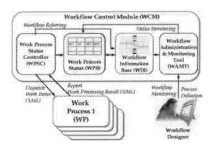

Fig. 2. Workflow control model

change the business process by modifying the business workflow with the graphical interface. The WAMT supports on-line workflow status monitoring. With the WAMT, the network operator can monitor the detail work processing status and can estimate the completion time of a certain workflow.

The Work Process Status Controller (WPSC) refers the WIB in order to decide on the current work process (WPcurrent) and the next work process (WPnext) to be executed and maintains the status of the WPcurrent and WPnext in the Work Process Status (WPS). In addition, the WPSC distributes WPcurrent to the appropriate Work Processes (WPs) that are distributed and executed in different machines. Each WP executes its own role in the delivery of the telecommunications services according to the activation request from the WPSC and reports the processing result to the WPSC via XML. On the other hand, the WPSC receives the processing result from the WP, and it changes the status of WPcurrent and moves the WPnext to WPcurrent.

4   Provisioning BGP/MPLS VPN Service

The IP virtual private network (VPN) feature for Multi-protocol Label Switching (MPLS) [9,10,11] allows a network service provider to deploy scalable IPv4 Layer 3 VPN backbone services [8]. IP VPN is the foundation used by companies for deploying or administering value-added services, including applications and data-hosting network commerce, and telephony services to business customers.

In this paper, we explore the issues surrounding the provision of MPLS VPNs, in particular, BGP/MPLS VPN known as MPLS-based Layer 3 VPNs or RFC 2547bis VPNs [8,9]. One of the services being investigated by many ISPs is that of BGP/MPLS VPNs. BGP/MPLS VPNs can form the center of an offering that provides substantial value to the customer in the form of increased simplicity, greater flexibility and outsourced routing, while representing a significant revenue stream for the ISP [8].

The BGP/MPLS approach to VPNs defines three roles for routers, which are as follows [8]:

– Customer Edge (CE) routers, which are associated with customer sites and usually managed by the customer.
– Provider Edge (PE) routers, which serve as the customer's entry and exit points for the VPN and are managed by the provider. Most of the VPN functionalities of the BGP/MPLS solution are concentrated in the PE routers.
– Provider (P) routers, which form the core of the provider network and are primarily concerned with forwarding VPN traffic that has been placed in MPLS frames by CE and PE routers. P routers are managed by the provider.

A BGP/MPLS VPN is formed when the provider associates the individual routing tables in the PE routers with MPLS Label-Switched Paths (LSPs) leading from that PE router to other PE routers that provide service to other CE routers on the same VPN [8]. A single PE router may serve many different

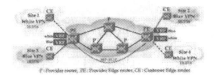

Fig. 3. An BGP/MPLS VPN model

CE routers and VPNs. The reference model for BGP/MPLS VPN is shown in Figure 3.

PE routers use an extended form of the Border Gateway Protocol version 4 (BGP4) routing protocol to share routing information and to populate the routing tables belonging to each VPN. The P routers, therefore, do not take part in the IP routing of customer traffic, but rather perform forwarding for the MPLS LSPs carrying that traffic. P routers do use a standard Interior Gateway Protocol (IGP) to maintain provider routing tables [9].

There are two broad configuration activities: one is for network configuration and the other is for VPN service configuration, as shown in Figure 4. The Network Configuration Activities (NCA) is for activating the signaling protocols for the management of LSPs in the MPLS network and the MP-BGP session configuration for the delivery of VPN information over the MPLS network. The Service Configuration Activities (SCA) is for creating VRF containing Route Distinguisher (RD)[9] and Route Target (RD) [9] and for activating the necessary routing protocol between CE and PE. The routing protocols between CE and PE can be OSPF, IS-IS, RIP and static.

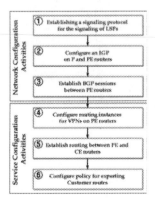

Fig. 4. Workflow for BGP/MPLS VPN configuration in terms of device activation

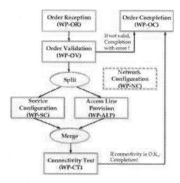

Fig. 5. A workflow for providing BGP/MPLS VPN service

Because the NCA can be a part of network provisioning and the SCA can be a part of service provisioning, this paper focuses on the SCA.

In this section, we describe the IP-VPN service delivery scheme using the workflow control scheme described in the previous section. In order to provide the Provider-Provided Virtual Private Network (PPVPN) using the proposed workflow control scheme, we define the various Work Processes (WPs) as follows:

Order Reception (WP-OR) - The WP-OR receives the service delivery order from the Customer Care and Billing System (CCBS) and notifies the CCBS of the completion of service delivery.

Order Validation (WP-OV) - The WP-OV validates the received orders from the CCBS. If the received order is not valid, WP-OV issues a completion notification with error to the CCBS.

Access Line Provision (WP-ALP) - The WP-ALP provides the facility for access lines between CE and PE. The access lines between CE and PE can be xDSL, ATM, Frame Relay, Ethernet, dedicated leased line and others.

Service Configuration (WP-SC) - The WP-SC activates the network elements, PE routers, for configuring BGP/MPLS VPN. In the WP-SC, it takes on the role of Service Configuration Activities (SCA), as defined in Figure 6.

Network Configuration (WP-NC) - The WP-NC takes on the role of network planning, as described in Figure 6. It does not participate in the service delivery process.

Connectivity Test (WP-CT) - The WP-CT provides the test functions for network operators to test the end-to-end connectivity or accessibility among the configured VPN sites.

Order Completion (WP-OC) - The WP-OC notifies the CCBS of the service delivery result, whether it is successful or not. In the WP-OC, the customer master is created.

There is a concurrent workflow. The WP-SC and WP-ALP are processed in current. The WP-CT is processed after the WP-SC and WP-ALP are finished.

5   Conclusion

This paper proposed an order management system and infrastructure to encompass and support all functions involved in the telecommunications service delivery process. We defined the generic workflow patterns that can simply manage the complicated and various telecommunications service orders and designed the workflow-based service delivery model for telecommunications services.

We also proposed the workflow patterns that can be applicable for telecommunications service delivery. This workflow patterns automate the entire service delivery process. Our workflow-based service delivery architecture is a highly scalable one adopting the web service. This architecture allows network service providers to manage their service orders. Our workflow-based service delivery architecture allows network service providers to simplify and accelerate the entire telecommunications service workflow.

In addition, we described a case study that applied the workflow-based service delivery architecture to the provision of MPLS provider-provided virtual private network (PPVPN).

With this workflow-based service delivery architecture, we can achieve the following:

- Differentiate the carrier as a provider of exceptional customer service, aggressively meeting the demand for new telecommunications services
- Increase service visibility to the customer, providing up-to-the-minute information on the status of orders and services.
- Provide customers with faster telecommunications services
- Expand markets and increase order-processing volumes without a simultaneous increase in human resource
- Increase productivity and reduce the number of personnel involved in the service delivery process

With the proposed workflow-based service delivery architecture, we can achieve low operating expenses while supporting the carrier's growth strategy.

Our further studies include solutions that enable customers to manage their own VPNs via web-based self-care, enable better tracking and feedback of real-time SLA conditions and rebates, and rapid deployment of additional products and services.

References

1. Carol Prior, Workflow and Process Management, Maestro BPE Pty, Austria, 2003.
2. WMFC, Workflow Management Coalition Terminology & Glossary, Document Number WFMC-TC-1100, Issue 3.0, February 1999.

3. TMForum, New Generation Operational Support Systems (NGOSS) - Architecture Overview, Member Evaluation Version 1.1, October 2000.
4. W.M.P. van der Aalst, A.H.M. ter Hofstede, B. Kiepuszewski, and A.P. Barros. "Advanced Workflow Patterns," 7th International Conference on Cooperative Information Systems (CoopIS 2000), volume 1901 of Lecture Notes in Computer Science, pages 18-29. Springer-Verlag, Berlin, 2000.
5. M. Dumas and A. ter Hofstede, "UML Activity Diagrams as a Workflow Specification Language," In Proc. of the International Conference on the Unified Modeling Language (UML), Toronto, Canada, October 2001. Springer Verlag.
6. B. Kiepuszewski, A.H.M. ter Hofstede, and W.M.P. van der Aalst, Fundamentals of Control Flow in Workflows, Acta Informatica, 39(3), pages 143-209, 2003.
7. Alfred Fent and Burkhard Freitag, "ULTRAflow - A Lightweight Workflow Management System," Proceedings of the International Workshop on Functional and Logic Programming (WFLP2001), Kiel, Germany, September 2001.
8. Matt Kolon, MPLS VPN Provisioning, Part Number: 552005-002, Juniper Networks, Inc., 2001.
9. Rosen and Rekhter, BGP/MPLS VPNs, IETF RFC 2547, March 1999.
10. Bates, Chandara, Katz, and Rekhter, Multiprotocol Extensions for BGP4, IETF RFC 2858, June 2000.
11. E. Rosen, A. Viswanathan, and R. Callon, Multiprotocol Label Switching Architecture, IETF RFC3031, 2001

One Round Identity-Based Authenticated Conference Key Agreement Protocol

Jeung-Seop Kim[1], Hyo-Chul Kim[2], Kyeoung-Ju Ha[3], and Kee-Young Yoo[1]

[1] Computer Engineering Department, E9-508, Kyungpook National University, 1370,
SanGyuk-Dong, Puk-gu, Tae-Gu, Republic of Korea
dambi@dittotec.com, yook@knu.ac.kr
[2] Computer Information Major, Keimyung College, 700, SinDang-Dong, DalSeo-gu,
Tae-Gu, Republic of Korea
khc@km-c.ac.kr
[3] Faculty of Multimedia, Daegu Haany University, 290, YuGok-Dong, Gyeongsan-si, Gyeong-
SangBukDo, Republic of Korea
kjha@dhu.ac.kr

Abstract. We propose an identity-based authenticated conference key agreement protocol for multi-party. Our protocol is a protocol meeting a lower
bound of only one round for multi-party contributory key agreement protocol.
The security attributes of our protocol are explained using the security model,
and the computational overheads are analyzed as well.

1 Introduction

Key agreement protocol is one of the basic cryptographic protocols. Most published
key agreement protocols are based on Diffie-Hellman key agreement protocol, which
enables two parties to establish a session key[1]. The basic Diffie-Hellman key
agreement protocol suffers from the man-in-the-middle attack because it does not
authenticate the two communication parties. This problem is solved using a digital
signature scheme, but this solution has another problem that the message lengths are
much greater than in the basic protocol[2,3,4,5].

The conference key is the secret key that three or more parties share. It requires multiple rounds of communications in order to share a secret key. But, a recent protocol
proposed by Joux for three parties is a one round authenticated tripartite protocol,
which requires a special setting[6]. And, a protocol by Boyd is a one round conference protocol, but does not provide forward secrecy[7]. In conference key agreement
protocol, the protocol round is one of important factors. One round means that all the
messages can be sent in parallel and are independent messages during the protocol.

Just like the basic Diffie-Hellman key agreement protocol, Joux's protocol is vulnerable to man-in-the-middle attack because it does not authenticate the three communicating parties. This problem is solved using certificates of the three parties, which are
issued by a certificate authority(CA), to bind a party's identity with his static keys[8].

M. Freire et al. (Eds.): ECUMN 2004, LNCS 3262, pp. 407–416, 2004.
© Springer-Verlag Berlin Heidelberg 2004

This solution requires a large amount of computing time and storage, because the entities must verify the certificate of the user before using the public key of a user. In 1984, Shamir asked for identity-based encryption and signature schemes to simplify key management in certificate-based public key infrastructure[9]. The basic idea of identity-based cryptosystem is that the identity information of a user functions as his public key. Recently, many identity-based cryptographic schemes using bilinear pairing have been proposed. There are Boneh-Franklin's identity-based encryption scheme, Smart's identity-based authenticated key agreement protocol, Zhang's ID-based one round authenticated tripartite key agreement protocol[10,11,12].

In 1998, Becker and Wille proposed several bounds on multi-party key agreement protocols[13]. The number of the rounds is the one of the bounds. No Diffie-Hellman generalization can meet this bound. But, in 2003, Boyd described a protocol, which meets the bound of Becker and Wille under the assumption that standard secure cryptographic primitives exist for encryption and signature, and using ideal hash functions (random oracles)[7]. The Boyd's protocol is simple and very efficient in comparison with previously published conference key agreement protocols. However, the Colin's protocol does not provide forward secrecy.

In this paper, we propose an identity-based authenticated conference key agreement protocol. We present a protocol using the identity information of a user functions as his public key, and meeting a lower bound of only one round for multi-party contributory key agreement protocol. In addition we show our protocol providing the forward secrecy and present the security attributes of our protocol and analyze the computational overheads.

The rest of the paper is organized as follows: The next section briefly explains the technical backgrounds about the bilinear pairing, the Diffie-Hellman problem and the communications model. Section 3 gives a detailed description of our protocol. In section 4, the security attributes and the analysis of the computational overheads are presented. Section 5 concludes this paper.

2 Technical Backgrounds

2.1 Bilinear Pairing

In this section, we shall briefly describe the properties of the bilinear pairings. The bilinear parings were used in negative meaning in cryptography, because the MOV attack using Weil pairing and FR attack using Tate pairing reduce the discrete logarithm problem on some elliptic curves to the discrete logarithm problem in a finite field[14,15]. Recently, the bilinear pairings have been used in positive meaning in cryptography, because they are important tools for construction of the identity-based schemes[6,10,16,17,18].

We let G_1 be a cyclic additive group generated by P, whose order is a prime q, and G_2 be a cyclic multiplicative group of the same order q. We assume that the discrete logarithm problem(DLP) in both G_1 and G_2 are hard. We let $\hat{e} : G_1 \times G_1 \rightarrow G_2$ be a pairing which satisfies the following properties:

1. Bilinear
$$\hat{e}(P_1 + P_2, Q) = \hat{e}(P_1, Q) \cdot \hat{e}(P_2, Q)$$
$$\hat{e}(P, Q_1 + Q_2) = \hat{e}(P, Q_1) \cdot \hat{e}(P, Q_2)$$
i..e., $\hat{e}(aP, bQ) = \hat{e}(P, Q)^{ab}$ where $a, b \in Z_q^*$, $P, Q \in G_1$.

2. Non-degenerate
   There exists $P \in G_1$ such that $\hat{e}(P, P) \neq 1$.

3. Computability
   There is an efficient algorithm to compute $\hat{e}(P, Q)$ for all $P, Q \in G_1$.

The non-degeneracy does not hold for the standard Weil pairing $e(P, Q)$, but it does hold for the modified Weil pairing $\hat{e}(P, Q)$. We note that the Weil and Tate pairings associated with supersingular elliptic curves or abelian varieties can be modified to create such bilinear maps. We refer to [10,16] for more details.

2.2 Diffie-Hellman Problem

In this section, we shall briefly explain the definition and the meaning about the Diffie-Hellman problem. We let the *Diffie-Hellman(DH) tuple* in G_1 be $(P, xP, yP, zP) \in G_1$ for some $x, y, z \in Z_q^*$ satisfying $z = xy \bmod q$.

Definition 1. The *Decision Diffie-Hellman(DDH) problem*: Given $P, xP, yP, zP \in G_1$, decide if it is a valid *DH tuple*. This can be solved in polynomial time by verifying $\hat{e}(xP, yP) = \hat{e}(P, zP)$.

Definition 2. The *Computational Diffie-Hellman(CDH) problem*: Given any three elements from the four elements in *DH tuple*, compute the remaining element.

CDH Assumption: There exists no algorithm running in expected polynomial time, which can solve the *CDH problem* with non-negligible probability.

Definition 3. The *Bilinear Diffie-Hellman(BDH) problem*: Given $P, xP, yP, zP \in G_1$, compute $\hat{e}(P, P)^{xyz} \in G_2$, where $x, y, z \in Z_q^*$. An algorithm is said to solve the *BDH problem* with an advantage of ε if

$$Pr[A(P, xP, yP, zP) = \hat{e}(P, P)^{xyz}] \geq \varepsilon.$$

BDH Assumption: There exists no algorithm running in expected polynomial time, which can solve the *BDH problem* in $<G_1, G_2, \hat{e}>$ with non-negligible probability.

2.3 Communications Model

We will prove the security of the protocol proposed in this paper using a communications model, which was proposed by Bellare and Rogaway incorporating later updates [19,20].

The protocol entities have an identifier U from a finite set $\{U_1, U_2, ..., U_n\}$ and these entities are modeled by oracles that represent an instance in a specific protocol. And these oracles keep transcripts that keep track of messages they have sent and received, and of queries they have answered. Oracle $\mathit{IT}_{I,J}$ represents the actions between two entities, I and J, in the protocol run indexed by integer n. The number of entity n is polynomial bound in the security parameter k. Each entity has a pair of identity-based long-term keys, where the public key is generated using the entity's identifier and the private key is computed and issued by a key generation center(KGC) at the start of the protocol. The oracle queries are the interactions with the adversary E, which is a probabilistic polynomial machine that controls all the communications that take place and has access to the entities' oracles. We describe each query by the adversary E.

1. *Send* : This query allows the adversary to send a message to the entity U, say $\mathit{IT}_{I,J}$, or to initiate a new protocol run between two entities, I and J, by sending a message.
2. *Reveal* : This query allows the adversary to ask a particular oracle to reveal the session key. If a session key K_s has been accepted by $\mathit{IT}_{I,J}$ then it is returned to the adversary, and the oracle is called opened.
3. *Corrupt* : This query allows the adversary to ask a particular oracle to reveal the oracle's internal state and set the long-term private key of the entity to be the value K chosen by the adversary. The entity is called corrupted.
4. *Test* : This query allows the adversary to check the oracle's accepted a session key K_s by distinguishing it from a test key. A random bit $b \leftarrow \{0, 1\}$ is chosen, if $b = 0$ then the session key K_s is returned while if $b = 1$ then a random string is returned to guess b. The advantage of the adversary E, denoted $advantage^E(k)$, is the probability that E can distinguish the queried oracle's session key K_s from a random string, and it is defined as:

$$Advantage^E(k) = |Pr[guess\ correct] - 1/2|.$$

And, an oracle's possible states are as follows.

1. *Accepted* : If an oracle decides to accept, holding a session key K_s, after receipt of properly formulated messages.
2. *Rejected* : If an oracle decides to reject holding a session key K_s.
3. *Opened* : If an oracle has answered a reveal query.
4. *Corrupted* : If an oracle has answered a corrupt query.

3   The Proposed Protocol

In this section, we present a one round identity-based authenticated conference key agreement protocol. Only one run of the protocol results in multiple conference key. The proposed protocol consists of two phases that are the setup phase and the key generation phase. First of all, we now describe the setup phase, which consists of the

KGC's setup operation, the entities' long-term public/private key setup, the entities' ephemeral public/private key setup, and a secure identity-based signature scheme. And then, we will explain the key generation phase.

3.1 Setup Phase

Let G_1 be a cyclic additive group generated by P, whose order is a prime order q, and G_2 be a cyclic multiplicative group of the same order q. The bilinear pairing is a map $\hat{e} : G_1 \times G_1 \rightarrow G_2$. We define three cryptographic hash functions $H : \{0, 1\}^* \rightarrow Z_q$, $H_1 : \{0, 1\}^* \rightarrow G_1$, and $H_2 : F_{q^k}^* \rightarrow \{0,1\}^*$. And, a secure identity-based signature scheme $\Sigma = (S, V)$ consists of two algorithms. The setup phase is as follows:

Step 1.   KGC's setup : The KGC chooses a random number $s \in Z_q^*$ and set $P_{pub} = sP$. The KGC publishes system parameter $params = \{G_1, G_2, P, P_{pub}, H, H_1, H_2\}$, and keep s as the *master-key*, which known only by itself.

Step 2.   Entities' long-term public/private key setup : Each entity submits his identity information ID_i to the KGC. The KGC computes the each entity's public key as $Q_i = H_1(Id_i)$, and returns $S_i = sQ_i$ to the entity as his private key, where $1 \leq i \leq n(n$ is the number of the entities). The pairs (Q_i, S_i) for ID_i serve as their long-term public, private key pairs.

Step 3.   Entities' ephemeral public/private key setup : Each entity randomly chooses an ephemeral private key $a_i \in Z_q^*$, and computes the value of the corresponding ephemeral public key a_iP. The pairs (a_iP, a_i) for ID_i serve as their ephemeral public/private key pairs.

Step 4.   Signing algorithm S setup : S takes a random integer $a \in Z_q^*$, and a plaintext message m. The signer A computes $P_A = aP$ and $T_A = H(P_A, m)S_A + am$. Then the signature of m is (P_A, T_A). Signer A has a long-term private key $S_A = sQ_A$, while the long-term public key is Q_A.

Step 5.   Verification algorithm V setup : V takes as input a message m and its corresponding signature (P_A, T_A). The verifier accepts the signature if and only the following equation holds:

$$\hat{e}(T_A, P) = \hat{e}(H(P_A, m)Q_A, P_{pub}) \cdot \hat{e}(m, P_A).$$

We note that the above signature scheme is not the normal signature scheme in the sense that the message m takes a special form of a multiple of P instead of any value. However, this signature scheme can be easily converted into a normal signature function $f : \{0, 1\}^* \rightarrow G_1$, that is $f(m) = H(m)P.$. For a signature scheme to be secure, we require that it be computationally impossible for any adversary to forge a signature on any message even under adaptive chosen message attacks. This signature scheme is secure against existential forgery under an adaptive chosen message attack in the random oracle model. The proof is similar to that of Scheme 3 in [21], and the security analysis of this signature scheme is referred in [12].

3.2 Key Generation Phase

We now define the proposed protocol. All parameter choices depend on a security parameter k. The protocol uses the set of n entities' identifiers $U = \{U_1, U_2, \ldots, U_n\}$ and a secure identity-based signature scheme. The KGC provides each user with an authentic copy of the entities' long-term public keys of all other users. The protocol is as follows:

$$U_i \rightarrow * \; : \; a_i P_{pub}, (P_i, T_i),$$

where $1 \le i \le n$, $P_i = aP$, $T_i = H(P_i, a_i P_{pub})S_i + aa_i P_{pub}$, $a \in Z_q^*$, and the asterisk is used to denote broadcasting messages.

In an implementation, there is no need for the message of some entity to be sent before the other messages, so it is perfectly possible for all messages to be sent together in only one round. The key generation phase consists of two steps that are the entity authentication step and the key generation step. After all entities do the entity authentication step, do the key generation step. In case of rejection in the entity authentication step, the entity cannot generate the conference key. The key generation phase is as follows:

Step 1.   Entity authentication step : All entities broadcast the $a_i P_{pub}$, (P_i, T_i) message and receive $n-1$ $a_j P_{pub}$, (P_j, T_j) messages, where $1 \le i, j \le n$ and $i \ne j(n$ is the number of the entities). The signature of $a_i P_{pub}$ is (P_i, T_i). The ith entity authenticates the $n-1$ entities using the signature (P_j, T_j), if and only the following equation holds:

$$
\begin{aligned}
\hat{e}(T_j, P) &= \hat{e}(H(P_j, a_j P_{pub})Q_j, P_{pub}) \cdot \hat{e}(a_j P_{pub}, P_j) \\
&= \hat{e}(H(P_j, a_j P_{pub})Q_j, sP) \cdot \hat{e}(a_j P_{pub}, aP) \\
&= \hat{e}(H(P_j, a_j P_{pub})S_j, P) \cdot \hat{e}(aa_j P_{pub}, P) \\
&= \hat{e}(H(P_j, a_j P_{pub})S_j + aa_j P_{pub}, P) \\
&= \hat{e}(T_j, P)
\end{aligned}
$$

Step 2.   Key generation step : The ith entity generates the session key K_i as follows, where $1 \le i \le n(n$ is the number of the entities), and H_2 is a one-way key derivation function, which will be modeled as a random oracle.

$$
\begin{aligned}
U_i \; : \; K_i &= \hat{e}(Q_1, a_1 P_{pub}) \cdot \ldots \cdot \hat{e}(a_i S_i, P) \cdot \ldots \cdot \hat{e}(Q_n, a_n P_{pub}) \\
&= \hat{e}(Q_1, a_1 P_{pub}) \cdot \ldots \cdot \hat{e}(a_i s Q_i, P) \cdot \ldots \cdot \hat{e}(Q_n, a_n P_{pub}) \\
&= \hat{e}(Q_1, a_1 P_{pub}) \cdot \ldots \cdot \hat{e}(a_i Q_i, sP) \cdot \ldots \cdot \hat{e}(Q_n, a_n P_{pub}) \\
&= \hat{e}(Q_1, a_1 P_{pub}) \cdot \ldots \cdot \hat{e}(a_i Q_i, P_{pub}) \cdot \ldots \cdot \hat{e}(Q_n, a_n P_{pub}) \\
&= \hat{e}(a_1 Q_1, P_{pub}) \cdot \ldots \cdot \hat{e}(a_i Q_i, P_{pub}) \cdot \ldots \cdot \hat{e}(a_n Q_n, P_{pub}) \\
&= \hat{e}(a_1 Q_1 + \ldots + a_i Q_i + \ldots + a_n Q_n, P_{pub})
\end{aligned}
$$
$$K_1 = \ldots = K_i = \ldots = K_n$$
$$K_s = H_2(K_1) = \ldots = H_2(K_i) = \ldots = H_2(K_n)$$

4 Security Analysis

4.1 Security Attributes

The principle of our protocol is that each entity's ephemeral key a_i determines the final session key $K_s = H_2(\hat{e}(a_1Q_1 + a_2Q_2 + \ldots + a_nQ_n, P_{pub}))$. The secrecy of the session key relies on the assumption of hardness of BDH problem, *i.e.*, given Q_i, P_{pub}, where $1 \leq i \leq n(n$ is the number of the entities), it is hard to determine $\hat{e}(Q_i, P_{pub})^{a_i}$. And, the authenticity of a_iP_{pub} is achieved by attaching the signature (P_i, T_i). The signature is generated using the entity's long-term private key S_i. And then, other entities verify the signature using the entity's long-term public key Q_i. Without knowledge of the long-term private key of the entity, an adversary can hardly impersonate that entity for a key agreement since he can hardly forge a signature of the broadcast message within a polynomial time with a nontrivial probability. And without knowledge of each entity's ephemeral private key a_i, it is impossible to know each entity's long-term private key S_i.

$$T_i = H(P_i, a_iP_{pub})S_i + aa_iP_{pub}$$
$$S_i = (T_i - aa_iP_{pub})H(P_i, a_iP_{pub})^{-1}$$

Below are security attributes of the proposed protocol.

1. **Known Session Key Security**: A protocol satisfies the *Known session key security*, if an adversary, having obtained some previous session keys, cannot get the session keys of the current run of the key agreement protocol. As with our protocol, suppose that the adversary knew the previous session key of a previous session key agreement protocol. In communications model, the adversary E is allowed to make the *Reveal* queries to any oracle except for $\Pi^n_{I,J}$ and $\Pi_{J,I}$. These *Reveal* queries do not help the adversary to obtain the session key between $\Pi^n_{I,J}$ and $\Pi_{J,I}$. Actually our protocol has the property of the known session key security because each run of the protocol produce a different session key and the adversary has to extract the ephemeral key a_i to know a session key. For example, to determine a_i from $\hat{e}(Q_i, P_{pub})^{a_i}$, is equivalent to solve the Discrete-Log problem in G_2. Without the ephemeral key, the adversary cannot extract the entity's long-term private key. In addition, the security of the signature scheme also prevents the adversary from forging a valid signature from entity and impersonating entity to get session key. Therefore the knowledge of past session keys does not allow the adversary to deduce the future session keys.

2. **Perfect Forward Secrecy**: A protocol satisfies the *Perfect Forward Secrecy*, if the compromise of the all long-term private keys of the all entities does not affect the security of the previous session keys. In communication model, the adversary E is allowed to make the *Corrupt* queries to any oracle. These *Corrupt* queries do not help the adversary to know the previous

session keys, because the session key is generated by the entity's ephemeral private key, long-term private key and P(a generator of G_1). Although the adversary knows entity's long-term private key and P, the adversary does not extract the entity's ephemeral private key. We can explain this problem as *CDH problem*. Let the *Diffie-Hellman tuple* be $(P, Z_i, a_i P, S_i)$ for the ith entity, where $\hat{e}(P, Z_i) = \hat{e}(a_i P, S_i) \in G_2$, $Z_i \in G_1$. Given P, Z_i, S_i, we cannot compute the $a_i P$. This problem is the *CDH problem* that is equivalent to solve the Discrete-Log problem in G_1. As with the proposed protocol, the adversary has to get the corresponding ephemeral keys to learn the previous session keys. And, suppose the adversary is able to compute $a_A P_{pub} = (T_A - H(P_A, a_A P_{pub})S_A)a^{-1}$ from A's the signature in broadcasting messages $a_A P_{pub}$, (P_A, T_A) for entity A, where $a \in Z_q^*$. However, given $a_A P_{pub}$, to extract a_A is equivalent to solve the Discrete-Log problem in G_1. Therefore our protocol has the property of the *Perfect Forward Secrecy*.

3. *No Key Control*: A protocol satisfies the *No Key Control*, if the session key is determined by the all entities and no one can influence the generation of the session key. In the entity A, the session key is generated using the A's information $\hat{e}(a_A S_A, P)$ and the received messages. In other words, the generation of the session key uses the information of the all entities. All entities broadcast the messages in same time, and receive the messages. So, no one can enforce the session key to fall into a pre-determined interval. In other words, no one can influence the generation of the session key. Therefore our protocol has the property of the *No key Control*.

4. *No Key-Compromise Impersonation*: A protocol satisfies the *No Key-Compromise Impersonation*, unless the compromise of one entity's long-term private key influences the compromise of the long-term private keys of other entities. The adversary E may impersonate the compromised entity in subsequent protocols, but he cannot impersonate other entities. In communication model, the adversary is allowed to make the *Test* queries to an oracle, which has been corrupted. Suppose that there exists an oracle $\Pi^t_{I,J}$(which has been corrupted) and the adversary impersonates another oracle $\Pi^s_{J,I}$(which has not been corrupted) using *Test* queries. If the adversary impersonates another oracle $\Pi_{J,I}$, the probability of the queried on $\hat{e}(a_J S_J, P)$ $\hat{e}(Q_J, a_J P_{pub})$ by the adversary is non-negligible. But, this contradicts the BDH assumption. And thus the probability of impersonating $\Pi_{J,I}$ to $\Pi^s_{I,J}$ must be negligible. Therefore our protocol has the property of the *No Key-Compromise Impersonation*.

5. *No Unknown Key-Share*: A protocol satisfies the *No Unknown Key-Share*, if the all entities do not share the session key with the adversary. If the adversary convinces a group of entities, they share some session key with the adversary, and this protocol suffers from unknown key-share attack. In our protocol, the adversary must know the long-term private keys, the ephemeral private keys of the all entities to share the session key with adversary. Otherwise the adversary cannot share the session key. Therefore our protocol has the property of the *No Unknown Key-Share*.

4.2 Computational Overheads

Let us consider the computational requirements for each entity. Each entity generates a message aP_{pub} and a signature (P_i, T_i), and then broadcasts and receives a message. After checking on n-1 signatures, each entity generates a session key using n-1 messages. The computational requirements are the same as for the two entities case. The computational overheads of the signing are needed three scalar multiplications, one scalar addition and one hashing over G_1. The case of the verifying is needed one scalar multiplication, one hashing and three pairings. When generating the session key, two elliptic curve multiplications, n pairings and one hashing are needed. The computations required are substantially less than in the proven secure generalized Diffie-Hellman protocols of Bresson et al.[22,23,24], which require U_i to perform i+1 exponentiations in addition to generating and verifying a signature.

5 Conclusion

We have described the key agreement protocol for multi-party, which can be completed in one round and provided an analysis of its security. Our protocol relates to identity based authenticated key agreement protocol. In the protocol presented, some party can share the key in the own party through broadcasting the message, which is composed of each entity's key information using the ephemeral private key and its signature. The session key can be generated using all entity's key information, and the signature is used for the security for the entity's key information. Our protocol has the security properties as *known session key security*, *perfect forward secrecy*, *no key control*, *no key-compromise impersonation*, and *no unknown key-share*. And, our protocol is more efficient than other Diffie-Hellman protocols in terms of bandwidth and computational overheads.

References

1. W. Diffie and M. Hellman, "New directions in cryptography," *In IEEE Transactions on Information Theory*, No.22, pp.644-654, 1976
2. T. Matsumoto, Y. Takashima, and H. Imai, "On seeking smart public-key distribution systems," *Trans. IECE of Japan*, E69, pp.99-106, 1986
3. L. Law, A. Menezes, M. Qu, J. Solinas, and S. Vanstone, "An efficient protocol for authenticated key agreement," *Technical Report CORR 98-05*, Department of C & O, University of Waterloo, 1988
4. S. Blake-Wilson and A. Menezes, "Authenticated Diffie-Hellman Key Agreement Protocols," *Proceedings of the 5th Annual Workshop on Selected Areas in Cryptography(SAC '98)*, LNCS 1556, Springer-Verlag, pp.339-361, 1999
5. B. Song and K. Kim, "Two-Pass Authenticated Key Agreement Protocol with Key Confirmation," *Proc. Of Indocrypt 2000*, LNCS 1977, pp.237-249, Springer-Verlag, 2000

6. A. Joux, "A one round protocol for tripartite Diffie-Hellman," *ANTS IV*, LNCS 1838, pp.385-394, Springer-Verlag, 2000
7. C. Boyd and J. M. G. Nieto, "Round-Optimal Contributory Conference Key Agreement," *PKC 2003*, LNCS 2567, pp.161-174, Springer-Verlag, 2003
8. S. Al-Riymi and K. Paterson, "Authenticated three party key agreement protocols form pairings," *Cryptology ePrint Archive*, Report 2002/035, 2002
9. A. Shamir, "Identity-based cryptosystems and signature schemes," *Advances in Cryptology-Crypto 84*, LNCS 196, pp.47-53, Springer-Verlag, 1984
10. D. Boneh and M. Franklin, "Identity-based encryption from the Weil pairing," *Advances in Cryptology-Crypto 2001*, LNCS 2139, pp.213-229, Springer-Verlag, 2001
11. N. P. Smart, "An Identity based authenticated Key Agreement protocol based on the Weil pairing," *Electron. Lett.*, Vol.38, No.13, pp.630-632, 2002
12. F. Zhang, S. Liu and K. Kim, "ID-based one-round authenticated tripartite key agreement protocol with pairings," *Cryptology ePrint Archive*, Report 2002/122, 2002
13. K. Becker and U. Wille, "Communication complexity of group key distribution," 5th Conference on Computer and Communications Security, pp.1-6, ACM Press, 1998
14. A. Menezes, T. Okamoto, and S. Vanstone, "Reducing elliptic curve logarithms to logarithms in a finite field," *IEEE Transaction on Information Theory*, Vol.39, pp.1639-1646, 1993
15. G. Frey and H. Ruck, "A remark concerning m-divisibility and the discrete logarithm in the divisor class group of curves," *Mathematics of Computation*, Vol.62, pp.865 874, 1994
16. D. Boneh, B. Lynn, and H. Shacham, "Short signature from the Weil pairing," *Advances in Cryptology-Asiacrypt 2001*, LNCS 2248, pp.514-532, Springer-Verlag, 2001
17. R. Sakai, K. Ohgishi, M. Kasahara, "Cryptosystems based on pairing," *SCIS 2000C20*, 2000
18. E. R. Verheul, "Self-blindable credential certificates from the Weil pairing," *Advances in Cryptology-Asiacrypt 2001*, LNCS 2248, pp.533-551, Springer-Verlag, 2001
19. M. Bellare and P. Rogaway, "Entry authentication and key distribution," *In Advances in Cryptology CRYPTO '93*, pp.232-249, Springer-Verlag, 1993
20. M. Bellare and P. Rogaway, "Provable secure session key distribution – the three part case," *In Proceedings of the 27th ACM Symposium on the Theory of Computing*, 1995
21. F. Hess, "Exponent group signature scheme and efficient identity based signature schemes based on pairings," *Cryptology ePrint Archive*, Report 2002/012, 2002
22. E. Bresson, O. Chevassut, and D. Pointcheval, "Provably authenticated group Diffie-Hellman key exchange – the dynamic case," *In Advances in Cryptology – Asiacrypt 2001*, pp.290-309, Springer Verlag, 2001
23. E. Bresson, O. Chevassut, D. Pointcheval, and J. Quisquater, "Provably authenticated group Diffie-Hellman key exchange," *In CCS'01*, pp.255-264, ACM Press, 2001
24. E. Bresson, O. Chevassut, and D. Pointcheval, "Dynamic group Diffie-Hellman key exchange under standard assumptions," *In Advances in Cryptology – Eurocrypt 2002*, Springer-Verlag, 2002

An Efficient Identity-Based Group Signature Scheme over Elliptic Curves

Song Han[1,2], Jie Wang[2], and Wanquan Liu[1,*]

[1] Department of Computing, Curtin University of Technology
GPO Box U1987, Perth 6845, Western Australia, Australia
[2] Department of Mathematics, Beijing University
Beijing, 100871, China

Abstract. Group signatures allow every authorized member of a group to sign on behalf of the underlying group. Anyone except the group manager is not able to validate who generates a signature for a document. A new identity-based group signature scheme is proposed in this paper. This scheme makes use of a bilinear function derived from Weil pairings over elliptic curves. Also, in the underlying composition of group signatures there is no exponentiation computation modulo a large composite number. Due to these ingredients of the novel group signatures, the proposed scheme is efficient with respect to the computation cost in signing process. In addition, this paper comes up with a security proof against adaptive forgeability.

Keywords: Group Signatures, Anonymity, Network Security, Weil Pairings, Security Protocol.

1 Introduction

Group signature is one of the most important security protocols in cryptography. A group signature scheme is a digital signature scheme in which an individual member of a group can generate a signature (for a document), and it can be verified by anyone, without revealing the signer's identity. At the same time, the signer can be identified later in case of disputes by a designated group manager. Moreover, the group manager can not forge valid group signatures for other group members. Therefore, the group signatures are suitable for the following scenario:

Several network policemen consist of a group NP in a network; Every one of them can find out what kind of documents are harmful and then sign them. Any user in this network who receives the above documents would check whether a signature exists on the documents and also it is validly signed by the group NP. And then, this user can be convinced the document is really harmful. How can we realize (or deploy) the network security of this scenario? A group signature protocol would be a better choice.

---

* Corresponding Author

M. Freire et al. (Eds.): ECUMN 2004, LNCS 3262, pp. 417–429, 2004.

The concept of group signatures was first introduced by Chaum and E.van Heyst [5] in 1991. Group signatures are also suitable in many applications: e-cash, bidding, voting, and so on. Recently in [2], Ateniese and B. de Medeiros presented a new application for group signatures: anonymous E-prescriptions in medical situations. Following [5], various group signature schemes have been proposed so far, for instance, [1,8,9,21,14,20,18]. Among these group signature schemes, the security of [5] and [8] relies on the difficulty of discrete logarithm problems and of the factoring problems. The security of [13] depends on the difficulty of both the discrete logarithm problem and the e-th roots computation. The security of [9] is based on the difficulty of discrete logarithm problems and on the security of Schnorr signatures [17] and of RSA signatures. And the security of [7] relies on the strong RSA assumption and the decisional Diffie-Hellman assumption. In [21] Wu and Varadharajan constructed new trapdoor one-way function by use of Boolean permutations and proposed some group signature schemes, whose security depends on the difficulty of solving nonlinear equations. In [14], Popescu proposed an identity-based group signature scheme, that makes use of bilinear pairings over elliptic curves and its security relies on the difficulty of elliptic curve discrete logarithm problem and Diffie-Hellman problem; also on the security of RSA signatures. However, the signing computations of [14] encompass the exponentiation computations modulo a large RSA modulus. Therefore, the performance is cost-inefficient. On the other hand, it should be noted that most group signature schemes for instance [9,1] whose security is based on the difficulty of discrete logarithm problems make use of the proofs of knowledge about discrete logarithms. Generally speaking, all these methods are used to prove the knowledge of holding secret keys. However, these methods resort to some proofs systems that are three-round protocols. Generally, these protocols can be shown to be honest-verifier zero-knowledge proofs of knowledge and need to be repeated sequentially many times in order to be zero-knowledge proofs. Hence, these methods are also a little inefficient in terms of security check. Therefore, it is still an open problem to construct a more efficient and provably secure group signature scheme that does encompass neither the exponentiation computations modulo a large RSA modulus nor the methods of proofs of knowledge in public key settings. Before the publication of [9], all the proposed group signature protocols, for instance [5,8,12] have the following undesirable properties:

(1) the length of a group signature and (or) the length of group's public key depend linearly on the number of underlying groups.

(2) it is necessary to modify at least the public key, when new authorized member joins the group.

Subsequent works on group signature schemes after 1997 for instance [9,21,1, 18,14] possess the desirable property: the length of a group signature and (or) the size of the group public key are independent of the number of group members.

In traditional group signature signing algorithms, the public keys of group members are essentially random bit strings from a given set. This leads to a problem of how the public keys being associated with the corresponding physical entities performing the signing computations. In order to solve this issue, the identity-based group signature scheme has been proposed with a trusted key generation center which gives each member a personalized smart card when the member first joins the network. The information embedded in this card enables each member to sign the documents she sends and verifies the documents she receives in a totally independent way, regardless of the identities of other members in this group. Previously issued cards do not have to be updated when new group member joins the network.

The concept of identity-based cryptography is due to A.Shamir [16]. Shamir's original motivation was to simplify certificate management in e-mail systems. When Alice sends mail to Bob at bob@hotmail.com, she simply encrypts her message using the public key string ' bob@hotmail.com '. There is no need for Alice to obtain Bob's public key certificate. Therefore, an identity-based cryptosystem is a system that allows a publicly known identifier (for instance email address, IP address) to be used as the public parameter or public key of a public/private key pair. In 1997, Park, Kim and Won presented the first identity-based group signature scheme [13], where the length of the group public key and signature are proportional to the size of the group. In 1998, Tseng and Jan proposed another identity-based group signature scheme [20], which is not secure against forgeability. Thereafter, Popescu brought forward a modification on [20]. In 2002, Popescu proposed a new identity-based group signature scheme [14], that makes use of the pairings over elliptic curves. However, the scheme in [14] made use of the RSA signatures in group signatures. It is known that at the same security level ECC-521 can be expected to be on average 400 times faster than 15,360-bit RSA [11].

In this paper, a novel identity-based group signature scheme is proposed. It makes use of the bilinear pairings over elliptic curves. The size of the group public key is independent of the size of the underlying group. Also, the length of a group signature is independent of the number of the underlying group. Different from [14], our signing computation does not encompass RSA signatures. Therefore, by [11] the new scheme is expected to be more efficient than [14]. Further, a security proof against adaptive forgeability is presented in this paper.

The new group signature scheme proposed in this paper has the following desirable properties:

(1) Provably secure against adaptive forgeability, which is not formally proved in [1,7,14].

(2) No exponentiation calculations during both the generation and the verification of signatures. Previous to this new scheme, some group signature schemes need to compute exponentiations modulo a large RSA modulus [14]. Therefore, the new scheme is efficient in terms of computation cost.

The rest of this paper is organized as follows. The next section comes up with the model for identity-based group signatures. In section 3, we present some preliminaries including the bilinear pairings over elliptic curves. Section 4 brings forward the descriptions of the details of our new identity-based group signature scheme. Subsequently, the security proofs and analysis are presented in section 5. The performance of the novel scheme is discussed in section 6. Finally in section 7, some conclusions are given.

2   The Model

In this section, the concept of an identity-based group signature scheme are presented as follows. This concept is based on an identity-based digital signature scheme [10].

Definition 1. (Identity-based Group Signatures) An identity-based group signature scheme is an identity-based digital signature scheme comprised of the following five procedures:

(1) Setup: An algorithm, executed by the group manager, takes a random security parameter l as input and generates some system parameters and a master key. The system parameters are publicly known as the initial group public key; while the master key is only known to the group manager.

(2) Extract: A protocol between the group manager and a member. We assume that communication between the member and the group manager is private and authenticated. As a result of the protocol, the member becomes an authorized member of this group. The output is a membership credential and a membership secret. Here the member's secret contains two parts: one is sent by the group manager, the other is chosen by herself; where the two parts are both authenticated.

(3) Sign: A probabilistic algorithm (with inputs as a group public key, a membership secret, and a message m) outputs a group signature of m.

(4) Verify: An algorithm for establishing the validity of an alleged group signature of a message with respect to the group public key.

(5) Reveal: An algorithm that, given a message, a valid group signature on it, a group public key and a group manager's master key, determines the identity of the actual signer.

A secure identity-based group signature scheme should satisfy all or part of the properties:

(1) Correctness: Group signatures produced by a group member using SIGN algorithm must be accepted by VERIFY algorithm.

(2) Unforgeability: Only group members are able to sign messages on behalf of the underlying group.

(3) Anonymity: Given a valid signature of a message, identifying the actual signer is computationally infeasible for everyone except for the group manager.

(4) Unlinkability: Deciding whether two different valid signatures were computed by the same group member is computationally hard.

(5) Exculpability: Neither a group member nor the group manager can sign on behalf of other group members.

(6) Traceability: The group manager is always able to identify the actual signer for a valid signature in case of disputes.

3   Preliminaries

3.1   Notations

In this subsection, we describe some notations used in this paper. Let q be a large prime, and Z_q^* be $Z_q \backslash \{0\}$. Let N be a positive integer. We write Z_N^* for the multiplicative group of integers modulo N. We denote $\varphi(n)$ as the Euler phi function. Let H and H_1 be two cryptographic hash functions: $H : \{0,1\}^* \to G_1$, and $H_1 : \{0,1\}^* \times G_1 \to G_1$.

3.2 Pairings over Elliptic Curves

Let p be a sufficiently large prime that satisfies: (a) $p \equiv 2\bmod 3$: (b) $p = 6q - 1$, where q is also a large prime. Consider respectively the elliptic curves E/F_p and E/F_{p^2} defined by the equation:

$$y^2 = x^3 + 1.$$

Let G_1 be an additive group of points of prime order q on an elliptic curve E/F_p and let G_2 be a multiplicative group of same order q of some finite field F_{p^2}. We assume the existence of a bilinear map, the modified Weil pairing,

$$e : G_1 \times G_1 \to G_2$$

such that the Elliptic Curve Discrete Logarithm (ECDL) problems are difficult in G_1 and the Computational Diffi-Hellman (CDH) problems and the Inversion of Weil pairing (IWP) problem are difficult in G_2.

The modified Weil pairings $e : G_1 \times G_1 \to G_2$ has the following properties:

(1) Bilinearity: $e(aP, bQ) = e(P, Q)^{ab}$ for every pair $P, Q \in G_1$ and for any $a, b \in Z_p$.

(2) Non-degenerate: there exists at least one point $P \in G_1$ such that $e(P, P) \neq 1$.

(3) Efficient Computable: there are efficient algorithms to compute the bilinear pairings e.

3.3 Elliptic Curve Discrete Logarithms

Definition 2. (Elliptic Curve Discrete Logarithm Problem) Given G_1 as above, choose P a generator from G_1, given xP, where x is an unknown random element of Z_q^*, the Elliptic Curve Discrete Logarithm (ECDL) problem is to find x.

Assumption 1. (ECDLP Assumption) Given xP and a generator P in G_1 with unknown $x \in Z_q^*$. There is no probabilistic polynomial algorithm to solve the Elliptic Curve Discrete Logarithm problem with non-negligible advantage [10].

3.4 Inversion of Modified Weil Pairings

Definition 3. (Inversion of Modified Weil Pairings Problem) Given G_1, G_2 and $e(\cdot, \cdot)$ as above, choose P a generator from G_1, given $e(P, *)$, where $*$ is an unknown point of G_1, the Inversion of Modified Weil Pairings (IWP) problem is to find $Q \in G_1$ such that

$$e(P, Q) = e(P, *).$$

Assumption 2. (IWP Assumption) Given G_1, G_2 and $e(\cdot, \cdot)$ as above, choose P a generator from G_1, given $e(P, *)$, where $*$ is an unknown point of G_1. There is no probabilistic polynomial algorithm to solve the Inversion of Modified Weil Pairings problem with non-negligible advantage.

4   New Identity-Based Group Signature Scheme

In this section the detail description of the new group signature scheme is presented.

4.1   SETUP

This is a system generation algorithm. The group manager (GM) executes the following procedures:

(1) Choose p, q, G_1, G_2 defined in subsection 3.2 [3].

(2) Choose two cryptographic hash functions:

$$H : \{0,1\}^* \mapsto G_1,$$

$$H_1 : \{0,1\}^* \times G_1 \mapsto G_1.$$

(3) Construct a bilinear function defined in subsection 3.2:

$$e : G_1 \times G_1 \longmapsto G_2.$$

(4) Select a generator element $P \in G_1$, therefore $e(P, P)$ is a generator element of G_2.

(5) Select an integer a from Z_q^* as the secret key of GM; Set $P_{pub} = aP$ as the public key of this group.

(6) Let a string $f \in \{0,1\}^*$ denoting an identifier (e.g. email address or IP address) of any group member. GM computes $Q_f = H(f)$ as the public key of this member. It is easy to see we may not confirm the real identity of some group member by her email address or IP address.

(7) Let $\{0,1\}^*$ (a set of strings of any length) be the message space. Therefore, the group public key of this group is: $PK = \{P, P_{pub}, e(\cdot, \cdot), H, H_1\}$ The master key of GM is $SK = a$.

4.2   EXTRACT

Suppose a new member U_i wants to be an authorized member of this group. GM will communicate with U_i through a secure channel (e.g. secure against tampering, intruding, intercepting):

(1) U_i sends her identifier f_i to GM;

(2) GM computes $sk_i = aQ_{f_i}$, and then sends it to U_i.

(3) U_i regards respectively private value b_i (secretly chosen by herself) and her identifier f_i as her personal secret key and personal public key. Suppose $b_i f_i \equiv 1 \bmod \varphi(n)$, where n is a product of two large prime numbers.

(4) U_i and GM simultaneously execute a Schnorr identification protocol (see [17]). Thereafter, GM obtains a credential t_i which is used to identify the membership of U_i.

(5) GM has a transcriptor: $trans = \{< f_i, t_i > |$ for every authorized group member $U_{f_i}\}$. This transcriptor is held only by GM.

(6) As a result of the communication, U_i becomes an authorized group member of this group. Her credential is t_i; her personal secret key is $\{b_i, sk_i\}$; and her personal public key is f_i. All these information are stored in a smart card held privately by U_i.

4.3 SIGN

This is a generation algorithm of group signatures. Suppose U_{f_i} is an authorized member of this group. Given a message $m \in M$, she performs the following algorithm:

(1) chooses randomly and uniformly x from Z_q^*, and sets $A = xP$.

(2) computes $B = x^{-1}sk_i + H_1(m, A)b_i$, where x^{-1} is the inversion of x in Z_q^*.

Therefore, the group signature on message m is $\{A, B, f_i\}$.

4.4 VERI

This is an algorithm of verification on alleged group signatures. Given a message m and its alleged group signature $\{A, B, f_i\}$, any verifier who holds public key can validate the validity of the group signature by carrying out the followings:

(1) computes $\alpha = e(f_i P_{pub}, Q_{f_i})$;

(2) computes $\beta = e(A, f_i B)$;

(3) computes $\gamma = e(A, H(m, A))$.

As a result, the verifier checks the equation:

$$\beta \overset{?}{=} \alpha\gamma \tag{1}$$

If the equality holds, then the verifier accepts: $\{A, B, f_i\}$ is a valid group signature on message m; otherwise, rejects it. On the one hand, by the group public key the verifier knows the signature coming from this group; On the other hand, by the personal public key the verifier knows this signature was generated by one of authorized members not by group manager.

4.5 REVEAL

This algorithm is only executed by the group manager GM. Given a message m and its corresponding valid group signature $\{A, B, f_i\}$, the group manager looks up the tanscriptor for the corresponding membership credential t_i. By the Schnorr identification protocol [17] and this group membership credential, the group manager can confirm the real identity of the group member.

Remark 1: In the above SETUP procedure, we can utilized the method of multiple trust authorities [6] to select and configure several GMs in order to avoid the centralization of the identities of group members into one GM. This will resist a denial of service possibly happened for one GM.

Remark 2: The proposed protocol is a group signature scheme which allows every authorized member to sign documents on behalf of the designated group. While the protocol of [10] is a general signature scheme, which implies each user can only sign documents on behalf of herself.

Remark 3: The proposed protocol is in the settings of supersingular elliptic curves while the setting in [8] is of a prime-order multiplicative subgroup of a finite field. On the other hand, the former did not utilize the proof of knowledge (see [1,9]), while the latter utilized this type of proof.

5   Security Proofs and Analyses

This section we will come up with the security proofs and analysis of the new identity-based group signature scheme. Specially we shall prove that the new identity-based is secure against adaptive chosen message attack. On the other hand, some properties in the definition of group signatures will also be analyzed here.

5.1   Correctness

Theorem 1. Given any message $m \in M$, if an authorized group member honestly computes the corresponding group signature $\{A, B, f_i\}$ on m by SIGN algorithm, then the VERI algorithm always accept it:

$$VERI(m, \{A, B, f_i\}) \equiv 1. \tag{2}$$

Proof Suppose $\{A, B, f_i\}$ is a group signature on message m honestly computed by an authorized member through SIGN algorithm, we shall prove it is valid. Therefore, it will always be accepted by VERI algorithm. In fact, $\{A, B, f_i\}$ has the following formula:

$$A = xP; \ B = x^{-1}sk_i + H_1(m, A)b_i. \tag{3}$$

Therefore,

$$
\begin{aligned}
\beta &= e(A, f_i B) \\
&= e(A, f_i x^{-1} sk_i + f_i)bH_1(m, A) \\
&= e(xP, f_i x^{-1} sk_i)e(xP, H_1(m, A)) \\
&= e(P, aQ_{f_i})^{x f_i x^{-1}} e(xP, H_1(m, A)) \\
&= e(f_i aP, Q_{f_i})e(xP, H_1(m, A)) \\
&= e(f_i P_{pub}, Q_{f_i})e(A, H_1(m, A)) \\
&= \alpha\gamma.
\end{aligned}
$$

Hence, VERI algorithm always accepts the group signature.

5.2   Security Against Adaptive Forgeability

Generally speaking, adaptive unforgeability in group signatures implies that: even if an adversary has oracle access [15] to the group signing algorithm which provides valid group signatures on messages of the adversary's choice, the adversary cannot create a valid group signature on a message not previously queried.

Theorem 2. Under the assumption of Elliptic curve Discrete Logarithm and the assumption of Inversion of Weil Pairings, the new identiy-based group signature scheme is secure against adaptive forgeability.

Proof. We shall prove that the new identity-based group signature scheme is secure against adaptive chosen message attacks.

Suppose **Adv** is a probabilistic polynomial time algorithm that will forge valid group signatures against the new identity-based group signature scheme.

First it is noted that **Adv** is not able to obtain the personal secret key of any authorized group member by observing the corresponding personal public key Q_{f_i} and the group public key PK. In fact,

(1) Due to the difficulty of Elliptic Curve Discrete Logarithm problems, **Adv** is not able to obtain a by solving $P_{pub} = aP$. Therefore, it is not able to work out $sk_i = aQ_{f_i}$.

(2) Due to the unknown factors of n, **Adv** is not able to figure out b by the relation of $f_i b_i \equiv 1 \bmod \varphi(n)$.

On the other hand, even though **Adv** can oracle access [15] to the Sign algorithm simulator, it is not able to return a valid group signature. In this case, we first assume that **Adv** is able to bring forward valid group signatures, then there will be a contradiction.

Adv would interact with GM, SIGN simulator, and hash oracle. The detailed descriptions of these interactions are as follows:
GM:

(1) **Adv** would choose freely a personal public key f_j of any authorized group member and interact with GM:

(2) GM randomly and uniformly selects a' from Z_q^* and sets $Q_{f_j} = a'P$, $sk_{f_j} = a'P_{pub}$, and then sets $H(f_j) = Q_{f_j}$.
SIGN simulator:

(1) Given any message m chosen by **Adv**, SIGN simulator will return a group signature with respect to f_j;

(2) By use of the results returned by interacting with GM, SIGN simulator computes

$$\{A = xP, B = x^{-1}sk_{f_j} + f_j^{-1}H_1(m, A)\},$$

where x is chosen by **Adv** from Z_q^*.
HASH ORACLE: For any message m chosen by **Adv** and the element A returned by SIGN simulator, HASH ORACLE defines $H_1(m, A) = gP$, where $g \in Z_q^*$. (It is known that P is a generator of G_1.)

In fact, we may regard **Adv**, GM, SIGN simulator and HASH ORACLE respectively as some probabilistic polynomial time algorithms. In the course of interacting with GM, SIGN simulator and HASH ORACLE, **Adv** would freely choose some messages and some personal public keys of authorized members. However, there is a limitation on the behavior of **Adv**: that is, as it forges a valid group signature for the corresponding message m_0, the message m_0 has to be not queried in the course of interactions by **Adv** to obtain its corresponding valid group signature.

By the descriptions of the above three probabilistic polynomial time algorithms, for any new message m not queried by **Adv**, due to the Theorem 1 in [15], we may make use of the random transcripts of GM and SIGN simulator respectively

$$\sigma \text{ and } \psi$$

as the auxiliary inputs, and then run the probabilistic polynomial time algorithm **Adv** twice. At the same time, we use the different values of hash function $H_1(m, \cdot)$: h_1 and h_2. Therefore, due to the assumption on **Adv** (that is, it is able to output valid group signatures.), we can obtain two different valid group signatures on message m with respect to public key f_i:

$$A_1, B_1, f_i \tag{4}$$

and

$$A_1, B_2, f_i \tag{5}$$

Since we used the different hash values,

$$B_1 \neq B_2.$$

Therefore, in the verification algorithm $VERI$, by the (4) we have:

$$e(A_1, f_i B_1) = e(f_i P_{pub}, Q_{f_i})\gamma_1; \tag{6}$$

By the (5) we have:

$$e(A_1, f_i B_2) = e(f_i P_{pub}, Q_{f_i})\gamma_2. \tag{7}$$

where

$$\gamma_1 = e(A_1, h_1);$$
$$\gamma_2 = e(A_1, h_2).$$

Therefore, we can obtain by equation (6) and (7) respectively :

$$\frac{e(A_1, f_i B_1)}{e(A_1, f_i B_2)} = \frac{e(f_i P_{pub}, Q_{f_i})\gamma_1}{e(f_i P_{pub}, Q_{f_i})\gamma_2};$$

Hence,

$$e(A_1, f_i(B_1 - B_2)) = \gamma_1 \gamma_2^{-1};$$

By the computations of γ_1 and γ_2, and the randomness of h_1 and h_2, we may understand:

$$\gamma_1 \gamma_2^{-1}$$

is a random element of the finite group G_2.

Therefore, given a point A_1 in the finite group G_1, for any element g in G_2, we may use a probabilistic polynomial time algorithm to find:

$$F = f_i(B_1 - B_2)$$

such that:

$$e(A_1, F) = g.$$

where $g = \gamma_1\gamma_2^{-1}$. Evidently, this contradicts the assumption of assumption of Inversion of Weil Pairings. Therefore, the theorem holds.

5.3 Anonymity

In identity-based group signatures, the anonymity implies that any user outside of the signing group cannot identify the membership of the original signer even though the user can check the validity of the group signature.

In this subsection, we discuss the anonymity property of the new identity-based group signature scheme. Given a valid group signature

$$\{A, B, f_i\},$$

since the group membership credential t_i is privately held by U_{f_i}, any user is not able to identify the real identification of U_{f_i}. On the other hand, because of the difficulty of elliptic curve discrete logarithm problem, any user is not able to work out t_i by use of the group public key PK.

5.4 Exculpability

Neither a group member nor the group manager can sign on behalf of other group members. In fact, due to the secure channel between authorized members and the group manager, $sk_i = aQ_{f_i}$ and $cred_i$ are secretly communicated. Additionally, the value b_i is privately held only by U_{f_i}. Therefore, for any authorized member U_{f_j}, she does not know the personal secret key $\{sk_i, b_i\}$ of the authorized member U_{f_i}. Hence, U_{f_j} cannot on behalf of U_{f_i} output a group signature A, B, f_i such that

$$e(A, f_iB) = e(f_iP_{pub}, Q_{f_i})e(A, H_1(m, A)) \tag{8}$$

At the same time, the group manager GM cannot represent or personate U_{f_i} to output valid group signatures. In fact, b_i is secretly chosen by U_{f_i}. Therefore, due to the difficulty of integer factor problem, GM is not able to work out b_i from $b_if_i \equiv 1\mathrm{mod}\varphi(n)$.

5.5 Traceability

The group manager is always able to open a valid signature and identify the actual signer in case of disputes. Given a valid group signature $\{A, B, f_i\}$. By the group membership credential t_{f_i} (related to f_i) and the committed property of the Schnorr identification protocol [17], GM can then identify the real identification of the corresponding authorized group member.

6   Performance

The performance of the new identity-based group signature scheme is dominated by the signing algorithm and the verification algorithm. As to the verification algorithm, there are two point multiplications, one modulus multiplication, one hash function evaluation, and two bilinear pairing computations. Moreover, the verification makes use of the bilinearity of the pairings over elliptic curves. As to the signing algorithm, there are three point multiplications and one hash function evaluation. There is no pairing evaluation in the signature generation.

We note that there is no exponentiation (specially RSA exponentiation) calculations during the generation of signatures. Additionally, there is no exponentiation calculations during the verification of signatures. Previous to this new scheme, some group signature schemes need to compute exponentiations modulo a large RSA modulus [14]. Therefore, the new scheme is efficient in terms of computation cost.

The group public key and group signatures are independent of the number of the authorized group members. Therefore, our scheme is suitable for large groups. Especially, the new group signatures may be applied in mobile communications.

For practical simulation, some researchers have provided useful tools [4,19], to deal with pairing evaluation, point multiplication or scalar multiplication, and hash function evaluation in the elliptic curve settings.

7   Conclusion

A novel identity-based group signature scheme is presented. It makes use of the bilinear pairings over elliptic curves. The size of the group public key is independent of the size of the underlying group. Also, the length of a group signature is independent of the number of the underlying group. In addition, the signing computation does not encompass RSA signatures. Therefore, the new scheme should be efficient. At the same time, the proof of security against adaptive forgeability is presented in this paper.

References

[1]   G.Ateniese, J.Camenisch, M.Joye & G.Tsudik, *A pratical and provably secure coalition-resistant group signature scheme*, Advances in Cryptology-CRYPTO 2000, Springer-Verlag, LNCS 1880, 255-270, 2000.

[2]   G.Ateniese & B. de Medeiros, *Anonymous E-prescriptions*, ACM Workshop on Privacy in the Electronic Society (WPES02), Sponsored by ACM SIGSAC, Washington DC, USA, 2002.

[3]   D.Boneh & M.Franklin, *Identity-based encryption from the Weil pairing*, Proceedings of CRYPTO 2001, Springer-verlag, LNCS 2139, 213-229, 2001.

[4]   P.S.L.M.Barreto, H.Y.Kim, B.Lynn & M.Scott, *Efficient algorithms for pairing-based cryptosystems*, Advances in Cryptology-Crypto 2002, Springer-Verlag, LNCS 2442, 354-368, 2002.

[5]   D.Chaum & E.Van Heyst, *Group signatures*, EUROCRYPT 1991, LNCS 547, Springer-Verlag, 257–265, 1991.

[6]   L.Chen, K.Harrison, N.Smart & D.Soldera, *Applications of multiple trust author-ities in pairing based cryptosystems*, Proc.InfraSec 2002, Springer, LNCS 2437, 260-275, 2002.

[7]   J.Camenisch & M. Michels, *A group signature scheme with improved efficiency*, Advances in Cryptology-ASIACRYPT 1998, Springer-Verlag, LNCS 1514, 160-174, 1998.

[8]   L.Chen & T.P.Pedersen, *New group signature schemes*, Proceedings of EURO-CRYPT 1994, Springer-Verlag, LNCS 950, 171-181, 1995.

[9]   J.Camenish & M.Stadler, *Efficient group signature schemes for large groups*, Pro-ceedings of CRYPTO 1997, Springer-Verlag, LNCS 1296, 410-424, 1997.

[10] F.Hess, *Efficient identity based signature schemes based on pairings*, K. Nyberg and H. Heys(Eds.), Selected Areas in Cryptography, SAC 2002 , Springer-Verlag, 310-324, 2003.

[11] K.Lauter, *The Advantages of Elliptic Curve Cryptography for wireless security*, IEEE Wireless Communications Magazine, IEEE Press, February 2004.

[12] H.Petersen, *How to convert any digital signature scheme into a group signature scheme*, Security Protocols Workshop 1997, 177-190, 1997.

[13] S.Park, S.Kim & D.Won, *Id-based group signature*, Electronics Letters, 33(19), 1616-1617, 1997.

[14] C. Popescu, *An efficient id-based group signature scheme*, Studia Universitatis Babes-Bolyai, Informatica, Vol. XLVII, November 2, 2002.

[15] D.Pointcheval & J.Stern, *Security arguments for digital signatures and blind sig-natures*, Journal of Cryptology, Springer-Verlag, 13(3), 361-396, 2000.

[16] A.Shamir, *Identity-based cryptosystems and signatures*, Proceedings of CRYPTO 1984, Springer-verlag, LNCS 196, 47-53, 1985.

[17] C.Schnorr, *Efficient signature generation by smart cards*, Journal of Cryptology, Springer-Verlag, 4(3), 239-252, 1991.

[18] D.Song, *Preatical forward-secure group signature schemes*, Proceedings of ACM Symposium on Computer and Communication Security, ACM Press, 225-234, 2001.

[19] N.P.Smart & E.J.Westwood, *Point multiplication on ordinary elliptic curves over fields of characteristic three*, Applicable Algebra in Engineering, Communication and Computing, Vol 13, 485-497, 2003.

[20] Y.M.Tseng & J.K. Jan, *A novel id-based group signature scheme*, Proceedings of Workshop on Cryptology and Information Security 1998, Tainan, 159-164, 1998.

[21] C.K.Wu & V.Varadharajan, *Many-to-one cryptographic algorithms and group signatures*, Australian Computer Science Communications, Proceedings of the Twenty Second Australasian Computer Science Conference (ACSC'99), Jenny Edward (Ed.), Springer, 432-444, 1999.

Handover-Aware Access Control Mechanism: CTP for PANA

Julien Bournelle[1], Maryline Laurent-Maknavicius[1], Hannes Tschofenig[2], and Yacine El Mghazli[3]

[1] GET/INT, 9 rue Charles Fourier, 91011 Évry, France
{Julien.Bournelle, Maryline.Maknavicius}@int-evry.fr
[2] Siemens AG, Otto-Hahn-Ring 6, 81739 Munich, Germany
Hannes.Tschofenig@siemens.com
[3] Alcatel, Route Nozay, 91460 Marcoussis, France
Yacine.El_Mghazli@alcatel.fr

Abstract. The PANA protocol offers a way to authenticate clients in IP based access networks. It carries EAP over UDP which permits ISPs to use multiple authentication methods. However, in roaming environments IP clients might change of gateways and new EAP authentication from scratch may occur. This can considerably degrade performance.

To enhance IP handover in mobile environments, we propose to use the Context Transfer Protocol. The aim is to recover from previous PANA Authentication Agent the PANA security context previously established. For this, we define some ways to trigger the transfer and the content of what we called a PANA context.

1 Introduction

Protocol for Carrying Authentication for Network Access (PANA) is a link-layer agnostic transport for the Extensible Authentication Protocol (EAP) to support network access authentication between clients and access networks. In IP based access network, a PANA server may be used as a front-end to a AAA architecture in order to authenticate users before granting them access to resources.

While roaming, the PANA client (PaC) might encounter change of gateway. Without extensions to PANA, the PaC has to restart a new PANA protocol exchange to authenticate itself to the network. In some cases, it is necessary to execute the EAP exchange from scratch while in other cases it might be possible to benefit from a state stored in the visited networks' AAA server.

In this paper, we propose a mechanism to avoid re-authentication from scratch. This is achieved by using CTP (Context Transfer Protocol) [1]. The security context established during the initial authentication and the key establishment procedure (using PANA) are transferred between the old and new points of attachment.

M. Freire et al. (Eds.): ECUMN 2004, LNCS 3262, pp. 430–439, 2004.

2   IP Access Networks

Access to IP networks is usually done from desktop hosts. Ethernet is widely used in companies and university networks, while personal access is often done using telephony networks (dialup). However new access methods are now emerging like ADSL, CaTV (Cable TV), GPRS, CDMA, 802.11a/b/g, providing end-users with high speed networks at low cost.

As a consequence, Internet operators now face new challenges in their access network as they must be able to provide these technologies to their subscribers. There are two kinds of operators: Network Access Providers (NAPs) and Internet Service Providers (ISPs). The former provides network access at the physical and link layers while the latter provides IP connectivity services for subscribers.

ISPs usually provide classical IP services such as packet forwarding, electronic mail, client IP stack parameters configuration (DHCP/DHCPv6/IPv6 Stateless autoconfiguration) and DNS service. In the near future, new services will be deployed like IP Telephony (VoIP), Mobility (Mobile IPv4/Mobile IPv6), Video on demand, etc.

From a security point of view, Internet operators will need:

- To authenticate their subscribers, to authorize them access to specific services and to collect accounting information.
- To provide per packet protection to their subscribers.
- Network access control mechanisms which may be combined with the protocol used to authenticate IP subscribers.

Legacy network access authentication protocols are link layer dependent (for example, approaches used for GSM based networks) or exploit other technologies (e.g. redirect web-logins for HTTP, Mobile IPv4 with its registration required flag). These protocols are difficult to extend and are useful in some specific usage environment whereas in other scenarios they are not suitable. PANA can be used as a single flexible and extensible mechanism in different environments. As will be shown in the next section, it supports various authentication methods, provides dynamic service provider selection, establishes a protected channel for the secure exchange of information between the end host and the network and takes roaming clients into account.

3   PANA

3.1   Benefits

PANA is a protocol currently under definition at the IETF: it can be used as a front-end to a AAA architecture and has the following benefits:

- It covers multiple usage scenarios (WLAN, ADSL, etc.). As such, operators do not need to maintain and deploy many authentication protocols.

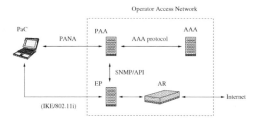

Fig. 1. Functional overview of PANA

- It is a link layer agnostic transport for EAP. Therefore, PANA can be used without considering specific link layer mechanisms. This seems to offer advantages in environments where different link layer technologies are used concurrently.
- It takes into consideration multiple IP address configurations: DHCP, static or stateless IP configuration.
- It supports separate NAP/ISP authentication.
- It offers a dynamic ISP selection.
- The use of EAP allows an easy interfacing with RADIUS or Diameter.
- It allows to bootstrap security associations for other protocols (cf. [2,3]). ·
- Based on the established PANA SA, information between the network and the user can be securely exchanged.

Given the above features, PANA appears to be an optimal choice amongst the available alternatives for future Access Networks.

3.2 Architecture

Figure 1 shows functional entities and protocols involved in an IP network access using PANA: PaC is the PANA client, the PAA is the PANA Authentication Agent and EP is the enforcement point where policy rules apply. Lastly, the AR is the access router; that is the default gateway for IP clients.

The PaC is the equipment which runs the client-side PANA software. It enables the subscriber to ask for network access authorization and to select the ISP. It contains an IP stack (IPv4 and/or IPv6) and has necessary credentials in order to be authenticated by the chosen ISP.

The PAA is positioned in the access network and must be one IP hop away from the PaC. It is in charge of getting credentials from the PaC and may be interfaced with a AAA infrastructure. Upon successful user authentication, the PAA configures the EP for the traffic of the PaC to pass through. When ESP or 802.11i is used for cryptographic data traffic protection, the PAA provides the EP with necessary information (e.g., session key).

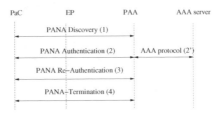

Fig. 2. PANA phases

The EP may be co-located with the PAA and if so, an API (*Application Programming Interface*) is used for communications. Whenever EP is a stand-alone device, the SNMPv3 protocol is used to communicate between PAA and EP [4]. This equipment provides layer 2 and IP filtering and according to operator's policy, implements 802.11i or IPsec ESP/IKE. The document [3] explains how IPsec can be bootstrapped with PANA.

From a protocol point of view, PANA is used between the PaC and the PAA. 802.11i or IKE/ESP is used between PaC and EP and supports packet's confidentiality, authentication and integrity. The PAA uses SNMPv3 to configure EPs. Finally, a AAA protocol such as RADIUS or Diameter may be used to carry EAP payloads between the PAA and the AAA servers to the EAP server to authenticate users and to exchange authorization and accounting information. In this case, EAP is used between the PaC and the AAA server and the PAA acts as a pass-through device. The PaC acts as an EAP peer and the AAA server as an EAP server.

3.3   PANA Overview

PANA can be divided into 4 phases: discovery, authentication, re-authentication and termination:

Discovery: To start the authentication phase, the PaC must discover a PAA. To do so, it sends a `PANA-PAA-Discover` (PDI) message to a well-known link-local address at a specific port. If multiple PAAs are on the same link, the first responder is selected. If PaC is configured with a static PAA's address, it sends a PDI message to this PAA. The network also has the ability to detect the PaC and trigger the authentication phase. To avoid Denial-of-Service, PAA sends a `PANA-Start-Request` (PSR) containing a cookie. The PaC must answer with a `PANA-Start-Answer` (PSA). This exchange bootstraps the authentication phase.

Authentication: During this phase, the PaC can select the ISP: PAA proposes a list of available ISPs in the first `PANA-Authentication-Request` (PAR) message. The PaC then selects the desired ISP and informs the access network of its choice through `PANA-Authentication-Answer` (PAN).

Fig. 3. Authentication stack

The main task of this phase is to carry EAP messages between PaC and the AAA infrastructure. The PAA uses its local AAA server as a relay/proxy or sends directly the AAA traffic to the home server. EAP [5] provides ISPs with the ability to use many authentication methods. Some of which provides key derivation: a first key MSK (*Master Session Key*) is derived at the PaC and at the AAA server (cf. figure 3). Then the AAA server derives another session key (AAA-Key) from the MSK which is sent to the PAA via the AAA infrastructure. This key transport mechanism is protected with RADIUS or Diameter specific security mechanisms. Further information concerning various aspects of this keying framework is available in [6]. If such a method is in use, PaC and PAA can protect their future PANA messages by appending a MAC generated thanks to the PANA security association shared between PaC and PAA.

The PaC is informed of the authentication's result through the message **PANA-Bind-Request**. This message also includes information such as the protection capability of the access network. The PaC answers with the **PANA-Bind-Answer** message.

Finally PAA configures the EP. According to the network policy, the PAA should give EP enough data to bootstrap IKE or the 4-way handshake of the 802.11i protocol [7]. In the former case, IKE configures an IPsec ESP Security Association between PaC and EP while in the latter case the 4-way handshake configures a security association at layer 2 between the same entities. If no packet encryption is required, the EP is just a packet filtering device.

Re-authentication: The protocol provides two levels of re-authentication thanks to PANA and EAP.

- *EAP:* the re-authentication may be triggered by the PaC or by the AAA server. In the former case, the EAP module in the PaC decides to perform a re-authentication of the network. It uses PANA Authentication messages (**PAR/PAN**) for this purpose. In the latter case, it is the EAP module located at the AAA server or PAA which decides to re-authenticate the PaC.
- *PANA:* only PANA modules are involved. The protocol provides two specific messages to handle PANA re-authentication:
  PANA-Reauth-Request(PRAR)/PANA-Reauth-Answer(PRAA). PANA specific reauthentication mechanisms are very efficient means to detect the aliveness of both the PAA and the PaC.

Termination: This phase allows to explicitly close a PANA session. If a PANA SA exists, this phase is protected.

3.4   PANA Context After a Successful Authentication

While authenticated, the PaC shares a session with a specific PAA. Its IP traffic is routed to an AR (Access Router) and is filtered by an EP. If IPsec is in use, the EP is co-located with the AR. The PAA is in the OPEN state and associated timers are set up.

The variables shared by PaC and PAA include:

- *Session-Id*: allocated by the PAA before the authentication phase and used to uniquely identify a session.
- *Device-Id of PaC*: identifier of the device used to handle traffic at the EP level. It is either the IP address or the layer 2 MAC address.
- *Device-Id of PAA*: identifier to whom PaC's PANA messages are sent.
- *Device-Id of EP*: identity of the Enforcement Point
- Initial transmitted sequence number `tseq` of PaC and PAA (ISN_pac/ISN_paa): these are values used in the PANA header.
- Last transmitted `tseq/rseq`. Basically `rseq` is the last `tseq` received.
- *Retransmission interval*.
- *Session lifetime*.
- *Protection Capability*: the protocol which has to be used to protect traffic.
- PANA SA Attributes: `AAA-Key`, `AAA-Key-identifier`, and `PANA_MAC_Key` (key used to integrity and replay protect PANA messages).

3.5   The Handover Problem

The PANA protocol can be used in multiple environments and to a large extent in Radio Access Network (e.g., 802.11a/b/g or GPRS). In that case, the PaC should be a mobile host and the user may move during his PANA session. As noted in section 3.2, the PAA must be one IP hop away from PaC: this means that a specific PANA module on a PAA is in charge of one IP network.

After PaC's IP handover, the PaC changes of IP network and the new PAA (nPAA) does not share a context with the PaC. This new IP access network will detect this new PaC and will trigger a new PANA authentication from scratch. A new authentication phase involving the AAA infrastructure will occur.

Whereas this PaC has already been authenticated, it is asked to prove its identity one more time. Clearly, this is not conceivable if the user is prompted to enter a *login/password*. In this paper, we propose to solve this specific issue by using the Context Transfer Protocol (CTP).

4   The Context Transfer Protocol (CTP)

From a more general perspective, an IP client may share some contexts with the access network equipment. For example, it may have a state information related to Quality of Service, header compression, security services such as AH/ESP. Those services need signaling messages to be established. Each time the IP client moves and changes of access router, those contexts need to be re-established from scratch. CTP [1] has been designed by the Seamoby Working Group at the IETF to solve this problem. The basic idea is to carry the context between Access Routers.

Transferring the Context: CTP provides two types of messages to carry context between access routers and to trigger transfer between IP client and access router. The first one is discussed in this section while the second one is discussed in the next section.

The transfer occurs between the previous (pAR) and the new access router (nAR). CTP provides four messages for this purpose:

- CTD *(Context Transfer Data)*: embeds the context to be transferred.
- CTDR *(CTD Reply)*: reply to CTD.
- CT-Request: request a transfer for a particular context.

All CTP messages must be protected by IPsec ESP. Thus pAR and nAR must be able to establish an IPsec Security Association or must already share one. From a key management point of view this does not cause problems since both entities are located within the same administrative domain.

Triggering the Transfer: The transfer takes place between the access routers but it must be triggered according to the movement of the client. Basically, the protocol considers two possibilities:

1. The client informs the network of its move by sending a CTAR (Context-Transfer-Activate-Request) message. If the MN has not yet performed the handover, it sends the message to the pAR. If the MN has already moved, it sends the message to the nAR. This message is signed by using a key that must be shared between the pAR and the client. This protects the access network against trivial DoS.
2. The network is able to anticipate client's move (cf. [8]).

5   Use of CTP for PANA

CTP provides a generic way to carry context between access routers. This avoids re-running some context signaling from scratch. We propose to use it to improve handover performance in IP access network where PANA is used.

The transfer takes place between PAAs and thus PAAs must share IPsec Security Associations for the protection of CTP messages. Our proposal takes into account the following scenarios: transfer before or after the handover.

5.1   Triggering the Transfer from the PaC

After the handover: Two possibilities are available to trigger the transfer at nPAA:

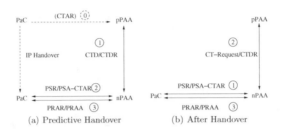

Fig. 4. CTP for PANA messages

1. **CTAR**: the PaC sends a CTAR message whose content is adapted to PANA.
2. **PSA-CTAR**: a CTAR message is embedded in a PSA message.

Basically, if a PaC moves to a new IP access network, it will be detected and the network will start a PANA authentication by sending a **PSR** message. Thus the second solution seems as the most appropriate. The PANA module at the nPAA receives the PSA message and delivers the CTAR message to the local CTP module.

Predictive handover: The PaC should use a classical CTAR message as it does not have to answer to a PANA message.

5.2   The Authorization Token

The CTAR message, which is used to trigger a context transfer, contains an authorization token. This token is a keyed hash over the CTAR content. It implies that PaC and pPAA share a key. For security reasons, it is better to have a unique and fresh session key for each session with each application usage: we propose to derive a new session key named **PANA-CTP-Key** from the **AAA-Key**.

5.3   CTP-PANA Operations at PAA and PaC

To support the context transfer extensions described in this proposal, the PAA and PaC software needs to implement two modules: the PANA module and the PANA-CTP module which provides the CTP functionality.

Predictive Handover. The PaC predicts its movement and initiates the context transfer procedure (cf. figure 4(a)). This allows to further improve performance. For this, the CTP-PANA module should be able to anticipate the move.

The pPAA knows that it must send a CTD message to the nPAA. The first possibility is that the PaC sends to pPAA a CTAR message containing data which permit to retrieve the nPAA's address (e.g., nAR's address). The second

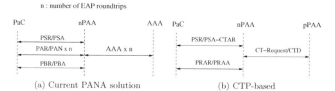

(a) Current PANA solution   (b) CTP-based

Fig. 5. Message exchanges after handover

possibility is that the network is able to anticipate the move of PaC. In this case, the network informs the pPAA of the nPAA's location and triggers the PANA context transfer. In both cases, the pPAA retrieves local information relative to PaC and sends the CTD message containing the PANA context to the nPAA. Then a timer is set up to delete the previous state of the PaC.

The nPAA receives the CTD message and creates state for the PaC; we propose to introduce a new state in the PAA state machines: PANA-CTP-WAITING. When receiving a PSA-CTAR message from the PaC (responding to a PSR while entering in the new access network), the nPAA checks CTAR's validity (cf. section 5.2) and then re-authenticates PaC with the specific PANA exchanges. A new Session-Id is allocated and nPAA can derive a new `PANA_MAC_Key`. Then, it configures the EP with necessary data.

After the Handover. The PaC enters in the network and receives a PSR from the nPAA (cf. figure 4(b)). It replies with a PSA embedding a CTAR message. This message contains data which permits the nPAA to recover the PaC context. The nPAA receives the message and decaspulates the CTAR message. It delivers it to its CTP-PANA module. From the CTAR message, the nPAA learns the pPAA's address and sends a CT-Request message with data provided by the PaC. It receives the PANA context in the CTD message and creates an entry in the state machines. Then it re-authenticates the PaC with the PANA procedure and configures the EP.

6   Theoretical Performance Improvement

Figure 5 presents the message exchanges after the handover as it is currently performed by PANA and as PANA combined with CTP enables to reach. We assume that the PaC has been previously authenticated and consider the handover already occured.

In the current solution, the network detects the PaC and triggers a classical PANA authentication. This approach could lead to a number of round-trips superior to eight messages between the PaC and the PAA depending on the EAP method. Moreover, the PAA needs to contact the AAA infrastructure:

depending on the AAA server location, a long delay may be introduced. The EAP method may also introduced a processing latency: heavy cryptographic computation.

In our approach, six messages are needed and less computational effort is required. The AAA infrastructure is not contacted. For these reasons, our proposal can enhance handover performance.

7   Conclusion

The PANA protocol appears as a technology of choice for ISPs. As explained in this article, it does not depend on the access technology and provides dynamic service provider selection. Moreover, it incorporates access control mechanism and provides bootstrapping method for per-packet protection.

However, PANA does not consider a handover between access networks of different PAA domains: each time a PaC changes access router, a new PANA authentication takes place from scratch. This can seriously degrade performances in mobile environments. To enhance handover performance, we propose to use the Context Transfer Protocol which was developed by the IETF. We define the considered PANA context and necessary operations for transfer.

References

1. Loughney, J., Nakhjiri, M., Perkins, C., Koodli, R.: Context Transfer Protocol. draft-ietf-seamoby-ctp-08.txt (2004) Work in progress.
2. Yegin, A., Tschofenig, H., Forsberg, D.: Bootstrapping RFC3118 Delayed DHCP Authentication Using EAP-based Network Access Authentication. draft-yegin-eapboot-rfc3118-00.txt (2004)
3. Parthasarathy, M.: PANA enabling IPsec based Access Control. draft-ietf-panaipsec-02.txt (2004) Work in progress.
4. El Mghazli, Y., Ohba, Y., Bournelle, J.: PANA: SNMP usage for PAA-2-EP interface. draft-ietf-pana-snmp-00.txt (2004) Work in progress.
5. Blunk, L., Vollbrecht, J., Aboba, B., Carlson, J., Levkowetz, H.: Extensible Authentication Protocol (EAP) (2004) Work in progress.
6. Aboba, B., Simon, D., Arkko, J., Eronen, P., Levkowetz, H.: Extensible Authentication Protocol (EAP) Key Management Framework. draft-ietf-eap-keying-02.txt (2004) Work in progress.
7. IEEE: Wireless Medium Access Control (MAC) and physical layer (PHY) specifications: Specification for Enhanced Security (2002) 802.11i/D3.0.
8. Arbaugh, W., Aboba, B.: Handoff extension to RADIUS. draft-irtf-aaaarch-handoff-04.txt (2003) Work in progress.

Authentication, Authorization, Admission, and Accounting for QoS Applications

Carlos Rabadão[1, 2] and Edmundo Monteiro[2]

[1] Superior School of Technology and Management
Polytechnic Institute of Leiria
Morro do Lena – Alto do Vieiro, 2411-901 Leiria, Portugal
crab@estg.ipleiria.pt
[2] Laboratory of Communications and Telematics
CISUC / DEI, University of Coimbra
Polo II, Pinhal de Marrocos, 3030-290 Coimbra Portugal
{crab, edmundo}@dei.uc.pt

Abstract. The main objective of the IETF Differentiated Services (DiffServ) model is to allow the support on the Internet of different levels of service to different sessions and information flows, aggregated in a few number of traffic classes. The flow classification is supported by some of the IP packet header fields. This approach shows some security limitations that are inherent to the DiffServ model. Being the edge routers (ER) the responsible for the admission and marking of packets, according to the class of service, they are the most vulnerable element to attacks. A security hole in ERs could be propagated to the entire domain, compromising the QoS of all the domain flows. To overcome these limitations, this paper proposes an architecture for Authentication, Authorization, Admission control and Accounting (AAAA) of QoS client applications with dynamic identification of sessions and flows. The proposal functionalities are described and analyzed in some detail, focusing the main modules and message exchange among modules. The paper ends with the discussion of the main advantages of the proposal over existing solutions.

1 Introduction

In communication systems, the expression "Quality of Service" (QoS) is used to characterize the capacity of the system to support data flows with service guaranteed parameters (e.g. bandwidth, delay, jitter, losses) in a more or less strict way. The QoS mechanisms impose priorities and restrictions in the access of flows to available communication system resources. In the case of the DiffServ model [1] this traffic prioritization is supported by the identification of Classes of Service (CoS) done according specific fields of the header of IP packets [2]. As discussed in [3, 4] this approach has some security limitations, namely authentication and authorization.

The IETF DiffServ working group has considered some methods to reduce the inherent security limitations of the DiffServ model [4]. These include auditing and IPSec [5, 6]. However the vulnerabilities to security attacks, such as man-in-the-middle and Denial of QoS (DQoS), remain open issues [7].

M. Freire et al. (Eds.): ECUMN 2004, LNCS 3262, pp. 440–449, 2004.

To overcome the security limitations of Diffserv model, this paper proposes an architecture for Authentication, Authorization, Admission control and Accounting (AAAA) of Quality of Service (QoS) client applications with dynamic identification of sessions and flows. Our proposal addresses the issues related with the secure negotiation of QoS, in an intra-domain scope, namely admission control at the edge devices of DiffServ domains and the processes of authentication of the customers and authorization of flows associated with the resource reservation procedures. The issues related with the confidentiality and integrity of information flows are relegated to other modules of the communication system.

Besides the present section, the paper has the following structure. Section 2 discusses relevant research work related with the improvement of security and management of QoS networks. Section 3 describes our proposal for dynamic QoS negotiation with authentication, authorization, admission control and accounting of sessions and flows. Section 4 will be dedicated to the evaluation of the proposed architecture. Finally, Section 5 will be devoted to conclusions and directions for future work.

2   Related Work

Data communication infrastructures are frequently exposed to attacks to the confidentiality, integrity and availability of the information in transit, and to the authenticity of the origin and destination of this information.

In communication infrastructures based in the DiffServ IETF model this situation is even more serious because they assure different levels of service to different flows of information, over TCP/IP communication infrastructures, contributing to increase the potential of occurrence of some of these attacks. On DiffServ networks the differentiation of the quality of the service provided to different customers is based on the value of the DSCP (DiffServ Code Point) field [2] included in the ToS (Type of Service) field of the IP header [8]. This approach presents some security limitations [3, 4] that could be explored by less scrupulous users trying to get better quality of service for its flows without however assuming the associated costs, leading to the theft of resources that, in extreme situations, may result on Denial of QoS (DQoS) of the active flows. The IETF DiffServ WG considered some methods to reduce the inherent security limitations of DiffServ model, such as auditing and the use of IPSec, but vulnerabilities to attacks such as man-in-the-middle and DQoS remain unsolved [7].

The ARQoS project [9] is one of the initiatives to improve the security in QoS networks, preventing and detecting control and data flow attacks on QoS mechanisms. The work *Preventing Denial of Service Attacks on Quality of Service* [10] addresses the use of the pricing paradigm in the process of resource allocation, that is, the price of the resources increases as the occupation of the resources increases and vice-versa. The process of allocation of network resources is always preceded by authentication and authorization procedures. For such, the proposal uses the *POLICY_DATA* object of RSVP (Resource ReSerVation Protocol) [11] proposed in the RFC2750 [12], to allow the use of the PBN (Policy Based Networks) model [13] and RSVP together. The authors also suggest the use of an authorization server, based on the SIP (Session Initiator Protocol) [14], to generate signed policy objects that grant the ability of users to pay a specific price for specific resources. Later, these objects will be carried to the admission control system, through RSVP messages.

A similar approach is proposed by the 3GPP (3rd Generation Partnership Project) to the UMTS *end-to-end* QoS architecture [15], to interoperate with external IntServ/RSVP networks. When external networks use RSVP, the signalling messages must contain one authentication token, when available, and the flow identification, issued by the SIP protocol. These elements are carried in the *POLICY_DATA* object.

The RSVP admission control mechanisms are often based on user or application identities [16]. The RFC3520 [17] specifies a new object aimed to add to the RSVP a mechanism for admission control, per session. This new object, named *Session Authorization Policy Element*, will transport the authorization of use of a set of specific resources, for a specific session. This object can include the information of the authorized resources (e.g. QoS parameters), identification of the flow and session, duration of the session and the identification of authorization entity.

Despite the above described RSVP extensions, providing additional versatility to the protocol, the RSVP was initially designed to support end-to-end QoS signalling in the Internet and targeted to multicast receiver oriented applications, resulting in a heavy message processing within network nodes. Moreover, there are many applications demanding different signalling services not supported by RSVP.

The IETF NSIS workgroup [18] is considering the decomposition of the overall signalling protocol suite into a generic lower layer, with separate upper layers for each specific signalling applications [19]. A proposal for an upper layer protocol was made by NSIS named *NSLP for Quality of Service Signalling* [20]. This protocol is similar to RSVP and uses soft-state peer-to-peer refresh messages as the primary state management mechanism. However, it supports both sender and receiver initiated reservations between arbitrary nodes, e.g. edge-to-edge, end-to-access, etc., and it doesn't support IP multicasting, making it a more flexible and light than RSVP.

A set of other analyzed works [21, 22, 23, 24] addresses the problematic of the security in infrastructures with QoS support, mainly on the subjects of admission control, proposing the adoption of the PBN architectures, and security aspects of inter-domain signalling between Bandwidth Brokers (BB). However, issues associated with the security in the peripheral networks, such as user authentication and dynamic authorization of resources, remain open.

3   AAAA Architecture for QoS Applications

In the DiffServ model flow authentication is carried out on a per packet basis, at the entrance of each domain. Flow classification is supported by some of the IP packet header fields. As said before this approach has some security limitations that are inherent to the DiffServ model.

Being the edge routers (ER) the responsible for the admission and packet marking according to flow's quality of service, they are the most vulnerable element to attacks and security holes.

To overcome these limitations, this paper proposes an architecture for QoS negotiation with authentication, authorization, admission control and accounting of client applications in a dynamic way, at the entrance of DiffServ domains. The proposed architecture will basically focus in questions of secure negotiation of QoS, in an intra-domain scope, addressing the questions related to admission control at the edge devices of DiffServ domains and with the procedures for customer authentication and resources reservation authorization.

3.1   Architecture Overview

The proposed architecture, shown in Figure 1, has seven main modules: Policy Repository, QoS Client, Authentication, Authorization, Admission Control, Accounting and Router PEP (Policy Enforcement Point). The AAAA modules interact with the Policy Repository module of the domain and with the QoS Client component located at each client with QoS capabilities. The Router PEP and the DiffServ Edge Router (of the network provider) are the responsible for the enforcement of the QoS definitions in the border between the user domain and the network provider.

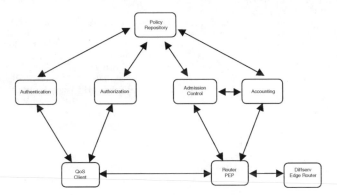

Fig. 1. AAAA Architecture for QoS applications

The QoS client will be initially authenticated in an Authentication module. This module will issue credentials to enable the client to contact the Authorization module. Subsequently, the client will request to the Authorization module the allocation of network resources with the characteristics needed for its new session. After that and according to the domain QoS policies the Authorization module will decide to authorize (or not) the new session, sending a *ticket* that contains, among other data, the resources approved for the new session and the identification of the session flows.

After the above procedures, the client has an authorization to use network resources, emitted by the Authorization module, based in QoS policies, but the needed resources are not yet reserved. To make this reservation, the client needs to issue a request to the Admission Control module, the entity that controls the network resources of the domain. This is achieved sending the *ticket* to the Admission Control module, protected against authenticity treads. This module, based in the domain re-sources availability and in the *ticket* information, will reserve the required resources, configuring the Edge Router and sending the DSCP to the client, to update the session description. After this, the client will mark packets and sends it to the network media. The authorization (*ticket*) has limited time so, before time-out, the QoS Client must refresh the ticket and send it to the Admission Control module.

Figure 2 shows a simplified QoS Client flow diagram for the session (or flow) set-up and teardown procedures. In both cases, the QoS client sends a request for a *ticket*. In the case of a session set-up, the Authorization module issues a *ticket* with informa-tion of the session identification, associated flows and authorized resources for each one. If the QoS Client needs to add a new flow to the session, the Authorization mo-dule will issue a new *ticket* with the same session identification and with information about all the flows belonging to this session (already authorized and new ones). In the case of a flow teardown, the issued *ticket* will include the session identifier, and all the flows to be supported. Finally, in the case of a session teardown, the issued *ticket* will include only the session identifier, being removed all the flows information.

3.2   Architecture Elements

As said before, there are seven fundamental entities in this architecture: Policy Re-pository, QoS Client, Authentication, Authorization, Admission control, Accounting and Router PEP. The functionalities of each of the modules are described bellow.

QoS Client - This module intercepts all the session/flow set-up and teardown re-quests, implicitly made by the application or on the behalf of an external protocol (like SIP signalling, for instance), and issues a resource request to the Authorization module indicating, among other information, the identification of the user, sessions and flows. The client must be authenticated before the request can be issued. If the resource request is authorized by the Authorization module, a *ticket* will be received, with similar capabilities of the gate [25], ticket [26] and token [27] proposals. With this *ticket*, the QoS Client will request resources reservation to the Admission Con-trol, sending the ticket inside a *POLICY_DATA* object [11, 12]. This object will be carried by a signalling protocol like RSVP or NSLP-QoS. This module is also respon-sible by the DSCP packet marking.

Authentication - This module is the responsible for the authentication of the QoS Client, enabling the access to the Authorization entity. This module will consult the Policy Repository in order to determine if user has administrative permission to ac-cess to the system.

Authorization - This module decides whether or not the client has administrative permissions to make an issued reservation and to generate the associated *ticket*. The ticket is based on the resource request, issued by the QoS Client, and on the domain QoS policies, defined by the network administrator and stored in the Policy Reposi-tory.

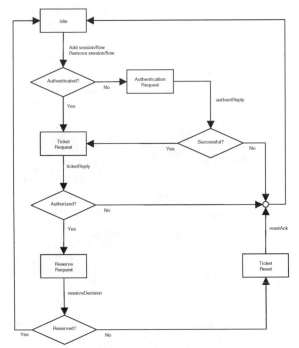

Fig. 2. QoS Client flow diagram for a session or flow set-up/teardown

Router PEP - This module is the responsible for the analysis of the resource request messages issued by the QoS Client. It will extract the relevant information from the received messages and forward this information to the Admission Control entity (e.g. COPS REQ message) [28, 29]. It should also configure the router interfaces according to the configurations received from the Admission Control entity, enabling the flows belonging to authorized sessions to be routed to its destinations, outside the local domain.

Admission Control - This module has as principal task the reception and analyse of the resources request issued by the QoS clients. After consulting the domain resources availability, and according to the resources requested by the QoS client, via *ticket*, this module will issue the decision to the client (for instance with a COPS DEC message [28, 29]) and will ask the PEP Router to configure the router interfaces. The module could also send *tickets* to the Accounting entity for accounting purposes.

Accounting - This module is the responsible to collect information concerning the characteristics and duration of the QoS resources used by the clients, using *tickets* received from the Admission Control entity and/or information received from the PEP Router.

Policy Repository - This module is the responsible to store all the domain policies, including security policies, QoS policies and accounting policies. These policies are defined by the network administrator and could be accessed by the Authentication and Authorization module, in a per session/flow set-up/teardown basis, and sporadically by the Accounting and Admission Control. The policies will reflect the treatment to give to each flow, according to its owner, user group it belongs and privileges, and according to the type of application/flow and its QoS requirements.

3.3   Example of a Session QoS Set-Up

In this section we will present and discuss an example of the use of the proposed architecture. The example will show the message exchange between the several network modules involved at the set-up of a session QoS reservation. Only the originating side flows are described for simplicity. The same concepts apply to the terminating side. The example is illustrated in Figure 3.

The QoS Client intercepts all the session/flow set-up requests, implicitly made by the application or on the behalf of an external protocol like SIP. After an interception, the QoS Client must verify if credentials to authenticate itself to the Authorization Server exists, and if not, get credentials (*authenRequest*) from the Authentication module. After this, the module issues a resource authorization request (*ticketRequest*) to the Authorization module, indicating, among other information, the identification of the user and flows to be authorized.

The Authorization Server queries the Policy Repository (*policyRequest*) in order to know if user has administrative permissions to make the reservation and to decide which resources should be authorized for each flow, according to the application and user privileges. After receiving and analysing the list of Policies (*policyList*), the Authorization entity sends a response to the QoS Client (*ticketReply*), possibly after modifying the parameters of the resources to be used. This response is personated by a *ticket* including, among other elements, the session identifier, the authorized flows and the approved resources.

After receiving the *ticket*, the QoS Client issues a request for reservation of the authorized resources (*resourceRequest*). The ticket provided by the Authorization module is included in this request.

The Router PEP intercepts the reservation request message and sends a policy decision request (e.g., COPS REQ message [12, 28, 29]) to the Admission Control entity (*provisionRequest*) in order to know if the resource reservation request should be allowed to proceed. The *ticket* is included in this above described request.

The Admission Control module uses the *ticket* to extract information about the user and about the services authorized by the Authorization module. The Admission Control, after examining the information, checks the availability of resources to supply the requested QoS and allows or rejects the reservation request. After this, it sends this decision (for instance with COPS DEC message [28, 29, 12]) to the Router PEP (`provisionDecision`). The parameters to configure the edge router interfaces and the DSCP values to apply to the flows of the new session are included in this message. The Admission Control sends the ticket (*accountInforme*) to the Accounting entity, for accounting of the network resources used by this session.

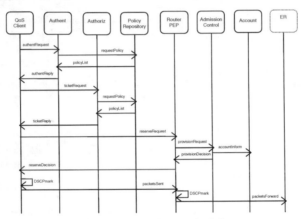

Fig. 3. Sequence diagram for the session QoS set-up

The Router PEP, after receiving the message, configures the router interfaces to allow the flows of the new session and sends a response (`reserveDecision`) to the QoS Client indicating that resource reservation is complete and the DSCP values to mark the packets of flows belonging to the authorized session.

After receiving the response, the QoS Client, allows the flow of the new session to proceed. All the packets of authorized sessions will be marked.

4   Evaluation

In the architecture proposed in this paper for QoS negotiation with AAAA, instead of resource allocation based on static rules, the resources are allocated based on the client and session identification, as well as in accordance with the defined policies, in a dynamic way, e.g. when new flows or sessions are requested. The requests are authorized by an authorization module issuing a *ticket* according to the security and QoS policies of the local domain. After the authorization, the admission control entity will analyse the *ticket* and, if resources are available, the edge router will be configured to enable the flow or the set of flows belonging to a session. With this procedure the contracts established between the users and network providers will not be violated and the resource allocation will be done in accordance with the domain policies.

Besides the above advantages, when the customer gets a *ticket* authorizing him to use a set of resources with QoS and the admission control module denies this reservation, an auditing process could be triggered to identify the causes of the fault (e.g. bad use of QoS policies, insufficient resources) enabling the system to take decisions to solve the malfunctioning of the system (e.g. to eliminate the cause of bad application of policies, to renegotiates the resources near the network provider).

The use of tickets enable the system to add and to remove new flows in an expedite way and to reduce the vulnerabilities to the attacks of theft of resources and consequent denial of QoS, because only clients who acquire one ticket could request resources reservation and only authenticated users with expressed permission will be able to acquire its. To reduce the vulnerabilities to theft or corruption of tickets, all the issued tickets will have a limited lifetime and will be authenticated by the authorization entity. These mechanisms reduce the vulnerabilities to man-in-the-middle attacks, aimed to adulterate ticket and consequently deny the QoS of the flows.

5   Conclusions and Future Work

In this paper, we proposed an architecture for QoS negotiation with Authentication, Authorization, Admission control and Accounting (AAAA), for client application session with QoS needs. The proposal seeks to overcome the security limitations of current DiffServ model and to manage the admission control of sessions in a dynamic way, according to the client profile and to the available resources.

The proposal addresses the issues of secure negotiation of QoS in an intra-domain scope, the issues related to admission control in the edge devices of DiffServ domains and the procedures of authentication of the clients and authorization of resources reservation for sessions establishment.

Future work (already ongoing) will address experimentation and evaluation of scalability and performance behaviour of the proposed architecture.

Acknowledgements. This work was partially financed by the PRODEP program supported by the Portuguese Government and the European Union FSE Programme.

References

[1]   S. Blake *et al*, *An Architecture for Differentiated Services*, RFC 2475, IETF, Dec. 1998.
[2]   K. Nichols *et al*, *Definition of the Differentiated Services Fields (DS Fields) in the IPv4 and IPv6 Headers*, RFC 2474, IETF, Dec. 1998.
[3]   C. Rabadão, E. Monteiro, *Segurança e QoS no Modelo DiffServ (Security and QoS in the DiffServ Model)*, 5th Conference on Computer Networks (CRC2002), Faro, Portugal, University of Algarve, 26-27 Sep. 2002.

[4] Zhi Fu *et al*, *Security Issues for Differentiated Service Framework*, Internet Draft (expired), Oct. 1999.

[5] S. Kent, R. Atkinson, *IP Encapsulating Security Payload (ESP)*, RFC 2406, Nov. 1998.

[6] R. Atkinson, *IP Authentication Header*, RFC 1826, IETF, Aug. 1995.

[7] A. Striegel, *Security Issues in a Differentiated Services Internet*, Proc. of Trusted Internet Workshop - HiPC, Bangalore, India, Dec. 2002.

[8] J. Postel, Editor, *Internet Protocol*, RFC 791, IETF, Sep. 1981.

[9] D. Maughan *et al*, *The ARQoS Project: Protection of Network Quality of Service Against Denial of Service Attacks*, http://arqos.csc.ncsu.edu/, State University of North Carolina, University of California and MCNC.

[10] E. Fulp *et al*, *Preventing Denial of Service Attacks on Quality of Service*, Proc. of DARPA Information Survivability Conference and Exposition (DISCEXII'01), IEEE Computer Society, 2001.

[11] R. Braden *et al*, *Resource ReSerVation Protocol (RSVP) - Version 1 Functional Specification*, RFC2205, IETF, Sep. 1997.

[12] S. Herzog, *RSVP extensions for policy control*, RFC2750, IETF, Jan. 2000.

[13] S. Hahn *et al*, *Resource Allocation Protocol*, IETF, http://www.ietf.org/html.charters/rap-charter.html.

[14] J. Rosenberg *et al*, *SIP: Session Initiation Protocol*, RFC 3261, IETF, Jun. 2002.

[15] *Access Security for IP-based Services*, Technical Specification 3GPP TS 33.203, Version 6.1.0, 3rd Generation Partnership Project, Dec. 2002.

[16] S. Yadav *et al*, *Identity Representation for RSVP*, RFC 3182, IETF, Oct. 2001.

[17] L-N. Hamer *et al*, *Session Authorization Policy Element*, RFC3520, IETF, Apr. 2003.

[18] J. Loughney *et al*, *Next Steps in Signaling (NSIS)*, IETF, http://www.ietf.org/html.charters/nsis-charter.html.

[19] R. Hancock *et al*, *Next Steps in Signaling: Framework*, Internet Draft (work in progress), IETF, Oct. 2003.

[20] S. Van den Bosch, G. Karagiannis and A. McDonald, *NSLP for Quality-of-Service Signaling*, Internet Draft (work in progress), IETF, Feb. 2004.

[21] G. Pujolle, H. Chaouchi, *QoS, Security, and Mobility Management for Fixed and Wireless Networks under Policy-based Techniques*, IFIP World Computer Congress 2002.

[22] E. Mykoniati *et al*, *Admission Control for Providing QoS in DiffServ IP Networks: The TEQUILA Approach*, IEEE Communications Magazine, Jan. 2003, pp. 38-44.

[23] A. Ponnappan *et al*, *A Policy Based QoS Management System for the IntServ/DiffServ Based Internet* , Proc. of 3rd International Workshop on Policies for Distributed Systems and Networks, POLICY'02, Monterey-California, 5-7 Jun. 2002.

[24] V. Sander *et al*, *End-to-End Provision of Policy Information for Networks QoS*, Proc. of 10th IEEE International Symposium of High Performance Distributed Computing, San Francisco-California, 07-09 Aug. 2001.

[25] *PacketCable Dynamic Quality of Service Specification*, CableLabs, Dec. 1999.

[26] J. Vollbrecht *et al*, *AAA Authorization Framework*, RFC 2904, IETF, August 2000.

[27] L-N. Hamer, B. Gage and H. Shieh, *Session Authorization Policy Element*, RFC3521, IETF, Apr. 2003.

[28] D. Durham, *The COPS (Common Open Policy Service) Protocol*, RFC2748, IETF, Jan. 2000.

[29] J. Boyle *et al*, *COPS usage for RSVP*, RFC2749, IETF, January 2000.

Implementation and Performance Evaluation of Communications Security in a Mobile E-health Service

Juan C. Sánchez, Xavier Perramon, Ramon Martí, and Jaime Delgado

Universitat Pompeu Fabra (UPF), Pg. Circumval·lació 8
E-08003 Barcelona, Spain
{jcarlos.sanchez,xavier.perramon,ramon.marti,jaime.delgado}@upf.edu

Abstract. This paper discusses the implementation and performance evaluation of the security services in a mobile healthcare system. The system consists of a number of sensors attached to the patient, which collect data that are sent via 2.5G or 3G mobile telephony to a hospital or healthcare centre for monitoring. Security is an essential requirement of the system due to the high sensitivity of the transmitted information. Since mobile devices are used, adding security introduces a deterioration in system and communications performance. For this reason, security implementation performance is a critical issue in mobile e-health services.

1 Introduction

Electronic healthcare or "e-health" is an IT service that demands security measures to protect the patient's personal data. If the service is enhanced by affording mobility to the patient by means of wireless communication devices, thus providing mobile healthcare or "m-health", the security requirements are stronger, since it is necessary to protect the information conveyed through the wireless links.

In this paper we present a specific implementation of an m-health system, designed for real-time monitoring of patient data from a hospital or healthcare centre via 2.5G or 3G (GPRS or UMTS) mobile telephony. This system allows patients with chronic diseases, such as diabetes or asthma, or in a rehabilitation process, to lead a normal life while their vital constants are measured and sent to the healthcare professionals, without the need for frequent visits to the hospital. Other possible applications of the system are monitoring of training programmes for athletes, etc.

Next, we discuss the security functionalities needed in this system. Some are common to most IT systems, and some are specific to mobile services. We will focus on those security requirements related to the communications in the m-health system.

Finally, we describe the solution that we have implemented to add security and privacy to the system, focusing on performance evaluation to assess the efficiency of our solution, and we present some conclusions from these results.

M. Freire et al. (Eds.): ECUMN 2004, LNCS 3262, pp. 450–459, 2004.
© Springer-Verlag Berlin Heidelberg 2004

2   The MobiHealth System

MobiHealth is a European project for developing a mobile healthcare system over GPRS and UMTS communications, with participation from several hospitals, mobile network operators, software companies and universities. Information about the project can be found in the MobiHealth website [1].

The core of the mobile part of the system is a Body Area Network (BAN) of devices worn by patients while carrying out their normal daily activities. The BAN design [2,3] is based on the novel concept of Personal Area Networks (PAN) [4]. The devices in the BAN are normally sensors that collect medical data about the patient, e.g., blood pressure, pulse, glucose level, electrocardiogram (ECG), etc. Apart from sensors, if necessary it would also be possible to include actuators, such as insuline pumps, in the BAN.

All sensors in a BAN are connected, following a star-like topology, to a front-end that acts as the hub of the network. Connections between sensors and the front-end will generally be wired, but it is also possible to have sensors with short-range radio links such as Bluetooth or Zigbee (IEEE 802.15.1 or 802.15.4). The front-end collects all data from the sensors and forwards them to the so-called Mobile Base Unit (MBU), a computing device running a BAN application for local processing of the data. The MBU will typically be a personal digital assistant (PDA), but it could also be a new generation programmable mobile phone.

The MBU sends the data to a server in the hospital or healthcare centre through GPRS/UMTS. If the MBU is a PDA, this communication can be done with an attached GPRS/UMTS jacket or, alternatively, the MBU can be connected to a mobile terminal via Bluetooth or Zigbee. Over this GPRS/UMTS communication, the MBU uses the Internet Protocol (IP) to access the server anywhere in the Internet.

In MobiHealth, the server is called the Back-End System (BEsys), and it is generally located in the hospital or healthcare centre where the end users (doctors, nurses or other healthcare professionals) can monitor the patient's data with a specific End-User Application (EUA). It is also possible for an external service provider (which could be the GPRS/UMTS network operator itself) to host the BEsys, or part of it, if there are not enough computing resources available at the hospital.

If the BEsys and the EUA are both at the same location, the communication between them will normally be through a Local Area Network (LAN). However, sometimes the caregiver may need to access the BEsys from outside the hospital, e.g., from a portable computer while visiting the patient at home. Therefore, there must also be provision for external EUA access to the BEsys. At the application layer, the communications with the BEsys make use of HTTP-based protocols exchanging XML-encoded messages.

There are three main components in the BEsys: the BAN Data Repository (BDR), the Surrogate Host (SH), and the Wireless Service Broker (WSB). The BDR provides persistent storage for data received from the BAN of each patient. The SH provides an interface based on distributed objects that allows the EUA to

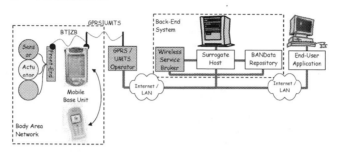

Fig. 1. Overview of the MobiHealth system

access the BAN data (the EUA interacts with the SH as though it were directly the BAN). The SH is implemented as a Jini system [5], which uses the Java Remote Method Invocation (RMI) technique for distributed object discovery and event notifications. Finally, the WSB is an optional component in charge of authenticating the mobile terminal in the BAN and authorizing its access to the BEsys.

Figure 1 presents an overview of the different components of the MobiHealth system.

3   Security and Privacy in MobiHealth

3.1   Security Requirements

As mentioned previously, there are a number of security requirements that are specific of e-health services, and further requirements in m-health services due to the use of mobile communications. We enumerate here some of the security functionalities that have been considered in the development of the MobiHealth system, and we will then concentrate on the implications of adding security to the communications, in particular between the MBU and the BEsys. A more exhaustive review of the threats and security services in MobiHealth can be found in [6,7].

As with most IT systems, threats to information security in an e-health service (apart from those of environmental or accidental origin) can have an impact on privacy, integrity, authenticity, and system performance. In addition, the application may require one or more of the following security services: traffic confidentiality (volume, frequency, origin and destination, etc.), non-repudiation, access control, secure data storage, and secure timestamping.

An m-health system like MobiHealth can be used in different scenarios, and with different types of applications: chronic disease, rehabilitation, physical activity monitoring, etc. Depending on the application, some security functions may be more necessary than others. For example, the recording of patient's data in local storage within the BAN (in the MBU) can be allowed at user's option, allowed temporarily only when the connection to the BEsys is not available (e.g., while out of GPRS/UMTS coverage), or disallowed at all due to the breach of privacy that this could represent.

3.2   Security in the Communications

In MobiHealth there are different communication interactions between its various components. In the development of the system we have considered basically the following communications: between sensors and front-end, between front-end and MBU/mobile terminal, between mobile terminal and GPRS/UMTS base station, between the Internet and the BEsys, and between the BEsys and the EUA.

In each of these communication segments, one or more of the following communication layers is involved: the data link layer, the network layer, the transport layer, and the application layer. Different techniques can be used for adding security at each layer.

- At the data link layer, all mobile communication technologies provide some form of encryption, to guarantee to a certain degree the privacy of the data, and authentication: Bluetooth, Zigbee, GPRS and UMTS, and also WEP (Wired Equivalent Privacy) if IEEE 802.11 Wireless LAN is used.
- At the network layer, the standard method for protecting IP traffic is the IPsec architecture, based on the AH (Authentication Header) and ESP (Encapsulating Security Payload) protocols.
- At the transport layer, one of the most implemented solutions is the use of the SSL/TLS (Secure Sockets Layer or Transport Layer Security) protocol. SSL/TLS provides a negotiation mechanism for exchanging encryption keys by means of public key cryptography, so that eavesdroppers can not discover the cryptographic parameters. This negotiation based on public keys also guarantees authenticity of the communicating parties.
- At the application layer, it is possible for applications to implement directly their own security mechanisms. However, it is more usual to benefit from the security provided by the lower layers, e.g., when an HTTP-based or an RMI-based application needs secure transport, HTTPS (i.e., HTTP over SSL/TLS) or secure RMI (i.e., Remote Method Invocation over SSL/TLS), respectively, can be used.

Therefore, if the communication between two system components makes use of several communication layers, it may be necessary to choose between different security protocols. For the MobiHealth system, we decided not to implement any application layer security mechanism, and to use always the data link layer protection as provided by the corresponding protocol (Bluetooth, Zigbee, GPRS, UMTS, WLAN), with an appropriate higher layer security protection on top of that. Thus, the choice of secure protocols was reduced to the network layer (IPsec) and/or the transport layer (SSL/TLS).

4   Security Implementation and Performance Evaluation in MobiHealth

4.1   The MobiHealth Security Implementation

Apart from the security requirements listed in the previous section, we also had to address the following requirements:

- Some components of the system may have a dynamically assigned IP address: typically the mobile terminal in the BAN will have an address provided by the network operator, and also the end-users accessing the system from outside the hospital may have a dynamic IP address. Therefore, the IPsec protocols are not well suited because node authentication is based on their IP address.
- The security modules should be as transparent to the rest of the system as possible, in order not to add too much complexity to the implementation of the other modules.
- The inclusion of security should not imply a significant overhead in the communications between the components of the system.

For these reasons, we decided to base the security implementation on the SSL/TLS protocols. This implies that every component that will act as an end-point of the application communication must have an X.509 certificate assigned, with its corresponding public key. In the case of the MBU, the associated private key is stored in encrypted form, and a personal identification number (PIN) is needed to decrypt it. This PIN is optional, since it may be difficult for some patients, in particular elderly people, to remember it or to enter it into the PDA.

Since the HTTP protocol is used in the communications between MBU and BEsys and between BEsys and EUA, this protocol is replaced by HTTPS. And since communications with the SH use Remote Method Invocation (RMI), this is replaced by secure RMI.

In addition, the goal of achieving maximum transparency is facilitated by the fact that SSL/TLS provides an interface that is very similar to the non-secure transport interface. Adaptations to existing libraries have been made so that the software in the MBU can simply be built with a secure version of the transport library (with SSL/TLS), without the need to modify the application code.

This has allowed an independent development of the security modules and the rest of the system. There has been no need to adapt the system to the security implementation, only the security modules have been adapted to the system in the integration phase.

Further adaptations to the transport library were necessary because of an internal characteristic of the BEsys: in order to facilitate the transmission of dynamically generated content, the BEsys makes use of the "chunked" transfer mode of HTTP/1.1, whereby a single HTTP message can be divided into chunks, each with its own length specification. The possibility of using chunking had to be considered in the HTTPS implementation.

We also adopted two measures for reducing the overhead introduced by the security functions. The first one was to modify the SSL/TLS library in order to negotiate only the use of fast symmetric encryption algorithms (in particular the RC4 algorithm). The other one was to reduce the length of the public keys in the certificates to 512 bits. This length is near the border of what is considered secure today, but we must take into account that the security of the system is not based exclusively on these public keys (e.g., there is additional encryption at the data link layer).

The possibility of adding or removing the security modules allows for an easier evaluation of their efficiency, because it makes it simpler to compare the performance of the system without security and with security, as shown in the next subsection.

4.2   Performance Evaluation

A number of trial scenarios have been defined in the MobiHealth project for testing the usability and performance of the system with real patients. One of these trials has also been selected for evaluating the security implementation (in this case with a simulated patient). This trial consists in supervising the rehabilitation of chronic respiratory patients by measuring pulse oximetry and ECG.

The configuration used in the security tests was a UMTS mobile phone connected via Bluetooth to a PDA where the BAN software was running. All sensors and the PDA were connected through wires to the front-end. The BAN was in Barcelona (Spain) and the BEsys was installed in Enschede (Holland).

It is worth noting that these tests were made with a static patient, due to the nature of the medical trial. In this case, mobility provides the possibility of taking measures from any location, but not necessarily while in motion. The available bandwidth for the UMTS communication and the distance to the base station were constant during the tests, but variability in these parameters would be another factor that would have an influence on the system performance.

The measurements were done by running the BAN for intervals of 3 minutes and capturing locally from the MBU the packets transmitted to the BEsys. The communication was restarted at every run, and the amount of data to be sent was the same for every interval.

Three communication modes were used in the tests for comparison: with HTTP and no WSB, with HTTPS and no WSB, and with HTTPS and WSB (see Sect. 2 above for an explanation of the role of the WSB). Table 1 summarizes the throughput achieved in each test, expressed in application bytes (i.e., not counting lower layer headers).

As can be seen from the table, the fact of using HTTPS (HTTP over SSL/TLS) causes a reduction of approx. 25% in the total throughput, due to the SSL/TLS headers overhead. The WSB also introduces an additional decrease of throughput.

The following figures present the instant throughput and the average accumulated throughput that was achieved during one of each type of tests. Figures 2

Table 1. Average performance in various tests

Test	Throughput
UMTS, HTTP, no WSB	2610 bytes/s
UMTS, HTTPS, no WSB	1935 bytes/s
UMTS, HTTPS, WSB	1631 bytes/s

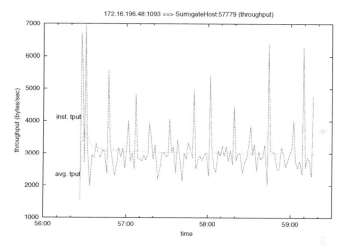

Fig. 2. Data throughput between MBU and BEsys with UMTS and without SSL (HTTP)

and 3 show the throughput average of the best performance for HTTP tests and the best performance for HTTPS tests without WSB, respectively. The two lines in each figure represent instant values and average values of the throughput.

It can be observed from the figures how both test cases differ not only in the end value of average throughput, but also in the establishment time of the connection due to the SSL/TLS negotiation.

The 25% of throughput reduction between both tests is due to two main reasons:

- addition of SSL/TLS header in data sent through the mobile telephony bandwidth-limited channel;
- additional resource requirements in encryption/decryption hardware.

SSL/TLS also adds an initial handshaking, which affects the amount of data and mainly the time to establish the connection. The SSL/TLS handshaking for

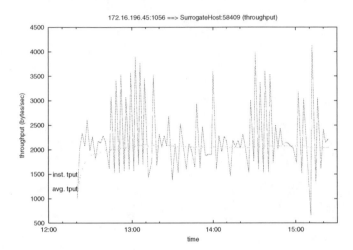

Fig. 3. Data throughput between MBU and BEsys with UMTS and with SSL (HTTPS)

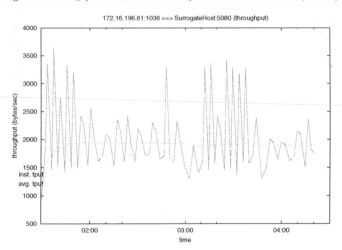

Fig. 4. Data throughput between MBU and BEsys with UMTS and with SSL (HTTPS) through the WSB

HTTPS connections in our setup takes 5 seconds. The modifications we have made to the SSL/TLS library have reduced to the minimum the handshaking while keeping the authentication, integrity and privacy.

Finally, Figure 4 shows the throughput for the HTTPS case including the WSB in the communications, which introduces additional overhead.

5   Conclusions

The main goal of the MobiHealth project is to prove the feasibility and the advantages of using 2.5G and 3G mobile telephony in a specific area of application, namely e-health services. But it also has other benefits, one of which is the implementation of the security functions that we have presented in this paper to enhance the quality of the service. This enhancement is aimed at increasing trust and acceptance of the users of e-health services, which will add to the social and economical advantages of mobile healthcare for an important number of citizens.

From the technical point of view, a secure communications implementation has been developed that fulfills the requirements of this type of system. The inclusion of security measures necessarily implies a decrease in overall system performance, but this decrease has been partially alleviated by using fast symmetric encryption algorithms, and adjusting the length of cryptographic keys. The observed performance lies within the reasonable expectations, given the resource limitations imposed by mobile devices, and is compensated by the enhanced quality of service that security provides.

Acknowledgements. The work presented here started in the implementation and performance evaluation of communications security in a mobile e-health service within the MobiHealth project, co-funded by the European Commission through the FP5 IST programme, and is being continued within VISNET, a European Network of Excellence (http://www.visnet-noe.org/) funded under the European Commission FP6 IST programme. In this Network of Excellence, our research is being centred on the development and profiling of an open security library for mobile and high-speed services.

References

1. MobiHealth website: www.mobihealth.org
2. V. Jones, R. Bults, D. Konstantas, P. AM Vierhout: Healthcare PANs: Personal Area Networks for Trauma Care and Home Care. In: Fourth International Symposium on Wireless Personal Multimedia Communications (WPMC), Aalborg, Denmark (2001)
3. K. van Dam, S. Pitchers, M. Barnard: From PAN to BAN: Why Body Area Networks? In: Proceedings of the Wireless World Research Forum (WWRF) Second Meeting, Helsinki, Finland (2001)
4. T. Zimmerman: Personal Area Networks (PAN): Near-field Intra-body Communication. IBM Systems Journal **35** (1996) 609–618

5. Jini Network Technology: `<http://www.sun.com/jini/>`
6. R. Martí, J. Delgado, X. Perramon: Security Specification and Implementation for Mobile e-Health Services. In: S.-T. Yuan, J. Liu (eds.): E-technology, e-commerce and e-service. IEEE Computer Society Press (2004)
7. R. Martí, J. Delgado, X. Perramon: Network and Application Security in Mobile e-Health Applications. In: Information Networking (ICOIN 2004). Lecture Notes in Computer Science. Springer-Verlag (to be published)

Inside BGP Tables: Inferring Autonomous System's Interconnectivity

Juan I. Nieto-Hipólito[1] and José M. Barceló

Computer Architecture Department, Technical University of Catalunya, Spain.
C/ Jordi Girona, s/n, Campus Nord UPC, D6-116, Barcelona, 08034, Spain
Tf: +34 93 401 7187, Fax: +34 93 401 7055
{jnieto,joseb}@ac.upc.es

Abstract. In this paper we address the problem of Autonomous System (AS) interconnectivity. We analyze the union of six BGP Tables taken from different places. The union of BGP tables from several AS's gives a richer perspective in terms of number of AS's and inter-domain links discovered. We infer AS relationships using well-known heuristics. Once we have the AS relationships we can obtain the AS's that form the regional area of the Internet eliminating the end customers. Pruning the heuristic we obtain AS's that form the core of the Internet. We, then, can study whether the set of AS's in the regional and core sub-graphs and the in-degree and out-degree distributions follow power-laws. Our results show that the In-degree CCDF fits well a power-law in all the graphs defined. However, the Degree CCDF only fits a power-law for the whole graph. For the regional and core sub-graphs the Degree CCDF is better fitted with a Go Model. The Out-degree CCDF for these sub-graphs are well fitted with a power-law only for the first hundred of neighbors.

1 Introduction

Recently, Faloutsos et al., [8], published that the Internet connectivity in terms of degree distributions at the Autonomous System (AS) level could be modeled by a power-law distribution of type $y=Ax^{-c}$, where c is a positive number. Other parameters such as the clustering coefficient and the AS path length are also analyzed when AS topologies are studied. The importance of understanding and modeling the Internet topology is concerned with traffic engineering, protocol performance evaluation, simulation of a variety of network problems, studies on protocols scalability specially those treating with routing, design of new protocols that take advantage of the Internet topology, adaptation of current deployed protocols, figure out the optimal location of servers and network resources and the evolution and growing of the Internet itself.
The above power-law has been verified in many works like [4], [5], [6], [7], [8] or [15] among others. Several authors like [4], [5], [8], [9] or [15] only used for their

---

[1] Juan I. Nieto-Hipólito belongs to the Autonomous University of Baja California (UABC), México. He is a PhD student at UPC, Spain, with a grant of the Mexican government. This work was also supported by the Ministry of Education of Spain under grant CYCIT TIC-2001-SGR-00226 and by CIRIT under grant CIRIT 2001-SGR-00226 and NoE EuroNGI.

M. Freire et al. (Eds.): ECUMN 2004, LNCS 3262, pp. 460–472, 2004.
© Springer-Verlag Berlin Heidelberg 2004

studies one source of data or BGP table obtained from Oregon Route Views project, [1]. Works such as [6], [7], [10] or [14] take several BGP tables from different ISP's and join the BGP table to enrich the adjacency matrix that represents the graph. Joining BGP tables allows the discovery of more AS's and links[2]. Authors from CAIDA, [11], propose to generate active probes to complete the AS interconnectivity. In many of the former works showing power-law behaviors, authors did not take into account that BGP is a protocol of business relationships, [13]. Authors of [9] defined five types of AS relationships and gave a heuristic to infer these relationships from a BGP table, and authors of [14] proposed a different heuristic to infer the AS relationships and furthermore divided the AS's as belonging to five hierarchical levels. They furthermore argued that tables taken from only one point such Oregon Route Views give a poor vision of the Internet.

We obtain BGP tables from several AS's (e.g.; repositories such as Route Oregon, [1], RIPE, [2], and Swinog) and from individual leaf AS's (e.g.; Exodus Comm. Europe, Exodus Comm. Asia and Opentransit) via telnet to its BGP routers, [3]. We then follow the approach of authors of [9] to classify peering relationships and the approach of [14] to join BGP tables from several points and eliminating end customers from the AS graph to obtain a sub-graph representing the AS's of the regional area of Internet that interconnect other AS's but that do not have peer to peer relationships among them. Iterating the algorithm we eliminate intermediate or regional AS's to obtain a new sub-graph representing the core of the Internet with AS's with peer to peer relationships. As an ongoing work of [10], we are interested in studying whether the power-laws are ubiquitous along BGP Tables. Or in other words, we want to know whether the degree distributions of the sub-graphs representing the AS's of the regional area and the core of the Internet follow the same distributions as the degree distributions of the whole AS graph. Furthermore, we will study the in-degree and out-degree distributions as in the whole AS graph as in the sub-graphs. This study may help to identify parameters in order to build Internet graph generators at AS level taking into account AS relationships and hierarchical levels.

Section 2 is devoted to explain the methodology used and to define the metrics elected and the heuristics used. Section 3 shows the results obtained applied to the whole BGP table, regional area and core sub-graphs. Section 4 finalizes with the conclusions of the work.

2   Methodology

We consider the AS level topology and define the graph G=(N,E), where N is the number of vertices or AS's and E the number of edges or links that connects the vertices. From now a link is an edge between two AS's regardless of the number of physical connections that exist between them. We define the adjacency matrix A as a symmetric matrix of size NxN with components $a_{ij}=1$ if node i has an edge joining node j and 0 elsewhere. The degree di of a node is calculated as $d_i = \sum a_{ij}$ for all j.

---

[2] Here we refer as link an inter-domain link or an edge between two neighbor AS's

Works such as [6], [7], [10] or [14] show that joining BGP tables from different AS's located in different geographical sites increment the number of links and AS's discovered. All BGP tables used in this work were collected over the same period of time (may/2003), and the difference in collection times between the first and the last one was 1 week. Of these tables Oregon and Ripe-rcc are Remote Route Collectors, which have a repository where the complete data can be obtained via an anonymous ftp. Ripe-rcc is a set of 9 remote route collectors, 7 deployed in Europe, 1 in Japan and 1 in the USA, see [3]. The rest of the data sources were downloaded using the CISCO or Zebra "sh ip bgp" command via telnet to the site. Swinog is a medium size AS with 41 neighbours and Exodus Comm. Europe, Exodus Comm. Asia and Open-transit are leaf ASs with only one neighbour.

From now on, data from Oregon Route Views repository will be referred as Oregon and the union of the seven sources of data will be referred as Union. With the union of data sources, the number of AS's and links discovered increase. We note an increase of 5.6% and 25.8% of AS's and links, respectively, between Oregon and the Union.

We will investigate degree's distribution and peering relationships at three levels: the whole AS graph, the regional area sub-graph (without end customers) and the core area sub-graph (without end customers and regional AS's).

2.1   Metrics

In this section we define the metrics selected to compare our sources of data:

Degree CCDF: This metric plots the fraction of nodes with degree greater than or equal to d versus the degree of the AS. In a probabilistic sense the $CCDF$ is defined as $F_d = Prob\{D \geq d\} = \Sigma f_i$ for $d \leq i < \infty$, where D is a random variable that indicates the number of incident neighbours upon an AS, i.e. its degree. F_d also follows a power-law with exponent α, $F_d \propto d^{\alpha}$.

Correlation coefficient R^2: is a normalized measure of the strength of the linear relationship between two variables. Uncorrelated data results in a correlation coefficient of 0 while equivalent data sets have a correlation coefficient of 1.

2.2   Inferring the AS's Relationships

L. Gao et-al, [9], give an heuristic to infer AS relationships into provider-to-customer (P2C), peer-to-peer (PEER), and sibling (SIB) relationships. Using Gao's heuristic we can obtain a directed provider-to-customer graph from which we define two more adjacency matrices: the in-degree adjacency matrix A^I and the out-degree adjacency matrix A^O, see Table 1. The in-degree adjacency matrix A^I has components $\{a^I_{ij}\}=1$ when node i is a customer of node j or node i is a peer or sibling of node j and $\{a^I_{ij}\}=0$ otherwise. The out-degree adjacency matrix A^O has components $\{a^O_{ij}\}=1$ when node i is a provider of node j or node i is a peer or sibling of node j and $\{a^O_{ij}\}=0$ otherwise.

The in-degree, d^I_i, of node i is the number of providers directly connected with it, so $d^I_i = \Sigma a^I_{ij}$ for all j. The out-degree, d^O_i, of node i is the number of customers directly connected with it, so $d^O_i = \Sigma a^O_{ij}$ for all j. Note that a peer to peer and a sibling relationship imply a provider to customer plus a customer to provider relationship. That means that $\Sigma d_i \neq \Sigma d^I_i + \Sigma d^O_i$.

Table 1. In-degree and out-degre in a direced provider-customer graph

AS	AS	Relationship	
i	j	Provider to Customer	$A_{ij}=1$ and $A_{ji}=0$ for out-degree
		$i\,(P) \rightarrow (C)\,j$	$A_{ij}=0$ and $A_{ji}=1$ for in-degree
i	j	Peer and Sibling	$A_{ij}=1$ and $A_{ji}=1$ for out-degree
		$i\,(P) \leftrightarrow (C)\,j$	$A_{ij}=1$ and $A_{ji}=1$ for in-degree

Figure 1 shows a very simple AS's Topology as an example. From 1a), that is, without taking into account AS relationships, each AS has degree equal to two. Taking into account AS relationships, 1b), AS_1 has in-degree $d^I_1 = 2$ and out-degree $d^O_1 = 0$, that is, AS_1 is client of AS_2 and AS_3. AS_2 has in-degree $d^I_2 = 1$ and out-degree $d^O_2 = 2$, since it is a peer of AS_3 and at the same time provider of AS_1. AS_3 has in-degree $d^I_3 = 1$ and out-degree $d^O_3 = 2$, since it is provider of AS_1 and a peer of AS_2. Besides AS_1 with an out-degree equal to zero does not collaborate in the routing process, because he is a customer of AS_2 and AS_3. An AS with $d^O_i = 0$ is defined as a leaf AS or end customer, see [14]. An AS with $d^O_i = 0$ and $d^I_i = 1$ is a stub AS and an AS with $d^O_i = 0$ and $d^I_i > 1$ is a multi-home AS.

Removing AS's with out-degree $d^O_i = 0$, end customers that do not participate in the routing process are eliminated. Iterating this process, the Internet AS topology can be split in an end customer area, a regional area and a core area. The customer area only includes end customers (AS's with out-degree initially equal to zero).

The regional area includes AS's (also called regional ISPs) that interconnect end customers with the core area (AS's that after removing the end customers have out-degree equal to zero). The core area includes AS's that form the core of the Internet and that are highly interconnected (AS's with an out-degree higher than zero). From these areas, see figure 2, we can consider the following graphs:

The whole AS graph that includes the customer, the regional and the core area
The regional area sub-graph that includes the regional area and the core area
The core sub-graph that only includes the core area

Most of the works presented in the references, like [4], [5], [6], [7] or [8] among others, have modeled degree distributions of the whole AS graph using power-laws. The works, [9] and [14], that introduced the heuristics to infer relationships have been centered in quantifying AS relationships and quantifying the size of the different hierarchical levels. We are here interested in knowing whether the regional and core areas have degree distributions following a power-law. The objective is to know whether Internet topologies can firstly be obtained building the core and then adding the end customers and regional AS's. Table 2 summarizes what is known about the degree, in-degree and out-degree distribution modeling of the different hierarchical levels:

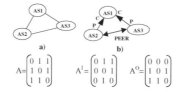

$$A=\begin{pmatrix} 0 & 1 & 1 \\ 1 & 0 & 1 \\ 1 & 1 & 0 \end{pmatrix} \qquad A^I=\begin{pmatrix} 0 & 1 & 1 \\ 0 & 0 & 1 \\ 0 & 1 & 0 \end{pmatrix} \qquad A^O=\begin{pmatrix} 0 & 0 & 0 \\ 1 & 0 & 1 \\ 1 & 1 & 0 \end{pmatrix}$$

Fig. 1. AS relationship: Degree, in-degree and out-degree adjacency matrices

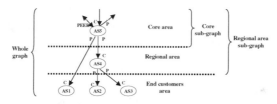

Fig. 2. Core, regional and end customer area

Table 2. Degree, in-degree and out-degree distributions

	Degree	In-degree	Out-degree
Whole AS graph	power-law	?	?
Regional area sub-graph	?	?	?
Core sub-graph	?	?	?

3   Results

In this section we will investigate the AS connectivity. We analyze the degree, in-degree and out-degree distributions for the whole AS graph, the regional and core area sub graphs and the relationships at each hierarchical level.

3.1   Relationship in the Whole Graph, Regional Area, and Core Area Subgraphs

We show in Table 3 some parameters obtained from analyzing the Oregon BGP table and the Union of Oregon with RIPE and the other AS's. We have added the analysis of Oregon in order to compare the increase of AS's and links discovered with the union of BGP tables. Oregon holds 15379 AS's with 34564 links, while the union holds 16244 AS's with 43483 links. That means an increase of 5.6% in AS's and 25.8% in links. Furthermore, the maximum degree increases from 2932 to 3407 (AS701) and the average degree increases from 4.49 to 5.35.

Table 3. Parameters obtained for the Oregon and Union of BGP tables.

	Number of AS's	Number of Links	Maxim Degree	Average degree	Max. In-degree	Maxim Out-degree
Whole Oregon	15379	34564	2932	4.4943	286	2487
Whole Union	16244	43483	3407	5.35373	380	3407

Table 4. Parameters obtained for the Union of BGP tables (whole graph, regional area and core sub-graph)

	Number of AS's	Number of Links	Maxim degree	Average degree	Max. In-degree	Maxim Out-degree
Whole AS graph	16244	43483	3407	5.35	380	3407
Reg. area sub-graph	3120	16660	773	10.67	380	772
Core sub-graph	1817	13174	516	14.5	380	515

Table 4 shows the same data after removing the AS's with out-degree equal to zero for the Union of data sources. The table shows the whole graph, the regional area sub-graph and the core area sub-graph. The adjacency matrix of the regional area sub-graph has 3120 AS's and 16660 links. This represents a sub-graph that contains 19.2% of the AS's and 38.3% of the links. The average degree in this sub-graph is now 10.67, showing that this area is highly interconnected. However, after removing the end customers there still are AS's with an out-degree $d^0=0$. These AS's are providers of the end customers and at the same time are customers of the high interconnected AS's of the core of the Internet, but without peer to peer relationships among them. We remove these AS's obtaining the core area sub-graph.

The core sub-graph has 11.18% of the AS's and 30.29% of the links. The average degree in the core sub-graph has increased to 14.5. These AS's are well interconnected among them forming a mesh topology with a large number of peer to peer and sibling relationships.

Table 5 shows the total number of AS's and link relationships and their distribution as end customers, regional AS's and core AS's. The table also shows the kind of relationships present at each area. From this table we can see that around 80.79% of the AS's are end customers with 61.6% of the links belonging to the whole graph. These AS's may have degree higher than 1 due to multi-homing. 8.02% of the AS's are regional AS's with 8.01% of the relationships. These AS's act as transit between the end customers and the core of the Internet but they do not have any peer-to-peer relationships among themselves. The rest of the AS's belong to the core area. The AS's present in the Internet core can further be divided in more sets as authors in [14] propose.

In the core sub-graph there are 19.06% provider to customer relationships, 9.93% peer to peer relationships and 1.3% siblings. These percentages are with respect the total number of relationships. If we calculate the percentage with respect the number of relationships in the core area we see that in the core area 62.94% are provider to customer relationships, 32.67% are peer to peer and 4.28% siblings. These high percentages of peering are the responsible of the high clustering coefficients, around 0.46, found in many works, see [10] as an example.

Table 5. Relationship obtained in the whole graph, the regional and core sub-graphs.

	Number (#) of ASs / (#) links	% of ASs / links	P2C (#) P2P (#) SIB (#)	P2C (%) P2P (%) SIB (%)
Total number of Ass / links	16244 / 43483	100 / 100	38601	88.71
			4318	9.93
			564	1.3
End customers Ass / links	13124 / 26823	80.79 / 61.6	26823	61.6
			0	0
			0	0
Regional ASs / links	1303 / 3486	8.02 / 8.01	3486	8.01
			0	0
			0	0
Core ASs/ links	1817 / 13174	11.18 / 30.29	8292	19.06
			4318	9.93
			564	1.3

Table 6. % AS's pruned in the regional area sub-graph

RANK	Degree	Regional area sub-grah Degree	%links pruned	% P2C	% PEERs
1- AS701	3407	698	5,05	79.51	20.48
2- AS1239	2762	773	3,71	72.01	27.99
3- AS7018	2074	432	3,06	79.17	20.83
4- AS3561	1319	479	1,57	63.68	36.32
5- AS1	1140	459	1,27	59.73	40.27
6- AS209	998	217	1,46	78.25	21.75
7- AS3356	734	404	0,62	44.95	55.05
8- AS3549	716	333	0,71	53.49	46.51
9- AS702	705	208	0,93	70.49	29.51
10- AS2914	660	292	0,69	55.75	44.25

Table 6 shows the 10 first ranked AS's after removing the end customers. The % of links pruned is computed as the percentage of end customers links removed from an AS with respect the total number of end customers links removed from the whole graph. The core sub-graph has 1817 AS's and 13174 links. Table 7 shows the 10 first ranked AS's after removing the regional AS's. The % of Regional links pruned is now computed as the percentage of regional links removed from an AS with respect to the total number of regional links removed from the regional area sub-graph. The % of total links pruned is computed as the percentage of end customers and regional links removed from an AS with respect to the total number of end customers and regional links removed from the whole graph. Figure 3 shows the degree, the number of end customer links removed, the number of regional links removed and the number of links with other core AS's for the 40 first ranked AS's. These 40 AS's are in the core area. These figures shows certain preferential attachment for the end customers and regional AS's with respect the high degree AS's. However, it is not clear that this preferential attachment is kept in the core area. To prove whether this behavior is presented we plot the degree distributions in the regional and core area sub-graphs.

Table 7. % of pruning and connectivity in su-graph core

RANK	Degree	Core Degree	% Regional links pruned	% total links pruned	% P2C	% PEER
1- AS701	3407	456	3,47	4,87	86.61	13.38
2- AS1239	2762	516	3,69	3,71	81.31	18.68
3- AS7018	2074	269	2,34	2,98	87.02	12.97
4- AS3561	1319	343	1,95	1,61	74.00	26.00
5- AS1	1140	325	1,92	1,34	71.50	28.50
6- AS209	998	130	1,25	1,43	86.97	13.02
7- AS3356	734	317	1,25	0,69	56.81	43.18
8- AS3549	716	275	0,83	0,73	61.59	38.40
9- AS702	705	165	0,62	0,89	76.59	23.40
10- AS2914	660	235	0,82	0,70	64.39	35.60

3.2   Degree, In-Degree, and Our-Degree CCDF

Degree CCDF measures the interconnectivity of the graph regardless the relationships among AS's: e.g.; customer to provider or peers, it gives the probability that an AS has a degree greater or equal to d. figure 4 shows the Degree CCDF of the whole graph for the union of data sources. The Degree CCDF fits a power-law with a constant of α=-1.2144 and a coefficient of correlation of R^2=0.97. This result has been shown in many papers, see [4], [5], [6], [7], [8] or [9] among others.

The In-degree and Out-degree CCDF measure the interconnectivity of the graph taking into account the relationships among AS's. The In-degree CCDF gives the probability for an AS to have more than d providers while the Out-degree CCDF gives the probability for an AS to have more than d customers. AS4513 has the maximum in-degree with 380 providers while AS701 has the maximum out-degree with 3407 customers. We can see, from figure 5 that the In-degree CCDF fits quite well a power-law with parameter α=-1.7519 and a coefficient of correlation of R^2=0.983. The same behaves with the Out-degree CCDF with a parameter α=-1.1729 and a coefficient of correlation of R^2=0.973. We have omitted values with d^O=0 since the X-axis is plotted in log-scale.

We observe an asymmetry between in-degree and out-degree distributions. Out-degree distributions are one order of magnitude greater than in-degree distributions. This is because of the in-degree curves plot only the contributions from peer to peer and siblings relationships while out-degree curves plot provider to customer, peer to peer and sibling relationships.

The following curves, Figure 6 and Figure 7 plot the Degree, In-degree and Out-degree CCDF after removing the AS's with out-degree equal to zero or end customers and therefore plotting the regional area sub-graph. The maximum degree in this graph is 773 and belongs to AS1239. The maximum in-degree is 380 and again belongs to AS4513 and the maximum out-degree is 772 and belongs to AS1239. The In-degree CCDF fits a power-law with a parameter α=-1.691 and a coefficient of correlation of R^2=0.974. Note that the slope is quite similar than that of the whole graph. This behavior is due to the fact that we have removed AS's that where customers (with d^O=0), leaving the peer to peer and sibling relationships. The Out-degree CCDF fits a

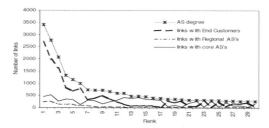

Fig. 3. % of pruning for the 30's top AS's in sub graph core

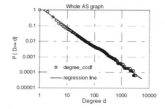

Fig. 4. Degree CCDF (R^2=0.9785, α=-1.2144) of the union of data sources

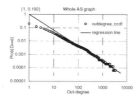 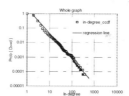

Fig. 5. In-degree CCDF (R^2=0.9837, α=-1.7519) and Out-degree CCDF (R^2=0.9731, α=-1.1729) of the union of data sources

perfect power-law for the first 400's out-degrees. However the tail of the distribution decays quite suddenly. The Degree CCDF, Figure 6, does not fit very well a power-law (R^2=0.906).

Figure 8 and figure 9 show the Degree, In-degree and Out-degree CCDF for the core area. We observe a low coefficient correlation factor for the Degree and Out-Degree CCDF as there were with the regional area. The In-degree CCDF again fits a power-law with a parameter α=-1.669 and a coefficient of correlation of R^2=0.968. The Out-degree CCDF fits the power-law with a coefficient of correlation of R^2=0.806 and the

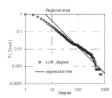

Fig. 6. Degree CCDF (R^2=0.9064, α=-1.6913) of the regional area

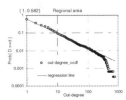 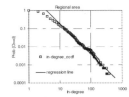

Fig. 7. In-degree(R^2=0.9743, α=-1.6913) , Out-degree CCDF (R^2=0.9952, α=-0.8217) regional area

Degree CCDF with a R^2=0.8512. In fact the regional area and the core area behaves quite similar.

We see that In-degree distributions are quite well fitted by power-laws. However, Degree and Out-degree distributions only are well fitted for the whole graph. Once the end customers are removed from the graph, the regional and the core areas are not well fitted by power-laws.

From figure 6 and figure 8 we note that the shapes of the Degree CCDF for the regional area and core area are very similar and only the tail decay heavily. In order to find a distribution that fits the data we have used a model proposed in [12] for SAR (synthetic Aperture Radar) images called Go-model. The Go-model probability density function is expressed as:

$$pdf = \frac{-2\alpha\gamma^{\alpha}x}{(\gamma+x^2)^{1-\alpha}} \quad \text{for -}\alpha, \gamma \text{ and } x > 0 \quad . \tag{1}$$

and a CCDF expressed as:

$$P\{X \geq x\} = \left(\frac{x^2}{\gamma}+1\right)^{-\alpha} \quad . \tag{2}$$

where α measures the slope or speed change among scales and γ is a form factor's.

Figure 10 and figure 11 show the regression line for the Go-model applied to the regional and core area sub-graphs respectively. We can observe that the Go-model fits the data with a high correlation coefficient (R^2=0.998), see figure 11, compared with a power law model (R^2=0.8512), figure 8, in the core sub-graph.

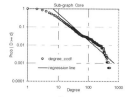

Fig. 8. Degree CCDF (R^2=0.8512) of the core sub-graph

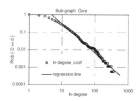 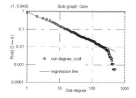

Fig. 9. In-degree(R^2=0.9686, α=-1.6693) and Out-degree CCDF (R^2=0.9945, α=-0.8061) of the core sub-graph

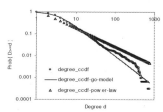

Fig. 10. Degree CCDF for the regional area sub-graph: Go-model (R^2=0.9977, γ= 11.6141 and α=0.685) and power-law (R^2=0.9064, α=-1.515)

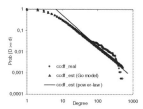

Fig. 11. Degree CCDF for the core sub-graph, Go-model (R^2=0.9986, γ=21.1591 and α=0.6598) and power-law (R^2=0.8512, α=-1.4724)

Table 8. Hierarcical level where is applicable a power-law and go models

Zone	Degree	Indegree	Outdegree
Whole	power-law	power-law	power-law
Regional area	Go model	power-law	power-law (d<400)
Core	Go model	power-law	power-law (d<400)

4   Conclusions

Using, well-known heuristics, we have found that 80% of the AS's are end customers with 61.6 % of the links. 8% of the AS's with 8% of the links are in a regional area serving as providers to the end customers and being customers of the more interconnected AS's that are in the core. The AS's in the core represent the 11.18% of the total AS's with 30.29% of the total links.

We have also shown that the higher ranked AS's have the higher number of end customers. This behavior shows a preferential attachment of end customers with respect to the big providers. However, this preferential attachment in not so clear in the core of the Internet, where AS's are more interconnected forming a mesh topology with high number of peer to peer relationships. All the peer to peer and sibling relationships are between AS's in this core area with only 11.18% of the total number of AS's and 30.29% of the total number of links. In fact, 32.67% of the core area relationships are peer to peer and 4.28% are siblings. This is the area that contributes to the high clustering coefficients found in many works on AS level topologies.

The last behavior is further studied analyzing the Degree, Out-degree and In-degree CCDF of the regional and core sub-graphs. We show as a summary in Table 8 whether the distributions fit a power-law to characterize their corresponding Degree/In-degree/Out-degree CCDF. We see that the Degree, In-degree and Out-degree CCDF for the whole graph fit well with a power-law. The regional and core subgraphs behaves quite similar. Their Degree CCDF does not fit a power-law and can be fit with a Go model and their Out-degree CCDF fit a power-law for the first hundreds of neighbours.

References

1. University of Oregon Route Views Project, http://www.antc.uoregon.edu/route-views/
2. RIPE Routing Information Service, http://data.ris.ripe.net
3. Public route server and looking glass list, http://www.traceroute.org
4. R. Albert, A.-L. Barabási, "Statistical Mechanics of Complex Networks", Reviews of Modern Physics, 74, 47 (2002)
5. T. Bu and D. Towsley, "On Distinguishing between Internet Power Law Topology Generators", In Proceedings of INFOCOM, 2002
6. H. Chang, R. Govindan, S. Jamin, S. J. Shrenker and W. Willinger, "On Inferring AS-Level Connectivity from BGP Routing Tables". In Proceeding of INFOCOM, 2002.
7. Q. Chen, H. Chang, R. Govindan, S. Jamin, S. J. Shenker and W. Willinger, "The Origin of Power Laws in Internet Topologies Revisited", INFOCOM 2002.

8. M. Faloutsos, P. Faloutsos, and C. Faloutsos, "On power-law relationships of the Internet topology", in Proceeding of ACM SIGCOMM'99, August 1999
9. L. Gao, J. Rexford, "On inferring Autonomous System relationships in the Internet", IEEE/ACM Transactions on Networking 2001, Vol 9, Dec 2001.
10. J. I. Hipolito-Nieto, J.M. Barcelo, "On the geographical properties of BGP routing tables", in Proceedings of HPSR2003, Torino, Italy, June 2003.
11. B. Huffaker, D. Plummer, D. Moore and K.C. Claffy, "Topology discovery by active probing", Symposium on Applications and the Internet (SAINT), 2002.
12. H. A. Navarrete, A. C. Frery, F. Sánchez and J. Antó, "Ultrasound Images Filtering Using the Multiplicative Model". Medical Imaging 2002: Ultrasonic Imaging and Signal Processing. 26-28 February 2002, San Diego, USA. Proceedings of SPIE, volume 4687.
13. Y. Rekhter and T. Li, "A border gateway protocol 4 (BGP-4)". Request For Comment – RFC 1771.
14. L. Subramanian, S. Argawal, J. Redxford, R. H. Katz, "Characterizing the Internet hierarchy from multiple vantage points", In Proceedings of INFOCOM 2002.
15. A.Vázquez, R. Pastor-Satorras, A. Vespignani, "Large-scale topological and dynamical properties of the Internet", Physical Review E., Vol 65, June 2002.
16. W. Willinger, R. Govindan, S. Jamin, V. Paxson, S. Shenker, "Scaling phenomena in the Internet: critically examining critically", Proceedings of the National Academy of Sciences, USA, Vol 99, supp 1, pp 2573-2580, February 2002.

Accelerating Computation Bounded IP Lookup Methods by Adding Simple Instructions

Hossein Mohammadi[1] and Nasser Yazdani[2]

[1] Computer Engineering Department, Islamic Azad University of Zanjan, Zanjan, Iran
hosm@ece.ut.ac.ir, http://www.hosmz.net
[2] Router Laboratories, ECE Department, University of Tehran, Tehran, Iran
yazdani@ut.ac.ir, http://web.ut.ac.ir/routerlab/members/yazdani

Abstract. Most of software-based IP lookup methods are memory bounded so their speed depends on memory technology. However, some of them are computation bounded, meaning that their memory time is less than computation. In this paper, we present general ideas to ask simple supports from hardware to accelerate an IP lookup method and as our main case we show that IP Lookup using DMP-Tree data structure is computation bounded and we accelerate this method by adding simple instructions to the running processor. We present hardware implementation of new instructions and resulted IP lookup speed.

1 Introduction

Performing fast IP address lookups is a serious challenge due to fast increase in line speeds and routing table sizes. In the most of the proposed methods, the major bottleneck is memory access time. In hardware-based solutions, this problem can be solved by increasing memory bus bandwidth on the lookup chip. Hardware-based methods are less flexible and scalable and software-based methods are better in these terms. But memory access overhead is a serious bottleneck in the most of software-based methods which we call them 'memory bounded' methods. A few software-based solutions exist in which computation time is more than memory access time. We call these methods 'Computation bounded'.

In this paper we take DMP-Tree IP lookup [1][3][21] and we show that this method can be computation bounded, and then we modify functional units of a RISC processor to add some simple and wisely selected instructions to take advantage of reducing time needed to perform computational jobs of the method. We show that adding the new instructions can help DMP-Tree software-based IP lookup. We also show that implementing these instructions is easy and feasible for high speed processors with a minimum overhead.

A brief review of proposed IP lookup methods comes in section 2. In section 3, we briefly discuss DMP-Tree software-based IP lookup. In section 4, we show that DMP-Tree software-based IP lookup can be computation bounded. In section 5, we propose the new instructions and do IP lookup using them. Simulation and hardware implementation results also come in this chapter. Finally section 7 concludes the paper and draws the future work path.

M. Freire et al. (Eds.): ECUMN 2004, LNCS 3262, pp. 473-482, 2004.

2 Related Work

Proposed methods for IP Lookup can be categorized based on the platforms and data structures. Hardware solutions like [12], [20], [14] are fast but less scalable and sometimes require a long update time [12], [13]. Flexibility and scalability to large routing tables are essential to lookup approach. We believe these are achievable in software. So we focus on these solutions. Proposed software lookups, which uses trie and its variations like LC-Trie [11], Patricia [8], Multibit-Trie [10], are slow due to multiple memory accesses. Lulea method [7] compresses trie efficiently regarding to common prefix lengths. This method is fast but it is restricted so it cannot scale to large routing tables while not supporting incremental updates, since each update requires building the whole trie structure. Some works like [6] combine hash with trie-based data structures. However, since there is no perfect hash function exists to do IP lookup efficiently for all possible routing tables, these methods are dependent to distribution of the prefixes. Tree-based methods like [9] seem to be better but it is difficult to accelerate them efficiently with hardware assistance because their computational tasks are too complex to be added as simple instruction.

Accelerating software-based IP lookup methods with hardware support maybe good if we have enough computational tasks comparing to memory access time. In such cases, two different approaches can be taken. Firstly, accelerating the lookup process by adding complex and method-specialized memory instructions to take advantage of overlapping memory access time with computation time. Secondly, adding simple and general computational instructions to speedup computational tasks and taking advantage of having computation bounded method by decreasing computation time using the new instructions [4][5]. We took first strategy in HASIL method [2] and the result was a scalable software-based IP lookup, which was accelerated by adding two instructions that modified memory unit and this approach lead to overlapping two computational tasks with a memory access. In [4] and [5] we have shown that augmenting general purpose processors may lead to efficient network processing. In this paper we take the second strategy and add some simple instructions by modifying functional units of the processor. Our method decreases computation time. This method is simpler than HASIL and the instructions are more general. Of course HASIL is faster since its instructions are specialized but our method in this paper is simpler and more general. These two methods are the same in terms of scalability because both of them use DMP-Tree software-based IP lookup.

3 IP Lookup Using DMP-Tree

DMP-Tree, proposed in [1][3][21], is a super set of the famous B-Tree data structure [15], which brings scalability of B-Tree to the string-matching problem in general and to LPM in particular. Our implementation of IP lookup with DMP-Tree shows that the height of this tree data structure is logarithmic with respect to the routing table size (see Figure 1). Therefore, the number of memory access, which is a major bottleneck in IP lookup, decreases sharply using this scheme. Another important point with DMP-Tree IP lookup is that this method is fairly scalable such that by increasing routing table size from 100K entries to 1M, height of tree just increases by one. Therefore, lookup time increases very slowly.

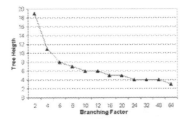

Fig. 1. Maximum Height of DMP-Tree for a 100K routing table. The figure tells that the height of DMP-Tree is similar to a logarithmic function of branching factor.

To build a B-Tree like data structure, first, we need a method to compare and sort items, here prefixes. To do this the following definition is proposed in [1][3][21] which is the basis of the DMP-Tree data structure.

Definition 1: Suppose $A = a_1...a_n$ and $B = b_1...b_m$ are two prefixes of $\{1, 0\}$. If $m=n$ then numerical values of prefixes are compared to determine which prefix is bigger. Otherwise, suppose $m<n$; numerical values of $a_1...a_m$ and $b_1...b_m$ are compared. Prefix with larger value is considered to be larger. If $a_1...a_m$ and $b_1...b_m$ are identical, then, a_{m+1} is checked. If it is 1 then A is considered to be bigger otherwise B is considered to be bigger.

To perform IP Lookup using DMP-Tree, we have to build DMP-Tree of the prefixes in the routing table according to definition 1. Building method is similar to B-Tree but with a special construction rule. The rule says that *'no prefix can stay in a higher level than its prefix'*.

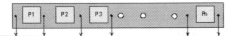

Fig. 2. General view of a bucket in a B-Tree like data structure like DMP-Tree.

LPM-Search method of DMP-Tree is expected to find LPM of the incoming packets destination IP address in order to find its next-hop. This method is similar to B-Tree search with just one more step. Fig.1 depicts a general view of a bucket in B-Tree-like data structures. A bucket is a node in the tree and contains sorted elements and pointers to the next level. There are n prefixes and $n+1$ pointers in a bucket. To do LPM-Search, we begin from the *root* bucket and compare incoming IP address with prefixes in the bucket to find P_i, such that $P_i < IP < P_{i+1}$. Then, we try to find LPM of IP in the current bucket. Finally, the pointer between P_i and P_j is followed to the next level bucket. This process continues until following pointer becomes null. (e.g.: in a leaf)

The following pseudo-code shows LPM-Search in DMP-Tree. *MaxMatch* holds the latest longest matched prefix. *CurrentBucket* is a pointer to the current bucket (node) of DMP-Tree.

LPM-Search (Input: IP Address)
/* Root is a pointer to the root of DMP-Tree – MaxMatch contains longest matching prefix found yet*/
Begin
CurrentBucket = Root;
MaxMatch = *;   */* Default Route */*
While CurrentBucket • Null **do**
Begin
Prf = first element in Bucket;
While PrefixCompare(IP,Prf)==Bigger
And Prf • NULL **do**   } Finding
Begin   place
Prf = Next Prefix in bucket;
End while;
Ptr = Left pointer of Prf;   }
For each Prf in bucket **do**   Longest
Begin   Matching
If Prf matches IP And Prf is Longer than MaxMatch **Then**   Prefix
              MaxMatch = Prf;
End for

CurrentBucket = Child pointed by Ptr;
End While
Return MaxMatch;
End LPM Search

Updating DMP-Tree is similar to B-Tree but it requires extra routines like Space Division, in order to satisfy DMP-Trees special construction rule. The update process is fast enough and it supports incremental updates. Details of DMP-Tree are beyond the scope of this paper and can be found [1][3][21].

4   Analysis of Software-Based Lookup Using DMP-Tree

In traditional trie-based methods, memory access time is the bottleneck since these methods require many memory accesses and their computational tasks are little. A few methods like [9] and [1][3][21] exist in which memory time is less than computation time. In this paper we use DMP-Tree software-based IP lookup since we have implemented it and we believe that this method has enough computational tasks and its computations are simpler than method used in [9].

To become clearer, we have implemented trie-based IP lookup as well and we have simulated running time of DMP-Tree software-based lookup and trie lookup with different memory delays. Figure 4 shows the results of the simulation. In this simulation we use a routing table of 100K entries for both trie and DMP-Tree and branching factor of DMP-Tree is set to 8 in order to fit each bucket in a cache line of our simulation platform.

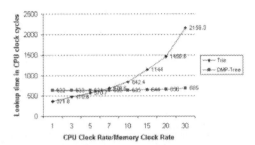

Fig. 4. Lookup time in clock cycles for trie and DMP-Tree using different memory configurations. The figure shows that when memory goes slower, Trie-based lookup requires a long time but DMP-Tree's lookup time remains almost the same.

In the simulation we have started memory delays from 1 meaning that in such a case memory can be accessed in one cycle of CPU's clock. This is important because in the case of very fast memory, almost whole time is computation and software control overhead time. As the figure shows, this value is 311 for trie and 632 for DMP-Tree. With increasing memory delay, trie lookup time increases exponentially but DMP-Tree software lookup time remains almost unchanged. This result implies that a large part of time needed to perform a lookup in trie is memory access time but in DMP-Tree, computations take more than memory access and can internally overlap with memory access. Therefore, increasing memory delay doesn't affect DMP-Tree lookup time.

5   New Instructions and Their Impact on Lookup Time

In this section, we decrease computation time of lookup methods by asking some simple supports from hardware to accelerate frequently executed operations. The first step is to find out these operations. In the second step we suppose that we have added the instructions and we perform IP lookup with them.

5.1   Identifying and Minimizing Supports Asked from Hardware

Generally, for most of IP lookup methods and especially for DMP-Tree software-based IP lookup, we can highlight these simple and frequent functions in the lookup-search procedure (see code of Figure 3):

1. Instructions to compare prefixes if the method requires prefix comparison.
2. Checking matching of an IP address with a prefix.
3. Extracting some bits from a word.
4. Extracting prefix length from represented format.
5. Converting (Value, Len) prefix to representation format.

6. Converting from representation format to (Value, Len).

7. Bit checking instructions.

Since adding an instruction requires complex hardware design process, it is wise to do our best to solve some of the above needs in software.

According to the LPM-Search method of DMP-Tree (Fig.3.); *Prefix Compare* and *Prefix Matching* are two main functions running many times for each lookup. For instance, in a non-optimized implementation of DMP-Tree of height 5 and branching factor of 16, each of these operations will be executed around 80 times (height * branching factor). Therefore, our first step is to optimize these two frequently used functions. Using ordinary prefix representation (Zero-filled prefix, Length), straight implementation of definition 1 to compare an IP address with a prefix, requires 13 instructions. Therefore, in our example, total prefix compares for each lookup will take 1040 clock cycles, because the IP address must be compared and matched with all prefixes in each bucket. Definitely, we need new ideas to do them faster.

In [2] we have solved comparison problem by defining a new prefix representation format called Hosm-Format. We have proposed a new format to represent prefixes. Using this format reduces instructions and time needed for *PrefixCompare* and *PrefixMatch* functions sharply.

Hosm-Format: Prefix can be represented by adding a zero to its tail, and then filling it with ones to make its length equal to the biggest possible prefix length plus one, here 32. (e.g.: 101* represents as 101<u>011....1</u>.)

Theorem 1: Numerical comparison between prefixes in *Hosm-Format* is equivalent to prefix-comparison using Definition 1.[1]

Since we use Hosm-Format to compare a prefix with an IP address, there are two concerns. First, what if prefix in Hosm-Format equals to IP. Second, what we can do with prefixes of length 32. The solution to the first concern is that, by assuming maximum prefix length to be 31, no prefix can equal to IP and in the case of equal Hosm-Format representations, IP should considered to be smaller. (Proof is trivial.). For the second concern, to keep prefix lengths smaller than 32, we can keep 32 bit prefixes in a simple jump table and check it first.

Using Hosm-Format, we can reduce the prefix compare time to just one clock cycle since it requires one integer-to-integer comparison, which is a common instruction in general-purpose processors. Therefore, we do not need any special hardware implementation for prefix compare.

Now we focus on prefix matching which is also a frequent job in the LPM-Search process.

Lemma 1: Assume *prf* is a prefix in Hosm-Format and *IP* is an IP address. If $prf \oplus IP < 2^{32-Length(prf)}$ holds then, *prf* matches *IP*.

Example: *prf=1101011..1 (1101*)* and *IP=1101xx...x. IP $\oplus$ prf= 0000yy...y < 000100...0*.

Now, we can compare prefixes in one clock and match them with IP using a straight implementation of Lemma 1, in up to 5 clocks. This is good for a software implementation but because of time-consuming match function, it is slower than hardware implementations with an order of magnitude. So we would like to accelerate this software-based lookup with some easy to implement hardware supports. Of course *Prefix Matching* is the first candidate to be implemented in hardware. While

---

[1] Proves are omitted due to space limitation.

using Hosm-Format, Implementation of *Prefix Matching* with Lemma 2 becomes straight forward. We discuss about implementation overhead in section 6.

5.2   New Instructions

By carefully examining results of section 5.1 and considering general requirements of packet processing applications (here, IP packet parsing) we can distinguish instructions presented in table 1 to be added to accelerate the lookup process. These instructions can generally accelerate every IP lookup process since they are general enough, but a lookup method will be accelerated if and only if it has enough computational tasks and it can fit its requirements to these instructions.

Table 1. New instructions proposed to be added to the processor to accelerate computations.

Inst.	Functionality	Example	Result
ebis	extracting bits	ebis Ra, #s, #l, Rr	Rr <= Ra & MASK[s,l]
ebia	extracting and adjusting	ebia Ra, #s, #l, Rr	Rr <= SHIFT_R((Ra & MASK[s,l]), s)
cbit	check bit	cbit R1 #b R2	If R1[b-th bit] is on R2 = 1 Else R2 = 0
cpr	Create prefix	cpr Rp, Rl, RPx	RPx <=CreatePref (Rp, Rl)
upr	prefix value	vpr RPx, Rv	Rv <= Value (RPx)
lpr	prefix length	lpr RPx, Rl	Rl<=Length (RPx)
mpr	match prefix	mpt RPx, Ra, ADDR	if Match (RPx, Ra) then jump to ADDR

cpr, upr, lpr and *mpr* instructions take Hosm-Format as their prefix representation format and do operations like prefix matching, prefix creation, etc. *ebis, ebia* and *cbit* are bitwise operations which do bit extraction, etc.

The philosophy behind selecting these operations to be implemented in hardware is that, in the most of packet processing applications, (lookup, parsing, classification, etc.) prefix operations executed many times. For example, in packet parsing, all we have to do is to extract some static or dynamic aligned fields from an IP packet. These tasks can be accelerated using our bit-wise operations. As another example we can consider lookup process in which prefix matching and prefix length extraction are frequently executed in the lookup-search method and prefix creation and decoding (converting from natural numbers to a specific prefix representation format and vice versa) is frequent in table-update routines. Therefore, Our Prefix unit instructions can generally help IP lookup.

5.3   Performing IP Lookup with New Instructions

Using the new instructions we can simplify lookup code and gain higher speeds. We have implemented DMP-Tree lookup method in assembly language and we have simulated its running time with different memory speed configurations. Using new instructions we can reduce code size by 13%. Since this 13% code is frequently executed during a lookup (matching instruction and length extraction from a prefix), lookup time is reduced by 27.44%. Figure 5 compares running DMP-Tree lookup method using mew instructions and without them. It is worth noting that for DMP-Tree lookup method we have used *mpr* and *lpr* and *ebia* instructions in matching phase of the LPM-Search code and other instructions may be used in other lookup methods or in packet processing applications like parsing.

We use *mpr* instruction to check matching of an IP address with a prefix and *lpr* to extract length of a prefix in Hosm-Format and *ebia* instruction to extract next-hop information from a tree element. As discussed this leads to 27.44% improvement in lookup speed.

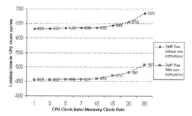

Fig. 5. DMP-Tree Lookup speed with and without new instructions. It gains about 30% in speed

As an example we can consider a 2.4GHZ processor with a 333MHZ DDR-Ram is this case memory is 7 times slower than the processor and according to figure 5, each lookup requires 457 clock cycles. It means that each lookup requires 190ns and consequently our method can forward 5.26 million packets per second. By assuming average packet size to be 2000bits this means supporting 10Gbps lines.

An important question is that why using these instructions can not reduce lookup time more sharply. The first reason is that when we develop a software program, control instructions like loop control branches and result checking instructions push a large overhead on system's behavior. Another reason is that in the case of DMP-Tree IP lookup, the new instructions is only used to accelerate matching part and the *while* loop in *finding place* part of code (see Figure 3) which finds place of the IP address in the bucket, remains unchanged. We have not added special instructions to help DMP-Tree IP lookup in all of its tasks because we would like our newly added instructions to remain general enough and useful for a verity of lookup methods just like DMP-Tree lookup. Of course adding more complex and specialized instructions can help DMP-Tree lookup. For examples instructions added in HASIL [2] can accelerate DMP-Tree lookup process with an order of magnitude but those instructions can not help other lookup methods.

6   Implementing the New Instructions in Hardware

In this section, we describe the overhead imposed of adding the new operations in terms of area and delay. We augmented a simple 32 bit ALU our extra operations to support prefix and parse instructions. We have implemented the ALU before and after being augmented with our new units in a .25u ASIC technology. Table 2 shows the result of this comparison in terms of area and delay.

Table 2. The results of the hardware implementation of Prefix and Parse unit (32 bit units)

32 bit units	Area	Delay
Prefix overhead on ALU	18%	0
Parse overhead on ALU	28%	0
Total overhead	46%	0

As the table shows, there is no delay overhead imposed from new operation. On the other side the area overhead is acceptable.

7   Conclusion

In this paper we climb the first stage and we have identified and added some bit-wise and some prefix instructions to accelerate IP packet processing methods in general and packet-parsing and IP lookup in particular. Also we showed that DMP-Tree based software lookup is a good candidate to simultaneous need to speed and scalability in IP lookup algorithms. Also this method can be computation bounded by selecting large enough branching factors (like 8) and can be accelerated efficiently with our new instructions. Our method can easily support routing tables with millions of entries because tree's height is logarithmic with respect to branching factor.

Future works for this research is to study the most famous and efficient mechanisms for packet classification and quality of service to identify their main computational bottleneck and complete the new instructions. This will lead to an efficient network processor architecture which is constructed by augmenting a RISC core.

References

1. Nasser Yazdani and Paul S. Min, "Fast and Salable Schemes for IP Lookup Problem," Proc. of IEEE Conf. on High Performance Switching and Routing, Heidelberg Germany, June 2000.
2. H. Mohammadi, N. Yazdani, B. Robatmili, M. Nourani, "HASIL: Hardware Assisted Software-based IP Lookup for Large Routing Tables", Proceeding of the 11th IEEE International conference on networks (ICON) 2003, pp. 99-105, Sydney – Australia
3. Nasser Yazdani, Hossein Mohammadi, "IP Lookup in Software for Large Routing Tables Using DMP-Tree Data Structure" Proc. of the 9th Asia Pacific Conference on Communications (APCC) 2003
4. H. Mohammadi, B. Robatmili, H.R. Ghasemi, N. Yazdani, M. Nourani, "Line-speed IP Lookup Using Improved Functional Units", Proc. Of the 9th Iranian Conference of Computer Science (CSICC) 2004, pp. 97-102, Tehran, Iran

5. B. Robatmili, H. Mohammadi, H.R. Ghasemi, N. Yazdani, "Augmenting General Purpose Processors for Network Processing" Proc. Of the 2nd Conf. of Field Programmable Technology, 2003, Tokyo, Japan
6. M.Waldvogel , G.Varghese, et.al."Scalable High Speed IP Routing Lookups", Proc. of ACM SIGCOM`97, pp. 25-35, Cannes, France.
7. M.Degermark, A.Brodnik, S.Carlsson and S.Pink, "Small Forwarding Tables for Fast Routing Lookups," Proceeding of ACM SIGCOM`97 Conf. , pp. 3-14 , Cannes , France.
8. W.Doeringer, G.Karjoth, M.Nassehi, "Routing On Longest Matching Prefixes," IEEE/ACM Trans. Net. vol.4, pp. 86-97, Feb. 1996.
9. B.Lampson , V.Srinivasan , and G.Varghese , "IP Lookups Using Multiway and Multicolumn Search," Proc. IEEE Infocom`98
10. Sartaj Sahni and Kun Suk Kim, "Efficient Construction of Variable-Stride Multibit Tries For IP Lookup," Proceedings IEEE Symposium on Applications and the Internet, SAINT, 2002.
11. S.Nilsson and G.Karlsson , "IP Address Lookups Using LC-Tries," IEEE JSAC, June 1999 ,Vol.17 , No. 6 , pp. 1083-1092.
12. Nasser Yazdani and Nazila Salimi, "Performing IP Lookup on Very High Line Speed," Proceeding of ICT 2002, Shiraz, Iran.
13. N. McKeown, P.Gupta, and S.Lin, "Routing Lookups in Hardware at Memory Access Speeds," Proceeding of IEEE Infocom`98 Conf. , pp. 1240-1247.
14. Chen, W.E.; Tsai, C.J. "A fast and scalable IP lookup scheme for high-speed networks, "Proceedings of IEEE ICON99.
15. T.Cormen, C.Leiserson, R. Rivest and Stein, "Introduction to Algorithms" , MIT Univ. Press , 2001
16. H.Y.Tzeng, "Longest Prefix Search Using Compressed Trees" , proceeding of IEEE GlobCom 98 Conf. , Sydney , Australia.
17. B.Lampson , V.Srinivasan , and G.Varghese , "IP Lookups Using Multiway and Multicolumn Search" , Proceeding of IEEE Infocom`98 Conf. , pp. 1247-1256 , San Francisco , CA.
18. Tzi-Cker Chiueh , Prashant Pradhan , "High Performance IP Routing Lookup Using CPU Caching" , Proceeding of IEEE Infocom`99.
19. Huan Liu, "Routing Prefix Caching in Network Processor Design", Proc. International Conference on Computer Communications and networks (ICCCN), Phoenix, AZ, 2001
20. N. McKeown, P.Gupta , and S.Lin , "Routing Lookups in Hardware at Memory Access Speeds" , Proceeding of IEEE Infocom`98 Conf. , pp. 1240-1247
21. Nasser Yazdani, Hossein Mohammadi, "DMP-Tree: A Dynamic M-way Tree Data Structure for String Matching" Submitted to Elsevier Journal of Algorithms

Using Lifecycles and Contracts to Build Better Telecommunications Systems

John Strassner

MDAPCE, 4790 Longwood Point, Colorado Springs, CO 80906 USA
john@mdapce.com

Abstract. Currently, network operation is divorced from how a business running the network operates. Concurrently, the complexity of both system design and business operation keeps increasing. Current approaches using best-of-breed applications present tremendous integration problems due to a lack of common information and an inability to share and reuse management data. This paper will describe a new approach in building next generation telecommunications components, systems and their management applications. It has three important parts – representing managed entities using the DEN-ng models, using contracts as the unit of interoperability, and modeling the lifecycle of the system and its components.

1 Introduction

Enterprises as well as Service Providers are under increasing pressure to do more with less. Policy based management, process management, web services, and other technologies have been proposed to improve the configuration and management of network services and resources. However, most of these approaches fail to take into account the needs of different constituencies involved in building a product or service offering. For example, consider a product that offers a VPN service. Many different types of people – business analysts, product and project managers, network administrators, system architects, and more – are required to work together to deliver this product.

Now consider what happens when a customer buys this product. The product was bought to fulfill one or more business purposes. However, if the business needs are not taken into account in the configuration of the VPN, then the VPN and the services that it provides cannot be adjusted to the changing demands of the environment and the users of the VPN.

If eBusiness and autonomic computing are to be realized, then the needs of the business must drive the services that the network provides.

Current approaches to network configuration and management are technology-specific and network-centric, and do not take business needs into account [1]. More specifically, current approaches lack the ability to define network services as a function of how a business operates as well as relate them to what a particular type of user needs at a given time. This prevents network services from being represented in a form that will enable them to be configured to adapt to the varying demands of a changing environment.

M. Freire et al. (Eds.): ECUMN 2004, LNCS 3262, pp. 483–497, 2004.

For example, SNMP and CLI focus on describing low-level technical aspects of devices and device interfaces. Their languages are currently unable to express business rules, policies and processes, which makes it impossible to use these technologies to directly change the configuration of network elements in response to new or altered business requirements. Instead, software must translate the business requirements to a form that is subsequently translated into SNMP or CLI [2]. This process disconnects the main stakeholders in the system (e.g., business analysts, who determine how the business is run, from network technicians, who implement network services on behalf of the business).

This problem is exacerbated by the proliferation of important management information that is only found in different, technology-specific forms, such as private MIBs and the many different dialects of CLIs. The latter is an especially difficult problem, due to the significant differences in the capabilities of a particular version of a network device's operating system. Fundamentally, these are all symptoms of a more strategic problem: common management information, defined in a standard representation, is not available. This missing piece prohibits different components from sharing and reusing common data, which leads to "stovepipe" or "silo" architectures that typify the Operational Support Systems (OSSs) and Business Support Systems (BSSs) of today. In addition, it requires a huge integration tax to be paid to integrate best-of-breed applications that were never designed to communicate with each other. This integration tax is the result of the one-to-one service integration forced on "stovepipe" architectures as the result of using incompatible non common data representations [2].

This paper describes a novel approach to fix these problems. The approach is rooted in the NGOSS (New Generation Operational Systems and Software) program of the TeleManagement Forum (TMF). This program consists of four coordinated efforts – the Shared Information and Data (SID) program, the Enhanced Telecom Operations Map (eTOM), the NGOSS Technology Neutral Architecture (TNA), and the NGOSS Lifecycle Methodology. These define information and data viewpoints, business processes, a distributed interface-oriented architecture, and a unique methodology for managing the lifecycle of NGOSS components, respectively.

The approach described in this paper uses three important concepts from the NGOSS programs: the NGOSS Lifecycle Methodology, the notion of a set of Contracts that are used by the Lifecycle to ensure that the solution is operating correctly, and the DEN-ng models to represent managed entities and management information.

The organization of this paper is as follows. Section 2 briefly describes the NGOSS programs in order to provide the reader with an appropriate context. Section 3 first provides a brief overview of DEN-ng, and then shows how its unique features are used in the context of NGOSS. Section 4 discusses the NGOSS contract model, while Section 5 describes the NGOSS lifecycle model. Section 6 provides an example of how these components are used. Finally, Sections 7 and 8 present conclusions and references, respectively.

2   The NGOSS Program

The NGOSS program is a coordinated set of four complementary efforts that describe a framework for building next generation Operation Support Systems (OSSs) and

Business Support Systems (BSSs). An important principle of the NGOSS program is to facilitate the rapid and flexible integration of best-of-breed components used in building OSSs and BSSs with legacy as well as next generation hardware, software and services.

The first, called the SID (Shared Information and Data) program [3], defines a comprehensive set of information and data models for representing entities of interest to the managed environment in a common way. DEN-ng was chosen as the basis of the SID due to a number of reasons that will be briefly described in Section 3. The most important of these are:

- DEN-ng is based on and extends UML's model and metamodel layers [4]
- DEN-ng defines a set of *views* that enable the same information to defined in different ways, in order to address the needs of a particular constituency
- DEN-ng was built to model the entire lifecycle (as opposed to just the current state) of a managed object.

The second program, called the eTOM [5], is a reference framework for categorizing the business process activities that a service provider will use. This enables progressively more information to be revealed about a process. These process elements can then be used to show organizational, functional and other relationships.

The third is the Technology Neutral Architecture [6] of NGOSS. This describes a distributed, contract-oriented architecture that realizes the principles of building a componentized OSS. This models a running system as a number of entities that together offer the functionality represented by a set of interfaces. Clients wishing to exploit this functionality bind to required interfaces and invoke operations over those bindings. A critical concept in this effort is the formal definition of the interface, the contract, which will be discussed in Section 4.

The fourth effort, the NGOSS Lifecycle and Methodology program [7], has as its goal the ensuring of consistent and common usage of the other elements of the NGOSS program. In effect, it is a unifying methodology that facilitates the holistic definition, creation, management, and deployment of a solution compliant with the NGOSS program. It will be discussed in more detail in Section 5.

The NGOSS program also has three other important efforts. First, the TMF sponsors a set of Catalyst programs, whose purpose is for groups of organizations (including Service Providers, Systems Integrators, and Systems Vendors) to work together to build practical implementations and live demonstrations of NGOSS solutions. These are showcased in the TeleManagement World shows sponsored by the TMF, twice a year. Second, the TMF is currently developing an industry compliance program to certify solutions and products for compliance to the NGOSS Specifications. Finally, the TMF is creating a *knowledge base* of documentation, models, code and training materials to support the TMF members in their own development of NGOSS-compliant components and solutions.

3   An Introduction to NGOSS

DEN-ng consists of a set of information and data models. It is the follow-on to the successful DEN [8] standard; however, work on DEN-ng is now being done primarily in the TMF. In fact, the "ng" was coined in recognition of the "NGoss" program.

The DEN-ng object-oriented information model provides a cohesive, comprehensive and extensible means to categorize and represent things of interest in a managed environment. Things of interest can include users, policies, processes, routers, services, and anything else that needs to be represented in a common way to facilitate its representation and management. The DEN-ng information model defines the static and *dynamic* characteristics and behavior of these managed entities as independent of any specific type of repository, software usage, or access protocol.

This was done in recognition of the fact that an OSS will require different types of repositories to hold management information, since that information have different characteristics and usage. A single information model can then be used to used to provide specific representations (e.g., data models) for each repository, as shown in Figure 1 below.

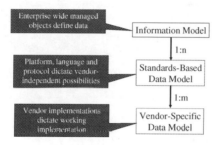

Fig. 1. The DEN-ng Framework

DEN-ng had several driving factors. First, it was realized that the many different stakeholders in building a product all have different perspectives on how a system works. This causes tremendous confusion, since it means that one concept might mean different things to different people. For example, when a business analyst looks at an SLA, that person thinks of contractual obligations and different options for realizing revenue. In contrast, the network technician responsible for implementing the SLA is at a loss – even though the SLA is a technical document, it doesn't contain the technical specifications needed by the network engineer to build the services that are being sold. Thus, the SLA is of little use to the network engineer in its form – it must be translated to a form that contains information suitable for the network engineer. DEN-ng made the decision that instead of trying to build a single "über-model" that was capable of representing these different concerns, it would instead build a *set* of models, each focused on a different constituency. This concept of views carries over into policy management, and gave rise to the policy continuum [9]. Both of these were incorporated into NGOSS.

The concept of views and building a set of models to *link knowledge together* is a unique characteristic of DEN-ng. This enables the needs of different constituencies to be associated with each other. Thus, each person in each constituency can use the terms and concepts that they understand to express their needs, yet have those needs clearly communicated to other personnel having different areas of expertise in the organization. DEN-ng enables these different concepts to be associated with each other.

DEN-ng is based on the Unified Modeling Language, the de facto international standard for defining information models. UML provides a robust metamodel and facilities that enable it to be extended and customized to meet application-specific needs. More importantly, UML can model the static and dynamic aspects of managed objects. For example, its metamodel defines the concepts of events and state machines. This is an important point, for as we will see, state machines are used in DEN-ng to model the lifecycle of managed objects. In contrast, previous approaches failed to adopt the facilities and approach specified by UML. For example, the DMTF has its own "metaschema" that is not compliant with UML and, at this time, has no real capability to model behavior. Other approaches, such as the ITU and IETF, don't use UML at all. Finally, the DMTF, IETF, and ITU don't address the *lifecycle* of managed entities.

DEN-ng is built as a framework of frameworks. This approach has two distinct advantages. First, it enables multiple modelers to work in parallel on different subject domains. Second, it enables information from other standards bodies and fora to be incorporated as is, or have their concepts and ideas mapped into the DEN-ng framework. Mapping is necessary if information is mined from an external standard or other document that isn't compliant with UML. In fact, information has already been mined from several different ITU Recommendations and IETF RFCs. The structure of the DEN-ng framework is shown in Figure 2 below.

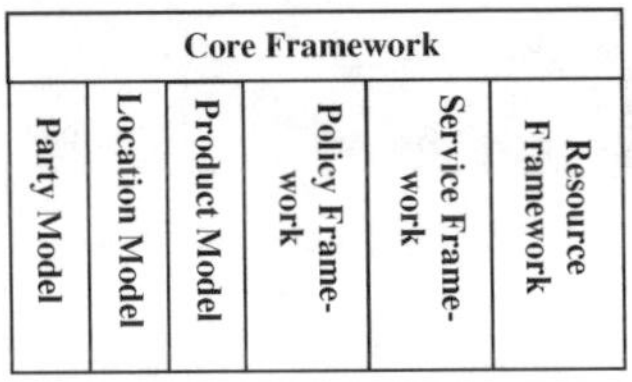

Core Framework					
Party Model	Location Model	Product Model	Policy Framework	Service Framework	Resource Framework

Fig. 2. The DEN-ng Framework

Figure 3 shows an operator trying to accomplish the same task on two routers made by two different vendors. As can be seen, the CLI is completely different. More importantly, the router on the left presents different configuration *modes*, which are absent in the router on the right. Thus, the *programming model* for these devices is different. This in turn means that the operator must be aware of these differences. This is analogous to requiring the network technician to speak multiple languages fluently. However, even that analogy falls short, because of the significant changes that can be introduced in a new version of an operating system of a device. These include new commands as well as changes to existing commands.

DEN-ng capabilities provide an intermediate level of abstraction, which enables the network technician to specify the desired functionality in a technology- and vendor-neutral way. Tools such as those provided by Intelliden [10] take this vendor-neutral specification and translate it into vendor-specific CLIs. In effect, DEN-ng sits as an intelligent lingua franca above the different methods (such as SNMP and CLI) that are used to control the device. This enables the network technician to use a single language to configure heterogeneous devices.

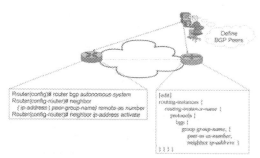

Fig. 3. Different Programming Models

Constraints define which capabilities can be used as a function of a particular context; context defines the current environment, objectives, obligations, and policies governing operation, and the desired behavior of the system. For example, consider a device that can perform different types of encryption. Each different type of encryption is modeled as a capability. Constraints, such as ITAR regulations, prohibit certain capabilities from being used in a given environment. The context is the overall environment that is being modeled. Thus, the combination of capabilities, constraints, and context enables the *behavior* of a system to be abstractly modeled. Since DEN-ng is built on UML (and thus uses OMG MOF), this combination enables code to be generated from the models, in a manner similar to the Model Driven Architecture (MDA) initiative of the OMG [11]. Please note that the combination of using capabilities, constraints and context as a methodology in developing and applying the information and data models is also unique.

DEN-ng is built using patterns [12]. These patterns frequently use other abstraction mechanisms, such as roles [13], to separate how an object is used from the basic definition of the object. This approach makes the model inherently extensible, and helps model the real-world application of the model by embedding in the model facilities that enable it to be used by practitioners.

DEN-ng, like DEN, is built around the use of a finite state machine (FSM) model to describe and control managed entities. DEN-ng defines three fundamental types of classes to support a FSM: (1) classes to model the current state of a managed entity, (2) classes to model the changing of state of a managed entity, and (3) classes to control when the state of a managed entity is changed.

Most other information models are "current state" models – they are limited to defining the behavior of a managed entity at a particular point in time. In contrast, DEN-ng models the static and dynamic characteristics of a managed object using different types of UML diagrams. In effect, static diagrams, such as class diagrams, define a set of building blocks that can be used to represent different features of a managed environment. Dynamic models, such as collaboration and sequence diagrams, use entities defined in these static diagrams to model the behavior of managed objects. This is shown conceptually in Figure 4 below.

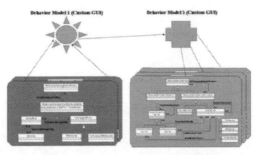

Fig. 4. DEN-ng Artifacts Used to Build a FSM

Figure 4 shows conceptually how different elements from different DEN-ng domain models can be combined to produce an FSM that describes the behavior of a solution that uses elements of these different models. Each FSM is populated with the contents of the appropriate DEN-ng models, which define the actors, elements, and transitions for a particular set of concepts and entities for a given domain. Thus, the DEN-ng models are conceptually a set of building blocks that can be put together to model application-specific behavior.

The above approach enables the knowledge of one constituency to be translated or mapped to a form that a different constituency can understand. For example, a business rule containing no networking terminology can thus be translated into a rule that contains basic networking terms, which can in turn be translated into lower-level networking concepts, until finally appropriate vendor-specific commands can be generated which can be used to change the configuration of a networking device. This process ensures that business concerns drive the (re)configuration of the network. More importantly, knowledge is used to characterize what is desired of the network, what the network can provide, and how the network is modified to achieve that goal.

The TMF uses DEN-ng to form the core of the SID models, as well as its policy, service and resource domains. Currently, the business view of the SID consists of over 1,600 pages of documentation, as well as detailed UML models.

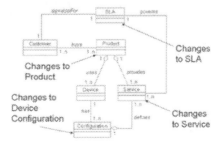

Fig. 5. Relating Different Abstractions in DEN-ng

Figure 5 is a simplified UML diagram showing some of the conceptual relationships that exist across the Product, Service, and Resource DEN-ng models. This provides a simple example of how different types of objects, and their views, can be combined. Conceptually, this figure shows that a Product, which is a business abstraction, contains Services and Resources. The power of this approach is that changes in these *business* entities (e.g., Product and SLA) can be directly mapped to changes in resource and service configuration. Thus, as a Product Offering changes to include new functionality, the underlying model changes to reflect the updated configuration of the services and resources making up the model.

4   The NGOSS Contract Model

The NGOSS Contract Model is based on the Design by Contract [14] approach pioneered by Bertrand Meyer and the concept of contractually defined interfaces developed by Bellcore in the INA project. The principal idea of this approach is that an object and its clients communicate using contracts. The client must guarantee certain conditions before calling a method, and the object must guarantee a set of properties after the method call. These are, of course, assertions, and are referred to as the pre-conditions and post-conditions of the contract. If these pre- and post-conditions can be realized, then the resulting code can guarantee how the two objects interact with each other. An important corollary of this approach is that no "hidden" obligations are present – everything that is required is specified in the pre- and post-conditions of the contract. It should be noted that such an approach does not guarantee a "successful" conclusion to a method, merely that the result of the method is predictable.

NGOSS extends and expands on this basic notion in several important ways. First, instead of focusing just on implementing code, NGOSS uses a contract to define how components *interact* with each other. Thus, an NGOSS contract is more than an interface or API description – it is an expression of behavior. This behavior takes the form of guaranteeing that a particular state was reached before a contract was invoked, and that a specified state or states will be reached when the contract has finished execution. The expression of state in a contract enables the DEN-ng models to be used to supply the entities defined in a contract. Thus, an NGOSS contract specifies how entities *interact* with each other (instead of simply specifying how they communicate).

In Design by Contract, class invariants were defined as an integral part of a contract. A class invariant describes a property that every instance of the class must satisfy whenever it is viewed by other entities. Conceptually, this property makes a class instance consistent, and can be linked to the state of the class instance. DEN-ng and NGOSS extend this to a more complete definition of state, so that the state is more than just a set of properties – it also means that a particular set of behaviors has been achieved and another (possibly different) set of behaviors is now possible to invoke.

Contracts are an important NGOSS artifact. It does no good to have a set of components that cannot interoperate with each other. Thus, NGOSS has defined a contract *template* that has five fixed parts, with room to extend its definition as required. This ensures that each contract will have the same look and feel. This is clearly a benefit to developers, but there are two even more fundamental reasons for using a template. The first is that contracts are one of the artifacts that can be put in the NGOSS knowledge base. Since components are required to communicate using contracts, the use of a common template enables a diverse set of contracts to be structurally consistent. This simplifies automation and usage, but has an even more important benefit – the collection of contracts has the same *engineering*. Put another way, it's not sufficient to define consistent interfaces and signatures. What is really needed is to define, in a precise and formal manner (so that code can be automatically generated), how the entities interact using the contract. Contracts enable this interaction to be specified in the form of obligations and benefits, which in turn enables policy management and process automation to be used.

The second benefit is that a contract is conceptually an intelligent container. This enables the contents of the contract to *morph* as a function of the particular view and/or stage in the lifecycle that is being examined. This concept will be explained in more detail in Section 5.

An NGOSS Contract consists of five parts – generic, functional, non-functional, management, and view-specific model. These parts change as the view changes; a brief description of the view-invariant fields is as follows.

The generic part of the contract consists of a standard header and a variable descriptive part. The former includes data such as the name of the contract, a unique identifier, its version, and the identifier of the organization that defined the contract. The latter contains the goals of the contract, search criteria (to facilitate finding contracts and contract instances), and description and comment fields.

The functional part of the contract defines the *capabilities* provided by the contract; the pre-conditions needed for correct operation, the post-conditions resulting from correct operation and a context for the capabilities. This includes fields such as pre- and post-conditions, result status, interaction roles, security, policies, processes, and context.

The non-functional part contains the obligations (required behavior), prohibitions (disallowed behavior), restrictions (limits on behavior) and permissions (allowed behavior) on the capabilities defined in the functional part of the contract. These include factors such as external influences (e.g., technological limitations, regulatory or legal limitations, and organizational limitations), as well as other considerations (e.g., cost and availability).

The management part defines the management capabilities needed to operate administer and maintain the capabilities of this contract. This includes defining the roles, activities, processes, and policies that are used to govern this contract.

Finally, the view-specific model part contains various types of models (UML and others, such as BPEL4WS [18] diagrams or workflows) tailored to support the specific view of the contract being represented. This will be explained more in Section 5.

As in the Eiffel programming language, NGOSS generalizes the notion of pre- and post-conditions into a set of obligations and a set of benefits. NGOSS also defines prohibitions, delegations, and other concepts mentioned above. In DEN-ng, policy can be viewed as defining when a state transition can occur. This enables policy management to be used to control which contracts are available to which clients when, why, and how.

5   The NGOSS Lifecycle and Methodology

DEN-ng was built to model not just the current state of a managed entity, but the entire lifecycle of that managed entity, using a set of FSMs. This concept is implemented by coordinating multiple FSMs, which enable the behavior of components and entire solutions to be represented. This forms the root of the DEN-ng lifecycle, which is model-driven and uses a combination of static and dynamic models to represent the overall behavior of the solution, using a set of FSMs.

The NGOSS Lifecycle and Methodology program takes the DEN-ng lifecycle model and extends it for use in communication applications. Significantly, NGOSS employs other tools in addition to models to define and implement its lifecycle.

The NGOSS Lifecycle is based on integrating a set of different views of the solution (and its components). This is based on a principle called *progressive grounding*, where conversation between the partners is used to form common understanding that develops over time [16].

Fundamentally, this approach quantifies an important fact: the processes of analyzing business and system requirements, modeling the design, implementing a solution and then deploying the solution *is iteself a lifecycle*. The NGOSS Lifecycle [17] uses the four views defined in DEN-ng (business, system, implementation and deployment) and formalizes them into a cohesive structure that can be used to define the characteristics and behavior of NGOSS components and solutions. More importantly, the Lifecycle is organized to enable flexible modeling of the solution. Sequential progress through the Lifecycle is, of course, supported; however, the NGOSS Lifecycle also enables organizations to progress through different aspects of the solution in whatever manner represents the way their organization works.

The Lifecycle provides traceability from the business definition through the deployment of a solution through defining views.

These views enable the interests of various stakeholders to be protected by representing the evolution of their interests as the solution progresses from the definition of its business concerns, through its mapping to a particular architecture and implementation, through its deployment. Specifically, the NGOSS Lifecycle contains two planes of interest: the Logical Plane, which is technology neutral, and the Physical Plane, which is technology specific. The Logical Plane describes the conceptual design of the solution from a business and a system perspective. The Physical Plane describes the actual implementation of the solution. The Logical and Physical Planes intersect two different perspectives, the Service Provider Perspective and the Service Developer Perspective. The Service Provider Perspective is concerned with the definition and operation of the business, while the Service Developer perspective focuses on the architecture and its implementation built to support the operation of the business.

The Business View is used to describe the goals, obligations, and policies that will be used to drive the services offered by the business. This is done using high-level, technology-independent terms. The eTOM and the SID are used to focus on the concerns of the business: processes, entities and business interactions. DEN-ng models these various tasks and functions as interaction diagrams in which contracts are used to specify how information is exchanged between collaborating entities.

The System View is primarily concerned with the modeling of processes and information that affect the overall architecture of a component or a solution in a technology neutral manner. The SID, eTOM, and the NGOSS Technology Netural Architecture are used in conjunction to focus on system concerns: managed objects, the specification of behavior, and associated computational interactions, again in a technology neutral manner. In addition, interactions are used; the difference is that in the system view, Contracts also specify how functionality is exchanged between collaborating entities.

The Implementation View puts the focus on *how* to build hardware, software, and firmware that will implement the system being designed. This View uses the particular NGOSS architectural style to map from the technology neutral specifications of the system to a selected target architecture. This drives the customization of an appropriate DEN-ng data model. Again, interactions are used; the difference is that contracts also contain implementation artifacts, such as code, APIs, or Web Service invocations.

Finally, the Deployment View is concerned with operating and actively monitoring the system to ensure that the observed solution behavior is what is expected – if not, then the behavior can be adjusted appropriately by using the NGOSS behavior and control mechanisms of process and policy based management. Once again, interactions are used to specify the details of the desired behavior. Thus, deployment contracts can contain a wide range of data (e.g., statistical performance data to evaluation the performance of the solution, policies to define specific behavior in response to particular stimuli, and so forth). The NGOSS Lifecycle and Methodology is shown below in Figure 6.

Fig. 6. The NGOSS Lifecycle Methodology

This approach enables business, technology, and product lifecycles of a given solution to be synchronized. The advantage of this approach is that it provides traceability from the business definition of the solution through its architecture, implementation and deployment. It also enables information and data entities to be further developed as part of the development process.

The NGOSS Lifecycle model uses Contracts as the "unit of interoperability". Contracts appear in each of the four views. This means that contracts are *always* used to provide a mechanism that links the business, system, implementation and deployment aspects of a solution together. However, the content of a Contract changes to suit the specific needs of a particular View, and the constituents of that View.

An important point is that the NGOSS Contract *morphs* as necessary to contain new and/or changed information. Morphing is necessary because, due to traceability concerns, we don't want to *lose* information. Yet, we want to be able to respond to changing needs. For example, a Contract in one View could contain the specification of a service to be delivered; that same Contract in a different View could specify information and code that implement the service. Thus, a Contract is more than a software interface specification – it also defines pre- and post-conditions, semantics for using the service, policies affecting the configuration, use, and operation of the service, and much more. In short, the Contract is a way of *reifying* a specification of a service, and implementing the functionality of the service (including obligations to other entities in the managed environment). An example is provided in Section 6.

6 An Example of Contract Morphing

As a simple example, consider an SLA. Initially, it will be written in business terms for business people to agree on – this is represented by the Business View Contract. Clearly, however, a link needs to be made to the other three Views. For example, the System View defines how the overall architecture is affected, and whether it can support the goals and objects defined in the Business Contract. Similarly, the Implementation View defines how the architecture is realized in hardware, software, and firmware for implementing this Contract. Finally, the Deployment View defines how the Contract can be monitored to ensure that the terms in the SLA are not violated (and if they are, what corrective action should be taken). The notion of Views enables the SLA (and its associated concepts) to be described using terms and concepts appropriate for a given constituency and, more importantly, be associated with each other.

These Views can be conceptually interconnected as shown in Figure 7 below.

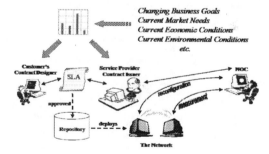

Fig. 7. The NGOSS Lifecycle Methodology

There are several different types of interactions present in the above diagram. The Issuer of the Contract that represents the Service Provider must understand the capabilities that are present in the network as well as the business needs of prospective Customers. (Note that there are many other types of stakeholders that would need to participate in this offering – Product Designers, Market Analysts, System Architects,

Programmers to mention a few – but this simply reinforces the need to enable different constituencies to speak a common lingua franca, which in turn emphasizes the need to enable knowledge describing the concerns of each constituency to be represented in a common form. That lingua franca is, of course, DEN-ng.)

When the Service Provider's Contract Issuer defines the SLA to be offered, prospective Customers will either accept it as is or try and negotiate some of its terms. Both of these (and other types of) actions need to be recorded and "follow" the SLA through its lifecycle. This function is represented by the Business Contract.

Once the SLA is agreed on, it is stored in a repository, so that it can be installed and deployed on the system supporting the Contract. At this point, several outcomes are possible; each can require a series of interactions between the Business, System, Implementation and Deployment Views. Three examples will be given to illustrate this complex set of interactions:

- The SLA is decomposed into other related forms (e.g., SLOs and SLSs) so that it can be implemented, tested and then offered to the Customer. If something goes wrong, then being able to relate what went wrong to entities representing the Contract in its various forms enables the process to be corrected. It also enables the Service Provider to build a knowledge base that learns from these problems, so they can be avoided in the future.
- The Customer asks for new features that the Service Provider agrees to add. The negotiation process needs to be recorded to protect both parties. However, being able to capture what specifically was changed in the service offering is important for marketing, product management, and future engineering. The new features may or may not have ramifications in the system design and architecture of the offering, and will certainly cause the service to be reconfigured to meet these new features.
- The environment changes, either because the Customer needs to use different resources and services, or because of external events (stock crash, more people than expected subscribing to different variations of the service, a DDoS attack, and so forth). The Customer doesn't want to renegotiate all aspects of the Service – he or she simply wants the "right" thing to happen. This autonomic behavior requires the system to know itself, its capabilities and constraints, and the environment in which it is operating in (i.e., its context). More importantly, it requires business goals and obligations to be "translated" into equivalent forms in the system, implementation and deployment views. This requires more than a common lingua franca – it requires a common representation of policy and process management, which can be used to govern how the solution morphs to suit the changing needs of the environment.

Thus, it is mandatory that associations between each of these four Views are supported, so that the process of offering the service may fluidly and dynamically be adjusted. Clearly, if Contracts are implemented as static entities, the above scenarios cannot be accommodated. However, it is equally important to realize that the above scenarios require policies, processes and data to change. Put another way, contracts establish the current context for delivering a Service, and enable that context to change as necessary. In reality, there is a continuum of Contracts, where each Contract is visible in one or more of the NGOSS Views.

To ensure that these changes do not compromise the overall solution, it is imperative that data, policies, processes, and most importantly, knowledge, do not "disappear" or become disconnected from their changed cousins. Hence, NGOSS Contracts are designed as dynamic containers that can morph and expand as needed to enable new and changed knowledge to be stored. This provides visibility and traceability, from business needs through the architecture, implementation, and deployment of the solution, as well as visibility and traceability describing how policies, processes, information processing entities, and other NGOSS Components interact to form a solution.

7   Conclusions and Future Work

Current network management approaches rely on stovepipe architectures that can only share data through complicated mediation services providing point-to-point integrations that are costly to design and even harder to maintain. Worse, they disconnect the various stakeholders from each other in two important ways. First, the lack of common information inhibits the sharing and reuse of management information. Second, the information models being used only describe the current state of the managed object. Thus, it is impossible to trace the evolution of the design, development, operation, and maintenance of a solution.

In direct contrast to this, the promise of future management architectures, such as autonomic computing, is that network services and resources can be quickly and efficiently reconfigured to meet changing environments and business needs. This demands the ability to associate business concepts with the architecture and components of a particular solution.

This paper has presented a new approach, based on the TMF NGOSS program, that solves the above problems by building componentized software that is inherently more maintainable and more reusable.

The DEN-ng models were developed in order to address the needs of all constituencies that design a solution, from product definition through retirement. This enables the business to define services as a function of the changing environment, and for the network to respond to these changing demands in an agile manner.

Two important foundations of DEN-ng are views and the use of finite state machines. Views enable different aspects of the same object to be described to best suit the needs of different constituencies. Finite state machines model the static and dynamic aspects of a set of collaborating entities. Other important features of DEN-ng include: (1) it is based on UML, and extends the metamodel of UML using UML approved techniques; (2) it is built as a framework of frameworks, which enables knowledge from external standard bodies and fora to be incorporated; (3) it uses a set of powerful abstractions – capabilities, constraints, context, and roles, and patterns – to make its models inherently extensible.

Contracts form a key foundation of the NGOSS program. They represent a mechanism that enables the interaction between entities to be specified. Contracts are strongly tied to the concepts of the DEN-ng model, and ensure that the state of the system (and hence its behavior) can be precisely specified. Policy is used to control when an entity can transition to a new state.

The NGOSS Lifecycle is a formal specification for defining the characteristics and behavior of NGOSS components. It also enables the interests of various stakeholders to be protected by representing the evolution of their interests as the solution progresses from the definition of its business concerns, through a mapping to a particular architecture and implementation, through its deployment.

References

[1]   Strassner, J. "A New Paradigm for Network Management – Business Driven Device Management". *In SSGRRs 2002 summer session.*
[2]   Strassner, J., and Reilly, J. 2003. "Learning the SID". *Day Tutorial for TMW 2003 Fall Conference*
[3]   TMF Forum, "Shared Information and Data (SID) model". *GB922 (and Addenda) and GB926 (and Addenda), v4.0.* Dec 2003
[4]   OMG, "Unified Modeling Language Specification", version 1.5, March 2003
[5]   TMF, "Enhanced Telecom Operations Map (eTOM) – The Business Process Framework", *GB921, v3.5, June 2003*
[6]   TMF, "The NGOSS Technology Neutral Architecture", *TMF053 (and Addenda),* Dec 2003
[7]   TMF, "The NGOSS Lifecycle and Methodology", GB927, v1.0, Jan 2004
[8]   J. Strassner, *Directory Enabled Networks,* Macmillan Technical Publishing, 1999, ISBN 1-57870-140-6
[9]   J. Strassner, "Policy-Based Network Management", Morgan Kaufman Publishers, ISBN 1-55860-859-1, Sep 2003
[10] Please see: www.intelliden.com
[11] Please see: www.omg.org/mda.
[12] See, for example, M. Fowler, "Analysis Patterns – Reusable Object Models", ISBN 0-201-89542-0
[13] In particular, variations of the role object pattern are used – please see http://www.riehle.org/computer-science-research/1997/plop-1997-role-object.pdf
[14] B. Meyer, "Object-Oriented Software Construction", Prentice-Hall, ISBN 0-13-629155-4, 1997
[15] TMF, "NGOSS Architecture Technology Neutral Specification – Contract Description: Business and System Views", TMF053B, Jan 2004
[16] Clark, H. H. *Using Language.* Cambridge University Press, 1996
[17] TMF, "The NGOSS Lifecycle Methodology", GB927, v1.0, Dec 2003
[18] Please see: http://www.oasis-open.org/committees/tc_home.php?wg_abbrev=wsbpel

Probabilistic Inference for Network Management

Jianguo Ding[1,2], Bernd J. Krämer[2], Yingcai Bai[1], and Hansheng Chen[3]

[1] Department of Computer Science and Engineering, Shanghai Jiao Tong University,
Shanghai 200030, P. R. China
Jianguo.Ding@sjtu.edu.cn
[2] Department of Electrical Engineering and Information Engineering,
FernUniversität Hagen,
D-58084 Hagen, Germany
Bernd.Kraemer@FernUni-Hagen.de
[3] East-china Institute of Computer Technology,
Shanghai 200233, P. R. China

Abstract. As networks grow in size, heterogeneity, and complexity of applications and network services, an efficient network management system needs to work effectively even in face of incomplete management information, uncertain situations and dynamic changes. We use Bayesian networks to model the network management and consider the probabilistic backward inference between the managed entities, which can track the strongest causes and trace the strongest routes between particular effects and its causes. This is the foundation for further intelligent decision of management in networks.

1 Introduction

With the increasing in the size, heterogeneity, pervasiveness, complex applications and services in networks, effective management of such networks becomes more important and more difficult. Uncertainty of the complex networks is an unavoidable characteristic which comes from hardware defects, software errors, the incomplete management information and the dependency relationship between the managed entities. An effective management system in networks should deal with the uncertainty and allow for partial automation in network maintain and recovering. In the past decade, a great deal of research effort has been focused on improving a network management in fault detection and diagnosis. Finite State Machines were used to model the concepts of fault propagation behaviour and duration [1]. Rule-base methods were proposed for fault detection [2]. Coding-based methods [3] and Case-Based methods [4] were used for fault identification and isolation. All these researches help to improve fault management efficiently in networks. But most of the technologies are sensitive to noise. That means they can not deal with the incomplete and uncertain management information efficiently. Probabilistic reasoning is another effective approach for fault detection in network management [5,6].

On the practical side, most of the current commercial management software, such as HP OpenView, IBM Tivoli, or Cisco serial network management

M. Freire et al. (Eds.): ECUMN 2004, LNCS 3262, pp. 498–507, 2004.

software, perform remote monitoring, provide fault alarm, and perform some statistics on management information. But they lack facilities for exact fault localization, or the automatic execution of appropriate fault recovery actions. A typical metric for on-line fault identification is 95% fault location accuracy and 5% faults can not be located and recovered in due time [7]. Hence for large network including thousands of managed components it may be rather time-consuming and difficult to resolve the problems in a short time by exhaustive search in locating the root causes of a failure.

We apply Bayesian networks (BNs) to model dependencies among managed objects and provide efficient methods to locate the root causes of failure situations in the presence of imprecise management information. In this paper a Strongest Dependence Route algorithm for backward-inference in BNs is presented. The inference algorithm will allow users to trace the strongest dependency route from some malicious effect to its causes, so that the most probable causes are investigated first. The algorithm also provides a dependency ranking of a particular effect's causes.

The application of BNs to network management is discussed in Section 2. The Strongest Dependency Route algorithm for probabilistic inference in BNs is presented in Section 3. Section 4 identifies directions for further research.

2   Bayesian Networks for Network Management

Bayesian Networks are effective means to model probabilistic knowledge by representing cause-and-effect relationships among key entities in a network. BNs can be used to generate useful predictions about future faults and decisions even in the presence of uncertain or incomplete information in network management.

2.1   Models of Bayesian Networks

Bayesian networks, known as Bayesian belief networks, probabilistic networks or causal networks, are an important knowledge representation in Artificial Intelligence [8,9]. BNs use DAGs (Directed Acyclic Graphs) with probability labels to represent probabilistic knowledge. BNs can be defined as a triplet (V, L, P), where V is a set of variables (nodes of the DAG), L is the set of causal links among the variables (the directed arcs between nodes of the DAG), P is a set of probability distributions defined by: $P = \{p(v \mid \pi(v)) \mid v \in V\}$; $\pi(v)$ denotes the parents of node v. The DAG is commonly referred to as the dependence structure of a BN. In BNs, the information included in one node depends on the information of its predecessor nodes. The former denotes an effect node: the latter represents its causes. Note that an effect node can also act as a causal node of other nodes, where it then plays the role of a cause node. An important advantage of BNs is the avoidance of building huge JPD (Joint Probability Distribution) tables that include permutations of all the nodes in the network. Rather, for an effect node, only the states of its immediate predecessor need to be considered.

BNs are appropriate for automated reasoning because of their deep repre-sentations and precise calculations. BNs have been applied in medical diagnosis [10], map learning [11] and language understanding [12].

2.2   Mapping Networks to Bayesian Networks

Fig. 1 shows a example of the campus network of the FernUniversität in Hagen.

When only the connection service for end users is considered, Fig. 2 illustrates the associated BN. The arrows in the BN denote the dependency from causes to effects. The weights of the links denote the probability of dependency between the objects. In this example, the component F and component E are the causes for component D. The annotation $p(\overline{D} \mid \overline{E}F) = 100\%$ denotes the probability of the non-availability of component D is 100% when component F is in order but component E is not. Other annotations can be read similarly.

The dependencies among network entities can be assigned probabilities to the links in the dependency or causality graph [13,14]. This dependency graph can be transformed into a BN with certain special properties [15].

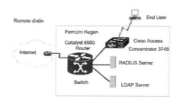

Fig. 1. Example of Campus Network

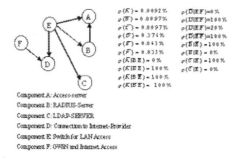

Fig. 2. Example of Bayesian Network for Fig. 1

When a network is modelled as a BN, two important processes need to be resolved:

(1) Ascertain the Dependency Relationship between Managed Entities.

When one entity requires a service performed by another entity in order for it to execute its function, this relationship between the two entities is called a dependency. The notion of dependencies can be applied at various levels of granularity. The functional model and the structural model are two useful models to get the dependency between cooperating entities in network [16].

(2) Obtain the Measurement of the Dependency.

When BNs are used to model networks, BNs represent causes and effects between observable symptoms and the unobserved problems, so that when a set of evidences is observed the most likely causes can be determined by inference technologies. Single-cause and multi-cause are two kinds of general assumptions to consider the dependencies between managed entities in network management. Single-cause means any of the causes must lead to the effect. While the multiple-cause means that one effect is generated only when more than one cause happens simultaneity. Management information statistics are the main source to get the dependencies between the managed objects in networks. The empirical knowledge of experts and experiments are useful to determine the dependency. Some researchers have performed useful work to discover dependencies from the application view in networks [17,16].

3   Probabilistic Inference for Network Management

The most common approach towards reasoning with uncertain information about dependencies in networks is probabilistic inference, which traces the causes from effects. The task of backward inference amounts to finding the most probable instances and the key factors that are related to the defect in networks.

3.1   Related Algorithms for Probabilistic Inference

There exist various types of inference algorithms for BNs. All in all they can be classified into two types of inferences: exact inference [9,18] and approximate inference [19]. Each resolution offers different properties and works better on different classes of problems, but it is very unlikely that a single algorithm can solve all possible problem instances effectively. For almost all computational problems and probabilistic inference, using general BNs has been shown to be NP-hard by Cooper [20].

However, Pearl's algorithm, the most popular inference algorithm in BNs, can not be extended easily to apply to acyclic multiply connected digraphs. A straightforward application of Pearl's algorithm to an acyclic digraph comprising one or more loops invariably leads to insuperable problems [21]. Another popular exact BN inference algorithm is the clique-tree algorithm [9]. It transforms a multiply connected network into a clique tree by clustering the triangulated moral graph of the underlying undirected graph first, and then performs message propagation over the clique tree. But it is difficult to record the internal nodes and the dependency routes between particular effect nodes and causes. In network management, the states of internal nodes and a key route, which connect the effects and causes, are important for management decisions. Moreover, the sequence of localization for potential faults is very useful reference to network managers. For system performance management, the identification of related key factors is also important. Few algorithms give satisfactory resolution for this case. In this paper the SDR algorithm presents resolutions for these tasks.

3.2   Backward Inference in Bayesian Networks

The most common approach towards reasoning with uncertain information about dependencies in network management is backward inference, which traces the causes from effects. We define E as the set of effects (evidences) which we can observe, and C as the set of causes.

Before discussing the complex backward inference in BNs, a basic model will be examined. In BNs, one node may have one or several parents (if it is not a root node), and we denote the dependency between parents and their child by a JPD.

Fig. 3 shows the basic model for backward inference in BNs. Let $X = (x_1, x_2, \ldots, x_n)$ be the set of causes. According to the definition of BNs, the following variables are known: $p(x_1), p(x_2), \ldots, p(x_n), p(Y|x_1, x_2, \ldots, x_n) = p(Y|X)$. Here $x_1, x_2, \ldots, x_n$ are mutually independent, so

$$p(X) = p(x_1, x_2, \ldots, x_n) = \prod_{i=1}^{n} p(x_i) \tag{1}$$

$$p(Y) = \sum_{X} [p(Y|X)p(X)] = \sum_{X} [p(Y|X) \prod_{i=1}^{n} p(x_i)] \tag{2}$$

by Bayes' theorem,

$$p(X|Y) = \frac{p(Y|X)p(X)}{p(Y)} = \frac{p(Y|X) \prod_{i=1}^{n} p(x_i)}{\sum_{X} [p(Y|X) \prod_{i=1}^{n} p(x_i)]} \tag{3}$$

which computes to

$$p(x_i|Y) = \sum_{X \backslash x_i} p(X|Y) \tag{4}$$

In Eq. (4), $X \backslash x_i = X - \{x_i\}$. According to Eqs. (1)-(4), the individual conditional probability $p(x_i|Y)$ can be achieved from the JPD $p(Y|X)$, $X = (x_1, x_2, \ldots, x_n)$. The backward dependency can be obtained from Eq. (4). The dashed arrowed lines in Fig. 3 denote the backward inference from effect Y to individual cause $x_i, i \in [1, 2, \ldots, n]$.

Fig. 3. Basic Model for Backward Inference in Bayesian Networks

In Fig. 2, when a fault is detected in component D, then based on Eqs. (1)-(4), we obtain $p(\overline{F}|\overline{D}) = 67.6\%, p(\overline{E}|\overline{D}) = 32.4\%$. This can be interpreted as follows: when component D is not available, the probability of a fault in component F is 67.6% and the probability of a fault in component E is 32.4%. Here only the fault related to connection service is considered.

3.3   Strongest Dependency Route (SDR) Algorithm for Backward Inference

Before we describe the SDR algorithm, the definition of the strongest cause and strongest dependency route are given as follows:

Definition 1. *In a BN let C be the set of causes, E be the set of effects. For $e_i \in E$, C_i be the set of causes based on effect e_i, iff $p(c_k|e_i) = Max[p(c_j|e_i)$, $c_j \in C_i]$, then c_k is the strongest cause for effect e_i.*

Definition 2. *In a BN, let R be the set of routes from effect $e_i \in E$ to its cause $c_j \in C, R = (R_1, R_2, \ldots, R_m)$. Let M_k be the set of transition nodes between e_i and c_j in route $R_k \in R$. Iff $p(c_j|M_k, e_i) = Max[p(c_j|M_t, e_i), t = (1, 2, \ldots, m)]$, then R_k is the strongest route between e_i and c_j .*

The detailed description of the SDR algorithm is described as follows:

Pruning of the BNs. Generally speaking, multiple effects (symptoms) may be observed at a moment, so $E_k = \{e_1, e_2, \ldots, e_k\}$ is defined as initial effects. In the operation of pruning, every step just integrates current nodes' parents into BN' and omits their brother nodes, because their brother nodes are independent with each other. The pruned graph is composed of the effect nodes E_k and their entire ancestor.

Strongest Dependency Route Trace Algorithm. After the pruning algorithm has been applied to a BN, a simplified sub-BN is obtained. Between every cause and effect, there may be more than one dependency routes. The SDR algorithm use product calculation to measure the serial strongest dependencies between effect nodes and causal nodes. Suppose $E_k \subset E, E_k = \{e_1, e_2, \ldots, e_k\}$. If $k = 1$, the BN will degenerate to a single-effect model.

Algorithm 3.2.1 (SDR)

Input: $BN = (V, L, P)$: $E_k = \{e_1, e_2, \ldots, e_k\}$: the set of initial effect nodes in BN, $E_k \subset V$.

Output: T: a spanning tree of the BN, rooted on E_k, and a vertex-labelling gives the probability of the strongest dependency from e_i to each vertex.

Variables: $depend[v]$: the probability of the strongest dependency between v and all its descendants: $p(v|u)$: the probability can be calculated from JPD of $p(u|\pi(u))$ based on Eqs. (1)-(4), v is the parent of u: $\varphi(l)$: the temporal variable which records the strongest dependency between nodes.

Initialize the SDR tree T as E_k;   // E_k is added as root nodes of T
Write label 1 on e_i //$e_i \in E_k$
While SDR tree T does not yet span the BN
    For each frontier edge l in BN
        Let u be the labelled endpoint of edge l,
        Let v be the unlabelled endpoint of edge l (v is one parent of u),
        Set $\varphi(l) = depend[u] * p(v|u)$:
    Let l be a frontier edge for BN that has the maximum φ-value;
    Add edge l (and vertex v) to tree T;
    $depend[v] = \varphi(l)$;
        Write label $depend[v]$ on vertex v;
    Return SDR tree T and its vertex labels:

The result of the SDR algorithm is a spanning tree T. Every cause node $c_j \in C$ is labeled with $depend[c_j] = p(c_j|M_k, e_i)$, $e_i \in E_k$, M_k is the transition nodes between e_i and c_j in route $R_k \in R$.

Proof of the Correctness of SDR Algorithm. Algorithm 3.2.1 gives a way to identify the strongest route from effect $e_i(e_i \in E_k)$ to $c_j(c_j \in C)$.

Lemma: when a vertex u is added to spanning tree T, $d[u] = weight(e_i, u) = -lg(depend[u])$.

Because $0 < depend[\delta_j] \leq 1$ so $d[\delta_j] \geq 0$. Note $depend[\delta_j] \neq 0$, or else there is no dependency relationship between δ_j and its offspring .

Proof: suppose to the contrary that at some point the SDR algorithm first attempts to add a vertex u to T for which $d[u] \neq weight(e_i, u)$.

See Fig. 4. Consider the situation just prior to the insertion of u and the true strongest dependency route from e_i to u. Because $e_i \in T$, and $u \in V \backslash T$, at some point this route must first take a jump out of T. Let (x, y) be the edge taken by the path, where $x \in T$, and $y \in V \backslash T$. We have computed x, so

$$d[y] \leq d[x] + weight(x, y) \tag{5}$$

Since x was added to T earlier, by hypothesis,

$$d[x] = weight(e_i, x) \tag{6}$$

Since $< e_i, \ldots, x, y >$ is sub-path of a strongest dependency route, by Eq.(8),

$$weight(e_i, y) = weight(e_i, x) + weight(x, y) = d[x] + weight(x, y) \tag{7}$$

Fig. 4. Proof of SDR algorithm

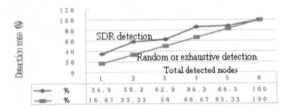

Fig. 5. Comparison between the SDR detection rate and random detection rate in a BN with 6 causal nodes. In the simulation experiment, we performed 500 tests.

By Eq. (7) and Eq. (9), we get $d[y] \leq weight(e_i, y)$. Hence $d[y] = weight(e_i, y)$.

Since y appears midway on the route from e_i to u, and all subsequent edges are positive, we have $weight(e_i, y) < weight(e_i, u)$, and thus $d[y] = weight(e_i, y) < weight(e_i, u) \leq d[u]$.

Thus y would have been added to T before u, in contradiction to our assumption that u is the next vertex to be added to T. So the algorithm must work. Since the calculation is correct for every effect node. It is also true that for multiple effect nodes in tracing the strongest dependency route.

Analysis of the SDR Algorithm. To determine the complexity of SDR algorithm, we observe that every link (edge) in BN is only calculated one time, so the size of the links in BN is consistent with the complexity. It is known in a complete directed graph that the number of edges is $n(n - 1)/2 = (n^2 - n)/2$, where n is the size of the nodes in the pruned spanning tree of BN. Normally a BN is an incomplete directed graph. So the calculation time of SDR is less than $(n^2 - n)/2$. The complexity of SDR is $O(n^2)$.

According to the SDR algorithm, we can observe multiple effect nodes From the spanning tree, the strongest routes between effects and causes can be obtained by Depth-First search. Meanwhile the value of $depend[c_j](c_j \in C)$ generates a dependency ranking of causes based on E_k. This dependency sequence is useful reference for fault diagnosis and system recovering.

Compared to other inference algorithms in BNs, the SDR algorithm belongs into the class of exact inferences and it provides an efficient method to trace the strongest dependency routes from effects to causes and to track the dependency sequences of the causes. It is useful in fault location, and it is beneficial for performance management. Moreover it can treat multiple connected networks modelled as DAGs.

In simulation experiment, detection rate represents the percentage of faults that were detected in a network by an algorithm. Using the simulation technique [22], we design a BN with 6 causal nodes. During the 500 tests, the simulation results are presented in Fig. 5. As stated in section 1, when a application system meets the 5% unlocated faults, generally the manager has to detect them randomly or exhaustively. Comparing with the random detection, the SDR algorithm provides more efficient approach to catch the causal nodes. When the set of causal nodes is larger the detection rate is more optimal.

4   Conclusions and Future Work

In network of realistic size and complexity, managers have to live with unstable, uncertain and incomplete information. Hence it is reasonable to use BNs to represent the knowledge about managed objects and their dependencies and apply probabilistic reasoning to determine the causes of failures or errors. Bayesian inference is a popular mechanism underlying probabilistic reasoning systems. The SDR algorithm introduced in this paper presents an efficient method to trace the causes of effects. This is useful for systems diagnosis and fault location and further can be used to improve performance management.

Most networks, however, dynamically update their structures, topologies and their dependency relationships between management objects. Due to the special requirements in network management, we need to accommodate sustainable changes and maintain a healthy management system based on learning strategies that allows us to modify the cause-effect structure and also the dependencies between the nodes of a BNs correspondingly. Also an effective prediction strategy, which takes into account the dynamic changes in networks, is important. This is related to discrete nonlinear time series analysis. Nonlinear regression theory [23] is useful to capture the trend of changes and give reasonable prediction s of individual components and the whole system.

References

1. C. Wang, M. Schwartz. Fault detection with multiple observers. IEEE/ACM transactions on Networking. 1993; Vol 1: pp48-55
2. M. Frontini, J. Griffin, and S. Towers. A knowledge-based system for fault localization in wide area networks. In Integrated Network Management, II. Amsterdam: North-Holland, 519-530, 1990.
3. S. A. Yemini, S. Kliger, E. Mozes, Y. Yemini, and D. Ohsie. High speed and robust event correlation. IEEE Communications Magazine, 34(5):82-90, 1996.

4. L. Lewis. A case-based reasoning approach to the resolution of faults in communication networks. In Integrated Network Management, III, 671-682. Elsevier Science Publishers B.V., Amsterdam, 1993.

5. R. H. Deng, A. A. Lazar, and W. Wang. A probabilistic Approach to Fault Diagnosis in Linear Lightwave Networks. IEEE Journal on Selected Areas in Communications, Vol. 11, no. 9, pp. 1438-1448, December 1993.

6. M. Steinder and A. S. Sethi. Non-deterministic diagnosis of end-to-end service failures in a multi-layer communication system. In Proc. of ICCCN, Scottsdale, AR, 2001. pp. 374-379.

7. The International Engineering Consortium. Highly available embedded computer platforms become reality.
http://www.iec.org/online/tutorials/acrobat/ha_embed.pdf.

8. J. Pearl. Probabilistic Reasoning in Intelligent Systems: Networks of Plausible Inference. Morgan Kaufmann, San Mateo, CA, 1988.

9. R. G. Cowell, A. P. Dawid, S. L. Lauritzen, D. J. Spiegelhalter. Probabilistic Networks and Expert Systems. New York: Springer-Verlag, 1999.

10. D. Nikovski. Constructing Bayesian networks for medical diagnosis from incomplete and partially correct statistics. IEEE Transactions on Knowledge and Data Engineering, Vol. 12, No. 4, pp. 509 - 516, July 2000.

11. K. Basye, T. Dean, J. S. Vitter. Coping with Uncertainty in Map Learning. Machine Learning, 29(1): 65-88, 1997.

12. E. Charniak, R. P. Goldman. A Semantics for Probabilistic Quantifier-Free First-Order Languages, with Particular Application to Story Understanding. Proceedings of IJCAI-89, pp1074-1079, Morgan-Kaufmann, 1989.

13. I. Katzela and M. Schwarz. Schemes for fault identification in communication networks. IEEE Transactions on Networking, 3(6): 733-764, 1995.

14. S. Klinger , S. Yemini , Y. Yemini , D. Ohsie , S. Stolfo. A coding approach to event correlation. Proceedings of the fourth international symposium on Integrated network management IV, p.266-277, January 1995.

15. D. Heckerman and M. P. Wellman. Bayesian networks. Communications of the ACM, 38(3):27-30, Mar. 1995.

16. A. Keller, U. Blumenthal, G. Kar. Classification and Computation of Dependencies for Distributed Management. Pro. of 5th IEEE Symposium on Computers and Communications. Antibes-Juan-les-Pins, France, July 2000.

17. M. Gupta, A. Neogi, M. K. Agarwal and G. Kar. Discovering Dynamic Dependencies in Enterprise Environments for Problem Determination. DSOM 2003, LNCS 2867, ISBN 3-540-20314-1, pp221-233, 2003.

18. J. Pearl. A constraint-propagation approach to probabilistic reasoning. Uncertainty in Artificial Intelligence. North-Holland, Amsterdam, pp357-369, 1986.

19. R. M. Neal, Probabilistic inference using Markov chain Monte Carlo methods. Tech. Rep. CRG-TR93-1, University of Toronto, 1993.

20. G. Cooper. Computational complexity of probabilistic inference using Bayesian belief networks. Artificial Intelligence, 42:393-405, 1990.

21. H. J. Suermondt and G. F. Cooper. Probabilistic inference in multiply connected belief network using loop cutsets. International Journal of Approximate Reasoning, vol. 4, 1990, pp. 283-306.

22. C. Wang. Bayesian Belief Network Simulation. Tech-Reprort, Department of Computer Science, Florida State University. 2003.

23. A. S. Weigend, and N. A. Gershenfeld. Time Series Prediction. Addison-Wesley, 1994.

A New Data Aggregation Algorithm for Clustering Distributed Nodes in Sensor Networks

Suk-Jin Lee[1], Chun-Jai Lee[1], You-Ze Cho[2], and Sung-Un Kim[3,*]

[1] Pukyong National University, 599-1 Daeyeon 3-Dong Nam-Gu, Busan,
608-737, Korea
{stone, leecjcpp}@mail1.pknu.ac.kr
[2] Kyungpook National University, 1370 San Kyok Dong, pook Gu,
Daegu, 702-701, Korea
yzcho@ee.knu.ac.kr
[3] Pukyong National University, 599-1 Daeyeon
3-Dong Nam-Gu, Busan, 608-737, Korea
kimsu@pknu.ac.kr

Abstract. The sensor nodes in sensor networks are limited in power, computational capacities, and memory. In order to fulfill these limitations an appropriate strategy is needed. Data aggregation is one of the power saving strategies in sensor networks, combining the data that comes from many sensor nodes into a set of the meaningful information. This paper proposes a new data aggregation algorithm named DAUCH (Data Aggregation algorithm Using DAG rooted at the Cluster Head) for clustering distributed nodes in sensor networks, combining the random cluster head election technique in LEACH with DAG in TORA. The proposed algorithm outperforms LEACH due to the less transmission power. Our simulation reveals that approximately a 4% improvement is accomplished comparing to the number of nodes alive with LEACH.

1 Introduction

Recent advances in MEMS-based sensor technologies, low-power analogy and digital electronics, and low-power RF designs have enabled the development of relatively inexpensive and low-power wireless micro sensors. These tiny sensor nodes can be used to a wide range of application areas such as health, military, home, and so on.

Realization of these sensor network applications requires wireless ad hoc networking techniques. But the sensor nodes in sensor networks are densely deployed, prone to failures, and are limited in power, computational capacities, and memory than the nodes in ad hoc networks. Moreover the topology of a sensor network changes very frequently, and sensor nodes mainly use a broadcast communication paradigm, whereas most of the ad hoc networks are based

---

* Corresponding Author

M. Freire et al. (Eds.): ECUMN 2004, LNCS 3262, pp. 508–520, 2004.
© Springer-Verlag Berlin Heidelberg 2004

on point-to-point communications. And Sensor nodes may not have global identification (ID) as well, because of the large amount of overhead and the large
number of sensors. Although many protocols and algorithms have been proposed
for traditional wireless ad hoc networks, they are not well suited to the unique
features and application requirements of sensor networks.

Many researchers are currently engaged in developing schemes that fulfill
these requirements. Especially in a view of the network, the power consumption
has been incurred to a critical issue, because the power efficiency is an important
performance metric, directly influencing the network lifetime in sensor networks
through.

Data aggregation is one of the power saving strategies in sensor networks,
combining the data that comes from many sensor nodes into a set of meaningful information. In LEACH, the useful data is aggregated to the cluster heads
that are randomly selected and allocate the time scheduling to their cluster
members[1]. But LEACH needs clustering formation overheads before performing the task and the nodes which are away from a cluster head consume much
more transmission batteries comparing to the nodes close to the cluster head. In
order to eliminate redundancy power consume in LEACH, TORA (Temporally-
Ordered Routing Algorithm) technique[2] which builds a DAG (Directed Acyclic
Graph) rooted at the destinations in ad hoc networks can be used. With DAG
each cluster head can construct an efficient cluster rooted at itself in sensor
networks.

This paper proposes DAUCH data aggregation algorithm for clustering distributed nodes in sensor networks, combining the random cluster head election
technique in LEACH with DAG in TORA. Using the efficient DAG, the proposed
algorithm saves the radio energy dissipations of the transmitter and the receiver
due to the short propagation distance. Our simulation reveals that approximately
a 4 percentages improvement is accomplished comparing to the number of nodes
alive with LEACH.

The rest of the paper is organized as follows: in section 2, we define the
merits and demerits of data aggregation methods in sensor networks and in ad
hoc networks, and show the properties of a new data aggregation method. In
section 3 we define the new data aggregation algorithm, and Section 4 presents
a model of the radio energy dissipation for the new data aggregation method
and an experiment result showing an effect of the power consumption comparing
with LEACH. We present our conclusion in section 5.

2   Preliminaries

2.1   Data Aggregation in Sensor Networks

Before starting the data aggregation techniques, we should investigate the routing models[3] that are assumed to consist of a single data sink attempting to
gather information from a number of data sources. Figure 1 is a simple illustration of the difference between simple models of routing schemes that use data

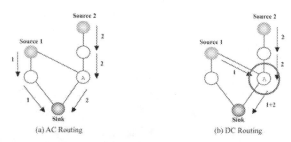

Fig. 1. Illustration of AC routing vs. DC routing

aggregation (which we term Data-Centric (DC)), and schemes that do not (which we term Address-Centric (AC)). They differ in the manner that the data is sent from a source to a sink. In the AC routing, each source independently sends data along the shortest path to the sink based on the route that the queries took (end-to-end routing), whereas in the DC routing the sources send data to the sink, but routing nodes on the way look at the content of the data and perform some form of aggregation and consolidation functions on the data originating at multiple sources.

In ad hoc networks, a routing model follows the AC routing, so each source sends its information separately to the sink like the figure 1(a). In sensor networks, a routing model follows the DC routing, so the data from the two sources are aggregated at node A, and the combined data is sent from node A to the sink like the figure 1(b). Therefore in sensor networks, the data aggregation technique is a critical factor different from ad hoc networks to save the power consumptions of the nodes in order to extend the sensor network lifetime.

In sensor networks, the data aggregation tree can be thought of as the reverse of a multicast tree. So optimal data aggregation is a minimum Steiner tree on the network graph. Instead of an optimal data aggregation, sub-optimal data aggregations are proposed to generate data aggregation trees that are aimed to diminish the transmission power. The figure 2 summarizes the properties and disadvantages of sub-optimal data aggregation methods.

The prevenient data aggregation methods [3] are efficient to the model where a single point in the unit square is defined as the location of an "event", and all nodes within a distance S (called the sensing range) of this event that are not sinks are considered to be data sources (which we term Event-Radius Model). In the model where some nodes that are not sinks are randomly selected to be sources, e.g. a temperature measurement and environment pollution detection (which we term Random-Source Model), it needs appropriate strategies for an efficient data aggregation.

In LEACH, all of the nodes in the field can be the source nodes in sensor networks, so this model can be considered Random-Source Model. The nodes in

Data aggregation method		Properties	Disadvantages
Optimal data aggregation method	Minimum Steiner Tree	The optimal number of transmissions required per datum for the DC protocol is equal to the number of edges in the minimum Steiner tree in the network.	The NP-completeness of the minimum Steiner problem on graphs
Sub-optimal data aggregation methods	CNS (Center at Nearest Source)	The source that is nearest the sink acts as the aggregation point. All other sources send their data directly to this source that then sends the aggregated information on to the sink.	The more great the gaps between the aggregation point and sources, the more the batteries consumptions.
	SPT (Shortest Paths Tree)	Each source sends its information to the sink along the shortest path between the two. Where these paths overlap for different sources, they are combined to form the aggregation tree.	The shorter the overlapped paths when the shortest route is established from each source to the sink, the more the batteries consumption.
	GIT (Greedy Incremental Tree)	At the first step the tree consists of only the shortest path between the sink and the nearest source. At each step after that the next source closest to the current tree is connected to the tree.	It takes some time for the identical data to arrive to the aggregation point and to aggregate the identical data from other source nodes.

Fig. 2. Comparison of the data aggregation methods

LEACH organize themselves into local clusters, with one node acting as the cluster head, which allocates the time slot to its cluster members. All non-cluster head nodes directly transmit their data to the cluster head, while the cluster head node receives data from all the cluster members, performs signal processing functions on the data (e.g., data aggregation), and transmits data to the remote BS (Base Station). If the cluster heads were chosen a priori and fixed throughout the system lifetime, these nodes would quickly use up their limited energy because being a cluster head node is much more energy intensive than being a non-cluster head node. Thus LEACH incorporates randomized rotation of the high-energy cluster head position among the sensors to avoid draining the battery of any one sensor in the network. In this way, the energy load of being a cluster head is evenly distributed among the nodes. But LEACH needs clustering formation overheads before performing the task, and the nodes which are away from the cluster head consume much more transmission batteries comparing to the nodes close to the cluster head. So it needs a strategy to eliminate the redundancy power consume in LEACH.

2.2   Data Aggregation in Ad Hoc Networks

Most of the ad hoc networks are based on point-to-point communications, so the data aggregation in ad hoc networks is not considered a critical issue except the multipath routing. In some routing protocols such as DSR [4], AODV [5], LMR [6], TORA, and so on, multi-paths can be established from the sources to the destination. In that case the data aggregation can be performed through the overlapped paths en route. But it depends on each routing technique, which is implemented in ad hoc networks. Amongst the multipath routing techniques, TORA builds a directed acyclic graph rooted at the destination in ad hoc networks. So using DAG all data in the field can be assembled at the destination node.

3   DAUCH Data Aggregation Algorithm

3.1   Overview

The new data aggregation algorithm is illustrated by the figure 3. Each cluster head that is elected randomly creates the DAG rooted at the cluster head. The nodes that have more than one uplink node aggregate the data arrived from the uplink nodes then transmit them to the downlink node. This manner is continued until all data arrive at the cluster head. The cluster heads receive and aggregate the data from the adjacent neighboring node, and then transmit them to BS.

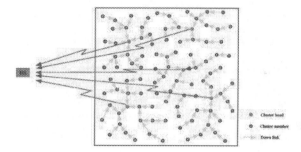

Fig. 3. Illustration of the new data aggregation algorithm

3.2   The Properties of DAUCH

In DAUCH, each cluster head that is elected randomly in the field creates a DAG centered at itself. Using DAG DAUCH creates the more effective cluster comparing to LEACH, and the nodes that are away from their cluster head save the data transmission power consumption using multi hop transmissions because of the shorter radio propagation distance. Moreover before the whole data get to the cluster head, some data aggregations are performed in the overlapped routes, so the task effort of data aggregations of the cluster head is distributed to other nodes that are not a cluster head.

But it is only effective supposed that all of the nodes are evenly distributed in the sensor field. Moreover there is a time delay in the case that the data of the nodes that are away from the cluster head arrive at the cluster head, and some overhead in the case that the links that connect the uplink with the downlink are unstable because of the dynamic movement of some nodes.

For the development of DAUCH, we made some assumptions about the sensor nodes and the underlying network model. For the sensor nodes, we assume that all nodes can transmit with enough power to reach the BS if need, that each node has the computational power to perform signal processing functions, and that all nodes are synchronized by each other in a sensor field. These assumptions are reasonable due to technological advances in radio hardware, low-power computing, and time synchronization techniques. For the network, we use a model where nodes always have data to send to the end user, nodes located close to each other have correlated data, all sensor nodes are evenly distributed and quasi static, and the message from the sender is correctly accepted by the receiver. Although DAUCH is optimized for this situation, it will continue to work if it were not true.

3.3   Requirements

The new algorithm uses five types of messages to communication.

DAG construction packet - When the cluster head creates DAG rooted at itself, the cluster head generates the DAG construction packet and broadcast the packet to the adjacent neighboring nodes. It contains the information of ClusterHeadID and DownlinkNodeID, where ClusterHeadID is the current cluster head's ID and DownlinkNodeID is the node that transmits the current DAG construction packet, and each DAG construction packet contains a cluster radius and a height. The former indicates the node hop number that is the farthest node of a cluster, and the latter express the hop number how far the current node is apart from the cluster head, respectively.

DAG deconstruction packet - When each round is ended, the cluster head generates a DAG deconstruction packet and sends it to all the cluster members in order to inform them of the end of a current round and the beginning of a new round. It contains ClusterHeadID and a cluster radius, and they inform the current cluster head's ID and the node hop number that is the farthest node of a cluster, respectively.

ACK packet - When a link breakage occurs, an uplink node sets a downlink flag and sends ACK packet to a downlink node whose height is the same as or less than itself in order to reconstruct the broken link. The ACK packet contains ClusterHeadID and TransmitterID. Herein TransmitterID presents the node ID that sends the ACK packet.

NonACK packet - When a node receives more than one message from its neighboring nodes and then collision occurs, it sends the packet to its neighboring node to retransmit. The NonACK packet contains ClusterHeadID and TransmitterID.

DATA packet - The nodes that have the data to send it to the cluster head use DATA packet to transmit the data. DATA packet is sent to the downlink node and on the way the packet is aggregated by the nodes that have more than one neighboring node. Finally all the data of a cluster are aggregated by the cluster head and sent to BS. The data packet contains SourceID, ClusterHeadID, TransmitterID, and Data. Herein SourceID presents the node that senses the event in a sensor field.

Each node caches the adjacent neighboring node IDs and stores their heights as well.

3.4   The Operation of DAUCH

The operation of DAUCH is divided into rounds and each round consists of five phases logically.

Cluster head selection phase - DAUCH's cluster heads are stochastically selected like LEACH. In order to select cluster heads, each node n determines a random number between 0 and 1. If the number is less than a threshold $T(n)$, the node becomes a cluster head for the current round. The threshold is set as follows:

$$\begin{cases} T(n) = \frac{P}{1-P(r \bmod \frac{1}{P})} & [\ \forall n \in G] \\ T(n) = 0 & [otherwise] \end{cases} \tag{1}$$

with P as the cluster head probability, r as the number of the current round, and G as the set of nodes that have not been cluster head in the last $1/P$ round. This algorithm ensures that every node becomes a cluster head exactly once within $1/P$ rounds.

DAG construction phase - After the cluster head selection phase, the cluster heads of each cluster broadcast the DAG construction packet to the neighboring nodes, as shown in figure 4(a). The nodes that receive the DAG construction packet set their heights that present the hop number how far the current node is apart from the cluster head, and then rebroadcast the packet to the neighboring nodes (figure 4(b)). If a node receives more than one DAG construction packet, it takes the message whose hop count is smaller than the others, and the messages that do not be taken are discarded. This manner is continued until the hop counter of the DAG construction packet reaches a cluster radius. After all, each cluster constructs DAG rooted at its cluster head.

DATA transfer phase - After a DAG construction is completed, the nodes that are farthest from the cluster head send the data to the neighboring downlink node, as shown in figure 5(a). After receiving the data, the node aggregates the data with own data, and then sends the aggregated data to the neighboring downlink node (figure 5(b), (c)). After all, every data within a cluster is assembled and aggregated at the cluster head, and then it is sent to BS, as shown in figure 5(d).

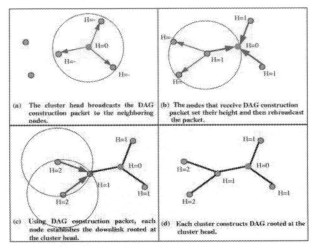

Fig. 4. The procedure of DAG construction

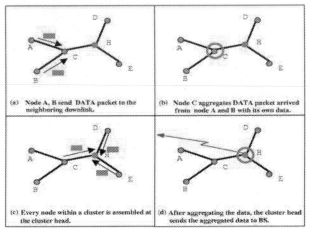

Fig. 5. The procedure of DATA transfer

Downlink failure phase - When a node's link that connects the neighboring two nodes is broken due to the node's movement or complete battery consumption, the uplink node sends ACK packet to reconnect a downlink path with the node which height is the same as or less than a previous downlink node with setting downlink flag. If there doesn't exist the node which height is the same as or less than a previous downlink node, or there isn't any respond to the ACK packet within TTL, the node directly sends its data message to the cluster head.

DAG deconstruction phase - Before the end of each round, the current cluster heads generate DAG deconstruction packets and send them to all the cluster members in order to inform the cluster members of the end of a current round and the beginning of a new round. And the packet is delivered until the nodes located in the end of the cluster, i.e. cluster radius.

4   Analysis and Simulation for DAUCH

4.1   Radio Energy Dissipation Model for the New Algorithm

For our experiment, we used a 100-node network where nodes were randomly distributed between (x=0m, y=0m) and (x=200m, y=200m) with BS at location (x=200m, y=200m). We assume a simple model[1] for the radio hardware energy dissipation where the transmitter dissipates energy to run the radio electronics ($E_{Tx-elec}$) and the power amplifier (E_{Tx-amp}), and the receiver dissipates energy to run the radio electronics ($E_{Rx-elec}$), as shown in figure 6. Therefore the radio energy dissipation of the transmitter is set as follows:

$$\begin{cases} E_{Tx}(k,d) = E_{Tx-elec}(k) \mid E_{Tx-amp}(k,d) \\ E_{Tx}(k,d) = E_{elec} \times k + \epsilon_{amp} \times k \times d^n \end{cases}, \tag{2}$$

and the radio energy dissipation of the receiver is set as follows:

$$\begin{cases} E_{Rx}(k) = E_{Rx-elec}(k) \\ E_{Rx}(k) = E_{elec} \times k \end{cases} \tag{3}$$

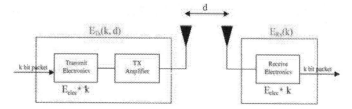

Fig. 6. Radio energy dissipation model

where d is a distance from a transmitter to a receiver, k is k-bit messages, and n is an exponential factor depending on the distance between the transmitter and the receiver. In the free space channel model, n is set to 2, and in the multipath fading channel model, n is set to 4 [7]. In our model, the distance between each cluster member and a cluster head is set to the free space channel model, and the distance between cluster heads and BS is set to the multipath fading channel model.

4.2   Analysis of the Radio Energy Dissipation Model

In order to analyze the different radio energy dissipation of LEACH and DAUCH, we apply the equation (2) and (3) to radio energy dissipation model for the data transmission and reception. Figure 7 presents the different data transfer model of LEACH and DAUCH. The inner and outer circles are implemented to easily compute the radio energy dissipation quantities based on the distance between each cluster member and a cluster head. We set the radius of the inner circle to d, and the radius of the outer circle to $2d$.

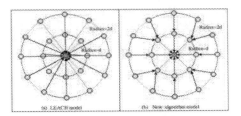

Fig. 7. Comparison LEACH model with the new algorithm model

In LEACH, the radio energy dissipations for the data transmission and reception can be computed using the figure 7(a). The energy dissipation of the inner nodes is set as follows:

$$8[E_{elec} \times k + \epsilon_{amp} \times k \times d^2 + E_{elec} \times k] \quad , \tag{4}$$

the energy dissipation of the outer nodes is set as follows:

$$12[E_{elec} \times k + \epsilon_{amp} \times k \times (2d)^2 + E_{elec} \times k] \quad , \tag{5}$$

and the total energy dissipation of the nodes is set as follows:

$$20[E_{elec} \times k] + 56[\epsilon_{amp} \times k \times d^2] + 20[E_{elec} \times k] \tag{6}$$

In the new algorithm, the radio energy dissipations for the data transmission and reception can be computed using the figure 7(b). The energy dissipation of the inner nodes is set as follows:

$$8[E_{elec} \times k + \epsilon_{amp} \times k \times d^2 + E_{elec} \times k] \quad , \tag{7}$$

the energy dissipation of the outer nodes is set as follows:

$$12[E_{elec} \times k + \epsilon_{amp} \times k \times d^2 + E_{elec} \times k] \quad , \tag{8}$$

and the total energy dissipation of the nodes is set as follows:

$$20[E_{elec} \times k] + 20[\epsilon_{amp} \times k \times d^2] + 20[E_{elec} \times k] \tag{9}$$

assuming that the total power consumption of the data aggregation is much smaller than the energy dissipation of the data transmission and reception, with comparing equation (6) with equation (9), we conclude that the new algorithm is superior to LEACH with regard to the radio energy dissipations for the data transmission and reception.

4.3   Simulation Results

For the presented simulations we use C++ programming. Each node is equipped with an energy source whose total amount of energy accounts for $2J$(Joule) at the beginning of the simulation. Every node transmits a 500 bytes message. The cluster head probability P is set to 0.05 [1].

Figure 8 and 9 illustrate simulation results of our sample network. According to our simulation results, the proposed method improves approximately 4% comparing to the number of nodes alive with LEACH. This is due that the new data aggregation uses smaller energy dissipation of the data transmission and reception than LEACH. Therefore the total amount of the new data aggregation algorithm's data received at the BS over time is better than that of LEACH due to the extended sensor network life time.

5   Conclusions and Future Work

This paper proposes a new data aggregation algorithm for clustering distributed nodes in sensor networks, combining the random cluster head election technique in LEACH with DAG which constructs an efficient cluster in TORA. The proposed algorithm outperforms LEACH by diminishing the radio energy dissipations for the data transmission and reception, and extending sensor network lifetime in comparison with LEACH. Even though our simulation result shows a good performance, the scheme is based on the assumption that the cluster head probability should be optimized like as to the case of LEACH. If we find a suitable cluster head number for the proposed algorithm, we can expect a more efficient outperforming result. This is our future work to make efforts to find an optimal cluster head probability.

In conclusion, this algorithm can be used in periodic data gathering application such as a temperature measurement and environmental pollution detection due to an efficient data gathering and low power consumption.

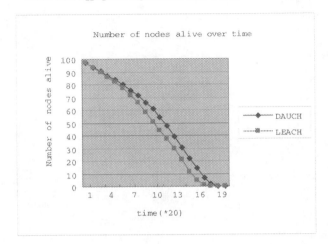

Fig. 8. Number of nodes alive over time

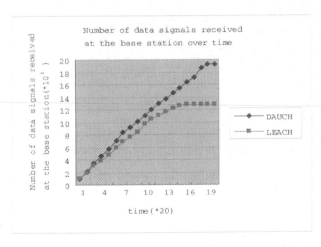

Fig. 9. Number of data signals received at the base station over time

Acknowledgment. This work was supported by Korea Research Foundation Grant (KRF-2003-002-D00243).

References

1. Wendi B. Heinzelman, Anantha P. Chandrakasan, and Hari Balakrishnan : An Application-Specific Protocol Architecture for Wireless Micro sensor Networks, IEEE TRANSACTIONS ON WIRELESS COMMUNICATIONS, Vol. 1, No. 4, Oct. 2002.
2. V. D. Park and M. S. Corson : A highly adaptive distributed routing algorithm for mobile wireless networks, INFOCOM '97. Sixteenth Annual Joint Conference of the IEEE Computer and Communications Societies. Proceedings IEEE, Vol. 3, pp.1405-1413, Apr. 1997.
3. Bhaskar Krishnamachari, Deborah Estrin, and Stephen Wicker : Modelling Data-Centric Routing in Wireless Sensor Networks, IEEE INFOCOM 2002.
4. David B. Johnson, David A. Maltz, and Josh Broch : DSR:The Dynamic Source Routing Protocol for Multihop Wireless Ad Hoc Networks, In Ad Hoc Networking, edited by Charles E. Perkins, Ch.5, Addison-Wesley, pp.139-172, 2001.
5. C.E. Perkins and E.M. Royer : Ad-hoc on-demand distance vector routing, Mobile Computing Systems and Applications, 1999. Proceedings. WMCSA '99. Second IEEE Workshop on, pp.90-100, Feb. 1999.
6. M. Scott Corson and Anthony Ephremides : A distributed routing algorithm for mobile wireless networks, Wireless Networks archive Vol. 1, Issue 1, pp.61-81, Feb. 1995.
7. T. Rappaport : Wireless Communications:Principles and Practice, Englewood Cliffs, NJ:Prentice-Hall, 1996.

On the Efficacy of PPPMux Within Radio Access Network

Jaesung Park, Beomjoon Kim, and Yong-Hoon Choi

LG Electronics Inc., LG R&D Complex, 533, Hogye-1dong, Dongan-gu, Anyang-shi,
Kyongki-do, 431-749, Korea, {4better,beom,dearyonghoon}@lge.com

Abstract. PPP Multiplexing (PPPMux) is agreed to be the most efficient solution for efficient use of transport resources in IP-based UMTS Terrestrial Radio Access Network (UTRAN). However, multiplexing inevitably introduces additional delay, which affects the overall performance of UTRAN. In this paper, we analyze the tradeoff of PPPMux between multiplexing efficiency and delay. In various operating environments considering maximum PPPMux frame size and input traffic rate, the performance of PPPMux is evaluated. It is shown that link utilization can be increased by using large PPPMux frame size, while multiplexing delay is tightly bounded below maximum waiting time.

1 Introduction

Efficient use of transmission resources becomes more important on a slow link such as E1/T1, which is common between Node-B and Radio Network Controller (RNC). Most of the traffic in current UMTS Terrestrial Radio Access Network (UTRAN) is voice and small sized control frames. Moreover, voice service would be the main revenue source of operators for several years to come. One of the characteristics of voice traffic is that the size of the traffic is small compared to the total protocol overheads imposed within UTRAN. Because the PDU size of voice traffic in frame protocol (FP) is too small to fill one cell of Asynchronous Transfer Mode (ATM), partially filled ATM cell causes waist of bandwidth. To solve this problem ATM Adaptation Layer Type 2 (AAL2) [8] has been used for efficient use of transmission resources in ATM-based UTRAN by trying to fully fill ATM cell payload. There have been a lot of works on the performance of AAL2 when it is applied to UTRAN [6], [7].

As networks are converged to all-IP, Third Generation Project Partnership (3GPP) started to work on the introduction of IP transport within UTRAN [2]. However, bandwidth efficiency is worse than that of AAL2/ATM case if the basic IP transport option is used, because the size of UDP/IPv4 header amounts to 28 bytes. To overcome this inefficiency and make IP transport option feasible within UTRAN, 3GPP agreed that header compression must be supported on slow PPP links, and no additional multiplexing layer/functionality shall be specified between UDP/IP and UTRAN FP because there is a PPP Multiplexing (PPPMux) [1] which can provide efficient multiplexing capabilities for PPP.

M. Freire et al. (Eds.): ECUMN 2004, LNCS 3262, pp. 521–530, 2004.
© Springer-Verlag Berlin Heidelberg 2004

On the contrary to the ATM transport option, PPPMux frame size can vary. One can exploit this feature to increase link utilization. However, a large PPP-Mux frame size also increases multiplexing delay and transmission delay. These delays must be handled carefully especially on slow links such as E1/T1 because UTRAN has very stringent delay requirements on a transport network [3]. It is not only for real time traffic but also for best-effort traffic because it is required that all the frames sent from RNC must arrive at Node-B within a predefined stringent time window [11]. Therefore, PPPMux should minimize the delays while maximizing link utilization. Engineering this trade-off is a key problem when PPPMux is used in the UTRAN.

There are a couple of works on the analysis of IP multiplexing schemes. The work in [9] analyzes IP multiplexer in terms of packetization delay and link utilization. The performance of Realtime Transport Protocol (RTP) multiplexing in terms of loss probability and the waiting time distribution of voice samples was evaluated in [5].

In this paper, a simple analysis on the performance of PPPMux within UTRAN is given. By considering various operating parameters of PPPMux, the performance is evaluated in terms of multiplexing delay and link utilization. We also analyze the maximum multiplexing efficiency, and identify the main factors which influence on the multiplexing efficiency. The performances are numerically evaluated for the 12.2 kbps Adaptive Multi-Rate (AMR) voice traffic in a various operating environments. Through the numerical evaluation, it is shown that PPPMux can support stringent delay requirements imposed by 3GPP while efficiently using transmission resources. In the evaluation, we focus on the Iub interface because it is usually composed of slow links where delay and efficient use of transport resources becomes more important problem. However, the same approach can be applied to other UTRAN interfaces exactly the same way.

2   Data Transport Within UTRAN

2.1   Radio Link Layer Operation

The UTRAN [10] is composed of a set of Radio Network Subsystems (RNS) connected to Core Network (CN) via Iu interface. A RNS consists of Radio Network Controller (RNC) and more than one Node-Bs. UTRAN is layered into a Radio Network Layer (RNL) and a Transport Network Layer (TNL). RNL protocols define UTRAN logical nodes and interfaces between them, and TNL provides services for transport of user, control, and management plane traffic.

In uplink direction, traffic from AMR source of an User Equipment (UE) goes through RLC, MAC, and WCDMA physical layer, and is transmitted into the air. Node-B receives a signal from air interface and extracts data from it, then transports the data to RNC via Iub in the form of FP frame by using transport network protocols [12]. When RNC receives a FP frame, corresponding RNL protocols process it according to the requested service. It should be noted that there is a stringent timing requirements between RNC and Node-B. Traffic sent

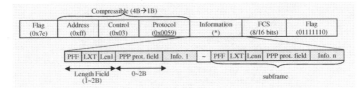

Fig. 1. PPPMux frame structure.

from RNC to Node-B over `Iub` will carry a reference to the specific radio frame when the data must be transmitted. If a FP frame arrives at a Node-B too late, then the information cannot be transmitted.

The characteristic of traffic changes when it goes through RNL protocols. For small sized voice traffic, RLC operates in Transparent Mode (TM). MAC layer selects an appropriate transport format (TF) for each transport channel. A transport format is selected from the transport format combination set (TFCS) negotiated at call setup time, and it is transmitted at each Transmission Time Interval (TTI). For voice traffic, TTI of 20 msec is used in general. FP layer adds a header to the MAC-PDU and sends it to the transport network layer protocol.

2.2   PPP Multiplexing Operation

Fig. 1 shows the frame structure of PPPMux [1]. It consists of High Level Data Link Control (HDLC) flag bits, a PPPMux header, a CRC field, and a number of sub-frames. HDLC header can be compressed to 1 bytes. Each sub-frame is composed of information field and sub-frame header, which indicates the protocol and length of each information field. Assuming sub-frame is less than 64 bytes, and each sub-frame has the same protocol number, sub-frame header becomes only 1 byte.

Initially, PPPMux negotiates working parameters such as maximum sub-frame size (`MAX_SF_LEN`), maximum frame size (X_{mux}), and last protocol ID (`Last_PID`). After setting up a connection, the transmitter side of PPPMux gets sub-frames from input buffer, and multiplexes sub-frames until one of the following stopping criteria is satisfied.

The stopping criteria of multiplexing are determined by the multiplexing delay and frame size. If there is no sub-frame to be served until maximum waiting time, PPPMux stops multiplexing and sends sub-frames multiplexed until that time by making a PPPMux frame with them (Stopping Criteria I). It also stops multiplexing if the sum of sub-frames is larger than maximum PPPMux size, even if maximum waiting time does not expire (Stopping Criteria II). In general voice samples are small, we can avoid the case where sub-frame size is larger than `MAX_SF_LEN` by setting it to have a reasonably large value. For example, because the maximum FP-PDU size for 12.2 kbps AMR source is 50 bytes [4] and maximum compressed UDP/IP header is 5 bytes, the above criteria can be avoided if `MAX_SF_SIZE` is set to be 60 bytes.

Both AAL2 and PPPMux have common features that they use a timer to limit the maximum multiplexing delay, and the size of multiplexed packets can vary. However, it should be noted that there is a fundamental difference between AAL2 and PPPMux in the method of multiplexing. AAL2 [8] is designed for fixed-sized ATM cell. The purpose of AAL2 is to fill 48 bytes ATM cell payload with as many data as it can by splitting a CPS packet into different ATM cells. The maximum number of AAL2 connections which can be multiplexed is limited by the 8-bit CID field. On the contrary, PPPMux does not segment a sub-frame to load it on two different PPPMux frames. Furthermore, multiplexing is ruled by the maximum size of PPPMux frame which can be set by a user. The variable frame size feature gives PPPMux an opportunity to increase link utilization by setting frame size large, especially when input traffic rate is high.

3   Analysis of PPP Multiplexing

3.1   Input Traffic to the PPPMux

FP-PDU of each connection is transmitted periodically at every TTI in the UTRAN. If we assume that compressed UDP/IPv4 provides only address information for Radio Access Bearer (RAB), this periodic pattern appears again when FP-PDU of a connection arrives to a PPP multiplexer. This is a valid assumption because most of the commercial systems are usually over-provisioned, so there is hardly a queuing effect. Node-B aggregates FP-PDUs of many connections and transports them to RNC via Iub interface. Therefore, for a given TTI period, the probability that FP-PDU arrives at PPP multiplexer may be random if there are a lot of traffic sources. Let N be the number of voice traffic sources and $TTI=20$ msec, the arrival process to the PPP multiplexer can be assumed to be *Poisson* process with mean $\lambda = N/TTI$ when N is large.

FP-PDU frame size is determined by the type of services and TFCS which are determined by UMTS signaling protocols at connection setup time. Frame size distribution is largely determined by the characteristics of traffic sources, but it is also affected by RNL protocol characteristics of UMTS system because control information is transported with data traffic as a Transport Format Set (TFS) at the same time. For example, 3 DCHs and one DCCH are used for the 12.2 kbps AMR voice service, and there can be 6 possible TFCSs [4].

3.2   PPP Multiplexer Model

Let T_w be the maximum waiting time of the multiplexer, and X_{mux} is the maximum PPPMux frame size excluding headers and CRC. As the multiplexing process repeats itself after it sends a PPPMux frame, PPPMux can be modeled as a renewal process with renewal time being the multiplexing delay. Let T_i be the inter-arrival time of sub-frames to the multiplexer, we can assume T_0 to be the arrival time of the first subframe after sending a PPPMux frame without loss of generality. Let M be the random variable which represents the number of arrivals. Denote k be the number of FP-PDUs arrived during $[0, t]$, and X_i be the size of subframe size at i-th arrival, then multiplexing delay (D) is given by

$$D = \begin{cases} t & \text{if } \sum_{i=1}^{k-1} X_i \leq X_{mux} \leq \sum_{i=1}^{k} X_i \text{ and } t < T_w \\ T_w & \text{if } t \geq T_w \end{cases} \quad (1)$$

where $t = \sum_{i=1}^{k} T_i$. Because the size of subframe and the number of arrivals are independent, the probability that multiplexing delay is equal to t can be obtained as follows:

$$P\{D = t\} = P\left\{ \left(\left(\sum_{i=1}^{k-1} X_i \leq X_{mux} \leq \sum_{i=1}^{k} X_i \right) \bigcap (t < T_w) \right) \mid M = k \right\} \\ = \sum_{k=1}^{\infty} \frac{(\lambda t)^k}{k!} e^{-\lambda t} \left(1 - F_X^{(k)}(X_{mux}) \right) F_X^{(k-1)}(X_{mux}) F_T^{(k)}(T_w) \quad (2)$$

where F_X is the cumulative distribution function (CDF) of subframe size and $F_X^{(n)}$ is the n-fold convolution of F_X. Similarly, F_T is the CDF of inter subframe arrival time and $F_T^{(n)}$ is the n-fold convolution of F_T. Since F_T is exponential, we have

$$F_T^{(k)} = 1 - e^{-\mu T_w} \sum_{i=1}^{k} \frac{(\mu T_w)^{i-1}}{(i-1)!} \quad (3)$$

where $\mu = 1/\lambda$. Let $Z = \sum_{i=0}^{k} X_i$, then $F_Z(x) = F_X^{(k)}(x)$. Therefore, average multiplexing delay and average PPPMux frame size can be obtained by

- Average Multiplexing Delay: $E[D] = \int_{t=0}^{t_w} t \cdot dF_D(t)$ where F_D is the CDF of D.
- Average PPPMux Size: $E[Z] = \int_{x=0}^{X_{mux}} t \cdot dF_X^{(k)}$

3.3   Multiplexing Efficiency

The efficiency is defined as the ratio of total payload size excluding all the overheads imposed by protocols to total frame size. Let OH_{RNL} is the total overhead from RNL protocols (RLC, MAC, FP), OH_{cUDP} is the size of compressed UDP/IP header, and H_{sub} is the header size of sub-frame of PPPMux. Then total overhead for one sub-frame would be $OH_{sub} = OH_{RNL} + OH_{cUDP} + H_{sub}$. If we denote the size of pure voice frame from codec by a, the number of multiplexed sub-frame by n, and the size of header and CRC of PPPMux frame by OH_{pppmux}, then the efficiency is defined as follows:

$$\begin{aligned} E_{PPPMux} &= \frac{a \cdot n}{(a + OH_{sub})n + OH_{pppmux}} = \frac{a}{a + OH_{sub} + OH_{pppmux}/n} < \frac{a}{a + OH_{sub}} \\ E_{PPP} &= \frac{a \cdot n}{(a + OH_{ppp})n} = \frac{a}{a + OH_{ppp}} \end{aligned} \quad (4)$$

As is evident above equations PPPMux can increase bandwidth efficiency by multiplexing a number of sub-frames. The maximum multiplexing efficiency is determined, not only by the number of multiplexed sub-frames, but also by the ratio of the size of pure voice frame to the total overhead of a sub-frame. These factors also determine the degree that PPPMux can achieve better efficiency over PPP.

We can also define TNL efficiency by considering FP-PDU as an input to TNL.

$$
\begin{aligned}
E_{TNL\_PPPMux} &= \frac{b \cdot n}{(b + OH_{cUDP} + H_{sub})n + OH_{pppmux}} \\
&= \frac{b}{b + OH_{cUDP} + H_{sub} + OH_{pppmux}/n} < \frac{b}{b + OH_{cUDP} + H_{sub}} \quad (5) \\
E_{TNL\_PPP} &= \frac{b \cdot n}{(b + OH_{cUDP} + OH_{ppp})n} = \frac{b}{b + OH_{cUDP} + OH_{ppp}}
\end{aligned}
$$

where $b = a + OH_{RNL}$. In this case, the number of multiplexed sub-frames and the ratio between overhead of sub-frame and overhead of PPP also determines the degree that PPPMux is superior to PPP in transmission efficiency.

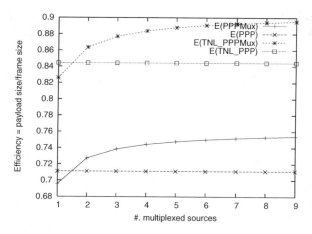

Fig. 2. Comparison of Efficiency between PPPMux and PPP.

Fig. 2 shows the efficiency of PPPMux and PPP. We assumed that the overhead of RNL protocols is 6 bytes [4], and overhead of cUDP/IP is 2 bytes, and sub-frame header size is 1 byte, respectively. As shown in Fig. 2, PPPMux is better than PPP in efficiency even though only two sub-frames are multiplexed. Compared to 84% of PPP efficiency, PPPMux can achieve efficiency of up to 90%, even if the frame size is small.

Fig. 2 also shows the efficiency of both RNL and TNL protocols. The efficiency is relatively low if the size of the traffic source is small compared to the total protocol overheads. We get efficiency of 78% when 32 bytes voice traffic arrives. However, efficiency increases as the size of traffic source becomes larger. It is noted that the difference between PPP and PPPMux in efficiency gets smaller if the size of traffic source becomes large. For example, the efficiency of PPP is 97.7% when 512 bytes of data traffic, which is the common TCP segment size, is transported. In case of PPPMux, the efficiency is 98.2%. This suggests that it is useful to use PPPMux for multiplexing only small sized traffic, and use PPP when the size of input traffic is large enough because multiplexing gain is minimal compared to protocol processing overhead when the size of input traffic is large.

4   Numerical Results and Discussion

In this section, we numerically evaluate the performance of PPPMux in terms of multiplexing delay and link utilization. We consider the transfer of 12.2 kbps AMR voice traffic between Node-B and RNC. We follow the RNL protocol layers and formats for `Iub` and `Iur` interfaces for a 12.2 kbps AMR voice traffic as defined in [4]. We also assume that compressed UDP/IP/PPPMux is used as TNL protocols [12], and a Radio Link (RL) is setup between a RNC and a Node-B.

4.1   Multiplexing Delay Distribution

Fig. 3 shows the distribution of multiplexing delay. PPPMux stops multiplexing mainly due to timeout when the input rate is low. On the other hand, maximum frame size dominates the multiplexing behavior when traffic arrives fast. In Fig. 3-(a), when the number of sources is 20, and X_{mux} size is 100 bytes, average multiplexed frame size is 68 bytes. This is because multiplexing stops due to Criteria I, and it explains the reason 75% of the multiplexing delay occurs due

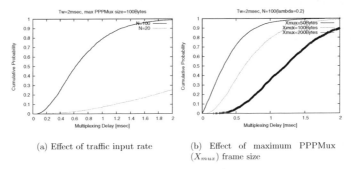

(a) Effect of traffic input rate   (b) Effect of maximum PPPMux (X_{mux}) frame size

Fig. 3. Distribution of Multiplexing Delay.

to timeout. However, in the same environments, if the input rate increases ($N = 100$), 95% of the multiplexing delay is 1.5 msec, and average PPPmux frame size is 80 bytes. Considering that average input data size is 22 bytes, this means multiplexing stops due to Criteria II.

Fig. 3-(b) shows the impact of X_{mux} on the multiplexing delay. Input rate relatively decreases as X_{mux} increases, because the more space there is, the more packets are needed to fill it in. Therefore, both multiplexing delay and multiplexing efficiency increases according to X_{mux}. For instance, when $N = 100$ and $T_w = 2$ msec, 95% of multiplexing delay was 1.5 msec and 0.96 msec when $X_{mux} = 100$ bytes and $X_{mux} = 50$ bytes respectively.

4.2   Impact of Maximum Wait Timer and Maximum Frame Size

We investigate the impact of maximum waiting timer T_w value on the multiplexing delay by varying input rate and X_{mux}. In general, multiplexing delay increases and multiplexing overhead decreases as T_w gets larger. This trade-off heavily depends on the X_{mux} and traffic input rate. Figs. 4-(a) and 4-(b) show the average multiplexing delay versus T_w when input rate is low and high respectively.

Because Stopping Criteria II dominates PPPMux, multiplexing delay decreases, and multiplexing efficiency reduces as X_{mux} decreases while traffic input rate is the same. (see Fig. 4-(a).) However, it is noted that the reduction rate of the efficiency is negligible because multiplexed packets are small compared to the total overhead when X_{mux} is small. For instance, in case of $N = 20$ and $T_w = 7$ msec, average multiplexing delay was 5.56 msec, 3.3 msec, and 1.99 msec while total efficiency was 0.73, 0.72, and 0.65, when $X_{mux} = 200$ bytes, 100 bytes, and 50 bytes, respectively.

It is also noted that X_{mux} and T_w do not influence on multiplexing delay for a reasonably high input rate. (see Fig. 4-(b).) It is because Criteria II determines

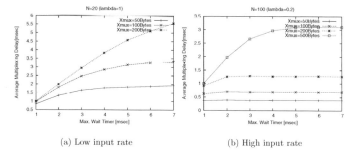

(a) Low input rate   (b) High input rate

Fig. 4. Average multiplexing delay vs. maximum wait timer.

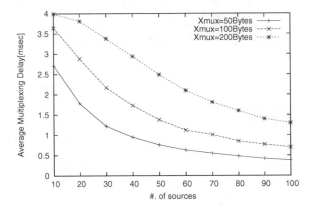

Fig. 5. Multiplexing delay vs. input traffic rate.

the multiplexing behavior if input rate is high enough. In this case we can increase multiplexing rate by increasing only X_{mux} size.

This is the fundamental difference between AAL2/ATM and IP/PPPMux. In case of ATM, one can not get more than 83% of link efficiency regardless of traffic input rate, because AAL2/ATM has a fixed cell size of 53 bytes. However, PPPMux can increases link utilization while limiting multiplexing delay to a given maximum wait timer by setting a frame size large, especially when traffic input rate is high. One thing to note is that the larger the size of PPPmux frame is, the more link efficiency one can get, but the transmission delay also increases with the size of PPPMux frame size. For example, it takes approximately 3 msec to transport 750 bytes frame on E1 links. Therefore the total delay of multiplexing and transmission should be considered together when determining maximum frame size. The safest policy for setting PPPMux parameters is that T_w is set to the maximum allowable multiplexing delay, and X_{mux} is set to the maximum allowable transmission delay. This policy leads to the maximum link utilization while limiting delays to a user defined maximum value. For example, if 12.2 kbps AMR sources are transported on an E1 link and the maximum allowable multiplexing delay is 1 msec, then by setting X_{mux} 100 bytes, one can bound multiplexing delay to 1 msec and limit the transmission delay of 0.4 msec for all input rates. (see Fig. 5.)

5   Conclusions

It is an important problem to use transport network resources efficiently while minimizing the delay introduced by an optimization technique when IP transport option is used within UTRAN. In this paper, we evaluate the performance of PPP Multiplexing protocol in terms of multiplexing delay and bandwidth utilization. It is shown that multiplexing delay can be bounded to maximum wait timer, and transmission delay can be limited by maximum PPPMux frame size.

In our efficiency analysis, it is suggested that it is useful to use PPPMux for multiplexing only small sized traffic, and use PPP when the size of input traffic is large because multiplexing gain over PPP is minimal compared to protocol processing overhead when the size of input traffic is large.

IP transport option within UTRAN is technically feasible, but there are a lot of things to make sure before deploying it in real network such as QoS, signaling in transport network, backward compatibility to R99 systems. For future works, we are planning to implement IP transport option within UTRAN, taking advantage of experiences of commercial ATM based UMTS system implementation. This will give us a comprehensive understanding of IP transport option.

References

1. R. Pazhyannur et al.: PPP Multiplexing. RFC 3153, (2001)
2. 3GPP TR 25.933 V5.3.0: IP Transport in UTRAN (Release 5). (2003)
3. 3GPP TR 25.853 V4.0.0: Delay Budget within the Access Stratum (Release 4). (2001)
4. 3GPP TS 34.108 V4.9.0: Common Test Environment for User Equipment (UE) Conformance Testing (Release 4). (2003)
5. Michael Menth: Carrying Wireless Traffic over IP Using Realtime Transport Protocol Multiplexing. In Proc. 12th ITC Spec. Seminar, (2000)
6. A. F. Canton, S. Tohme, D. Zeghlache, and T. Chahed: Performance analysis of AAL2/ATM in UMTS Radio Access Network. In Proc. PIRMC'2002, (2002)
7. S. Nananukul et.al.: Some Issues in Performance and Design of the ATM/AAL2 Transport in the UTRAN. In Proc. IEEE WCNC'2002, (2002)
8. ITU-T: B-ISDN ATM Adaptation Layer Specification: Type 2 AAL ITU-T Recommendation I.363.2. (1997)
9. A. Samhat and T. Chahed: Performance Evaluation of IP in the UMTS Terrestrial Radio Access Network. ITC'18, (2003)
10. 3GPP TS 25.401 V5.3.0: UTRAN Overall Description (Release 5). (2003)
11. 3GPP TS 25.402 V5.3.0: Synchronization in UTRAN Stage 2 (Release 5). (2003)
12. 3GPP TS 25.426 V5.3.0: UTRAN Iur and Iub Interface Data Transport & Transport Signaling for DCH Data Stream (Release 5). (2003)

HAIAN: Hash-Based Autoconfiguration by Integrating Addressing and Naming Resolution for Mobile Ad Hoc Networks*

Namhoon Kim, Saehoon Kang, and Younghee Lee

Information and Communications University, 119, Munji-Ro, Yuseong-Gu, Daejeon,
305-714, KOREA
{nhkim, kang, and yhlee}@icu.ac.kr

Abstract. Addressing is a basic step for communications among network nodes, and a naming service can help users easily and conveniently use their applications. To the best of our knowledge, addressing is studied regardless of naming service and vice versa. In this paper, we suggest an autoconfiguration scheme that allocates an address using a hash value of node's name. As we bound an address and the name at the joining phase, users use the naming service on their application without an additional name resolution step. Moreover, the total number of messages can be reduced by integrating addressing and naming resolution.

1 Introduction

A Mobile Ad hoc Network (MANET) is an arbitrary network consisting of a group of nodes without any infrastructure or administration. To communicate between nodes of a MANET, every node has a different address; however, because a MANET lacks an infrastructure, such as DHCP server, it required a distributed address autoconfiguration method. Usually, users want to use a name instead of an address on their application, such as ftp, telnet, e-mail, and web service. However, we cannot consider a DNS server in the MANET environment. Thus, we also require a name resolution scheme for the MANET.

To the best our knowledge, address configuration and name resolution has been dealt with separately in previous researches. In the study of address autoconfiguration, some researchers suggest that one node can act as a DHCP server in that the server node maintains the address pool and is responsible for the allocation of addresses [1][2][3]. Other researchers have proposed distributed address autoconfiguration mechanisms that may avoid the single point of failure problem: however these mechanisms have difficulty handling the address pool and guaranteeing the uniqueness of the addresses[4][5]. The previous autoconfiguration researches do not mention naming resolution because their

---

* This work is supported in part by the Korea Science and Engineering Foundation under grant No. R01-2003-000-10562-0, and by the Institute of Information Technology Assessment under grant No. A1100-0300-0004.

M. Freire et al. (Eds.): ECUMN 2004, LNCS 3262, pp. 531–539, 2004.

main concern is how to allocate addresses in the MANET. On the other hand, in the study of naming resolution, some researchers classify the name resolution mechanisms as partially centralized approaches and fully distributed approaches based on a communication model[6][7][8]. Other researchers suggest a naming architecture and structure for a MANET environment[9][10]. In addition, the previous naming resolution researches also do not describe the addressing because name resolution mechanisms assume each node already has its address. The IETF Zeroconf working group also divides addressing and naming services into distinct factors because the name relates to the application layer, while the address relates to the IP layer.

If a user knows the name of the node that the user wants to communicate, the user may not take interest in the address of the node although the network layer require the address. Of course, some users are adept at using the address instead of the name on their applications. However, it is generally regarded as common sense that users want to use a name instead of an address. For example, when we want to access to Yahoo website, we are likely to type *www.yahoo.com* instead of *216.109.118.76* on our web browser.

Such a convenient naming service can be used based on the network address. That is, naming depends on addressing. Therefore, if we want to use a naming service, we must pay additional overhead for that service. Is it possible, therefore, to achieve the addressing and naming resolution together? To answer this question, in this paper we suggest an autoconfiguration mechanism that allocates addresses by name of nodes. In other words, a node can get its address by using its name. Our main concept can help users to use the naming service without additional overhead for name resolution.

This paper is organized as follows: in Section 2, we describe our proposed mechanism and, in Section 3, we show the evaluation of our scheme. In Section 4, we mention related works; and finally, we present our conclusion in Section 5.

2   HAIAN: Proposed Autoconfiguration Scheme

A node to join a MANET must get an address to communicate, and users prefer to use the name of nodes instead of their address. However, existing autoconfiguration mechanisms or naming resolution mechanisms are located in different research areas in order to solve their respective problems. In other words, autoconfiguration mechanisms do not consider naming resolution, and name resolution mechanisms assume that every node already possessed its own address.

However, HAIAN tries to solve addressing and naming at the same time. HAIAN can help users quickly use the naming service. Moreover, message overhead can be reduced by integrating addressing and naming resolution.

2.1   Joining

The proposed scheme handles the address autoconfiguration and naming resolution at the same time. We assume each node has a hash function, such as MD5,

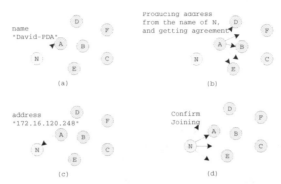

Fig. 1. Joining Process: This figure shows the process until a new node gets its address using its name. The Node N, a white circle, means a new node. The gray circle means MANET node; and in particular, Node A means the agent node for a new node. We have abbreviated the process of a new node finding its agent.

SHA, and HAS-160. A new node must find an entrusting agent node that helps the new node to obtain an address because the new node is unable to communicate at first. Following that, the new node sends its name to the agent node. When the agent node receives the name of the new node, it generates an address from the name using its hash function. Before the agent node responds to the new node with permission of the name and address, it must confirm whether there is a conflict between the address of new node and other existing MANET nodes' addresses. In that case, the agent node broadcasts the message, each node replies with negative confirms when it finds the conflict between the new address and its own address. For example, in Figure 1, a new node N wants to join the MANET. Node N finds an agent in order to get an address among its neighbors. We assume node N selects node A as its agent. Node N sends its name, David-PDA, to Node A to get an address. And then, Node A produces an address using its hash function, 172.16.120.248. Before Node A gives the address to Node N, it must confirm whether the address is already being used. If the address is not already being used, Node A gives the address to Node N. However, if the address is already being used for an existing MANET node, Node A sends a fail message to Node N. Node N then chooses another name, and it again sends its name to Node A.

After the joining is complete, the new-comer must send a verifying message including its address and name. When a node receives the message, it must update its name resolution table. The new node should receive the resolution table from its agent and add itself to the table. This table is used for address-to-name resolution. Instead of a resolution table, we can directly contact a node in order to resolve address-to-name.

2.2   Address Allocation

In the proposed scheme, the address allocation depends on the name of a node. Thus, we assume every node has one network interface for a MANET. A new node can obtain its address from the hash function using its name; however, every new node must ask an agent node to confirm whether its address is duplicated. If the address is already occupied by another node, the new node must select another name.

Actually, when the agent node receives the name of a new node, it first checks if the name is already being used on its resolution table. If the name is already bing used in the MANET, the agent node immediately sends a negative message to the new node. Otherwise, the agent node produces an address using the name, and gets agreements for the address from other existing MANET nodes.

2.3   Name Resolution

Usually, the naming resolution is completed through the following processes: Name Request, Name Response, and Name Bind. In our scheme, when we get a unique address, we also obtain the formal name in the MANET. Moreover, because each node maintains the resolution table, we can easily resolve name-to-address or address-to-name. The resolution table must update when a new node joins the MANET.

3   Evaluations

The proposed scheme combines address allocation and name resolution. Thus, it does not require additional overhead of name resolution because the allocation phase is a basic necessary step, and the proposed scheme finishes name resolution during the address allocation phase.

For example, we assume the use of an allocation method like MANETconf [5] or Strong DAD [4]. In order to get an address, we need many broadcast messages even if there is no failure. First, the agent node chooses a random address and sends a message including the address to check whether or not the address is already being used. The MANETconf needs two broadcasts, while the Strong DAD needs three broadcasts in checking duplicated addresses. In addition, after the agent node confirms the address is not being used, it broadcasts the message in order to inform that the address has been allocated to the new node. If the address is already being used, the agent node must retry the allocation phase using another random address. Thus, if there is a failure in getting agreements, additional broadcast messages are required.

In the naming resolution phase, the number of necessary messages differs according to the architecture, such as whether it is a centralized or distributed scheme. The centralized scheme must periodically broadcast messages to search unregistered nodes. Otherwise, every node should find the centralized server to register its name when it joins the network. And, when a node wants to use

Table 1. Comparison of the number of messages: B means broadcast message and U means unicast message. P means the period of beaconing. The naming resolution consists of registering and resolving. N means number of nodes. Constants - α, β, and δ - mean the probability of failure due to conflict of name or address in each scheme. RAA means Random Address Allocation, CN is Centralized Naming, and DN is Distributed Naming.

Schemes	Addressing	Naming Resolution
HAIAN	$(2B+2U)*\alpha$	None
RAA with CN	$(2B + 2U)*\beta$	$(P*B + U)*\delta + 2U$
RAA with DN	$(2B + 2U)*\beta$	$N*(2B)*\delta + 2U(\text{ or } 2B)$

the naming service, it must contact the server node. On the other hands, the distributed scheme also needs to broadcast messages to resolve name-to-address: name registration and name resolution. Unlike the centralized scheme, in the distributed scheme each node should broadcast in order to inform its name. And if the name is already being used, the node wanting to register must retry with another name. Also, when a node wants to use the naming service, it broadcasts the message including the address of destination. The destination node then responds using a broadcasting message including its name because it doesn't know the path from it to the source.

However, HAIAN doesn't require a separate name resolution phase. That is, addressing and naming resolution is completed at once. Thus, if the number of messages for the proposed scheme is similar to that of the existing address allocation method during the addressing phase, we conclude HAIAN is better in terms of communication overhead. The number of messages depends on the number of retrials caused by name or address conflicts. Our scheme must retry the allocation phase when a new-comer selects an existing name on the network or the hash function generates the same address although the name is different. However, in actuality, we need not broadcast messages on the same name case because if a new node chooses a duplicate name, the agent node immediately replies with a negative message. Therefore, when the new node chooses the same name, HAIAN needs two unicast messages. In producing a duplicate address from a hash function, we need broadcast messages in order to retry the allocation phase. While on the other hand, the existing address allocation scheme must retry the allocation phase when the agent node selects a duplicated address.

We summarize the number of messages for addressing and naming; Table 1 shows a comparison of the number of messages.

Intuitively, the proposed scheme is likely to often experience conflicts because HAIAN considers two factors: one is the name conflict, and the other is the address conflict. However, the name conflict probability is much lower than the address conflict because the name doesn't have any restriction but the address range is fixed, such as at 16 bits. Moreover, the name conflict also occurs in the existing name resolution scheme and it must retry to register with another name. Thus, we are mainly concerned with the case where the hash function generates the same address from different names.

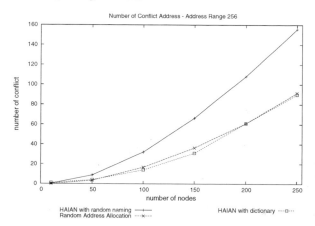

Fig. 2. Shows the number of conflict addresses when address bit is 8 bits. We compared HAIAN with random naming and the dictionary to the random address allocation method. In random naming, we randomly select strings with from 4 to 15 characters for the name of nodes. In the dictionary case, we used the word list of dictionary for the name of nodes.

To show the conflict ratio, we created a program with C. The random address allocation scheme and the proposed scheme are simulated to compare the address conflict according to the number of nodes. In simulation, node mobility and message loss were not considered. After the address allocation for the previous node has been completed, the next new node arrives. All nodes stay in the network until the end of the simulation and the number of nodes always increases. We selected MD5 hash function for the simulation. Actually, MD5 generates 128 bits word, but we lessen the word according to the address range.

As we can see in Figure 2, the number of conflicts in HAIAN is lower than that of the random address allocation method. That means the hashed value from different names is lower than the random selection from the fixed address range. And, we can see that the HAIAN with random naming and HAIAN with dictionary are almost the same. This means there are a few conflict names in random naming. From Figure 2, we concluded that it is practical to use HAIAN until the number of nodes reaches 50.

When the address range is 65535, the result is similar to Figure 2. From Figure 3, we suggest that the HAIAN can apply to a small scale network – until the number of nodes reaches 1000 – without consideration of the conflict problem.

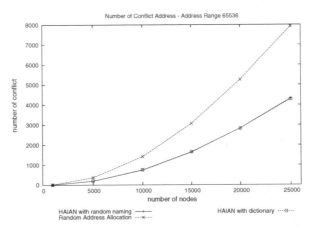

Fig. 3. Shows the number of conflict addresses when address bit is 16 bits. We compared HAIAN with random naming and the dictionary to the random address allocation method. In random naming, we randomly select strings with from 4 to 15 characters for the name of nodes. In the dictionary case, we used the word list of dictionary for the name of nodes.

From the simulation, the hashed value from the name of nodes is not much affected by the conflict problem on a small scale network. Moreover, although the number of nodes increases, the conflict ratio is still lower than that of the random address allocation. In conclusion, HAIAN can effectively support an addressing and naming service on a relatively small scale MANET. The main advantage of HAIAN is lower communication overhead for the naming service because preparations were made for the naming service during the addressing phase.

4   Related Works

4.1   Address Autoconfiguration for MANET

The Centralized Schemes. In the centralized schemes, there is only a server node which is responsible for allocating and maintaining addresses for a MANET. We see instances of centralized schemes in [1][2][3]. The server node refers to an agent node [1], a leader node [2], or an address authority [3], in each scheme, respectively, but the operation of the server nodes is similar in all cases: the server node maintains an address pool and is responsible for the address allocation. When two or more MANETs merge, the number of server nodes is reduced to one.

In these methods, the duplicated addresses by network merger can be simply detected using the server node's address pool. However, the server node becomes burdened and may be a single point of failure. The larger the number of hop counts from a new node to the server node, the longer these mechanisms take to find the server node and to get an address.

The Distributed Schemes. In contrast to centralized schemes, the distributed schemes operate in such a way that every node must communicate with each other to get an address.

C. Perkins et al. proposed a distributed Duplicated Address Detection (DAD) scheme called Strong DAD [4]. A new node randomly selects an IP address and examines whether it is used in a MANET. If the chosen address is already being used, it retries until it gets an unused address. The new node uses a temporary address for communication among other MANET nodes. The address allocation time increases in proportion to the number of failures. The maximum number of hop counts in a MANET also affects the allocation time.

S. Nesargi et al. suggested an agent-based distributed address autoconfiguration, MANETconf, using the distributed agreement concept [5]. Unlike Strong DAD, a new node, which is called a requestor, asks for an address to one of the neighbors in a MANET, which is called an initiator. The initiator then randomly selects an address and gets agreements from all other nodes in the MANET and assigns the address to its requestor. To acquire agreements, the initiator uses a modified DAD, receiving not NACK but ACK, which may result in an ACK explosion. Every node manages the list of all nodes in order to count the ACK messages to be received and decides the limit of the waiting time. Because this scheme manages the list of all nodes and collects response messages from all nodes, it can easily decide the success or failure of acquiring agreements. The node list can be used to find the conflict addresses when two or more networks are merged. It may take a longer time to allocate an address when the initiator fails to get agreements.

4.2   Name Resolution for MANET

P. Engelstad et al. classified the name resolution mechanism as a partially centralized, fully distributed, and hybrid approach [6][7][8]. The partially centralized scheme introduced a Name Coordinator that acts as a DNS server. It is responsible for resolving the name-to-address and address-to-name. Although this mechanism is effective in avoiding name collision, the name coordinator is a single failure point. On the other hand, in the fully distributed mechanism each node must resolve the name for itself. To avoid name collision, it requires much communication overhead during name resolution. In order to reduce the communication overhead, they suggest combining an on-demand routing protocol and the name resolution mechanism.

M. Aoki et al. explained design issues of name resolution that include name space, name assignment, and name management, and analyzed them to find what fits for a MANET [9]. On the basis of their analysis, they proposed a distributed name resolution scheme, ANARCH, which has a flat name space and a distributed name management method for the small scale MANET.

J. Jeong et al. suggested a name directory service mechanism, NDR, to provide mobile users with the name resolution service and give them the ability to exchange the user information that is needed to identify neighbors in a connected MANET [10]. NDR also has a mechanism that generates a unique domain name using a network device identifier. However, it takes too long to use the generating name for real applications.

5   Conclusion

The proposed scheme, HAIAN, provides address allocation and name resolution simultaneously for a MANET. Thus, HAIAN results in reducing total messages in performing address allocation and name resolution. Moreover, users can easily and quickly use the name service due to early binding between name and address. There are cases where we cannot use a favorite name due to generating the duplicate address by the hash function. However, if the address range is larger and the number of nodes is small, we can neglect that problem in this approach.

References

1. M. Gunes and J. Reibel, An IP address configuration Algorithm for Zeroconf. Mobile Multi-hop Ad-hoc Networks, Proc. of the International Workshop on Broadband Wireless Ad-Hoc Networks and Services, September 2002.
2. S. Toner and D. O'Mahony, Self-Organizing Node Address Management in Adhoc Networks, Springer Verlag Lecture Notes in Computer Science 2775, Springer Verlag, 2003, pp 476-483.
3. Y. Sun and E. Belding-Royer, Dynamic Address Configuration in Mobile Ad hoc Networks, UCSB Technical Report 2003-11, June 2003.
4. C. Perkins, J. Malinen, R. Wakikawa, E. Belding-Royer and Y. Sun, IP Address Autoconfiguration for Ad Hoc Networks, draft-ietf-manet- autoconf-01.txt, November 2001.
5. S. Nesargi and R. Prakash, MANETconf Configuration of Hosts in a Mobile Ad Hoc Network, Proc. of IEEE INFOCOM, June 2002.
6. P. Engelstad, D. Thanh, T. Jonvik, Name Resolution in Mobile Ad-hoc Networks, 10th International Conference on Telecommunications, ICT'2003. Vol. 1, 23 Feb.-1 March 2003, pp 388 - 392
7. P. Engelstad, D. Thanh, T. Jonvik, Name resolution in on-demand MANETs and over external IP networks, IEEE International Conference on Communication, ICC'2003. Vol. 2, 11-15 May 2003, pp 1024 - 1032.
8. P. Engelstad, D. Thanh, T. Jonvik, Name Resolution in On Demand MANET, Internet Draft, draft-engelstad-manet-name-resoultion-00.txt
9. M. Aoki, M. Saito, H. Aida, H. Tokuda, ANARCH: A Name Resolution Scheme for Mobile Ad Hoc Networks, Advanced Information Networking and Applications, AINA'2003. 27-29 March 2003, pp 723 - 730
10. J. Jeong, J. Park and H. Kim, "NDR: Name Directory Service in Mobile Ad-Hoc Network", ICACT 2003, Korea, January 2003.

Simulation Analysis of Teletraffic Variables
in DCA Cellular Networks

Antonietta Spedalieri, Israel Martín-Escalona, and Francisco Barcelo

Dept. d'Enginyeria Telemàtica de la Universitat Politècnica de Catalunya,
C/ Jordi Girona 1-3 (Mod. C3), 08034 Barcelona
{aspedali, imartin, barcelo}@entel.upc.es

Abstract. This paper deals with the characterization, by means of their statistical distributions, of several random variables related to cellular mobile telephony in a network that uses Dynamic Channel Allocation (DCA). In order to achieve this characterization, a simulation tool has been developed. The simulator implements the performance of a UMTS network as an example of DCA network, taking into account issues such as power control and admission control. The simulator provides data about channel holding time, time between two consecutive handoffs (handoff traffic characterization) and handoff duration, along with the information on the quality of service about the simulated environment. The effects of two different user mobility patterns have been analyzed.

1 Introduction

The progressive and intensive cellular division process has introduced several changes in the teletraffic analysis, which are strictly connected to modern mobile network design in opposition with the study carried out in fixed network and even in the firsts mobile cellular networks based on larger cells. Smaller cell size, which characterizes latest mobile networks, has the increase of the number of handoffs (HO) produced by a call as direct consequence. New variables such as channel holding time, the time between two consecutive handoff processes, the handoff duration, the time within the handoff area, the number of request of a single handoff process, etc, bound to the new concept of mobility and cellular division, must be introduced and deeply studied in the teletraffic analysis. All these variables are useful for a better planning.

The channel holding time has been widely studied using several approaches. Thus, [1] proposes an analytical study of this variable, while [2] provides a characterization of it using a simulation procedure and [3] does the same by means of a field-data analysis. On the other hand, variables related with the handoff process have been rarely considered in the recent teletraffic research [4, 5, 6]. However, their importance increases since the handoff rate becomes higher along with the reduction of the cell size. To the authors' knowledge, the existing studies related with these variables consider only Fixed Channel Allocation (FCA) networks [1, 2, 3, 4, 5, 6].

M. Freire et al. (Eds.): ECUMN 2004, LNCS 3262, pp. 540–553, 2004.
© Springer-Verlag Berlin Heidelberg 2004

This paper is aimed at jointly characterizing three teletraffic variables in DCA networks: channel holding time, time between two consecutive handoffs and handoff duration (i.e. time elapsed from handoff first request to handoff finalization). In addition to this, the study shows results of QoS (i.e. call blocking probability, forced interruption probability). The differences that came out form the fact of considering two different types of user mobility pattern are analyzed.

2   Variables Under Study

The variables under study are gathered into two groups: QoS and teletraffic. The following subsections provide a complete definition for both groups of variables.

2.1   QoS Variables

The first group of variables studied in this work aims at providing information about the QoS offered by the system. Two QoS figures have been considered in this study: the call blocking probability (BP) and the probability of forced termination of a call, i.e. due to handoff failure (FP). The reason after studying these two figures is that both of them are clearly perceived and can be measured at the user plane.

BP is the probability that a fresh call cannot be accepted due to the lack of available resources in the set of Base Stations (BS) that are able to serve the call. In the simulator it is computed as:

$$BP = \frac{Number\ of\ blocked\ fresh\ calls}{Number\ of\ fresh\ call\ attempts}. \tag{1}$$

FP is the forced termination probability, i.e. the probability that an ongoing call is interrupted due to a handoff failure. The handoff failure occurs when a call looses coverage from the serving BS without having seized a channel in a new one. The simulator obtains this figure as:

$$FP = \frac{Number\ of\ interrupted\ calls}{Number\ of\ established\ calls} \tag{2}$$

2.2   Teletraffic Variables

As a Mobile Station (MS) moves thought-out the network during a communication, it will be cross a number of cells and, as a direct consequence, changes the communication channel. Different BSs attend the communication whilst the service is being provided. Figure 1 shows this scenario for the case of two MSs setting up two different calls. In this figure, letters indicate time marks and the lines trace the geographical path followed by each MS while the call is in process. Throughout this study, we have not sketched the specific geographic boundaries that make the MS change serving cell: the changes are determined by constraints on the signal's strength. This results in an overlapping area between the source and target cells, known as the handoff or degradation area. This area is defined as the region in which a handoff may be requested and achieved. For example, in Figure 1, MS1 starts in *Cell 1* at time *A*. At time *B*, MS1 detects that the signal strength provided by the BS in *Cell 1* falls below the handoff threshold. This leads to the launch of a handoff attempt to *Cell 2*. However, due to traffic overload, *Cell 2* cannot accept the handoff request until *C*. Finally, MS1 ends at time *F* and MS2 at time *J*.

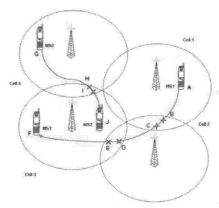

Fig. 1. Timestamps during the lifetime of a call.

Three teletraffic variables are investigated in this paper: the channel holding time, the time between two consecutive handoff arrivals and the handoff delay. These variables are defined as follows:

• *Channel holding time (CHT)*. This variable gathers the time periods in which a channel is allocated to a call, starting the call as a fresh one (i.e. a new originated call) or as an accepted handoff. Under the conditions presented in Figure 1, channel holding time samples include $(A - C)$, $(C - E)$, $(E - F)$, $(G - I)$ and $(I - J)$ time values.

• *Time between two consecutive handoff arrivals (TBH)*. These variable stores the time elapsed between two consecutive handoff arrivals at the same cell. Notice that this variable only considers the first request made by each handoff process received in the cell, i.e., the reattempts associated to handoff processes that were not immediately served are not taken into account. With reference to the communications shown in Figure 1, these samples include the $(D - H)$ value, under the assumption that attempt *D* took place before *H*.

• *Handoff duration (HD)*. This random variable saves the time elapsed since an MS requests a handoff until the handoff is supplied or dropped. For example, in the scenario proposed in Figure 1, this sample includes the $(B - C)$, $(D - E)$ and $(H - I)$ values.

3   Simulator Description

3.1   Introduction

The simulator presented in [6] has been adapted to include the particularities of DCA networks. The main changes are summarized in the following subsections.

Some hypotheses have been made in order to achieve an optimum execution speed vs. complexity trade-off. The following points describe these assumptions:

- The simulation layout is composed by NxN cells.
- Statistics are taken only from the inner cells in order to overcome the edge effect. Single wrapping has been used: the MS path is reflected when reaching the cell boundaries instead of appearing at the opposite edge.
- Fresh call arrivals follow a Poisson process and appear uniformly spread over the simulation area (i.e. uniform user density).
- All cells have equal properties both for radio conditions and for offered traffic. This is the homogeneity assumption; hot spots are not studied.
- The initial direction of a mobile station (MS) follows a uniform distribution.
- The speed of the mobile users is Gaussian-distributed.
- The MSs change their speed and direction while the communication is going on. The time between two consecutive changes follows an exponential law.
- Handoff re-attempts are allowed until a maximum dwell time in the handoff zone (an input parameter) is reached.
- No trunk reservation is used to prioritize handoffs.
- Call Duration follows an Exponential distribution.

3.2   Call Admission Control (CAC)

A CAC algorithm is applied every time that a fresh call arrives to the system. According to [7], the CAC scheme must be able to adapt the number of active calls in each cell according to interference level and power availability. This algorithm aims at verifying that accepting a new call does not increase the Signal to Interference Ratio (SIR) of the ongoing calls further than a certain quantity (1 dB in this study). If so, the new call is blocked. Obviously, this increases the Call Blocking Probability (*BP*) and contributes to obtain better figures in the Interrupted Call Probability (*FP*).

The *SIR* of each Mobile Station (MS) is computed as (see [8]):

$$SIR = \frac{P_r \times SF}{\alpha_{sir} I_{intra} + I_{inter} + P_n}. \tag{3}$$

where SF is the spreading factor of the physical connection, P_r is the received power, P_n is the thermal noise, I_{inter} is the sum of signal powers received from others cells, I_{intra}

is the sum of the signal power received from the other users in the same cell and α_{ser} is the loss-of-orthogonality factor due to the multipath.

3.3 Power Control

In this work an Adaptive-Step Power Control (ASPC) is used. This algorithm helps to improve network performances by increasing its capacity. The simulator continuously sets the power of each ongoing call to the minimum necessary to maintain the communication. This power control is carried out once per second. The algorithm followed to decide whether it is necessary to update the power on any MS has been borrowed from [9]. Table 1, where notation is also taken from [9] gathers the values of the parameters used with this algorithm in the simulations.

Table 1. Parameters of the power control algorithm (notation from [9])

Parameter	Value
Δ_0	1 dB
$\mu_{cp}, v_{cp}, \lambda_{cp}$	10
$n_0 = n_1 = n_{01}$	2

In order to estimate the power received by a MS and to take into account a correlated shadow model, the equation proposed in [7] has been improved using [10], redefining the A_s parameter to the following formula:

$$a = \varepsilon_D^{vT/D},$$
$$A_s(n+1) = A_s(n) \cdot a + LN(\sigma_s^2) \cdot (1-a). \qquad (4)$$

where A_s is the coefficient that considers that shadow for a MS is different at every simulation step, n indicates the time instant, a is the correlation factor of two consecutive time values, LN computes a lognormal distribution and σ is the standard deviation of the lognormal shadow and ε_D is the correlation between two points separated by distance D.

3.4 Handoff Management

In the simulator, a handoff process is started each time the *SIR* in the downlink or uplink falls below some defined thresholds (*Threshold$_{1,i}$*). Once started, the handoff process is held as long as a new channel is not achieved and the following conditions are satisfied:

- The *SIR* in downlink or uplink does not decrease below the sensibility thresholds (*Threshold$_{2,i}$*), where $\forall i$, Threshold$_{1,i} \geq$ Threshold$_{2,i}$.
- The duration of the handoff process does not reach a maximum threshold.
- The call is not voluntarily terminated by the user.

3.5   Modeling the Environment: Traffic, Mobility, Cell Layout

The whole-call duration and the fresh call arrival process are the only traffic inputs to the simulator. No other consideration has been taken regarding channel holding time or handoff process: those variables are a result of the traffic, type of environment, cell size, mobility pattern, signal power received by each station, etc. This provides the simulator a very realistic scope. The simulated scenario emulates a UMTS network operating in a suburban environment. The Base Station (BS) layout is based on hex-agonal-shaped cells and each cell is composed of three sectors. The scenarios consid-ered in this study use the parameters gathered in Table 2, to which an offered load of 50, 70 and 90% has been applied. Most parameters have been borrowed from similar studies and handbooks [6, 11].

Two kinds of mobility patterns have been used in simulations to study the impact ok users speed on the system: high mobility (HM) and low mobility (LM).

Table 2. Propagation, mobility and cell layout parameters

Power Parameters			
Max power emitted by BS	43 dBm	Max power emitted by MS	21 dBm
BS power gain	18 dB	MS power gain	0 dB
Min E_b/N_0 in the uplink	5 dB	Min E_b/N_0 in the downlink	4 dB
Mobility Parameters			
High Mobility Pattern (HM)		**Low Mobility Pattern (LM)**	
Average speed	25 m/s	Average speed	10 m/s
Standard deviation of the speed	7.5 m/s	Standard deviation of the speed	0 m/s
Mean time between speed changes	15 s	Mean time between speed changes	0 s
Other			
Distance between BS	600 m	Max handoff duration	15 s
Mean call duration	120 s		

In order to calculate the fresh (new) call arrival rate for each load needed as an in-put parameter to the simulator, we start by computing the maximum offered traffic per cell as:

$$M \approx \frac{G_P}{\dfrac{E_b}{N_0 + I_0}} \cdot \frac{1}{1 + \beta} \cdot \alpha \frac{1}{\nu} \cdot G_S \text{ [calls/cell]}, \tag{5}$$

where, as pointed out in [13], G_p is the processing gain, E_b is the bit energy, N_0 and I_0 represent noise spectral density and interference respectively, β is a penalty factor due to the interference generated by neighboring cells, α is a factor derived from power control, ν is the activity factor and G_s represents the gain due to cells division. Now the traffic and arrival rate (of fresh calls only) offered to each cell can be computed respectively as:

$$A = \rho M \tag{6}$$
$$\lambda = A/\tau \,$$

where ρ is the load offered to each channel and τ the mean call duration.

4 QoS Results

Table 3 gathers QoS results for both HM and LM scenarios. The ratio of around 14 handoffs per call for HM and 6 handoffs for LM could have been roughly estimated in advance. To obtain this estimation for HM we can proceed as follows: since the distance between BSs is 600 m. and there are 3 sectors per cell, 200 m. is a rough estimation of the length that a MS runs within a sector. At 25 m/s the time within a sector is around 8 s. and a call that takes 120 s. involves 120/8=15 sectors, hence 14 handoffs. Same procedure can be used for LM case. The random nature of the radio path including shadowing comes to increase this ratio.

Table 3. QoS Results for HM and LM scenarios

Load	HO to fresh call ratio (HM/LM)	BP% (HM/LM)	FP% (HM/LM)
50 %	13.95 / 5.34	8.05 / 5.17	0.05 / 0.01
70 %	14.30 / 5.83	14.95 / 11.78	0.70 / 0.11
90 %	13.81 / 5.97	21.25 / 17.23	1.70 / 0.93

Notice how a good balance is maintained between the blocking and interruption probability. The impact of increasing the load is more noticeable on the former, while the later is always bounded to low percentages. This is desirable and is achieved by means of the strict power control and CAC that favors continuity of service and thus decreases the interruption probability to the detriment of the call blocking probability. If the bound for the CAC changed from 1 dB to 2 dB, one should expect lower blocking probabilities but more interrupted calls. While the main objective of the CAC is to avoid the quality degradation of the conversation, a side impact is the desirable control of the balance between *BP* and *FP*.

As expected, the HM case shows worse performance figures. This is known among planning practitioners as the "mobility cost". On the one hand, *FP* is higher since each call involves more handoffs on average: for the same probability of a handoff being dropped the probability of having a dropped handoff increases. On the other hand, handoffs are prioritized over fresh calls since handoffs continuously retry while fresh calls are rejected if not immediately accepted. Hence when the handoff ratio increases the probability of a new call being rejected (i.e. *BP*) also increases.

5 Teletraffic Results

5.1 Characterization Methodology

First, the simulation is run and all data collected. Next, the mean and the squared coefficient of variation (*SCV*) are obtained from the samples. These statistical parameters are computed to forecast which theoretical distribution functions could fit the samples. Once the candidate function list is built, the parameters for each theoretical distribution are estimated using the Maximum Likelihood Estimation (MLE) method. MLE is used because this estimation method achieves better significance values than the ones obtained through any other technique, as long as the results of the fitting are evaluated by means of a Goodness-Of-Fit (GOF) test [3]. The notation is borrowed from [3, 6]. Finally, the distribution functions are evaluated through the Kolmogorov-Smirnov GOF test.

5.2 Channel Holding Time

An important variable that has been analyzed in this study is channel holding time. As it can be seen in Table 4, experimental results for this variable have a *SCV* for both HM and LM patterns lower than one. This suggests that the exponential distribution does not provide a good fitting of the data set. Moreover, the faster speed of users has as direct consequence on the decrease of the average and SCV of channel holding time. Faster speed means more cells crossed per call on average: the whole call holding time (exponential with SCV=1) is split into more pieces.

Table 4. Channel Holding Time Results

Offered Load	Average (HM/LM)	SCV (HM/LM)
50 %	8.60 / 15.07	0.66 / 0.86
70 %	8.39 / 14.68	0.66 / 0.81
90 %	8.68 / 13.29	0.68 / 0.91

The best fitting distributions are the Hyper Erlang and the Lognormal ones, as shown in Table 5, where the two best fitting for each load is highlighted. Traditionally, the Lognormal-2 distribution has been acknowledged as the one providing the best fit. However, the results herein presented show that, for HM scenarios, the best performances are achieved by the Hyper Erlang family distributions among which the best suited distribution is deemed to be the Hyper Erlang-K one, which gives quite identical results to the Hyper Erlang-JK in terms of goodness of fit, while requiring the estimation of a minor number of parameters. The best fitting distributions in LM are the lognormal distributions, among which the best performance is achieved by Lognormal-3 one.

Figure 2 shows the fitting results of Lognormal-3 in the 70%-loaded scenario for HM pattern while Figure 3 shows the same results in a LM environment.

Table 5. Fitting results (K-S distance) of the Channel Holding Time

Distribution	Offered load		
	50 % (HM/LM)	70 % (HM/LM)	90 % (HM/LM)
Exponential	6.45 / 8.35	6.70 / 5.04	6.02 / 5.17
Erlang-JK	3.42 / 3.69	3.79 / 3.34	4.48 / 4.32
Hyper Erlang-K	**2.84** / 4.78	3.02 / ----	**2.59** / ----
Hyper Erlang-JK	**2.40** / 4.46	**2.73** / 3.59	**2.59** / 4.89
Lognormal	6.15 / 7.91	5.84 / 5.80	5.44 / 5.14
Lognormal-2	3.02 / **2.28**	**3.01 / 2.44**	2.61 / **3.89**
Lognormal-3	3.50 / **1.64**	3.57 / **2.31**	3.20 / **4.01**
Weibull	3.82 / 4.98	3.43 / 5.04	2.71 / 4.69

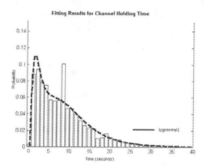

Fig. 2. Fitting results for Channel Holding Time for 70% load and HM scenario – Lognormal-3

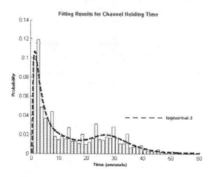

Fig. 3. Fitting results for Channel Holding Time for 70% load and LM scenario – Lognormal-3

5.3 Handoff Traffic

This random variable represents the time between two consecutive handoffs, coming from different MSs, to the same cell (i.e. sector in a BS). This variable was studied both considering and neglecting simultaneous handoff requests. Let P_o be the probability that two handoff requests from different MSs are simultaneously submitted to the same cell. Notice that, in the simulated scenarios, simultaneous handoff requests are highly probable events. This is because cells are divided into 120° sectors; this fact, along with the high user mobility (i.e. high speed vs. small areas), leads to a high rate of handoffs per call. Thus, the probability density function has a high peak at the origin, due to the high frequency of simultaneous handoffs. For fitting purposes it is easier to fit separately the simultaneous and non-simultaneous handoffs than trying to fit a distribution for all.

As a consequence, the characterization of this random variable was conducted in two steps. First, the probability of two or more handoff requests arriving simultaneously at the same cell is calculated. Simultaneous handoffs will thus be eliminated from the empirical sample prior to the second step, which consists in calculating the optimal parameters of the fitting distribution after having discarded the zero values. The combination of these two distributions completely defines the variable under study.

Table 6 shows the achieved results, where $(Average_1, SCV_1)$ and $(Average_2, SCV_2)$ represent the mean and the squared coefficient of variation of the variable including and not including simultaneous handoffs and P_o provides the probability of the time between consecutive handoffs being zero. Notice how, both in HM and LM, the average time between handoffs decreases and P_o increases when the load increases: this was expected since more load involves more ongoing calls, hence more handoff attempts received by each cell.

As expected, the probability of contemporary arrival is higher in the case of HM than in the one of LM. This is because in HM the handoff rate is higher than in low mobility case, therefore the number of cells changing gets higher and increases the probability that two calls can arrive at the same time in the same cell.

Table 6. Time between two non-simultaneous handoffs results

Offered load	Average₁ (HM/LM)	SCV₁ (HM/LM)	P₀ (%) (HM/LM)	Average₂ (HM/LM)	SCV₂ (HM/LM)
50 %	0.06 / 0.97	3.36 / 1.75	64.20 / 40.81	1.16 / 1.65	0.24 / 0.51
70 %	0.05 / 0.59	2.20 / 2.65	69.50 / 55.44	1.09 / 1.37	0.11 / 0.48
90 %	0.04 / 0.29	2.00 / 2.93	72.23 / 63.71	1.06 / 1.30	0.09 / 0.32

Table 7 gathers the fitting results obtained from the simulation of the proposed scenarios.

Figure 4 shows the fitting in the 50%-loaded in HM scenario of time between non-simultaneous handoffs using Gamma function, which is one of the distributions that provide better fitting. For LM the best fitting distributions belong to the Erlang and

Hyper Erlang families, among which the best performance is achieved by the Erlang-JK distribution

Table 7. Fitting results of the time between two non-simultaneous handoffs

Distribution	Offered load		
	50 % (HM/LM)	70 % (HM/LM)	90 % (HM/LM)
Exponential	18.99 / 23.72	18.18 / 21.83	17.61 / 25.24
Erlang-K	14.84 / 17.89	**15.04** / 17.08	14.91 / 19.07
Erlang-JK	**14.62 / 17.30**	------- / **16.41**	------- / **18.76**
Hyper Erlang-JK	**14.70 / 17.37**	------- / **16.46**	------- / **18.90**
Lognormal	17.04 / 20.74	15.65 / 19.49	15.56 / 22.49
Weibull	20.80 / 32.86	18.33 / 26.70	18.33 / 29.77
Gamma	14.73 / 20.02	**14.28** / 17.88	**14.28** / 18.91

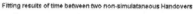

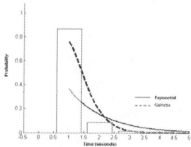

Fig. 4. Fitting results for Handoff traffic for 50% load and HM scenario – Exponential and Gamma

5.4 Handoff Duration

In the analysis and dimensioning of mobile cellular networks, the duration of handoff process is also of fundamental importance. In characterizing this variable, it must be taken into account that, if a system is properly dimensioned, the probability of immediately obtaining a new channel at the first handoff attempt is high. This is due to the fact that, since the CAC only applies to fresh call arrivals and fresh calls do not automatically reattempt in a very short period, handoffs have a higher priority in the competition for available channels.

Thus, the probability density function for this variable is given by a delta function at zero delay, plus a probability density function that only considers delays longer than zero. In this work, these two components were separately studied: we considered separately the probability of immediately serving a handoff and the characterization of the delayed handoffs.

Table 8 shows the results for the handoff duration (in seconds), where $Average_1$ and $Average_2$ stand for the average handoff duration including and not including the handoffs that are immediately accepted. The SCV in this table refers to the handoff duration without considering the immediately-served handoffs and it is lower than unity. Notice that the longer average can be expected as a consequence of the heavier load: the number of handoffs and the probability of a handoff not being attended by the system also increase. Such behavior occurs for all mobility kinds and can be observed in both HM and LM scenarios.

In HM scenarios the average duration of handoff is longer than in LM scenarios, while the probability of immediate handoff serving is lower. This can be explained by taking into account the values of interruption probability in the two cases: if the FP increases, the probability that an handoff has to finalize correctly decreases and increases the dwelling time in handoff zone too.

Table 8. Handoff duration results

Offered Load	$Average_1$ (HM/LM)	$P_0(\%)$ (HM/LM)	$Average_2$ (HM/LM)	SCV (HM/LM)
50 %	0.26 / 0.03	85.90 / 97.41	1.90 / 1.25	0.57 / 0.38
70 %	0.56 / 0.25	74.73 / 86.27	2.24 / 1.81	0.75 / 0.51
90 %	0.96 / 0.51	65.07 / 79.87	2.57 / 2.14	0.89 / 0.71

Table 9 shows the K-S statistical distances associated with the delayed handoffs duration. The best fitting for HM scenarios is achieved by the Hyper Erlang distributions in agreement with [12]. Erlang-JK function also provides great results, even more if it is taken into account the smaller number of parameters to fit. For LM scenarios the distribution that gives the best fitting is the Hyper Erlang-JK one.

Figure 5 shows the fitting results of Hyper Erlang-JK and exponential distribution functions in the 70%-loaded scenario for HM pattern.

Table 9. Fitting results of the delayed handoffs

| Distribution | Offered load to each channel | | |
	50 % (HM/LM)	70 % (HM/LM)	90 % (HM/LM)
Exponential	21.24 / 22.70	19.31 / 22.31	18.50 / 12.03
Erlang-K	**14.69 / 17.10**	15.46 / 16.32	16.90 / 12.24
Erlang-JK	**14.86** / 17.25	**13.91** / 16.34	14.24 / **9.46**
Hyper Erlang-K	14.92 / -----	14.06 / **16.21**	**14.04** / 9.81
Hyper Erlang-JK	14.91 / **16.95**	**13.87 / 16.21**	**13.84 / 9.31**
Lognormal	17.85 / 20.14	16.75 / 19.54	16.52 / 10.94
Gamma	16.05 / 17.23	16.30 / 19.07	17.14 / 14.33

6 Conclusion

In this work, simulations results have been analyzed in order to model DCA cellular networks in suburban scenarios with different loads and mobility patterns. The software simulator was used to obtain empirical data sets that allow characterizing teletraffic and QoS variables in UMTS-like networks.

The study of the QoS variables shows the impact of the mobility cost and that the CAC allows a good control of the balance between the blocking and interruption probabilities.

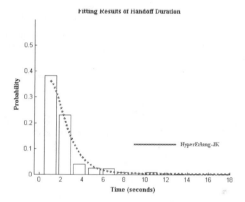

Fig. 5. Fitting Results for Handoff duration 90% Load and HM scenarios – Hyper Erlang-JK.

For channel holding time, the best fitting was reached trough Hyper Erlang JK for HM and Lognormal-3 for LM. It is important pointing out the bad result given by the exponential function. Regarding the time between two consecutive handoffs to the same cell, the exponential proved to be a bad choice, while the best characterization is achieved by the Gamma distribution in HM scenarios and Erlang-JK in LM scenarios. Results obtained in this work for the handoff duration shows that the best fitting is obtained with the Hyper Erlang-JK distribution for both HM and LM cases.

More details on the impact of the CAC on the balance between *BP* and *FP*, results on Manhattan scenarios and the impact of lognormal call holding time are left for further study.

Acknowledgments. This work has been funded by Telefonica Moviles España S.A. through Plan de Promoción Tecnológica, FEDER and the Spanish Ministry of Science and Technology through CICYT Project TIC 2003-01748.

References

1. D. Hong, S.S. Rappaport, "Traffic model and performance analysis for cellular mobile radio telephone systems with prioritized and nonprioritized handoff procedures", *IEEE TVT. 35(3)*, pp. 77-92, August 1986.
2. E. Chlebus, T. Zbiezek, "A novel approach to simulation of mobile networks", *12th ITC Specialist Seminar on Mobile Systems and Mobility*, March 2000.
3. F. Barceló, J. Jordán, "Channel Holding Time Distribution in Public Telephony Systems (PAMR and PCS)", *IEEE Trans. Veh. Tech.*, Vol. 29 No. 5, pp. 1615-1625, September 2000.
4. F. Barceló, J. I. Sánchez, "Probability distribution of the Inter-Arrival time to Cellular Telephony Channels", *IEEE 49th Vehicular Technology Conference (VTC 1999 Spring)*, pp. 762-766.
5. M. Ruggieri, F. Graziosi, F. Santucci, "Modeling of the Handoff Dwell Time in Cellular Mobile Communications Systems", *IEEE Trans. Veh. Tech.*, Vol. 47, No 2, pp. 489-498, May 1998.
6. I. Martin-Escalona, F. Barcelo, J. Casademont, "Teletraffic simulation of cellular networks: modeling the handoff arrivals and the handoff delay", *IEEE 13th IEEE Int. Symp. on Personal, Indoor and Mobile Radio Communications (PIMRC02)*, pp. 2209-2213, 2002.
7. A.Capone, S. Redana, "Call Admission Control Techniques for UMTS", *IEEE 54th Vehicular Technology Conference (VTC 2001 Fall)*, Vol: 2, pp. 925-929, October 2001.
8. A. Capone, M. Cesana, G. D'Onofrio y L. Fratta, "Mixed traffic in UMTS downlink", *IEEE Microwave and Wireless Components Letters*, Vol: 13, Issue: 8, pp. 299-301, August 2003.
9. L. Nuaymi, P. Godlewski, X. Lagrange, "Power allocation and control for the downlink in cellular CDMA networks", *12th IEEE International Symposium on Personal, Indoor and Mobile Radio Communications*, Vol 1, pp. C-29 - C-33, 30 Oct. 2001.
10. M. Gudmundson, "Correlation Model For Shadow Fading In Mobile Radio System", *Electronic Letters*, Vol.27 No. 23, November 1991.
11. H. Holma y A. Toskala, *WCDMA for UMTS*, John Wiley & Sons, 2000.
12. V. Pla y Casares-Giner, V., "Analytical-numerical study of the handoff area sojourn time", IEEE GLOBECOM 2002, Vol.1, pp. 886 – 890, 17-21 Nov.2002.
13. A.J. Viterbi, "The Orthogonal – Random Waveform Dichotomy for Digital Mobile Communications", *IEEE Personal Communications*, Vol 1, Is 1, pp. 18 -24, 1994.

Author Index